Biological Physics

Biological Physics

Edited by Keira O'Donnell

SYRAWOOD
PUBLISHING HOUSE

New York

Published by Syrawood Publishing House,
750 Third Avenue, 9th Floor,
New York, NY 10017, USA
www.syrawoodpublishinghouse.com

Biological Physics
Edited by Keira O'Donnell

International Standard Book Number: 978-1-64740-097-2 (Hardback)

Cataloging-in-Publication Data

Biological physics / edited by Keira O'Donnell.
     p. cm.
Includes bibliographical references and index.
ISBN 978-1-64740-097-2
1. Biophysics. 2. Biology. 3. Physics. I. O'Donnell, Keira.
QH505 .B56 2022
571.4--dc23

TABLE OF CONTENTS

PREFACE

Biological physics or biophysics is an interdisciplinary science which applies the principles and methods of physics to study biological phenomena. Biophysics encompasses the study of all scales of biological organization, by directly observing, modeling and manipulating the structures and interactions of molecules and molecular complexes. A number of techniques are used for the study of biological systems. Fluorescent imaging techniques include X-ray crystallography, electron microscopy, small-angle scattering, etc. to visualize structures of biological significance. Dual polarization interferometry and circular dichroism are used to measure conformational change in structure, and optical tweezers or AFM are used to directly manipulate molecules. Research in biophysics is interdisciplinary in its approach and often draws from the fields of physiology, molecular biology, biochemistry, nanotechnology, bioengineering, biomechanics, systems biology, among others. Different approaches, evaluations, methodologies and advanced studies on biological physics have been included in this book. There has been rapid progress in this field and its applications are finding their way across multiple industries. This book aims to equip students and experts with the advanced topics and upcoming concepts in biophysics.

Significant researches are present in this book. Intensive efforts have been employed by authors to make this book an outstanding discourse. This book contains the enlightening chapters which have been written on the basis of significant researches done by the experts.

Finally, I would also like to thank all the members involved in this book for being a team and meeting all the deadlines for the submission of their respective works. I would also like to thank my friends and family for being supportive in my efforts.

Editor

Theoretical model and characteristics of mitochondrial thermogenesis

Jian-Sheng Kang[1]✉

[1] The First Affiliated Hospital of Zhengzhou University, Zhengzhou 450052, China

Abstract   Based on the first law of thermodynamics and the thermal diffusion equation, the deduced theoretical model of mitochondrial thermogenesis satisfies the Laplace equation and is a special case of the thermal diffusion equation. The model settles the long-standing question of the ability to increase cellular temperature by endogenous thermogenesis and explains the thermogenic characteristics of brown adipocytes. The model and calculations also suggest that the number of free available protons is the major limiting factor for endogenous thermogenesis and its speed.

Keywords   Mitochondria, Thermogenesis, Thermal physical model, Brown adipocyte, Electrochemical potential energy

INTRODUCTION

Mitochondria are the main intracellular sites for thermogenesis, especially the mitochondria of brown adipocytes (BA), which have been targeted for therapy to reduce obesity. However, long-standing critique (Baffou et al. 2014) and debates (Baffou et al. 2015; Chrétien et al. 2018; Kiyonaka et al. 2015; Lane 2018; Suzuki et al. 2015) exist on the ability to increase cellular temperature by endogenous thermogenesis, and a good theory of intracellular thermogenesis and temperature is lacking. In this work, based on the first law of thermodynamics and the thermal diffusion equation, the thermal physical model of a mitochondrion is deduced. We found that mitochondrial thermogenesis is a special case of the thermal diffusion equation, which satisfies the Laplace equation ($\nabla^2 T = 0$).

CELL AS A THERMAL PHYSICAL SYSTEM

The cell is a membrane-enclosed grand canonical ensemble of systems that exchanges both heat and particles with its surroundings. We can use the first law of thermodynamics to show that

$$U = -Q + W, \tag{1}$$

where U is the cellular internal energy; Q is the heat dissipated to the surroundings, and W is the work added to the system. In this equation, the negative sign means that heat flows out of the cell.

For a differential change, the relation in Eq. 1 is given by the following:

$$dU = -\,dQ + dW. \tag{2}$$

For thermogenesis of BA, a cell with such a small size has limited sources for energy extraction or delivery compared with the amplitudes of heat (dQ) and work (dW). Thus, it is acceptable to claim that the change in cellular internal energy (dU) can be neglected.

$$dU \approx 0. \tag{3}$$

Therefore, the relation (Eq. 2) is reduced to the following:

$$dQ = dW. \tag{4}$$

We can write the cellular work (dW) as the sum of various forms, such as kinetic energy and potential energy:

✉ Correspondence: kjs@zzu.edu.cn (J.-S. Kang)

$$dW = -pdV + Fdx + \sum_i \mu_i dN_i + \sum_j \varphi_j dq_j, \qquad (5)$$

where pdV is the work performed by the volume change (dV) under pressure (p), and Fdx is the mechanical energy used to move a distance (dx) under force (F). In addition to kinetic energy, potential energy contains chemical potential (μ) and electric potential (φ) in the case of changing the numbers of particles (dN) or charges (dq).

THERMOGENESIS USING ELECTROCHEMICAL POTENTIAL ENERGY

Thermogenesis in BA is executed at the mitochondrial level. A single BA contains numerous mitochondria, which show minimal volume change and almost no motility in such a crowded space (Xie *et al.* 2017a, b). It is acceptable to consider that both dV and dx are equal to zero such that we can ignore the changes in kinetic energy and only consider the changes in potential energy. Thus, the relation (Eq. 5) is reduced to the following:

$$dW = \sum_i \mu_i dN_i + \sum_j \varphi_j dq_j. \qquad (6)$$

A mitochondrion with a large negative membrane potential has the proton-motive force (pmf) for ATP synthase as well as motive forces (mf) for other particles, such as Ca^{2+}, among others. Thus, we can write the relation (Eq. 6) as follows:

$$dW = pmf \cdot dH^+ + mf_{Ca^{2+}} \cdot dCa^{2+} + \cdots. \qquad (7)$$

With aforesaid relations (Eqs. 4 and 7), we also ignore the transient changes in mitochondria, such as $[Ca^{2+}]$ (Xie *et al.* 2017b), for sustained thermogenesis such that the following applies:

$$dQ = pmf \cdot dH^+. \qquad (8)$$

This equation matches the fact that the co-stimulation of neurotransmitters norepinephrine (NE) and ATP can effectively convert the electrochemical potential energy stored in the mitochondrial proton gradient into heat via the mitochondrial uncoupling protein-1 (UCP1) in BA (Xie *et al.* 2017b).

TEMPERATURE GRADIENT

According to Fourier's law, the relation between heat flux (J, heat per unit time per unit area, J/s·m) and temperature gradient (∇T, K/m) is written as follows:

$$J = -\kappa \nabla T. \qquad (9)$$

Fourier's law is also stated as follows:

$$dQ = JAdt = -\kappa A \nabla T dt, \qquad (10)$$

where dt is the time interval and A is the area. Equations 10 and 8 together give the following:

$$\nabla T = -\frac{pmf \cdot dH^+}{\kappa A dt}. \qquad (11)$$

We can consider that a mitochondrion with a spherical shape and radius (r) has an area of $A = 4\pi r^2$, and dH^+/dt is clearly the proton current (I_{H^+}) of the mitochondrion. The thermogenic proton current is directed inward and is mediated by UCP1 (I_{UCP1}) after its activation. These statements mean that we can rewrite the gradient expression (Eq. 11) for BA thermogenesis as follows:

$$\nabla T = -\frac{pmf}{4\pi\kappa r^2} I_{H^+} = -\frac{pmf}{4\pi\kappa r^2} I_{UCP1}. \qquad (12)$$

TEMPERATURE AS A FUNCTION OF TIME

After determining the equation for the temperature gradient, we can deduce the relation between temperature and time by applying the thermal diffusion equation with a heat source (Blundell and Blundell 2010):

$$\frac{\partial T}{\partial t} = D\nabla^2 T + \frac{H}{C}, \qquad (13)$$

where $D = \kappa/C$ is the thermal diffusivity (m²/s); κ is the thermal conductivity (W/m·K); C is the volumetric heat capacity (J/K·m³), and heat is generated at a rate H per unit volume (W/m³, $H = P/V$, P is the power, and V is the volume).

In spherical polar coordinates (Blundell and Blundell 2010), we write

$$\nabla^2 T = \frac{1}{r^2}\frac{\partial}{\partial r}\left(r^2 \frac{\partial T}{\partial r}\right), \qquad (14)$$

$$\frac{\partial T}{\partial r} = \nabla T. \qquad (15)$$

Because pmf and I_{H^+} are not functions of radius (r) for a single mitochondrion, Eqs. 14 and 15 together with Eq. 12 state that the thermogenesis of a mitochondrion satisfies the Laplace equation

$$\nabla^2 T = 0. \qquad (16)$$

Thus, the thermal diffusion Eq. 13 for a spherical mitochondrion reduces to the following:

$$\frac{\partial T}{\partial t} = \frac{H}{C} = \frac{P}{VC}. \tag{17}$$

Dividing both sides of Eq. 8 by a dt time, we write

$$P = \frac{dQ}{dt} = pmf \cdot \frac{dH^+}{dt} = pmf \cdot I_{H^+}. \tag{18}$$

Equations 17 and 18 yield the following:

$$\frac{\partial T}{\partial t} = \frac{pmf \cdot I_{H^+}}{VC}. \tag{19}$$

DISCUSSION

Steady state versus thermogenic state

In the resting state of BA, without stimulation of sympathetic transmitters, UCP1 is inactivated by purine nucleotides. The BA or mitochondrion has a steady state described according to Eq. 19 as follows:

$$I_{H^+} = I_{UCP1} = 0, \tag{20}$$

$$\frac{\partial T}{\partial t} = 0. \tag{21}$$

In the thermogenic state, it is clear that the proton current is not zero and is mediated by the activated UCP1 such that Eq. 19 states the following:

$$I_{H^+} = I_{UCP1} \neq 0, \tag{22}$$

$$\frac{\partial T}{\partial t} \neq 0. \tag{23}$$

Using the steady state to discuss the thermogenic state leads to an $\sim 10^{-5}$ gap between Baffou's model and well-known facts (Baffou et al. 2014). In our previous paper (Xie et al. 2017a), we noted Baffou's mistakes and properly applied Eq. 17 for theoretical estimation of the maximum rate of mitochondrial temperature change. The theoretical estimation matched well with the experimental result (Xie et al. 2017a).

Thermogenic rate and capacity

After constructing the thermogenic model as a function of time (Eq. 19), we can further discuss the thermogenic characteristics of BA, such as the thermogenic capacity of the mitochondrion and the limiting factors for BA thermogenesis.

To estimate the temperature profiles of mitochondria, we must know $pmf \cdot I_{H^+}$ in Eq. 19. Mitchell's chemiosmotic theory states the following:

$$pmf = \Delta\psi - \frac{2.3RT}{F} \cdot \Delta pH, \tag{24}$$

where $\Delta\psi$ is the electrical gradient; ΔpH is the proton gradient; R is the gas constant; T is the temperature in Kelvin, and F is the Faraday constant. The mitochondrial pmf is ~ 200 mV. For a single mitochondrion of BA under thermogenesis, the inward thermogenic proton current is the current of the mitoplast, which is mediated by UCP1 (I_{UCP1}). It is known that mitoplasts typically have membrane capacitances of 0.5–1.2 pF and proton current (I_{UCP1}) densities of 60–110 pA/pF (Bertholet et al. 2017).

If defining the change rate of mitochondrial temperature ($\frac{\partial T}{\partial t}$) as a measurement of the thermogenic capacitance in BA, by taking the proton current of mitochondrion as 100 pA and the mitochondrial volume as 1 μm^3, we obtain a theoretical rate of mitochondrial $\frac{\partial T}{\partial t}$ of ~ 4.8 K/s based on Eq. 19.

The maximum experimental thermogenic capacitance of BA is comparable to 10 $\mu mol/L$ CCCP-induced thermogenesis (Xie et al. 2017b). However, the measured maximum rate (Xie et al. 2017a) of mitochondrial $\frac{\partial T}{\partial t}$ is ~ 0.06 K/s, which suggests that the proton current (I_{H^+}) is a limiting factor for BA thermogenesis. For the maximum transient rate of mitochondrial $\frac{\partial T}{\partial t}$, an initial transient $[Ca^{2+}]$ change in mitochondria evoked by stimulation of sympathetic transmitters (Xie et al. 2017b) should be counted (Eqs. 6 and 7), which also makes a comparable contribution.

Proton pool is a limiting factor

A proton current of 100 pA means that a single mitochondrion consumes 6.24×10^8 protons per second and that a single BA with ~ 1000 mitochondria requires 6.24×10^{11} proton (~ 1 pmol) per second. Clearly, free cellular protons are the major limiting factor for thermogenesis, which was experimentally supported by the cytosol alkylation during BA thermogenesis (Xie et al. 2017b).

In Eq. 3, the change in cellular internal energy is claimed to be negligible. For verification, we calculated the numbers of free available protons, which are $\sim 6.3 \times 10^2$ in a mitochondrion and $\sim 10^5$ in a BA with a diameter of 20 μm and a cytosol pH of 7.4. Thus, we indeed confirmed that dU can be neglected for sustained thermogenesis. Additionally, $dU \approx 0$ suggests that the increased mitochondrial or cellular temperatures must be balanced and compensated by selected intra-mitochondrial or intracellular energy changes, such as exergonic reactions of NADH (52.6 kcal/mol) and $FADH_2$ (43.4 kcal/mol), which were also experimentally supported by NADH and $FADH_2$ consumption during BA thermogenesis (Xie et al. 2017b).

Consequently, the gap between the maximum experimental $\frac{\partial T}{\partial t}$ and the theoretical $\frac{\partial T}{\partial t}$ suggests that thermogenesis of BA uses less than 1% of its thermogenic capacity. In addition, as illustrated in Fig. 1, the results demonstrated that the overall averaged $\frac{\partial T}{\partial t}$ was less than ~ 0.005 K/s. In reality, a single BA might only consume $\sim 10^{-3}$–10^{-2} pmol proton per second for sustained thermogenesis in BA (Fig. 1). Furthermore, depolarization of the mitochondrial membrane potential and cytosol alkylation during BA thermogenesis (Xie *et al.* 2017b) suggest that the value of $pmf \cdot I_{H^+}$ is a factor of self-restriction for thermogenesis.

In summary, BA and its mitochondria are heat-producing micro-machines with a high efficacy limited by free proton pools. The thermogenic model (Eq. 19) and calculations suggest that BA thermogenesis relies on hydrogen and energy sources such as glucose, water, fatty acid, NADH, and FADH$_2$. One mol glucose and 6 mol water together can supply 24 mol protons in the tricarboxylic acid cycle. Even if glucose is supplied at a rate (Zamorano *et al.* 2010) of 0.18 pmol/h·cell without tens or hundreds of times the glucose uptake in BA under stimulation (Orava *et al.* 2011; Vallerand *et al.* 1990), it is sufficient to sustain thermogenesis in BA.

A generalized thermogenic model

$$\frac{\partial T}{\partial t} = \eta \frac{pmf \cdot I_{H^+}}{VC}. \tag{25}$$

Finally, the thermogenic model of mitochondria (Eq. 19) can be generalized as shown in Eq. 25 by multiplying the thermogenic efficiency (η). The thermogenesis of BA under NE and ATP co-stimulation is a special case ($\eta = 1$) for Eq. 25. In general, of the potential free energy in glucose, approximately 40% is conserved in ATP in mitochondrial oxidative phosphorylation, and thus, the value of η is approximately 0.6 for all other cell types except erythrocytes ($\eta = 0$), which lack mitochondria. Interestingly, NE stimulation alone activates the proton-pumping ATPase function of mitochondrial complex V in BA (Xie *et al.* 2017b), so that NE-stimulated BA show a variety of responses (heating, constant temperature or occasionally cooling) and a low efficacy of thermogenesis (Xie *et al.* 2017b). Consequently, the phenomena and Eq. 25 restate that the limiting factor for the capability of intracellular thermogenesis is the net proton current (I_{H^+}) of proton outflow by proton pumps and proton leakage by UCP1 or other uncoupling factors.

METHODS

The thermogenic model (Eq. 19) overcomes obstacles related to the ability to increase cellular temperature by endogenous thermogenesis. Therefore, quantification is needed, which has not been performed in our previous works. Thus, the temperatures were calculated and converted from our previous data (Xie *et al.* 2017a, b). The calculation was based on the relation (Eq. 26) between temperature (T) and the normalized intensity ratio (nr) of thermosensitive and thermoneutral mitochondrial dyes (Xie *et al.* 2017a).

$$\frac{1}{T} - \frac{1}{T_{ref}} = -\frac{k_B}{E_a} \cdot \ln nr, \tag{26}$$

where k_B is the Boltzmann constant, and E_a is the measured activation energy (~ 6.55 kcal/mol) (Xie *et al.* 2017a).

Acknowledgements The author thanks Dr. Xiao-Feng Liu and Dr. Tao-Rong Xie for discussions.

Compliance with ethical standards

Human and animal rights and informed consent This article does not contain any studies with human or animal subjects performed by any of the authors.

References

Baffou G, Rigneault H, Marguet D, Jullien L (2014) A critique of methods for temperature imaging in single cells. Nat Methods 11:899–901

Baffou G, Rigneault H, Marguet D, Jullien L (2015) Reply to: 'Validating subcellular thermal changes revealed by fluorescent thermosensors' and 'The 105 gap issue between calculation and measurement in single-cell thermometry'. Nat Methods 12:803

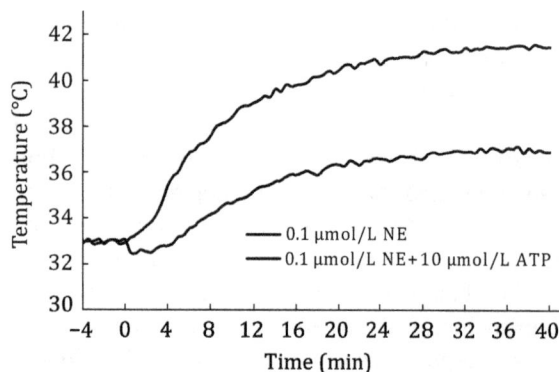

Fig. 1 Change profiles of mitochondrial temperature in BA under stimulation starting from 0 min of 0.1 μmol/L NE without (*black line*) or with (*red line*) 10 μmol/L ATP-induced thermogenesis in BA (Xie *et al.* 2017b)

Bertholet AM, Kazak L, Chouchani ET, Bogaczynska MG, Paranjpe I, Wainwright GL, Bétourné A, Kajimura S, Spiegelman BM, Kirichok Y (2017) Mitochondrial patch clamp of beige adipocytes reveals UCP1-positive and UCP1-negative cells both exhibiting Futile creatine cycling. Cell Metab 25:811–822e.4

Blundell S, Blundell KM (2010) Concepts in thermal physics. OUP, Oxford

Chrétien D, Bénit P, Ha H-H, Keipert S, El-Khoury R, Chang Y-T, Jastroch M, Jacobs HT, Rustin P, Rak M (2018) Mitochondria are physiologically maintained at close to 50 °C. PLoS Biol 16:e2003992

Kiyonaka S, Sakaguchi R, Hamachi I, Morii T, Yoshizaki T, Mori Y (2015) Validating subcellular thermal changes revealed by fluorescent thermosensors. Nat Methods 12:801–802

Lane N (2018) Hot mitochondria? PLoS Biol 16:e2005113

Orava J, Nuutila P, Lidell ME, Oikonen V, Noponen T, Viljanen T, Scheinin M, Taittonen M, Niemi T, Enerbäck S, Virtanen KA (2011) Different metabolic responses of human brown adipose tissue to activation by cold and insulin. Cell Metab 14:272–279

Suzuki M, Zeeb V, Arai S, Oyama K, Ishiwata S (2015) The 105 gap issue between calculation and measurement in single-cell thermometry. Nat Methods 12:802–803

Vallerand AL, Perusse F, Bukowiecki LJ (1990) Stimulatory effects of cold exposure and cold acclimation on glucose uptake in rat peripheral tissues. Am J Physiol 259:R1043–R1049

Xie T-R, Liu C-F, Kang J-S (2017a) Dye-based mito-thermometry and its application in thermogenesis of brown adipocytes. Biophys Rep 3:85–91

Xie T-R, Liu C-F, Kang J-S (2017b) Sympathetic transmitters control thermogenic efficacy of brown adipocytes by modulating mitochondrial complex V. Signal Transduct Target Ther 2:17060. https://doi.org/10.1038/sigtrans.2017.60

Zamorano F, Wouwer AV, Bastin G (2010) A detailed metabolic flux analysis of an underdetermined network of CHO cells. J Biotechnol 150:497–508

Accelerating electron tomography reconstruction algorithm ICON with GPU

Yu Chen[1,2], Zihao Wang[1,2], Jingrong Zhang[1,2], Lun Li[1,3], Xiaohua Wan[1], Fei Sun[2,4,5✉], Fa Zhang[1✉]

[1] Key Laboratory of Intelligent Information Processing, Institute of Computing Technology, Chinese Academy of Sciences, Beijing 100190, China

[2] University of Chinese Academy of Sciences, Beijing 100049, China

[3] School of Mathematical Sciences, University of Chinese Academy of Sciences, Beijing 100049, China

[4] National Key Laboratory of Biomacromolecules, CAS Center for Excellence in Biomacromolecules, Institute of Biophysics, Chinese Academy of Sciences, Beijing 100101, China

[5] Center for Biological Imaging, Institute of Biophysics, Chinese Academy of Sciences, Beijing 100101, China

Abstract   Electron tomography (ET) plays an important role in studying *in situ* cell ultrastructure in three-dimensional space. Due to limited tilt angles, ET reconstruction always suffers from the "missing wedge" problem. With a validation procedure, iterative compressed-sensing optimized NUFFT reconstruction (ICON) demonstrates its power in the restoration of validated missing information for low SNR biological ET dataset. However, the huge computational demand has become a major problem for the application of ICON. In this work, we analyzed the framework of ICON and classified the operations of major steps of ICON reconstruction into three types. Accordingly, we designed parallel strategies and implemented them on graphics processing units (GPU) to generate a parallel program ICON-GPU. With high accuracy, ICON-GPU has a great acceleration compared to its CPU version, up to 83.7×, greatly relieving ICON's dependence on computing resource.

Keywords   Acceleration, Electron tomography, GPU, ICON, Missing wedge restoration

INTRODUCTION

Electron tomography (ET) plays an important role in studying *in situ* cell ultrastructure in three-dimensional space (Yahav *et al.* 2011; Fridman *et al.* 2012; Rigort *et al.* 2012; Lučić *et al.* 2013). Combining with a sub-volume averaging approach (Castaño-Díez *et al.* 2012), ET demonstrates its power in investigating high-resolution *in situ* conformational dynamics of macro-molecular complexes. Due to limited tilt angles, traditional ET reconstruction algorithms including weighted

back projection (WBP) (Radermacher 1992), simultaneous iterative reconstruction technique (SIRT) (Gilbert 1972), direct Fourier reconstruction (DFR) (Mersereau 1976), iterative non-uniform fast Fourier transform (NUFFT) reconstruction (INFR) (Chen and Förster 2014), *etc.*, always suffer from the "missing wedge" problem, which causes density elongation and ray artifacts in the reconstructed structure. Such ray artifacts will blur the structural details of the reconstruction and weaken the further biological interpretation (Lučić *et al.* 2005).

In recent years, many algorithms have been proposed to deal with the "missing wedge" problem. Some of them apply prior constrains to the reconstructed tomogram to compensate the missing wedge, such as filtered

Yu Chen and Zihao Wang have contributed equally to this work.

✉ Correspondence: feisun@ibp.ac.cn (F. Sun),
zhangfa@ict.ac.cn (F. Zhang)

iterative reconstruction technique (FIRT) (Chen *et al.* 2016), discrete algebraic reconstruction technique (DART) (Batenburg and Sijbers 2011), and projection onto convex sets (POCS) (Sezan and Stark 1983; Carazo and Carrascosa 1987). These constraints include density smoothness, density non-negativity, density localness, *etc.* Others try to solve the reconstruction problem as an underdetermined problem based on a theoretical framework called "compressed sensing" (CS) (Donoho 2006). Compressed sensing electron tomography (Saghi *et al.* 2011, 2015; Goris *et al.* 2012; Leary *et al.* 2013) demonstrated certain success for the data with a high signal to noise ratio (SNR) (*e.g.*, material science data or resin-embedded section data). To cope with the low SNR case (*e.g.*, biological cryo-ET data, in which a low total dose of electron is used to avoid significant radiation damage), Deng *et al.* proposed iterative compressed-sensing optimized NUFFT reconstruction (ICON) by combining CS and NUFFT together (Deng *et al.* 2016). With a validation procedure, ICON not only restores the missing information but also measures the fidelity of the information restoration. ICON demonstrated its power in the restoration of validated missing information for low SNR biological ET dataset.

However, the convergence process of ICON is time-consuming. The huge computational demand has become a major problem for the application of ICON. The traditional solution to cope with the high computational cost has been the use of supercomputers and large computer clusters (Fernández *et al.* 2004; Fernández 2008), but such hardware is expensive and can also be difficult to use. Graphics processing units (GPU) (Lindholm *et al.* 2008) can be the attractive alternative solution in terms of price and performance. In this work, we developed the parallel strategies of ICON and implemented a GPU version of ICON, named ICON-GPU. Experimental results based on a Tesla K20c GPU card showed that ICON-GPU exhibits the same accuracy and a significant acceleration in comparison with the CPU version of ICON (ICON-CPU).

RESULTS AND DISCUSSION

Reconstruction precision

First, we evaluated the numerical accuracy of ICON-GPU using the root-mean-square relative error (RMSRE) ε as Eq. 1. To avoid dividing 0 when calculating the RMSRE, we first normalized the reconstructed slices into [0,1] using Eq. 2.

$$\varepsilon = \sqrt{\frac{\sum_{i=1}^{N}\left(\frac{Pnorm_i - Cnorm_i}{Cnorm_i}\right)^2}{N}}, \tag{1}$$

where N is the size of one slice; $Cnorm$ is the normalized slice reconstructed by ICON-CPU; $Cnorm_i$ is the value of the ith pixel in $Cnorm$; $Pnorm$ is the normalized slice reconstructed by ICON-GPU; $Pnorm_i$ is the value of the ith pixel in $Pnorm$.

$$Pnorm = \frac{P - minP}{maxP - minP} + c, \tag{2}$$

where $Pnorm$ is the normalized slice; P is the originally reconstructed slice; $minP$ is the minimum value of P; $maxP$ is the maximum value of P; c is a small constant to avoid 0 in $Pnorm$, in this work, $c = 10^{-7}$.

The RMSRE of ICON-GPU increases slowly with the image size; they are in the range of $(6 \times 10^{-7}, 4 \times 10^{-6})$ yielding a reasonable numerical accuracy for the float format data (Fig. 1).

Then, we evaluated the reconstruction accuracy by investigating the reconstructed tomograms. The *XY*-slices reconstructed by ICONs (Fig. 2B, C) show better SNR than that by WBP (Fig. 2A), yielding a better contrast to discriminate the cellular ultrastructures. Besides, ICON-CPU and ICON-GPU are identical with each other and the normalized cross-correlation (NCC) between them is 1. The *XZ*-slices reconstructed by ICONs (Fig. 2E, F) are also identical with each other and the ray artifacts in ICONs are significantly reduced in comparison with WBP (Fig. 2D). To be noted that, to eliminate any suspicion on the gray-scale manipulation (which could enhance the visual advantage), all images were normalized and displayed based on their minimum and maximum value.

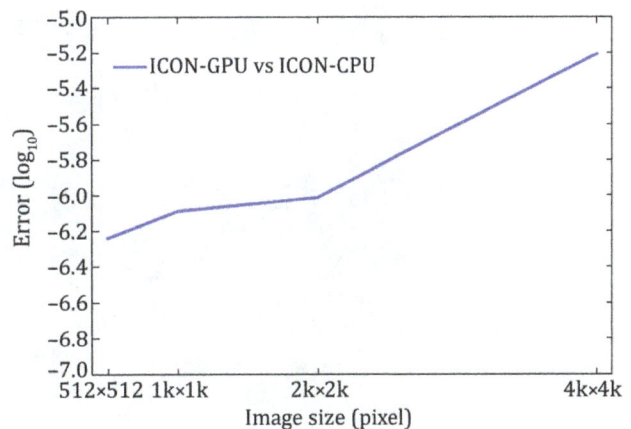

Fig. 1 The RMSREs of ICON-GPU

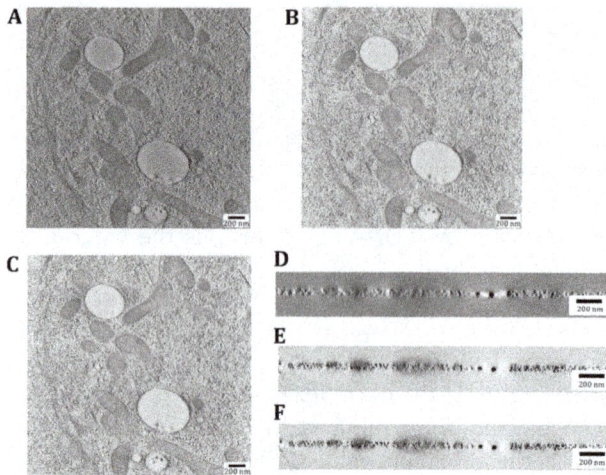

Fig. 2 Evaluate ICON-GPU by investigating the reconstructed tomograms. **A–C** The *XY*-slices of the tomograms reconstructed by WBP, ICON-CPU, and ICON-GPU, respectively; **D–F** The *XZ*-slices of the tomograms reconstructed by WBP, ICON-CPU, and ICON-GPU, respectively

We further investigated the reconstruction accuracy by the pseudo-missing-validation procedure (Deng *et al.* 2016). Here, the $-0.29°$ tilt (the minimum tilt) projection was excluded as the omit-projection ("ground truth") (Fig. 3A). We re-projected the reconstructed omit-tomograms at $-0.29°$. The re-projections of ICONs

(Fig. 3C, D) are identical with each other and the NCC between them is 1. The re-projections of ICONs are clearer in detailed structures and close to the "ground truth", compared to that of WBP (Fig. 3B). Such visual assessments were further verified quantitatively by comparing the Fourier ring correlation (FRC) curves between the re-projections and the "ground truth". The FRCs of ICONs coincide with each other, and they are better than that of WBP (Fig. 3E). The coincident FRCs of ICONs further demonstrate the accuracy of ICON-GPU from the perspective of restoring missing information.

Speed up

We evaluated the acceleration of ICON-GPU by comparing the running time of reconstructing one slice under 200 iterations. We reconstructed the datasets with sizes of 512×512, $1 \text{ k} \times 1 \text{ k}$, $2 \text{ k} \times 2 \text{ k}$, $4 \text{ k} \times 4 \text{ k}$, respectively. The acceleration of ICON-GPU improves when the slice size increases (Fig. 4; Table 1). The maximum speedup is $83.7\times$ in the reconstruction of a $4 \text{ k} \times 4 \text{ k}$ slice. With the efficient acceleration, the reconstruction time of one $4 \text{ k} \times 4 \text{ k}$ slice is reduced from hours to minutes, which greatly relieves ICON's dependence on computing resource.

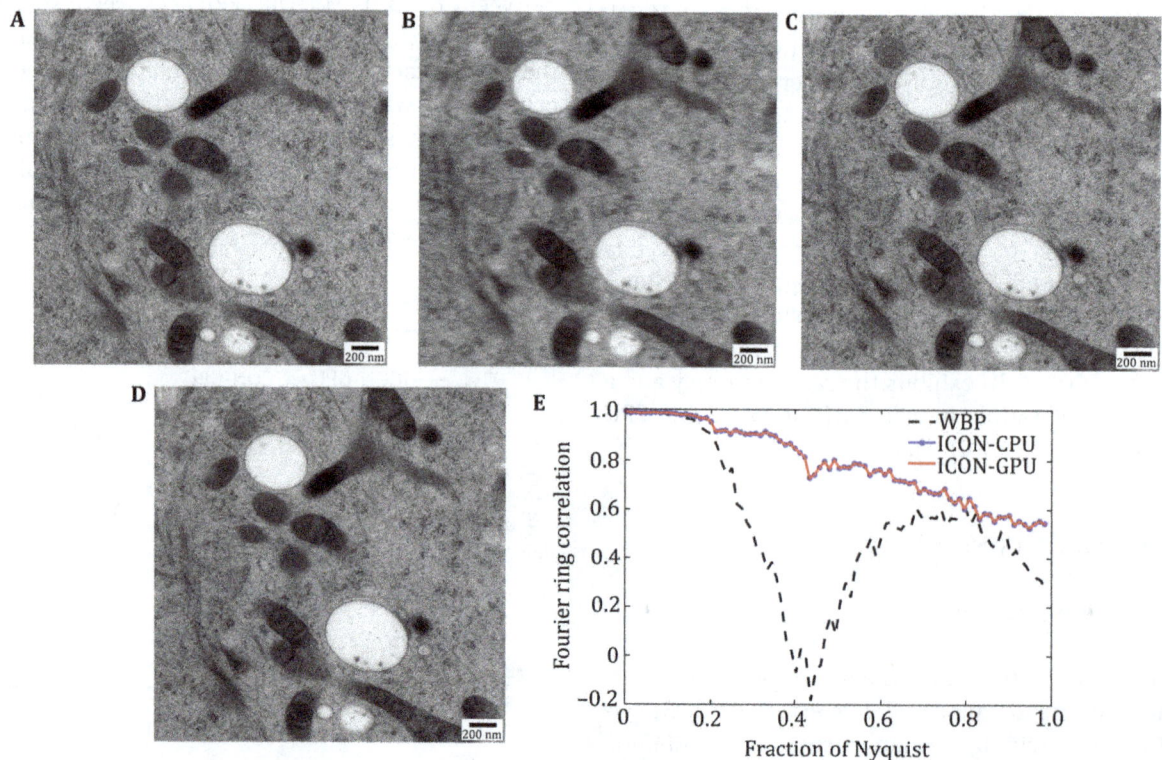

Fig. 3 Evaluate ICON-GPU by the pseudo-missing-validation procedure. **A** The omit-projection ("Ground truth"); **B–D** The re-projections of the omit-tomograms reconstructed by WBP, ICON-CPU, and ICON-GPU, respectively; **E** The pseudo-missing-validation FRCs of WBP, ICON-CPU, and ICON-GPU

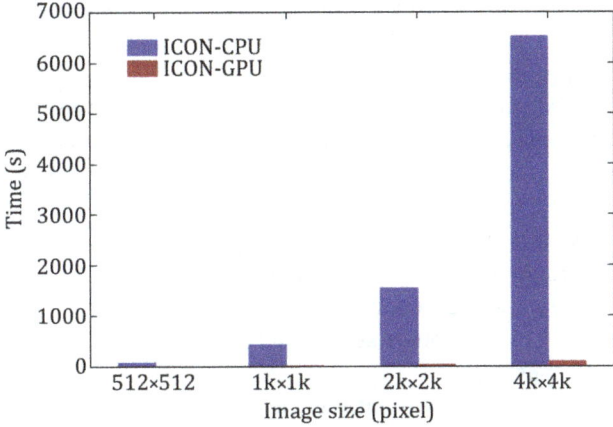

Fig. 4 The comparison of time-consuming of ICON-CPU and ICON-GPU

Table 1 The speedups of ICON-GPU compared to ICON-CPU

Image size	Speedup
512^2	8.7×
1024^2	41.5×
2048^2	61.2×
4096^2	83.7×

CONCLUSIONS

In the present work, we analyzed the iterative framework of ICON and classified the operations of ICON reconstruction into three types. Accordingly, we designed parallel strategies and implemented them on GPU to generate a parallel program ICON-GPU. We tested ICON-GPU on a resin-embedded ET dataset of MDCK cell section. The RMSRE between ICON-GPU and ICON-CPU is about $10e^{-6}$, yielding a reasonable numerical accuracy of ICON-GPU compared to ICON-CPU. In addition, ICON-GPU has the same ability of restoring missing information with ICON-CPU. In addition, ICON-GPU has a great acceleration, up to 83.7× in the reconstruction of one 4 k × 4 k slice in comparison with ICON-CPU.

To be noted that, ICON-GPU can also run on multiple GPU system such as TIANHE-2, a supercomputer developed by China's National University of Defense Technology, which is based on multi-core and many-core architectures (Liao *et al.* 2014).

The software package of ICON-GPU can be obtained from our website (http://feilab.ibp.ac.cn/LBEMSB/ICON.html or http://ear.ict.ac.cn).

MATERIALS AND METHODS

Iterative compressed-sensing optimized NUFFT reconstruction (ICON)

ICON is an iterative reconstruction algorithm based on the theoretical framework of "compressed sensing" and is designed to restore missing information caused by limited angular sampling (Deng *et al.* 2016). ICON is formulated as Eq. 3.

$$argmin \parallel Px \parallel_{L_0}, \; subject\;to: \parallel A^h WAx - A^h Wf \parallel_{L_2} < \varepsilon, \tag{3}$$

where x is the two-dimensional (2D) reconstructed slice; W follows INFR's description (Chen and Förster 2014) and contains the weights that account for the non-uniform sampling in the Fourier space (similar to the ramp filtering in WBP); A is the projection operation, defined as a non-uniform Fourier sampling matrix, which performs Fourier transform on the non-integer grid points (NUFFT); A^h stands for the conjugate transpose of A; f is the Fourier transform of acquired projections; $\parallel \cdot \parallel_{L_2}$ is an operator that calculates the Euclidean norm (L_2-norm); ε is a control parameter that is determined empirically according to the noise level; $\parallel \cdot \parallel_{L_0}$ stands for the operator that calculates the number of the non-zero terms. P is a diagonal sparse transformation matrix, whose diagonal element $\emptyset$ is defined as in Eq. 4.

$$\emptyset x = \emptyset(x) \stackrel{def}{=} \begin{cases} 0, (if \quad x < 0) \\ 1, (if \quad x \geq 0) \end{cases}. \tag{4}$$

The complete workflow of ICON can be divided into four steps (Deng *et al.* 2016):

Step 1   Pre-processing. Align tilt series and correct contrast transfer function (CTF).

Step 2   Gray value adjustment. Subtract the most frequently appeared pixel value in the micrographs, which is given from the embedding material (*e.g.*, resin or vitrified ice).

Step 3   Reconstruction and pseudo-missing-validation. Reconstruct tilt series into a 3D volume with an iterative procedure of fidelity preservation and prior sparsity restriction, and evaluate the restored information with pseudo-missing-validation.

Step 4   Verification filtering. Exclude the incorrectly restored information.

A series of tests showed that Step 3 accounts for at least 95% of the execution time of ICON. Thus, the major task for accelerating ICON is parallelizing Step 3 effectively on GPU. Since the procedures of "reconstruction" and "pseudo-missing-validation" are similar, only the parallelization of "reconstruction" will be discussed in this paper.

The major steps of "reconstruction" can be briefly described as followed.

Step 3.1: Fidelity preservation step. In this step, steepest descent method (Goldstein 1965) is used to calculate the subject function of Eq. 3 as follows:

$$r = A^h WA x^k - A^h Wf, \tag{5}$$

$$\alpha = \frac{r^T r}{r^T A^h WA r}, \tag{6}$$

$$y^{k+1} = x^k - \alpha r, \tag{7}$$

where x^k is the 2D reconstructed slice of the kth iteration, r is the residual, α is the coefficient used to control the step of updating, y^{k+1} is the intermediate updating result of the $(k+1)$th iteration.

Step 3.2: Prior sparsity restriction step. The diagonal sparse transformation matrix P can be re-formulated as a "hard threshold"-like operation as in Eq. 8:

$$x^{k+1} = H(y^{k+1}) = \begin{cases} 0, & \text{if } y^{k+1} < 0 \\ y^{k+1}, & \text{if } y^{k+1} \geq 0 \end{cases}, \tag{8}$$

where y^{k+1} is the intermediate updating result of the $(k+1)$th iteration. $H(\cdot)$ is a thresholding function, x^{k+1} is the 2D reconstructed slice of the $(k+1)$th iteration.

We classified the operations of these two steps into three types: (1) the summation of a matrix; (2) element-wise operations of matrices; (3) the NUFFT and the adjoint NUFFT. For a fast summation of matrix, we took advantage of the API function *cublasSasum* from the standard CUDA library cuBLAS (NVIDIA Corp, 2007). For type 1 and 2, parallel strategies are proposed in the following sections.

Parallelizing element-wise operations of matrices

GPU is a massively multi-threaded data-parallel architecture, which contains hundreds of scalar processors (SPs) (Lindholm *et al.* 2008). NVIDIA provides the programming model on GPU called CUDA. The CUDA program running on GPU is called Kernel, which consists of thousands of threads. Thread is the basic running unit in CUDA programming model and it has a three-level hierarchy: grid, block, thread. Besides, CUDA devices use several memory spaces including global, shared, texture, and registers. Of these different memory spaces, global memory is the largest but slowest in data accessing. CUDA provides API function *cudaMemcpy* to transfer data between host memory and device

memory; the time-consuming of such transfer sometimes is non-negligible especially for an iterative procedure like ICON reconstruction.

Since micrographs in ET are usually large (*e.g.*, 2 k × 2 k or 4 k × 4 k in float or short format) and exceed the limitation of most types of CUDA device memory (*e.g.*, 16 or 48 KB for shared memory), data in ICON-GPU are restored in global memory using float format. In order to cut down the time-consuming of memory transfer, we parallelized all operations of ICON on GPU even though some operations may have negligible speedups.

To deal with element-wise operation, ICON-GPU uses a 2D distribution of threads with a fixed block size of 32×32 and a fixed grid size of 4β, β is a parameter to be determined according to the matrix size N. ICON-GPU assigns the operation of one element to one thread according to the index of element. Pseudo codes for calling a kernel function and the operations inside a kernel function are shown in Fig. 5.

Parallelizing NUFFT and adjoint NUFFT

First, we give a brief description of NUFFT. Given the Fourier coefficients $\hat{f}_k \in \mathbb{C}, k \in I_N$ and $I_N = \left\{ k = (k_t)_{t=0,\dots,d-1} \in \mathbb{Z}^d : - \frac{N_t}{2} \leq k_t < \frac{N_t}{2}, t = 0, \dots, d - 1 \right\}$ as input, NUFFT tries to evaluate the following trigonometric polynomial efficiently at the reciprocal points $x_j \in \left[-\frac{1}{2}, \frac{1}{2} \right)^d$, $j = 0, \dots, M-1$:

$$f_j = f(x_j) = \sum_{k \in I_N} \hat{f}_k e^{-2\pi i k x_j}, \quad j = 0, \dots, M-1. \tag{9}$$

Correspondingly, the adjoint NUFFT tries to evaluate Eq. 10 at the frequency k.

$$\hat{h}_k = \sum_{j=0}^{M-1} f_j e^{2\pi i k x_j}. \tag{10}$$

Call of Kernel function:

operationFunc«<$dim3\left(4, \frac{\left(\frac{N-1}{1024}\right)}{4} + 1\right)$, $dim3(32,32)$»>(*input, output*);

- -

Operation inside Kernel function:

input, output: global memory
//find the index of element (thread)
blockid=blockIdx.x+blockIdx.y×gridDim.x
threadid=threadIdx.x+threadIdx.y×blockDim.x
index=threadid+blockid×(blockDim.x×blockDim.y)
//operation
if (index<N)
    output[index]=operation(input[index])

Fig. 5 Pseudo codes for calling a kernel function and the operations inside a kernel function for element-wise operations

NFFT3.0 (Keiner *et al.* 2010), a successful and widely used open source C library, is used in ICON-CPU for NUFFT and adjoint NUFFT. Yang *et al.* proposed a different theoretical derivation of NFFT and demonstrated the high efficiency of GPU acceleration of NFFT (Yang *et al.* 2015, 2016). To make ICON-GPU consistent with ICON-CPU, in this work, we parallelized the NUFFT and the adjoint NUFFT based on the algorithms described in NFFT3.0 and the algorithm of 2D NUFFT is displayed in Algorithm 1.

Algorithm 1: NUFFT

Input : $M, N = \{N_1, N_2\}, \sigma = \{\sigma_1, \sigma_2\}, m, x_j \in \left[-\frac{1}{2}, \frac{1}{2}\right]^2, j = 0, \dots, M-1, \hat{f}_k \in \mathbb{C}, k \in I_N,$

$$n = \sigma N = \{n_1, n_2\} = \{\sigma_1 N_1, \sigma_2 N_2\}.$$

1: For $k \in I_N$ compute

$$\hat{g}_k = \frac{\hat{f}_k}{|I_n| c_k(\hat{\varphi})},$$

$$c_k(\hat{\varphi}) = \hat{\varphi}(k_1)\hat{\varphi}(k_2).$$

2: For $l \in I_n$ compute by 2-variate FFT

$$g_l = \sum_{k \in I_N} \hat{g}_k e^{-2\pi i k (n^{-1} \odot l)}.$$

3: For $j = 0, \dots, M-1$ compute

$$f_j = \sum_{l \in I_{n,m}(x_j)} g_l \tilde{\psi}(x_j - n^{-1} \odot l),$$

$$I_{n,m}(x_j) = \{l \in I_n : n \odot x_j - m\mathbf{1} \leq l \leq n \odot x_j + m\mathbf{1}\},$$

$$\tilde{\psi}(x_j - n^{-1} \odot l) = \varphi(x_{1j} - \frac{l_1}{n_1})\varphi(x_{2j} - \frac{l_2}{n_2}).$$

$\varphi(x)$ and $\hat{\varphi}(k)$ are the window functions. In this work, the (dilated) Gaussian window functions (Eqs. 11, 12) are used.

$$\varphi(x) = (\pi b)^{-\frac{1}{2}} e^{-\frac{(nx)^2}{b}} \quad \left(b = \frac{2\sigma}{2\sigma - 1}\frac{m}{\pi}\right), \tag{11}$$

$$\hat{\varphi}(k) = \frac{1}{n} e^{-b\left(\frac{\pi k}{n}\right)^2}, \tag{12}$$

where x is a component of the reciprocal points x, k is a component of the frequencies k, σ is a component of the oversampling factors σ with $\sigma > 1$. In this work, $\sigma = 2$, n is one component of $n = \sigma N$, $m \in \mathbb{N}$ and $m \ll n$. In this work, $m = 6$.

The operations in 2D NUFFT and 2D adjoint NUFFT can be classified into three types: (1) element-wise operations of matrices; (2) 2D FFT; (3) calculation of window functions $\varphi(x)$ and $\hat{\varphi}(k)$. The parallel strategy of type 1 is the same as the strategy described in Section "Parallelizing element-wise operations of matrices." For type 2, to achieve a high performance FFT, we took advantage of the NVIDIA's FFT library, CUFFT (NVIDIA Corp 2007). Since ICON is an iterative algorithm, 2D NUFFT and 2D adjoint NUFFT will be repeated many

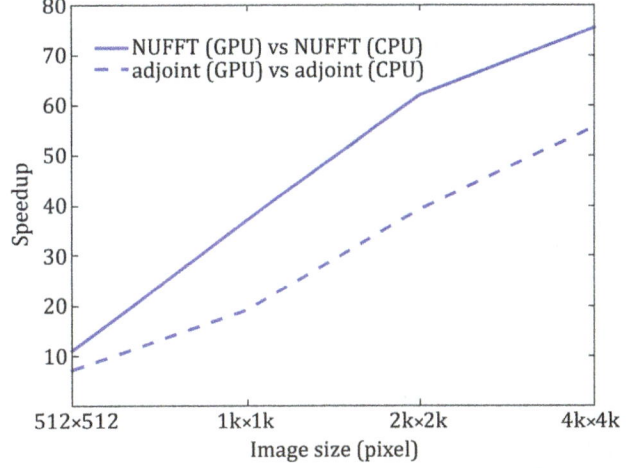

Fig. 6 The speedups of parallel NUFFTs compared to NFFT3.0

times. To cut down the time of calculation and memory transfer, we pre-computed the window functions for once and stored them in the device memory.

Parallel NUFFTs were tested using a resin-embedded ET dataset (see "Resin embedded ET Dataset" for details). Here, all CPU programs ran on one core (thread) of an Intel® Xeon™ CPU E5-2620 v2 @ 2.1 GHz (six cores per CPU) and all GPU programs ran on a NVIDIA Tesla K20c (2496 CUDA cores and 5 GB device memory). The accelerations of parallel NUFFTs improve when the image size increases and are up to 75.4x for NUFFT and 55.7x for adjoint NUFFT in the transform of one 4 k × 4 k image (Fig. 6).

Resin-embedded ET dataset

We tested ICON-GPU using a resin-embedded ET dataset of MDCK cell section. The tilt angles of the dataset originally ranged from −68° to +68° with 1° increment. In order to verify ICON-GPU's ability of restoring missing information, we extracted every other projection from the original dataset to generate a new tilt series with 2° increment for the following experiments. The tilt series were aligned using atom align (Han *et al.* 2014). The original image size is 4 k × 4 k with a pixel size of 0.72 nm. We also compressed the tilt series with factors of two, four, eight to generate datasets with smaller sizes of 2 k × 2 k, 1 k × 1 k, and 512 × 512, respectively.

Acknowledgements We would like to thank Prof. Wanzhong He (NIBS, Beijing) for providing the resin-embedded ET dataset. All the intensive computations were performed at TIANHE-2 supercomputer in National Supercomputer Center in Guangzhou and at the high performance computers in Center for Biological Imaging, Institute of Biophysics, Chinese Academy of Sciences (http://cbi.ibp.ac.cn). This work was supported by the National Natural Science Foundation of China (U1611263, U1611261, 61232001, 61472397, 61502455, 61672493), Special Program for Applied Research on Super Computation of the NSFC-Guangdong Joint Fund (the second phase), the Strategic Priority Research Program of Chinese Academy of Sciences (XDB08030202), and the "973" Program of Ministry of Science and Technology of China (2014CB910700).

Compliance with Ethical Standards

Conflict of interest Yu Chen, Zihao Wang, Jingrong Zhang, Lun Li, Xiaohua Wan, Fei Sun, Fa Zhang declare that they have no conflict of interest.

Human and animal rights and informed consent All institutional and national guidelines for the care and use of laboratory animals were followed.

References

Batenburg K, Sijbers J (2011) Dart: a practical reconstruction algorithm for discrete tomography. IEEE Trans Image Process 20(9):2542

Carazo JM, Carrascosa JL (1987) Restoration of direct Fourier three-dimensional reconstructions of crystalline specimens by the method of convex projections. J Microsc 145(Pt 2):159–177

Castaño-Díez D, Kudryashev M, Arheit M, Stahlberg H (2012) Dynamo : a flexible, user-friendly development tool for subtomogram averaging of cryo-em data in high-performance computing environments. J Struct Biol 178(2):139

Chen Y, Förster F (2014) Iterative reconstruction of cryo-electron tomograms using nonuniform fast Fourier transforms. J Struct Biol 185(3):309–316

Chen Y, Zhang Y, Zhang K, Deng Y, Wang S, Zhang F, Sun F (2016) Firt: filtered iterative reconstruction technique with information restoration. J Struct Biol 195(1):49–61

Deng Y, Chen Y, Zhang Y, Wang S, Zhang F, Sun F (2016) Icon: 3d reconstruction with 'missing-information' restoration in biological electron tomography. J Struct Biol 195(1):100

Donoho DL (2006) Compressed sensing. IEEE Trans Inf Theory 52(4):1289–1306

Fernández JJ (2008) High performance computing in structural determination by electron cryomicroscopy. J Struct Biol 164(1):1–6

Fernández JJ, Carazo JM, García I (2004) Three-dimensional reconstruction of cellular structures by electron microscope tomography and parallel computing. J Parallel Distrib Comput 64(2):285–300

Fridman K, Mader A, Zwerger M, Elia N, Medalia O (2012) Advances in tomography: probing the molecular architecture of cells. Nat Rev Mol Cell Biol 13(13):736–742

Gilbert P (1972) Iterative methods for the three-dimensional reconstruction of an object from projections. J Theor Biol 36(1):105–117

Goldstein AA (1965) On steepest descent. J Soc Ind Appl Math 3(1):147–151

Goris B, Broek WVD, Batenburg KJ, Mezerji HH, Bals S (2012) Electron tomography based on a total variation minimization reconstruction technique. Ultramicroscopy 113(1):120–130

Han R, Zhang F, Wan X, Fernández JJ, Sun F, Liu Z (2014) A marker-free automatic alignment method based on scale-invariant features. J Struct Biol 186(1):167–180

Keiner J, Kunis S, Potts D (2010) Using NFFT 3—a software library for various nonequispaced fast Fourier transforms. ACM Trans Math Softw 36(4):1–30

Leary R, Saghi Z, Holland PAMDJ (2013) Compressed sensing electron tomography: theory and applications. Ultramicroscopy 131(8):70–91

Liao X, Xiao L, Yang C, Lu Y (2014) Milkyway-2 supercomputer: system and application. Front Comput Sci 8(3):345–356

Lindholm E, Nickolls J, Oberman S, Montrym J (2008) NVIDIA tesla: a unified graphics and computing architecture. IEEE Micro 28(2):39–55

Lučić V, Förster F, Baumeister W (2005) Structural studies by electron tomography: from cells to molecules. Annu Rev Biochem 74(1):833

Lučić V, Rigort A, Baumeister W (2013) Cryo-electron tomography: the challenge of doing structural biology in situ. J Cell Biol 202(3):407–419

Mersereau RM (1976) Direct Fourier transform techniques in 3-d image reconstruction. Comput Biol Med 6(4):247

NVIDIA Corp (2007) CUDA CUFFT Library

Radermacher M (1992) Weighted back-projection methods. In: Frank J (ed) Electron tomography. Springer, Berlin, pp 91–115

Rigort A, Villa E, Bäuerlein FJB, Engel BD, Plitzko JM (2012) Chapter 14—integrative approaches for cellular cryo-electron tomography: correlative imaging and focused ion beam micromachining. Methods Cell Biol 111:259–281

Saghi Z, Holland DJ, Leary R, Falqui A, Bertoni G, Sederman AJ, Gladden LF, Midgley PA (2011) Three-dimensional morphology of iron oxide nanoparticles with reactive concave surfaces. A compressed sensing-electron tomography (CS-ET) approach. Nano Lett 11(11):4666–4673

Saghi Z, Divitini G, Winter B, Leary R, Spiecker E, Ducati C, Midgley PA (2015) Compressed sensing electron tomography of needle-shaped biological specimens—potential for improved reconstruction fidelity with reduced dose. Ultramicroscopy 160:230–238

Sezan MI, Stark H (1983) Image restoration by convex projections in the presence of noise. Appl Opt 22(18):2781

Yahav T, Maimon T, Grossman E, Dahan I, Medalia O (2011) Cryo-electron tomography: gaining insight into cellular processes by structural approaches. Curr Opin Struct Biol 21(5):670–677

Yang SC, Wang YL, Jiao GS, Qian HJ, Lu ZY (2015) Accelerating electrostatic interaction calculations with graphical processing units based on new developments of ewald method using non-uniform fast Fourier transform. J Comput Chem 37(3):378

Yang SC, Qian HJ, Lu ZY (2016) A new theoretical derivation of NFFT and its implementation on GPU. Appl Comput Harmon Anal. doi:10.1016/j.acha.2016.04.009

3

The application of CorrSight™ in correlative light and electron microscopy of vitrified biological specimens

Xiaomin Li[1], Jianlin Lei[1,2][✉], Hong-Wei Wang[2][✉]

[1] Technology Center for Protein Sciences, Ministry of Education Key Laboratory of Protein Sciences, School of Life Sciences, Tsinghua University, Beijing 100084, China
[2] Beijing Advanced Innovation Center for Structural Biology, School of Life Sciences, Tsinghua University, Beijing 100084, China

Abstract   Correlative light and electron microscopy is a powerful technique for identification and determination of the structures of interested macromolecules *in situ*. Combined with sample vitrification, it would be much easier to preserve the native state of macromolecule complexes and distinguish them from the crowded structure environment. In this article, we present a detailed process for the application of the CorrSight system, a light microscope equipped with a cryo module, in combination with a cryo-electron microscope. A relatively long course of up to 7–8 h for cryo module preparation and multichannel light microscopy imaging of vitrified specimen can be sustained. Correlation of light and electron microscopy images at both grid levels to locate squares and square level to locate target particles, and verification of target particles can be performed with the help of AutoEMation software. Cryo-electron tomography is used for obtaining the three-dimensional structure information.

Keywords   CorrSight system, Vitrified biological specimens, Light microscope, Cryo-electron microscope

INTRODUCTION

Vitrification of biological specimens in liquid nitrogen (LN2) temperature has been proved to be a powerful technique to preserve the native structures of macromolecules either *in vitro* or *in situ*. In combination with the most recent hardware and software breakthroughs, cryo-fixed macromolecule complexes isolated from cells can be determined at near atomic resolution using the single particle reconstruction method (Ma *et al.* 2017; Peng *et al.* 2016; Wang 2015; Wang *et al.* 2017a). However, it is still very challenging to identify and determine the structures of interested macromolecules *in situ* with cryo-electron microscopy (cryo-EM) (Lucic *et al.* 2008; Oikonomou and Jensen 2017; Plitzko *et al.*

2009). The major hurdle lies in the fact that many different kinds of macromolecules are crowded within the cell and therefore are hard to be distinguished from each other simply by their shapes revealed in cryo-EM (Bauerlein *et al.* 2017; Zhang 2013). Furthermore, the large dimension of a cell and a rather small viewing area of EM at a high magnification render it more difficult to localize relatively sparse molecules in a specific cellular structure environment (Plitzko *et al.* 2009). Light microscopy (LM), more specifically, fluorescence microscopy (FM) is well developed to label specific molecules and localize them in a large viewing area with a resolution of several hundreds of nanometers. The advantages of specific labeling and localization in LM and high spatial resolution in EM can be combined, leading to the development of a powerful technique as the correlative light and electron microscopy (CLEM) (Anderson *et al.* 2017; Compera *et al.* 2015; Faas *et al.* 2013; Koning *et al.* 2014; Sjollema *et al.* 2012).

✉ Correspondence: jllei@tsinghua.edu.cn (J. Lei), hongweiwang@tsinghua.edu.cn (H.-W. Wang)

CLEM technique was first reported in the early 1960s for adenovirus study with conventional chemical fixation EM specimen preparation (Godman *et al.* 1960; Morgan *et al.* 1960). While the chemical fixed specimens are still the majority of targets investigated by CLEM (Hellstrom *et al.* 2015; Kobayashi *et al.* 2016; Kong and Loncarek 2015; Loussert Fonta and Humbel 2015; Mourik *et al.* 2014), vitrified biological specimens become more and more popular to avoid the potential artifacts and damages caused by the chemical fixation. Cryo-CLEM is thus developed in order to visualize vitrified specimens by cryo-EM (Bykov *et al.* 2016; Mahamid *et al.* 2016; Wolff *et al.* 2016). Currently, two types of methods are used for correlation: (1) correlative LM/EM with freezing after FM imaging (Briegel *et al.* 2010). FM imaging of samples at room temperature could be done with any typical FM instrument. FM can have better resolution with high-powered, oil-immersion lenses with large numerical apertures. After FM imaging, the sample needs to be frozen for further cryo-EM imaging. Various factors can be introduced to the specimen between the LM imaging and specimen vitrification, causing more difficulties for accurate correlation. Therefore, it works better for specimen that is relatively stable and the fluorescently marked structure would not change in the seconds to minutes before vitrification. (2) Correlative LM/EM with freezing before FM imaging (Briegel *et al.* 2010). The sample is plunge frozen or cryo-sectioned first, then the grid is visualized under LM and EM sequentially with little disturbance of its structure, thus making the correlation more consistent. Such a practice needs the cryo-fixed specimen to be maintained always in a humidity-free environment and below ~ -140 °C during the whole CLEM process to prevent ice contamination on the specimen surface or recrystallization in the specimen. Special specimen stages designed for the related instruments are essential to preserve the cryo-fixed specimens (Li *et al.* 2018; Liu *et al.* 2015; Plitzko *et al.* 2009; Rigort *et al.* 2010; Schorb and Briggs 2014; Schorb *et al.* 2017; Zhang 2013). While the first type of CLEM method provides convenience of use and better resolution of LM, the second type guarantees more accurate correlation and is more favored (Briegel *et al.* 2010).

For the second type of CLEM method, two different designs are commercially available. One is the integrated (Agronskaia *et al.* 2008; Faas *et al.* 2013) and the other is the separated. The first integrated CLEM system called Tecnai with iCorr (Thermo Fisher Scientific Inc.) allows the stage to be tilted 90° to switch between the LM and EM mode for direct correlation of a specimen *in situ*, preventing the specimen from potential contamination and damage. The experiments efficiency is greatly increased without repeated grid transfers (Agronskaia *et al.* 2008). Within the iCorr system, however, only small light objective lens as $15\times/0.5$ NA can be integrated due to the limited space between the pole pieces of a standard Tecnai Spirit TEM. The iCorr system thus can only have fluorescence signals at the green channel with a low resolution of about 500 nm (Wang *et al.* 2017b). In comparison, separated CLEM systems such as the CorrSight system equipped with a cryo module from Thermo Fisher Scientific Inc. (Arnold *et al.* 2016) and Leica cryo-CLEM System (Hampton *et al.* 2017) allow the observation of a cryo-fixed specimen in a more advanced LM at better resolution and in an EM with high acceleration voltage. Different image processing software packages have been developed for high-precision correlation. The study of virus-infected or transfected mammalian cells by using Leica Cryo system has been reported. Here we present a detailed process for the application of the CorrSight system in combination with a cryo-TEM. We used plunge-frozen technique to fix the biological specimen and cryo-electron tomography (cryo-ET) for obtaining the three-dimensional (3D) structure information.

EXPERIMENT PROCEDURES

Sample preparation

In our CLEM study, biological samples were vitrified and assembled into EM Autogrids as below:

(1) GiG R3.5/1 200 mesh grids (GiG C200F1, Changshu Zhongke Xinghua Technology) with indexed letters and numbers were glow discharged by means of a basic plasma cleaner PDC-32G-2 (Harrick Plasma).

(2) After biological sample was applied onto the grids, cryo-EM grids were prepared using Vitrobot Mark IV (Thermo Fisher Scientific Inc.) at 20 °C and with 100 percent humidity. The blot force is set as −2 and the blot time is set as 2–2.5 s. The vitrified specimen on cryo-EM grids was stored in LN2 dewars for further examination.

(3) The assembly of Autogrids is executed inside a regular Cryo Transfer Workstation (Thermo Fisher Scientific Inc.) for a TEM with autoloader. Each grid is placed inside a C-clip ring (Thermo Fisher Scientific Inc.) with the sample-application side facing down and then clipped with a C-clip (Thermo Fisher Scientific Inc.) from the top to form a sandwich-type Autogrid.

(4) In the cryo shuttle and Transfer Box Assembly Workstation (Thermo Fisher Scientific Inc.) of the

CorrSight system, the Autogrids are mounted on a cryo shuttle with the C-clip ring side facing up (Fig. 1). The metal spacing cylinder is then put on top of the Autogrids, and the Plexiglas lid is closed immediately. Two Autogrids can be loaded into one cryo shuttle. The cryo shuttle can be stored in the transfer box.

The steps (2)–(4) should all be performed at LN2 temperature. The Autogrids or transfer boxes are all stored in LN2 dewars for further cryo-LM or cryo-EM imaging.

Fluorescence imaging with the CorrSight system

The flowchart of fluorescence imaging with the Corr-Sight system is illustrated as shown in Fig. 2. Before LM imaging of cryo samples, purging of the objective (chamber) with dry nitrogen gas and cooling down of the cryo module with liquid nitrogen are processed for a total time of 2–3 h (Fig. 3). Because of the continuous objective purging demand during the whole process, a pipeline supply of dry nitrogen gas is preferred or a self-pressurized LN2 dewar is the second choice. In this experiment, a few dry nitrogen gas cylinders were used, which can sustain as long as 8 h, from the objective purging to system heating. High-pressure gas purging on the cover glass must be done for de-icing, after the temperature of sample and chamber reaches a range from −185 to −195 °C (Fig. 4C). Live imaging mode of the system can be used to check whether the cover glass is clean enough for light source to penetrate, especially at the location of the grids. When the system is stable, the transfer box can be transferred into the cryo module chamber in a far end, and the cryo shuttle is placed in the center of the bottom, with the screw facing toward the operator (Fig. 3B). The fluorescence imaging steps are detailed as follows:

(1) Check the intactness of carbon film, flatness of grid, and thickness of vitreous ice for the Autogrids (Fig. 4A), using the 5×/0.16 NA or 10×/0.3 NA objective lenses. Fluorescence signal can be captured with the 20×/0.8 NA or 40×/0.9 NA objective lenses. Screen the cryo grids to choose the best one (called finder grid) for further image acquisition.

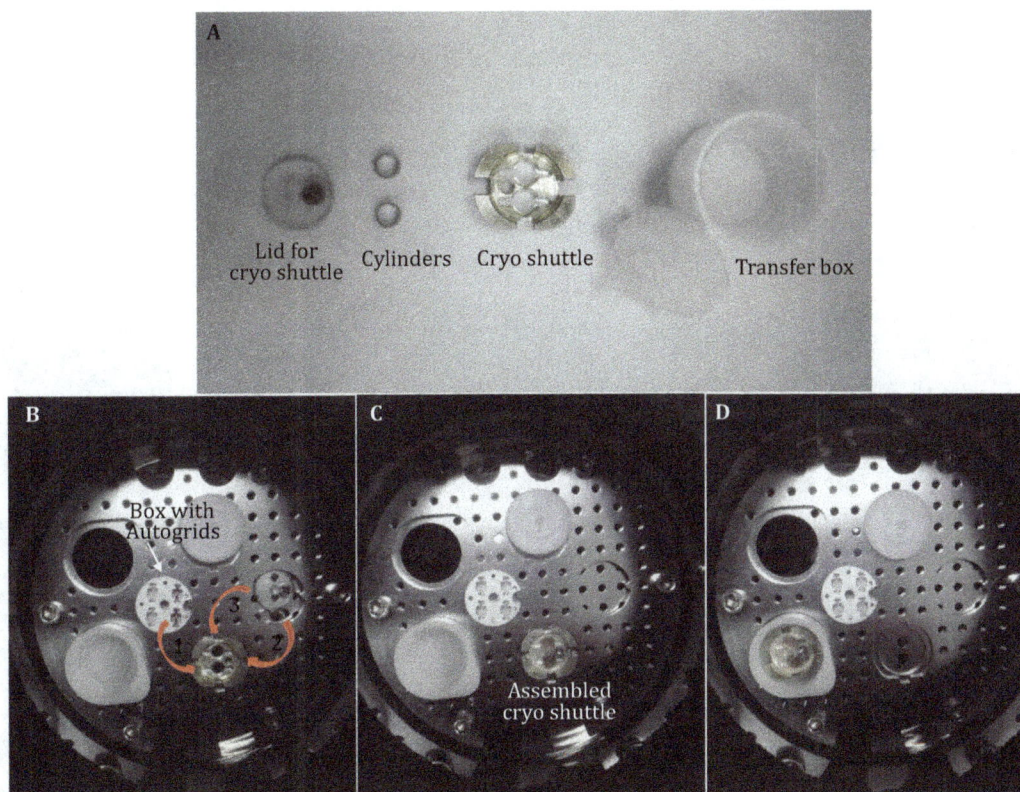

Fig. 1 Cryo shuttle and transfer box assembly for CorrSight. **A** The display of cryo shuttle components and transfer box. Lid and cylinders are used for fixing Autogrids. **B–D** The process of assembly. Cryo shuttle is assembled after mounting Autogrids on cryo shuttle (1), putting cylinders (2) and then putting lid (3) on the grids as the numbers correspond to the sequence illustrated in **B**. The assembled cryo shuttle in **C** is eventually loaded into transfer box in **D**

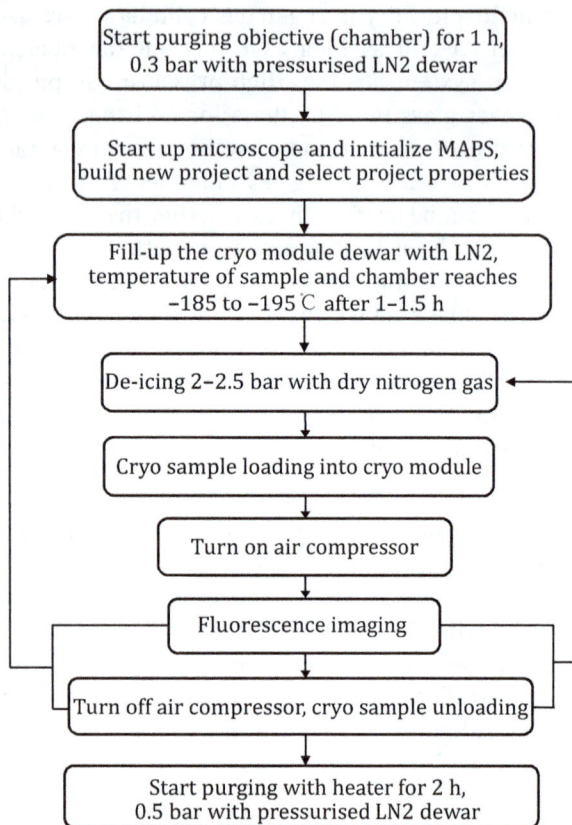

Fig. 2 Flowchart of fluorescence imaging with CorrSight

(2) Acquire the montaged map of the finder grid at a magnification 5×/0.16 NA or 10×/0.3 NA with the cryo trans-illumination channel and autofocus strategy (Fig. 5A). Stitch tiles to minimize the gap.

(3) Multichannel images of interested squares can be acquired at a magnification of 40×/0.9 NA as exampled in Fig. 6 shows clear fluorescence signals. Acquire images with z-stack mode if desired. Wide-field or spinning disk acquisition mode can be chosen according to the fluorescence intensity, the imaging resolution, and so on. For Wide field fluorescence, there is Xe-lamp light source and the oligochrome has three lines (405, 488, and 561 nm). For spinning disk confocal, four laser line combiner (405, 488, 561, and 640 nm) is provided. The Maps software (Thermo Fisher Scientific Inc.) provided with CorrSight allows the z-stack MIP (maximum intensity projection) averaging and adjustment of the correlation parameters, such as the contrast and transparency. Overlay of the transillumination and fluorescence images make available the identified information for EM correlation. Adjust the fine alignment between 5×/0.16 NA and 40×/0.9 NA images manually to locate good squares within the finder grid. This alignment can be done only once at the beginning, unless the cryo shuttle is moved. Good squares are normally the ones coated with a complete layer of carbon film and free from indexed letters and numbers as shown in Fig. 5A. Note that during the process of fluorescence imaging, the cover glass should be purged repeatedly to prevent ice growth, especially in a humid environment. Every time during the purging, the cryo shuttle should be put back into the transfer box to protect grids from temperature increase during the purging. Keep live transillumination

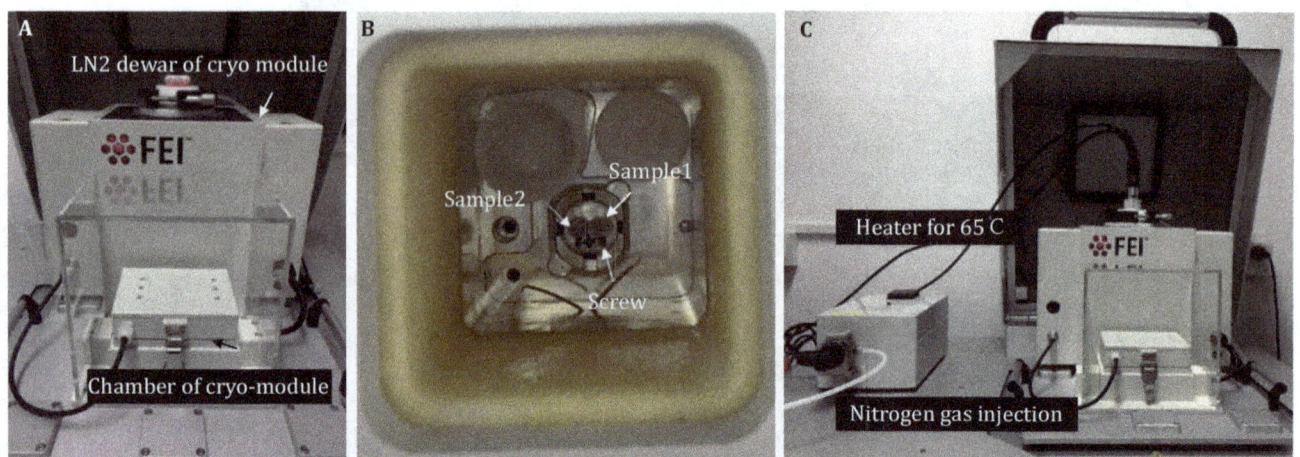

Fig. 3 The use of the cryo module of CorrSight. A Purging and cooling of the cryo module chamber before LM imaging. B The overhead view of cryo module chamber. There are locations with similar shape of transfer box and cryo shuttle. The screw of cryo shuttle should be facing toward the user. The right sample numbered 1 would be viewed as the left one in Maps software, while the left sample numbered 2 would be viewed as the right one in Maps software. C Warm up of the cryo module chamber at 65 °C with the heater and nitrogen gas injection after LM imaging

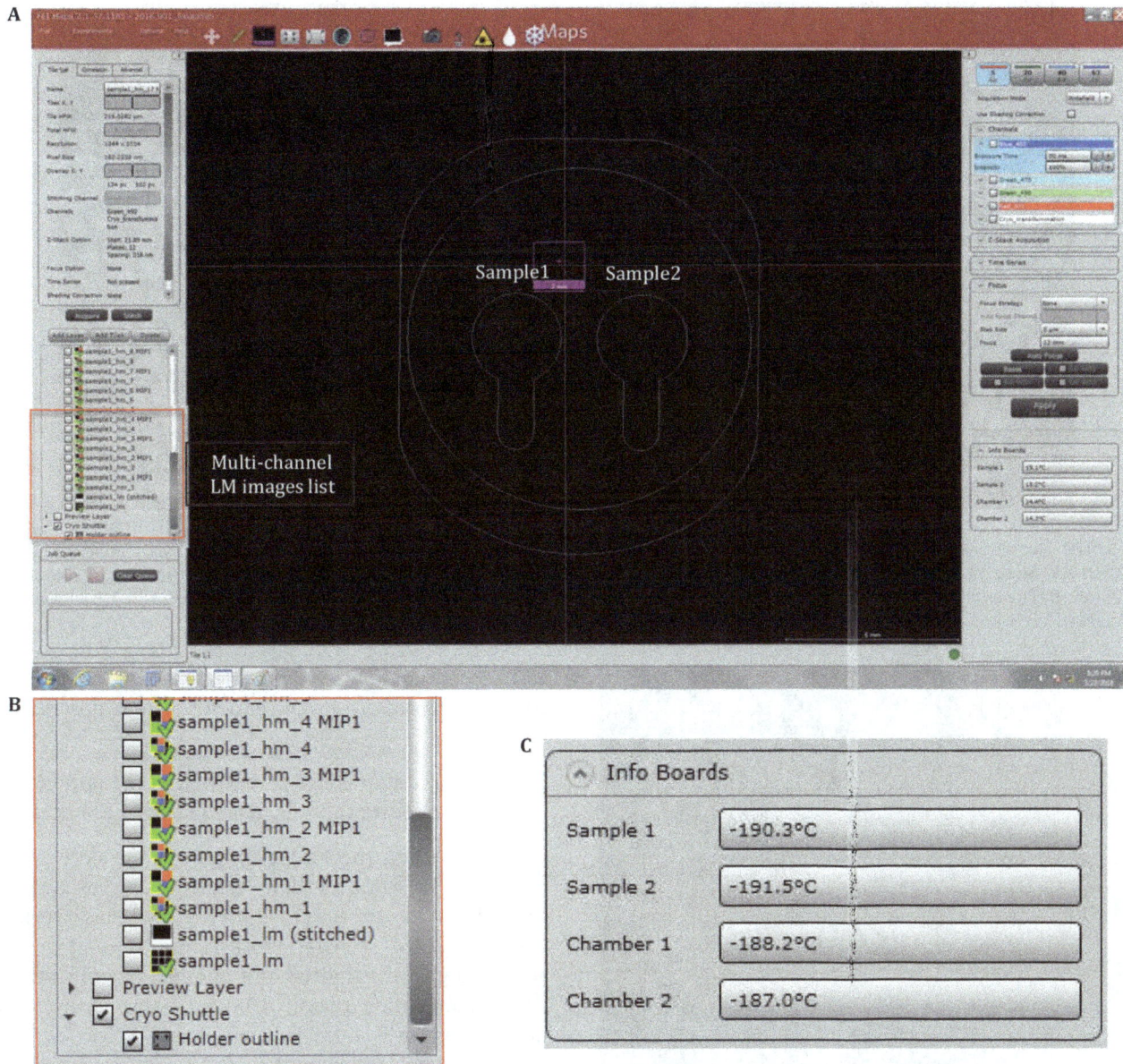

Fig. 4 The acquisition of multichannel LM images with Maps software. **A** The samples 1 and 2 correspond to the ones described in Fig. 3B. **B** Layer of sample 1. Each sample can have a layer which may contain one low magnification grid map and many high-magnification fluorescence square images numbered in sequence. **C** The temperature of sample and chamber should be kept in the range from −185 to −195 °C during imaging process

imaging by focusing on the cover glass, the crystal ice would be blown away gradually. The temperature changes of the samples and specimen chambers should be closely monitored, especially after continuous imaging for more than 2 h, to ensure that all liquid nitrogen in the cryo module dewar is not used up (Fig. 3A).

Note: Turn on the air compressor only at the beginning of the multichannel image acquisition and turn it off before the cryo sample transfer to prevent severe vibration of CorrSight. Once the imaging is completed, the cryo samples should be unloaded and put back in the LN2 tank immediately.

(4) After the image acquisition, put the cryo shuttle back to the LN2 storage tank. The cryo module should be heated continuously at 65 °C for 2 h and purged at 0.5 bar with pressurized LN2 tank at the same time (Fig. 3C).

(5) Overlaid multichannel images can be adjusted and saved manually for EM correlation. Within a good grid, more than 60 squares could be imaged and marked (Figs. 4B, 5A). This is the step for rough and quick screening of squares. Find squares with

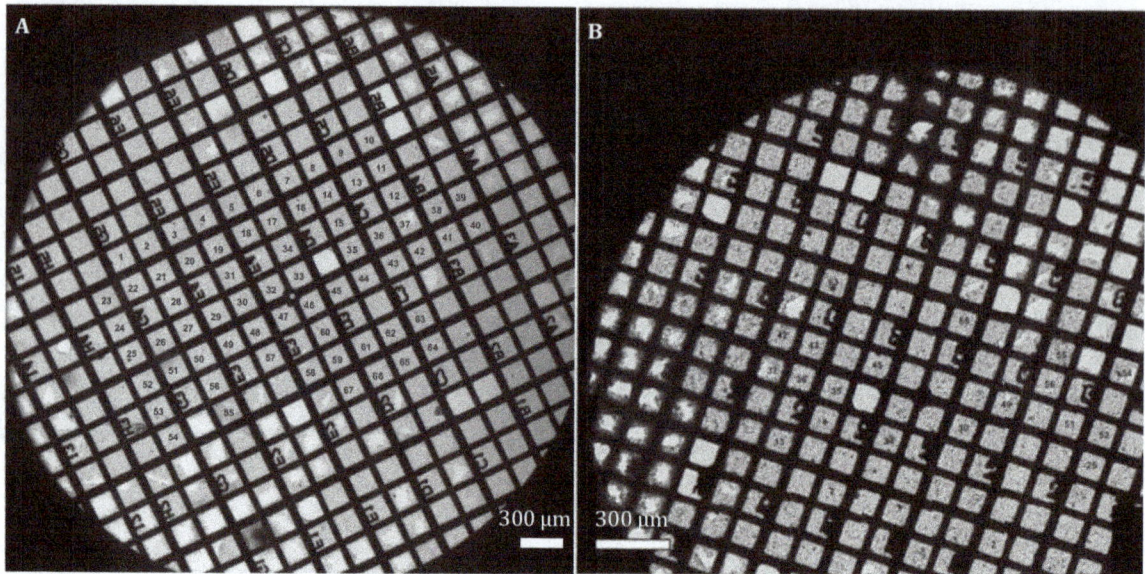

Fig. 5 The correlation of LM and EM grid map. **A** The montaged LM trans-illumination image of the grid at a magnification of 5×/0.16 NA. More than 60 squares are multichannel imaged and labeled with sequence number, which corresponds to the number in Fig. 4B. **B** The montaged EM image of the grid at a nominal magnification of 40×. The squares with targeted fluorescence signal are marked after the correlation of LM and EM grid map with AutoEMation

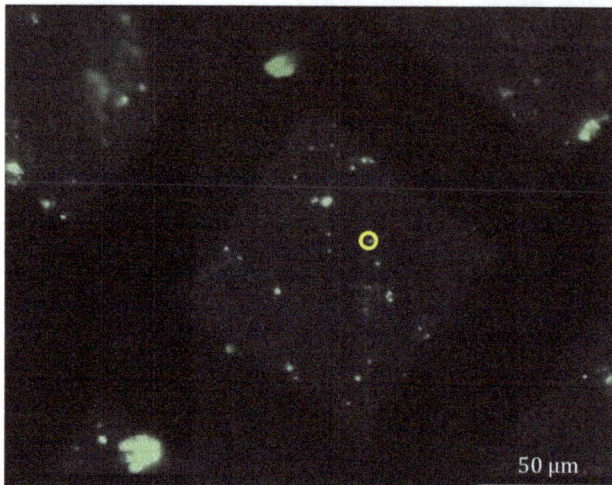

Fig. 6 A typical LM image of a square at a magnification of 40×/0.9 NA showing clear fluorescence signals

interested fluorescence signal and label the locations within the overlaid images (Fig. 5A) for further cryo-EM imaging.

Correlation of cryo-LM and cryo-EM images

Two functions in the upgraded version 2.0 of AutoEMation software (Lei and Frank 2005) are used for acquisition of montaged EM atlas (Fig. 7A), correlation of LM and EM images (Fig. 7B).

To locate and verify target particles for cryo-ET data collection, there are three sequential steps described as follows:

(1) The first step is the correlation of LM and EM grid map to locate the selected squares in EM grid map.

After loading the finder grid from the cryo shuttle into a cryo transmission electron microscope, for example, a Cs-corrected Titan Krios in this work, by using the Cryo Transfer Workstation, a montaged EM atlas of a grid is acquired at a nominal magnification in low magnification range (40× for Gatan Orius SC200 CCD camera, 100× for Falcon camera, 175× for Gatan GIF K2 camera) as exampled in Fig. 5B. The acquisition usually starts from the stage center and stage movement is used.

The conversion matrix between the LM and EM grid maps should be determined, which can be derived from two reference positions, *i.e.*, indexed letters and numbers, recognizable in both the LM and EM maps. For each reference position, the coordinate in the LM map is obtained by using any graphic software, while the coordinate in the EM map is obtained and converted to the corresponding EM stage position via a precalibrated matrix between the camera and stage by using AutoEMation.

With the help of the conversion matrix, the coordinates of those marked squares in the LM map (Fig. 5A) can be converted to their corresponding coordinates in

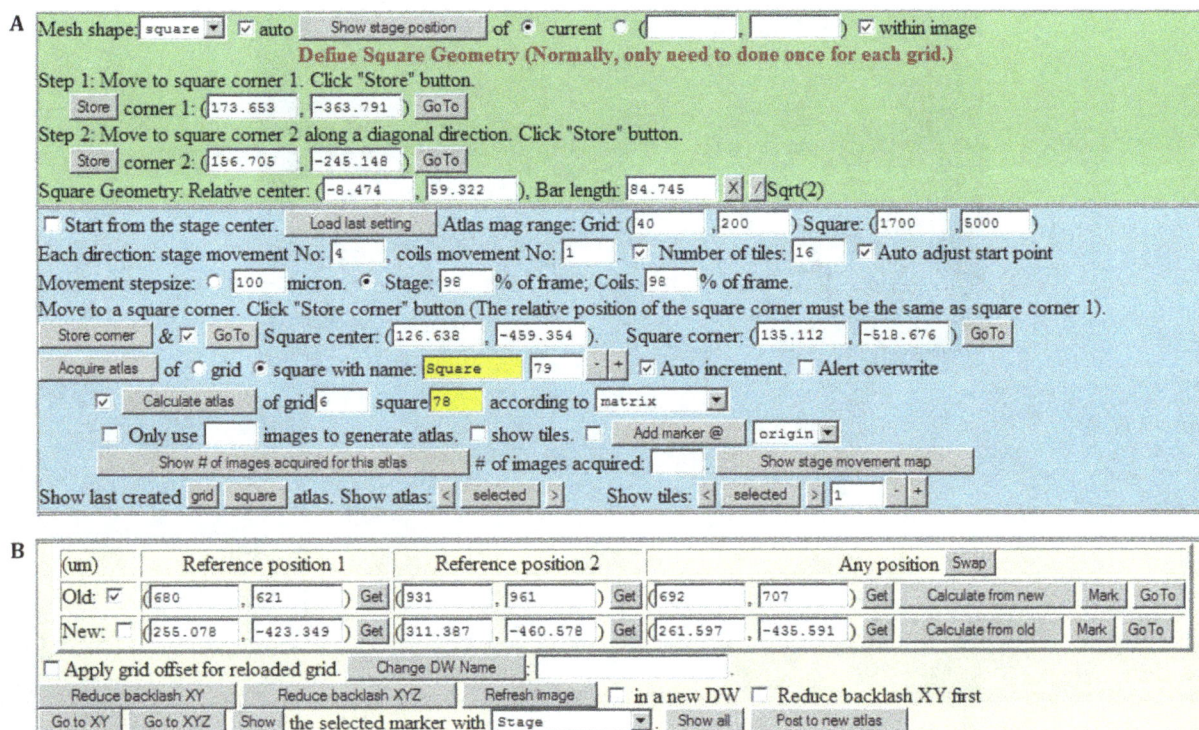

Fig. 7 Two functions in AutoEMation used for this CLEM study. **A** Acquisition of montaged EM atlas of grid and square. **B** Correlation of LM and EM images

the EM map marked (Fig. 5B) by clicking the "Mark" button in Fig. 7B, and the EM stage positions which can be reached by clicking the "GoTo" button near to the "Mark" button directly or the "Go to XYZ" button (Fig. 7B) if the marker in the EM map is selected (Fig. 5B).

(2) The second step is the correlation of LM and EM square map to locate target particles in EM square map.

For each marked square, a montaged EM atlas of the square (Fig. 8C) is acquired at the lowest nominal magnification in the SA range (1700× for both Gatan Orius SC200 CCD camera and Falcon camera, 2250× for Gatan GIF K2 camera). The acquisition starts locally after reaching the stage position of the square in the above step and stage movement or a combination of stage movement and image-beam shift is used.

Similarly, the conversion matrix between the LM and EM square maps should be determined, which can be derived from two reference positions, *i.e.*, square corners or any features, recognizable in both the LM and EM maps.

With the help of the conversion matrix, the coordinates of fluorescence target in the LM map (Fig. 8A, B) can be converted to their corresponding coordinates in the EM map (Fig. 8C, D) marked by clicking the "Mark" button in Fig. 7B, and the EM stage positions which can

be reached by clicking the "GoTo" button near to the "Mark" button directly or the "Go to XYZ" button (Fig. 7B) if the marker in the EM map is selected. The location of the target particle can be finely tuned with the combination of "Refresh" and "Go to XY" buttons (Fig. 7B) as the low-dose alignment between this magnification and a much higher magnification for verification can be set precisely with the inaccuracy of less than 0.02 μm by using AutoEMation.

Sometimes it is useful to verify the target particle visually. In Fig. 8B and D, three vesicles near the fluorescence signal are used as references for verification.

(3) The third step is the verification of the target particles at a much higher magnification to ensure they are suitable for cryo-ET data collection.

The target particles selected from the above two steps may not be suitable for cryo-ET data collection as images acquired at the above magnification cannot provide enough details. Therefore, it is worth imaging the target particles at a much higher magnification (*i.e.*, nominal magnification of 18,000× for Falcon camera, 2–3 e/Å^2 dose, −8 to −10 μm defocus) for verification as further cryo-ET data collection takes very long time. The imaging for verification can be simply done by clicking the "Refresh" button (Fig. 7B) after switching to the preset high-magnification mode.

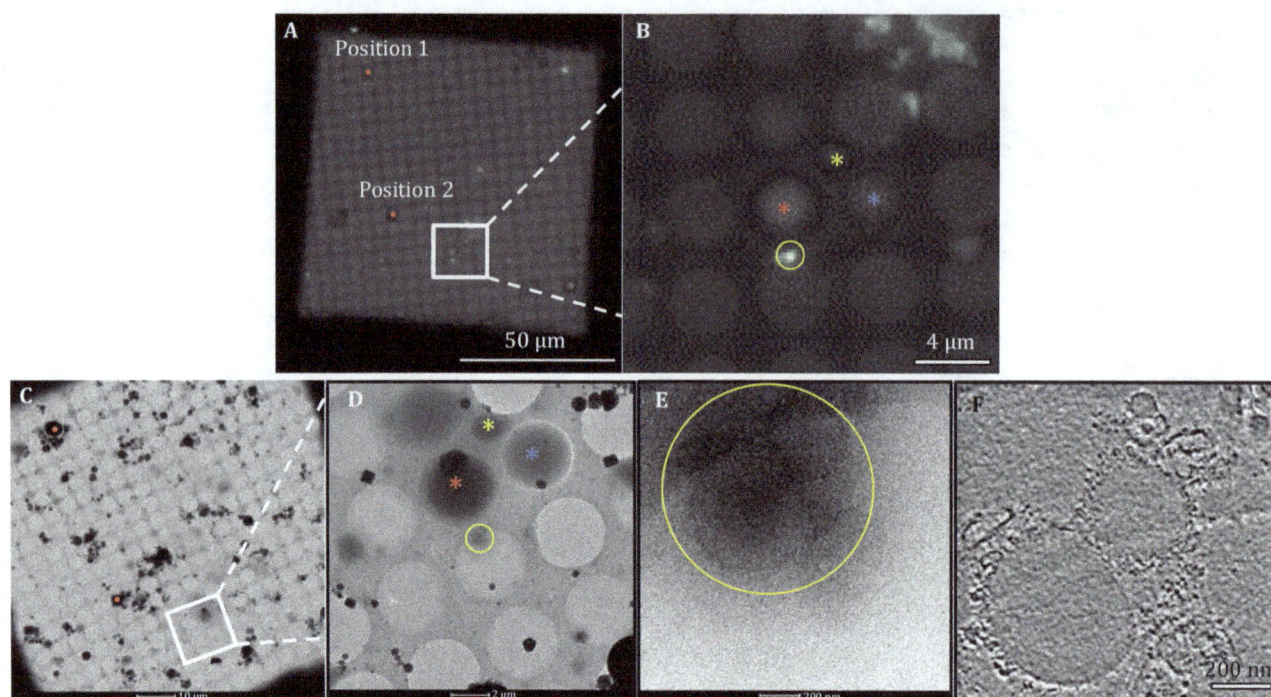

Fig. 8 The correlation of LM and EM square images and target identification. **A** The LM image of a square at a magnification of 40×/0.9NA. **B** Zooming in on the area highlighted in **A**. Three vesicles can be identified and marked by asterisks in *red, blue,* and *yellow,* respectively, which are close to the fluorescence signal, highlighted by yellow circle. **C** The montaged EM image of a square at a nominal magnification of 1700×. **D** Zooming in on the area highlighted in **C**. The target can be located via correlation by means of AutoEMation and verified with the three neighboring vesicles identified accordingly. **E** Low-dose EM imaging for verification of the target at a nominal magnification of 18,000×. **F** Tomogram slice of the target after tomographic data collection and reconstruction

Cryo-ET data collection and reconstruction

Cryo-ET data collection and reconstruction is usually performed by using free software packages and/or packages from EM vendors.

Set batch tomography data collection positions for all targets

(1) Set up FEI TEM tomography and build a new specimen session by checking the low-dose and batch options. Select file format and output folder.
(2) Add every target with separate defocus settings, ranging from −5 to −8 μm. Set the focus and tracking areas away from the exposure area. Determine the exposure time of data collection based on the limit of 1–2 e/$Å^2$ for each tilted image.
(3) As a typical setting, acquisition usually starts at 0° and single-axis tilt series are collected with 1.5° increment between −65° and 65° for frozen samples. Autofocus is performed before image acquisition periodically. The general periodicity switch angle is 30° as focus would be checked every five images when the tilt angle is lower than 30° and

every two images when the tilt angle is higher than 30°. In order to track the targets during the tilting, "tracking before" is usually performed when the tilt angle is lower than 10° and "tracking after" is performed for other tilt angles.

Data alignment and reconstruction

For the tomographic data with fiducial gold, use IMOD software package to do alignment and reconstruction. Execute the Protomo software package (Winkler and Taylor 2006) for the alignment and reconstruction for tomographic data without gold particles. Low-pass and high-pass filtering might be used for the noisy images, which is particularly helpful for the cross-correlation calculation during alignment. With Protomo, four or five iterations is enough for the 2 k × 2 k images and the weighted back-projection algorithm is used to generate the final 3D reconstruction from raw images. Apply median filtering or nonliner anisotropic diffusion filtering distributed in IMOD to enhance the contrast and signal-to-noise ratio of the tomograms. Further analysis of subvolume averaging and segmentation can be performed in Dynamo and Amira, respectively.

DISCUSSION

The CorrSight system can sustain cryo-LM imaging for a long duration of up to 5 h. The Maps software can save all the acquired data automatically in the project and avoid data loss due to unexpected interruption such as software crashing. More importantly, the packed Auto-grids matches well with cryo-SEM, cryo-TEM with autoloader, and Cryo Transfer Workstation. The application of AutoEMation software makes the correlation much easier. For the manual verification during the correlation, fluorescent beads can be used if there are no distinguishable particles. Verification with high-magnification imaging is necessary before cryo-ET data collection although the target is preexposed with a small amount of dose, as acquisition of each tilt series would take nearly 1 h.

The protocol in this study applies mainly to relatively thin samples, which can be prepared by plunge freezing by Vitrobot or similar apparatus and imaged with cryo-TEM directly. If the samples are too thick for electron beam to penetrate, ultrathin sectioning under LN2 environment needs to be executed before the sample can be loaded into the cryo-TEM. Sometimes high-pressure freezing should be performed instead of fast frozen for large cells or tissues, to get vitreous ice for the whole volume (Mahamid *et al.* 2015).

Besides the cryo- sectioning by diamond knife (Cortese *et al.* 2013), milling by cryo-focused ion beam (cryo-FIB) recently becomes the priority to get artifact-free, thin, frozen-hydrated lamella via fabrication (Arnold *et al.* 2016; Beck and Baumeister 2016; Lucic *et al.* 2013). After cryo-LM imaging, the locations with fluorescence signal can be specifically milled by FIB to generate the thin lamella, followed by cryo-EM high-resolution data collection (Fukuda *et al.* 2014). In addition, the resolution of LM imaging needs to be improved by integrating a super-resolution LM platform and a well-designed cryo module, which would be very powerful for the correlation of light and electron microscopy. In the future, more integrated and auto-mated system to bring all the above-mentioned steps in a seamless pipeline would make the correlative cryo-EM more powerful and robust.

Acknowledgements We thank the Tsinghua University Branch of the China National Center for Protein Sciences (Beijing) for providing facility support. We thank Dr. Peng Li and Dr. Xuchao Lv for providing the test samples, and Dr. Qiang Zhou for technical suggestions. This work was supported by the National Key R&D Program of China (2016YFA0501100, 2017YFA0503500).

Compliance with Ethics Standards

Conflict of interest Xiaomin Li, Jianlin Lei, and Hong-Wei Wang declare that they have no conflict of interest.

Human and animal rights and informed consent This article does not contain any studies with human or animal subjects performed by any of the authors.

References

Agronskaia AV, Valentijn JA, van Driel LF, Schneijdenberg CT, Humbel BM, van BergenHenegouwen PM, Verkleij AJ, Koster AJ, Gerritsen HC (2008) Integrated fluorescence and transmission electron microscopy. J Struct Biol 164:183–189

Anderson KL, Page C, Swift MF, Hanein D, Volkmann N (2017) Marker-free method for accurate alignment between correlated light, cryo-light, and electron cryo-microscopy data using sample support features. J Struct Biol 201:46–51

Arnold J, Mahamid J, Lucic V, de Marco A, Fernandez JJ, Laugks T, Mayer T, Hyman AA, Baumeister W, Plitzko JM (2016) Site-specific cryo-focused ion beam sample preparation guided by 3D correlative microscopy. Biophys J 110:860–869

Bauerlein FJB, Saha I, Mishra A, Kalemanov M, Martinez-Sanchez A, Klein R, Dudanova I, Hipp MS, Hartl FU, Baumeister W, Fernández-Busnadiego R (2017) *In situ* architecture and cellular interactions of PolyQ inclusions. Cell 171(179–187):e110

Beck M, Baumeister W (2016) Cryo-electron tomography: can it reveal the molecular sociology of cells in atomic detail? Trends Cell Biol 26:825–837

Briegel A, Chen S, Koster AJ, Plitzko JM, Schwartz CL, Jensen GJ (2010) Correlated light and electron cryo-microscopy. Methods Enzymol 481:317–341

Bykov YS, Cortese M, Briggs JA, Bartenschlager R (2016) Correlative light and electron microscopy methods for the study of virus–cell interactions. FEBS Lett 590:1877–1895

Compera D, Entchev E, Haritoglou C, Mayer WJ, Hagenau F, Ziada J, Kampik A, Schumann RG (2015) Correlative microscopy of lamellar hole-associated epiretinal proliferation. J Ophthalmol 2015:450212

Cortese K, Vicidomini G, Gagliani MC, Boccacci P, Diaspro A, Tacchetti C (2013) High data output method for 3-D correlative light-electron microscopy using ultrathin cryosections. Methods Mol Biol 950:417–437

Faas FG, Barcena M, Agronskaia AV, Gerritsen HC, Moscicka KB, Diebolder CA, van Driel LF, Limpens RW, Bos E, Ravelli RB, Koning RI, Koster AJ (2013) Localization of fluorescently labeled structures in frozen-hydrated samples using integrated light electron microscopy. J Struct Biol 181:283–290

Fukuda Y, Schrod N, Schaffer M, Feng LR, Baumeister W, Lucic V (2014) Coordinate transformation based cryo-correlative methods for electron tomography and focused ion beam milling. Ultramicroscopy 143:15–23

Godman GC, Morgn C, Breitenfeld PM, Rose HM (1960) A correlative study by electron and light microscopy of the development of type 5 adenovirus. II. Light microscopy. J Exp Med 112:383–402

Hampton CM, Strauss JD, Ke Z, Dillard RS, Hammonds JE, Alonas E, Desai TM, Marin M, Storms RE, Leon F, Melikyan GB, Santangelo PJ, Spearman PW, Wright ER (2017) Correlated fluorescence microscopy and cryo-electron tomography of virus-infected or transfected mammalian cells. Nat Protoc 12:150–167

Hellstrom K, Vihinen H, Kallio K, Jokitalo E, Ahola T (2015) Correlative light and electron microscopy enables viral replication studies at the ultrastructural level. Methods 90:49–56

Kobayashi S, Iwamoto M, Haraguchi T (2016) Live correlative light-electron microscopy to observe molecular dynamics in high resolution. Microscopy 65:296–308

Kong D, Loncarek J (2015) Correlative light and electron microscopy analysis of the centrosome: a step-by-step protocol. Methods Cell Biol 129:1–18

Koning RI, Celler K, Willemse J, Bos E, van Wezel GP, Koster AJ (2014) Correlative cryo-fluorescence light microscopy and cryo-electron tomography of Streptomyces. Methods Cell Biol 124:217–239

Lei J, Frank J (2005) Automated acquisition of cryo-electron micrographs for single particle reconstruction on an FEI Tecnai electron microscope. J Struct Biol 150:69–80

Li S, Ji G, Shi Y, Klausen LH, Niu T, Wang S, Huang X, Ding W, Zhang X, Dong M, Xu W, Sun F (2018) High-vacuum optical platform for cryo-CLEM (HOPE): a new solution for non-integrated multiscale correlative light and electron microscopy. J Struct Biol 201:63–75

Liu B, Xue Y, Zhao W, Chen Y, Fan C, Gu L, Zhang Y, Zhang X, Sun L, Huang X, Ding W, Sun F, Ji W, Xu T (2015) Three-dimensional super-resolution protein localization correlated with vitrified cellular context. Sci Rep 5:13017

Loussert Fonta C, Humbel BM (2015) Correlative microscopy. Arch Biochem Biophys 581:98–110

Lucic V, Leis A, Baumeister W (2008) Cryo-electron tomography of cells: connecting structure and function. Histochem Cell Biol 130:185–196

Lucic V, Rigort A, Baumeister W (2013) Cryo-electron tomography: the challenge of doing structural biology in situ. J Cell Biol 202:407–419

Ma C, Kurita D, Li N, Chen Y, Himeno H, Gao N (2017) Mechanistic insights into the alternative translation termination by ArfA and RF2. Nature 541:550–553

Mahamid J, Schampers R, Persoon H, Hyman AA, Baumeister W, Plitzko JM (2015) A focused ion beam milling and lift-out approach for site-specific preparation of frozen-hydrated lamellas from multicellular organisms. J Struct Biol 192:262–269

Mahamid J, Pfeffer S, Schaffer M, Villa E, Danev R, Cuellar LK, Forster F, Hyman AA, Plitzko JM, Baumeister W (2016) Visualizing the molecular sociology at the HeLa cell nuclear periphery. Science 351:969–972

Morgan C, Godman GC, Breitenfeld PM, Rose HM (1960) A correlative study by electron and light microscopy of the development of type 5 adenovirus. I. Electron microscopy. J Exp Med 112:373–382

Mourik MJ, Faas FG, Valentijn KM, Valentijn JA, Eikenboom JC, Koster AJ (2014) Correlative light microscopy and electron tomography to study Von Willebrand factor exocytosis from vascular endothelial cells. Methods Cell Biol 124:71–92

Oikonomou CM, Jensen GJ (2017) Cellular electron cryotomography: toward structural biology in situ. Annu Rev Biochem 86:873–896

Peng W, Shen H, Wu J, Guo W, Pan X, Wang R, Chen SR, Yan N (2016) Structural basis for the gating mechanism of the type 2 ryanodine receptor RyR2. Science 354:aah5324

Plitzko JM, Rigort A, Leis A (2009) Correlative cryo-light microscopy and cryo-electron tomography: from cellular territories to molecular landscapes. Curr Opin Biotechnol 20:83–89

Rigort A, Bauerlein FJ, Leis A, Gruska M, Hoffmann C, Laugks T, Bohm U, Eibauer M, Gnaegi H, Baumeister W, Plitzko JM (2010) Micromachining tools and correlative approaches for cellular cryo-electron tomography. J Struct Biol 172:169–179

Schorb M, Briggs JA (2014) Correlated cryo-fluorescence and cryo-electron microscopy with high spatial precision and improved sensitivity. Ultramicroscopy 143:24–32

Schorb M, Gaechter L, Avinoam O, Sieckmann F, Clarke M, Bebeacua C, Bykov YS, Sonnen AF, Lihl R, Briggs JAG (2017) New hardware and workflows for semi-automated correlative cryo-fluorescence and cryo-electron microscopy/tomography. J Struct Biol 197:83–93

Sjollema KA, Schnell U, Kuipers J, Kalicharan R, Giepmans BN (2012) Correlated light microscopy and electron microscopy. Methods Cell Biol 111:157–173

Wang H (2015) Cryo-electron microscopy for structural biology: current status and future perspectives. Sci China Life Sci 58:750–756

Wang HW, Lei J, Shi Y (2017a) Biological cryo-electron microscopy in China. Protein Sci 26:16–31

Wang S, Li S, Ji G, Huang X, Sun F (2017b) Using integrated correlative cryo-light and electron microscopy to directly observe syntaphilin-immobilized neuronal mitochondria in situ. Biophys Rep 3:8–16

Winkler H, Taylor KA (2006) Accurate marker-free alignment with simultaneous geometry determination and reconstruction of tilt series in electron tomography. Ultramicroscopy 106:240–254

Wolff G, Hagen C, Grunewald K, Kaufmann R (2016) Towards correlative super-resolution fluorescence and electron cryo-microscopy. Biol Cell 108:245–258

Zhang P (2013) Correlative cryo-electron tomography and optical microscopy of cells. Curr Opin Struct Biol 23:763–770

4

The advent of structural biology *in situ* by single particle cryo-electron tomography

Jesús G. Galaz-Montoya[1], Steven J. Ludtke[1][✉]

[1] National Center for Macromolecular Imaging, Verna and Marrs McLean Department of Biochemistry and Molecular Biology, Baylor College of Medicine, Houston, TX 77030, USA

Abstract   Single particle tomography (SPT), also known as subtomogram averaging, is a powerful technique uniquely poised to address questions in structural biology that are not amenable to more traditional approaches like X-ray crystallography, nuclear magnetic resonance, and conventional cryoEM single particle analysis. Owing to its potential for *in situ* structural biology at subnanometer resolution, SPT has been gaining enormous momentum in the last five years and is becoming a prominent, widely used technique. This method can be applied to unambiguously determine the structures of macromolecular complexes that exhibit compositional and conformational heterogeneity, both *in vitro* and *in situ*. Here we review the development of SPT, highlighting its applications and identifying areas of ongoing development.

Keywords   Cryo-electron tomography, Single particle tomography, Subtomogram averaging, Direct detection device, Contrast transfer function

INTRODUCTION: THE NEED FOR SINGLE PARTICLE TOMOGRAPHY

Over the last five years, thanks to the development of direct detection devices (DDDs) (Milazzo *et al.* 2005), cryo-electron microscopy (cryoEM) single particle analysis (SPA) has transitioned from being an established, but limited, technique to being at the forefront of structural biology (Eisenstein 2016; Nogales 2016). SPA can now achieve resolutions comparable to those of typical X-ray crystal structures while maintaining the specimen in a solution-like environment, thereby avoiding dehydration and crystallization artifacts. While a very powerful technique, SPA still suffers from two primary limitations: first, it is sometimes unable to unambiguously resolve reliable structures of macromolecules exhibiting continuous conformational flexibility; second, it cannot be directly used to study macromolecules within cells or other unique structures *in situ*.

Single particle tomography (SPT), also known as subtomogram averaging (STA), offers a solution to both of these problems. Indeed, with per particle 3D data, it is easier to unambiguously discriminate between changes in particle orientation versus changes in particle conformation, addressing the first issue above. Furthermore, the most impactful application of SPT lies in the cellular milieu. Since tomograms are a 3D representation of the imaged specimen, with SPT it is possible to isolate individual macromolecules from a cellular tomogram. These individual "subtomograms" can then be subjected to SPA-like 3D alignment and averaging, making true *in situ* structural biology at nanometer resolution feasible.

The difficulty faced by SPA when studying particles undergoing large-scale continuous conformational change is the ambiguity produced by making projections of 3D objects. As conceptually outlined in Fig. 1, extremely different 3D structures can theoretically yield one or more indistinguishable projections, particularly given the high noise levels present in typical CryoEM images, both *in vitro* and *in situ*. With continuous

✉ Correspondence: sludtke@bcm.edu (S. J. Ludtke)

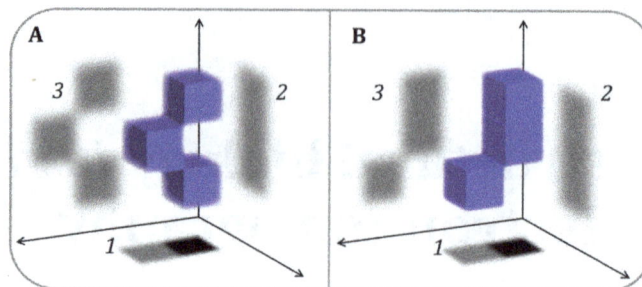

Fig. 1 2D versus 3D imaging for particle classification. A simple conceptual demonstration that conformationally different particles (**A** and **B**) can yield multiple projections that are extremely similar (even if orthogonal), as shown by projections *1* and *2*, particularly given the high levels of noise typical of cryoEM micrographs of either *in vitro* or *in situ* specimens. To push structures to high resolution, it is critical to distinguish between projections of structurally different molecules that might otherwise be erroneously classified together and aligned based on overall low-resolution feature similarity. Collecting more than two images, as done in tomography, possibly from a different axis (as shown by projection *3* in the figure), can help to distinguish conformational differences among particles

conformational variability, it can thus be mathematically impossible to unambiguously distinguish changes in particle orientation from changes in particle conformation with only a single 2D image for each particle. These limiting factors, namely high levels of noise and conformational variability, are exacerbated for macromolecules *in situ*.

For macromolecules *in vitro*, the "tilt validation" method (Henderson *et al.* 2011, 2012) partially addresses the issue of confounding orientation and conformation by imaging isolated particles from two different directions to assess the reliability of orientation assignment in SPA reconstructions. However, this is a validation method and not a tool for initial analysis. The random conical tilt (RCT) (Radermacher *et al.* 1986) and orthogonal tilt reconstruction (OTR) (Leschziner and Nogales 2006) methods make use of this concept to reconstruct challenging structures, but have a number of other limitations. Furthermore, all these methods image the specimen from only two angles about the same axis and, again, are only applicable to isolated complexes.

In addition to allowing for the computational isolation of macromolecular complexes from cells, single particle tomography (SPT; also known as subtomogram averaging, or STA), can be viewed as an extension of the tilting concept of RTC and OTR by collecting multiple tilted views of single particles. This is literally "tomography of single particles," as the inherent goal is to produce a tomographic 3D view for each individual particle in a system, be it *in vitro* or *in situ*. The particles are then processed through a pipeline akin to that of SPA cryoEM. That is, the tomographic single particles are (in simplified terms) aligned, classified by composition and/or conformation, and averaged as part of a standard pipeline that is applicable to both particles *in vitro* and *in situ*.

Although SPT has opened the window to increasing the resolution of structural biology *in situ* by averaging repeating features in cellular tomograms, this method also suffers from significant limitations. SPT builds on cryo-electron tomography (cryoET), which historically has been considered a low-resolution technique. In cryoET, a set of images (*i.e.*, a "tiltseries") is collected for each specimen area by tilting the specimen stage through a range of angles, usually about a single axis (*e.g.*, $\pm 60°$ in increments of $1°$–$5°$ or more). Very high cumulative electron dose (~ 50–120 e/A^2) has been the norm to obtain sufficient contrast in each image of a tiltseries to permit accurate alignment, with the side effect of destroying high-resolution information progressively through the series. Each tiltseries can then be computationally reconstructed into a 3D tomogram representing the 3D structure of the imaged area (Fig. 2). The resolution of raw tomograms is highly anisotropic and remains somewhat ill-defined (Cardone *et al.* 2005), but is generally estimated to range from roughly 50–150 Å (depending on the specimen and on data collection parameters). While this is sufficient to resolve cellular organelles and identify large macromolecular complexes, it is not sufficient to resolve macromolecular structure in detail. Nonetheless, each individual subtomogram does contain some high-resolution information, which, upon averaging with other ostensibly identical particles, can be recovered to yield structures at much higher resolution, depending on imaging conditions.

The recent development of direct electron detectors, phase plate technology, and improved contrast transfer function (CTF) correction methodologies for cryoET have made it possible to achieve images with higher contrast and resolution, respectively, using much lower dose. Additionally, when averaging subtomograms, fewer images may be collected for each tiltseries,

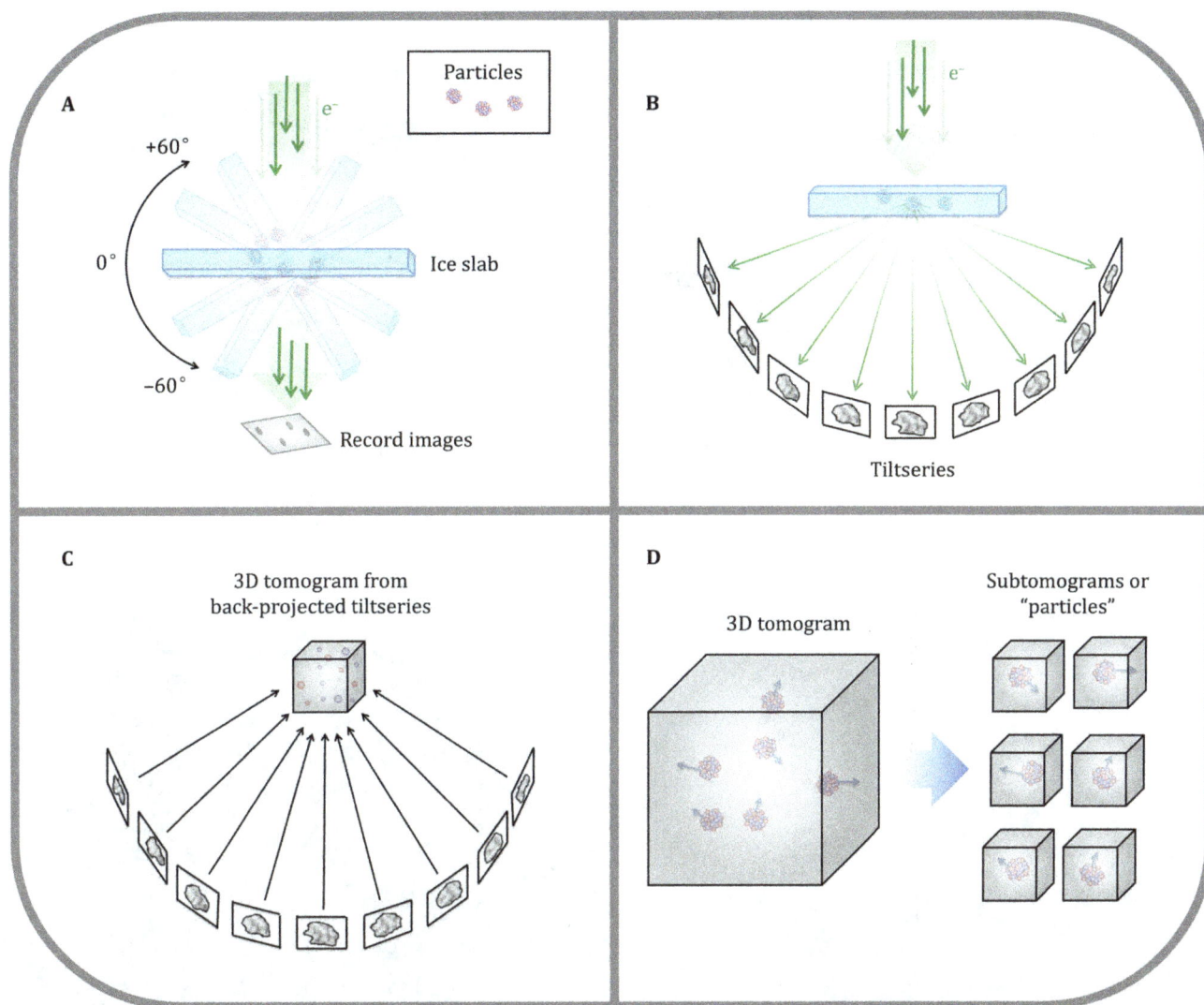

Fig. 2 CryoET schematic. **A** In cryoET, the ice-embedded specimen, typically shaped as a slab, is tilted through a wide range of angles in the electron microscope and an image is recorded at each angle. **B** This collection of images around a common axis constitutes a "tiltseries." **C** The images in a tiltseries can be computationally aligned to their common axis and reconstructed into a 3D tomogram by weighted back-projection or other methods. **D** Subtomograms representing a 3D view of individual macromolecules can be extracted from the reconstructed tomogram, then aligned and averaged (Fig. 7). (Partially inspired by Grünewald *et al.* 2002)

permitting the use of higher dose per image without impacting total dose. Indeed, these improvements have led to several recent cryoSPT studies approaching (Cassidy *et al.* 2015; Galaz-Montoya *et al.* 2016a; Khoshouei *et al.* 2016; Kudryashev *et al.* 2016; Li *et al.* 2016) or achieving (Bharat *et al.* 2015; Pfeffer *et al.* 2015a, b; Schur *et al.* 2013, 2015a, b; Mattei *et al.* 2016) subnanometer resolution, with the highest-resolution structure being solved to ∼4 Å (Schur *et al.* 2016). Simplistic CTF correction (not accounting for defocus gradients) of cellular data has also demonstrated measurable improvements through SPT of microtubules *in situ* (Grange *et al.* 2017). Some of the most impressive results in cryoSPT during the last few years are shown in Fig. 3. Perhaps the greatest remaining limiting factor for *in situ* experiments is specimen thickness, which limits electron penetration, making it impractical to study specimens thicker than roughly 0.5–1.0 μm. The study of thicker eukaryotic cells requires significant physical manipulation, such as slicing the specimen into thin sections (Al-Amoudi *et al.* 2004).

Continued development of computer technology has also played a critical role in improving the resolution achievable by cryoSPT. A single cellular tomogram (including those with repeating features that can be averaged) can exceed 64 GB in size at full resolution; therefore, the reconstruction of hundreds of tomograms is extremely computationally intensive, particularly if

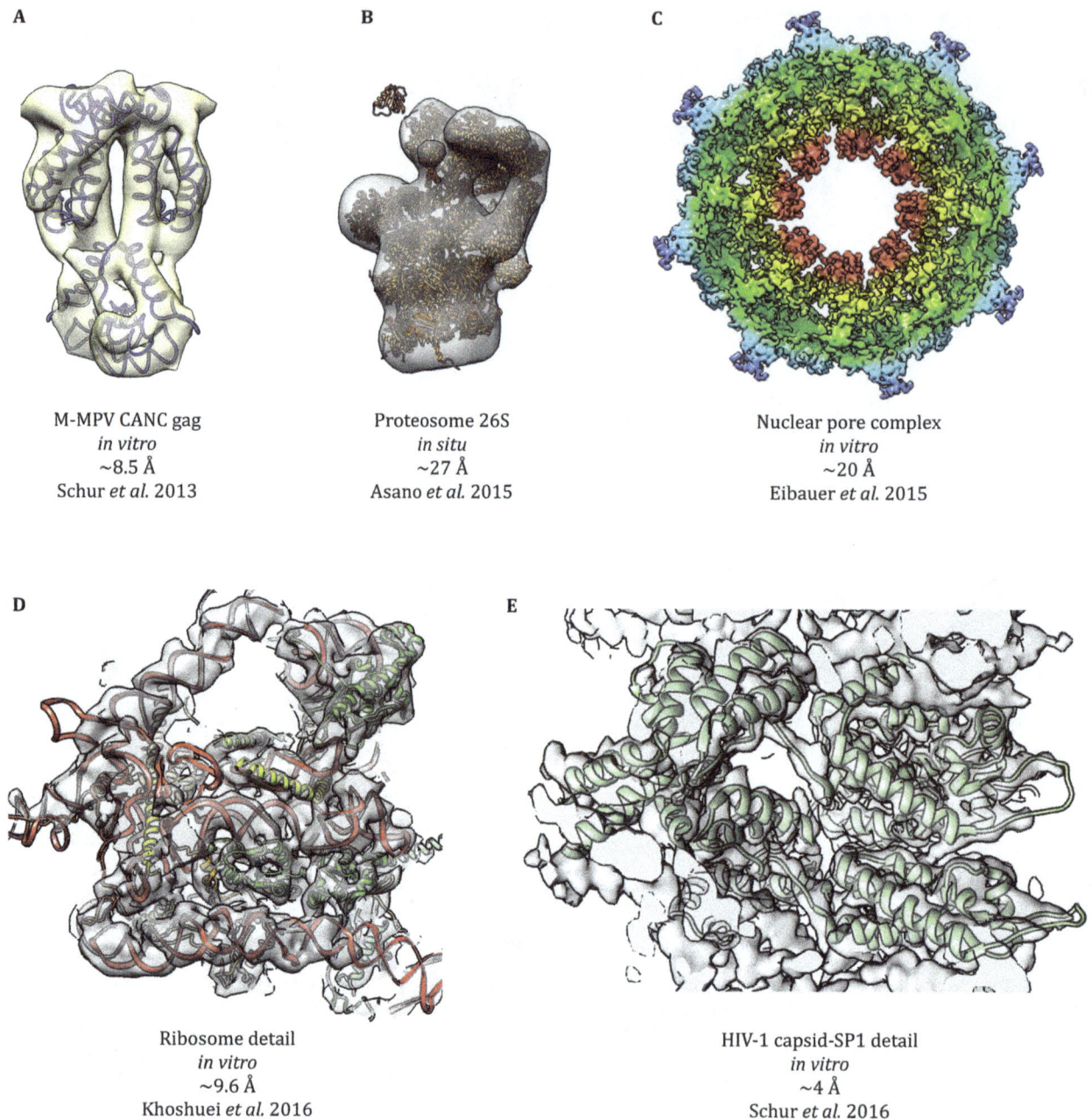

A

M-MPV CANC gag
in vitro
~8.5 Å
Schur *et al.* 2013

B

Proteosome 26S
in situ
~27 Å
Asano *et al.* 2015

C

Nuclear pore complex
in vitro
~20 Å
Eibauer *et al.* 2015

D

Ribosome detail
in vitro
~9.6 Å
Khoshuei *et al.* 2016

E

HIV-1 capsid-SP1 detail
in vitro
~4 Å
Schur *et al.* 2016

Fig. 3 Sampling of notorious cryoSPT studies. This figure is a sampling of cryoSPT structures published at different resolutions, prepared directly from the corresponding deposited (emdatabank.org) maps and models. **A** The structure of M-MPV CANC Gag (EMD-2488) was the first one solved to subnanometer resolution by cryoSPT. **B** Asano *et al.* undertook the study of proteasomes inside intact neurons, resolving multiple states for the 26S proteasome (EMD-2830 is shown), making use of a Volta phase plate (VPP). **C** Nuclear pore complexes (NPCs) are some of the most challenging specimens studied by cryoSPT due to their large size, extreme conformational flexibility, and the need for a lipid environment. Eibauer *et al.* solved the structure of the *X. laevis* NPC at unprecedented resolution for this specimen (EMD-3005). A comparable resolution was recently achieved for another nuclear pore complex (not shown; Kosinski *et al.* 2016). **D** A recent proof-of-concept study demonstrated that the VPP could be used in cryoSPT experiments to solve the structure of particles without any symmetry, such as the ribosome, to subnanometer resolution, using a relatively small number of particles ($N = 1400$; EMD-3418). **E** The highest-resolution structure by cryoSPT to date is that of the immature HIV-1 CA-SP1 lattice (EMD-4015), which allowed building an atomic model (PDB-5L93)

iterative reconstruction methods are used without downsampling. Per-particle computational costs (*i.e.*, preprocessing, alignment, classification, and averaging of subtomograms) are several orders of magnitude greater for SPT than for SPA. However, Moore's Law (Schaller 1997) has finally caught up with this field, and it is now practical to compute hundreds of tomographic reconstructions and average tens of thousands of subtomograms making use of algorithms that previously would have been untenable (Agulleiro *et al.* 2012).

HISTORY AND APPLICATIONS OF SINGLE PARTICLE CRYO-ELECTRON TOMOGRAPHY

The mathematical foundations underlying 3D image reconstruction date to 1917 when Johann Radon demonstrated that a function could be precisely reconstructed from an infinite set of its projections (Hawkes 2007). Since then, many mathematical techniques have been devised to reconstruct a 3D model from a set of 2D projection images, for medical imaging and a wide range of other applications, including transmission electron microscopy (TEM) tomography.

The theory underlying electron tomography (ET) and its application to study biological specimens (De Rosier and Klug 1968a, b; Hart 1968; Hoppe *et al.* 1968), including individual metal-stained macromolecules (Hoppe *et al.* 1974) and averages of a few subvolumes (Knauer *et al.* 1983; Oettl *et al.* 1983), were first demonstrated decades ago. However, there were many experimental and computational barriers to widespread adoption of ET at that time.

The development of cryo-electron microscopy (cryoEM) was a major breakthrough that demonstrated that biological specimens, including cells and 'single particles,' could be better preserved in vitrified water solutions, free from crystallization and staining artifacts (Dubochet *et al.* 1981, 1988), in a close-to-native state. CryoEM was first applied to 2D protein crystals (Taylor and Glaeser 1974) and was demonstrated for isolated particles (viruses) a decade later (Adrian *et al.* 1984). It took many more years before Walz *et al.* completed the first cryoSPT experiments in 1997, in their study of thermosomes *in vitro*. They showed that the structure of this chaperonin could be determined without missing wedge artifacts by computing tomograms of the specimen in solution, extracting volumes (*i.e.*, subtomograms) containing individual thermosomes, and averaging them after correct alignment. Since then, the publication rate of studies using SPT has been accelerating. Figure 4 shows the number of yearly structures solved by SPT and deposited in the Electron Microscopy Data Bank (EMDB) from 2004 to 2016, as well as the best resolution achieved in each of those years. A few structures that were not deposited to the EMDB were solved in 1997, 1998, and 2003, as noted in previous reviews (Schmid 2011; Kudryashev *et al.* 2012). The resolution averages presented in Fig. 4 exclude structures for which no resolution was reported (pink line). Of note, averages of at least two particles should always

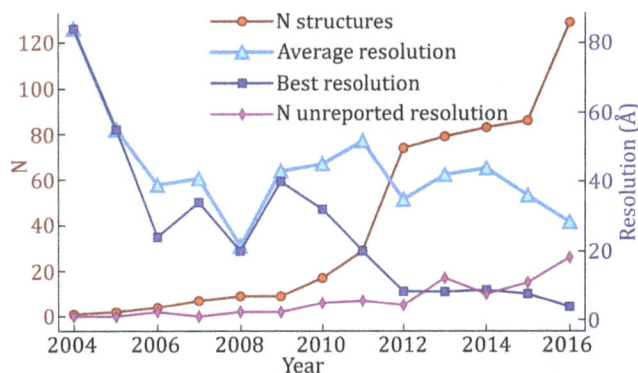

Fig. 4 Increase in yearly structures deposited in the EMDB by SPT from 2004 to 2016 and resolution improvement trend. While the highest resolution achieved each year has continued to improve, the average resolution is improving only very gradually

report an estimate of the resolution, however modest it might be. This will facilitate the interpretation of features in the structures being published and in the figures being displayed.

Early SPT relied largely on undocumented *ad hoc* scripts (Winkler and Taylor 1999; Pascual-Montano *et al.* 2002; Förster *et al.* 2005; Nicastro *et al.* 2006; Schmid *et al.* 2006; Winkler 2007; Schmid and Booth 2008). However, the methodology has gradually become more accessible with the surge of open-access software for SPT, such as AV3 (Förster and Hegerl 2007), PEET (Nicastro *et al.* 2006) (available through IMOD (Kremer *et al.* 1996)), Jsubtomo (Huiskonen *et al.* 2010), PyTom (Hrabe *et al.* 2012), Dynamo (Castaño-Díez *et al.* 2012), EMAN2 (Galaz *et al.* 2012; Murray *et al.* 2014; Galaz-Montoya *et al.* 2015), and RELION (Bharat *et al.* 2015).

The evolution of cryoET can be followed through reviews in this discipline over the last couple of decades, (*e.g.*, Koster *et al.* 1997; Baumeister 2002; Förster *et al.* 2005; Crowther 2010; Fernández 2012). CryoET can now be routinely applied to study macromolecules in solution (Medalia *et al.* 2002a) and in their native cellular context (Medalia *et al.* 2002b; Ortiz *et al.* 2006, 2010; Brandt *et al.* 2010; Schwartz *et al.* 2012). Challenging specimens whose structures have been solved by cryoSPT include complexes that exhibit extensive structural heterogeneity, such as carboxysomes (Schmid *et al.* 2006), dynein interacting with microtubules along axonemes (Nicastro *et al.* 2006), and pleomorphic viruses (Harris *et al.* 2006, 2013; Huiskonen *et al.* 2010; Schmid *et al.* 2012). CryoSPT has also been used to study viruses infecting their host cells (Hu *et al.* 2013; Peralta *et al.* 2013; Sun *et al.* 2014; Riedel *et al.* 2017; Murata *et al.* 2017), even in transient conformations along their assembly pathway inside cells (Dai *et al.* 2013). Much smaller complexes, such as the proteasome in different conformational states, have also been visualized *in situ*, inside neurons (Asano *et al.* 2015) (Fig. 3). Other complex systems whose structures have been best characterized using SPT are flagellae (Koyfman *et al.* 2011; Carbajal-González *et al.* 2013; Zhao *et al.* 2013), polysomes *in situ* (Brandt *et al.* 2010), membrane-bound ribosomes (Pfeffer *et al.* 2012, 2015a, b), nuclear pore complexes (Stoffler *et al.* 2003; Maimon *et al.* 2012; Eibauer *et al.* 2015; Kosinski *et al.* 2016), and other membrane-bound complexes (Davies *et al.* 2011; Eibauer *et al.* 2012; Dalm *et al.* 2015; Nans *et al.* 2015; Briegel *et al.* 2016; Sharp *et al.* 2016), as well as amyloid protein aggregates interacting with chaperones (Shahmoradian *et al.* 2013; Darrow *et al.* 2015), among others. A recent review (Asano *et al.* 2016) and a book chapter (Wan and Briggs 2016) describe in detail the technical aspects of carrying out SPT analyses.

CHALLENGES IN SINGLE PARTICLE TOMOGRAPHY

While cryoSPT is now being successfully applied to many biological problems that could not be addressed in near-native conditions a few years ago, it is also important to understand its current limitations and their underlying sources.

Radiation damage

A fundamental limitation in any cryoEM/ET study is the unavoidable fact that the specimen is being destroyed as it is being imaged. Thus, the permissible dose is limited to preserve detailed features (Glaeser 1971; Grubb 1974). This problem is exacerbated in tomography, since many tilt images of the same specimen area must be collected, and yet each tilt image must contain sufficient information for accurate tiltseries alignment to yield a high-fidelity reconstruction. How to optimally allocate the total cumulative dose among all the images of a tiltseries depending on the particular goals of a study (*i.e.*, "dose fractionation") has been a longstanding problem in cryoET (McEwen *et al.* 1995) and is still being actively researched (Hagen *et al.* 2017). A recent clever technique to turn the radiation sensitivity of biological specimens into an advantage is the concept of "bubblegrams" (Cheng *et al.* 2012; Wu *et al.* 2012) and "tomo-bubblegrams" (Fontana *et al.* 2015), in which the varying radiation sensitivity of different molecular species can be used to localize and identify substructures, while still preserving high-resolution detail in the early portion of the exposure. Aside from such unorthodox tricks, however, radiation damage remains

the primary limiting factor in cryoEM and cryoET (Cosslett 1978; Glaeser and Taylor 1978; Baker and Rubinstein 2010). Indeed, radiation damage is a complex phenomenon that remains under active investigation, since it depends not only on the chemical composition of the specimen but also on multiple data collection parameters such as cumulative dose, imaging temperature (Comolli and Downing 2005; Iancu *et al.* 2006; Bammes *et al.* 2010), and dose rate (Chen *et al.* 2008; Karuppasamy *et al.* 2011), among others. The recent development of DDDs has reduced the impact of radiation damage by permitting the recovery of a larger fraction of information at lower cumulative doses owing to the detector's higher detective quantum efficiency (DQE) and improved modulation transfer function (MTF) (Milazzo *et al.* 2010).

Dose fractionation, ice thickness, and beam-induced specimen motion

The SNR and contrast of an electron micrograph also depend on the thickness of the ice in which the specimen is embedded. In cryoET, the effective ice thickness scales with the secant of the tilt angle (Fig. 5). Images from tilt angles any higher than ±65° are often unusable and can decrease the quality of the tomographic reconstruction if included. The inability to collect a complete tomographic tiltseries (tilting through ±90°) causes the so-called "missing wedge" artifact. This term refers to the wedge-shaped region of Fourier space that is empty due to missing tilt images. This artifact produces anisotropic resolution in tomograms

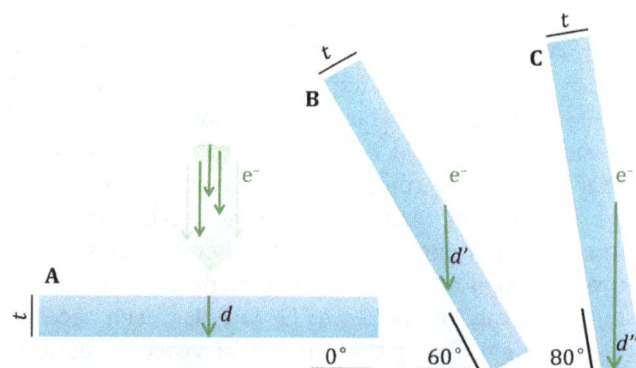

Fig. 5 The effective ice thickness of slab-shaped specimens increases with tilt angle. Even for relatively thin specimens (**A**), in cryoET the path of the electron beam through the slab-shaped specimen (*green arrows*, labeled d, d', and d") increases dramatically with tilt angle (**B, C**). This limits the thickness t of cellular material that can be studied by cryoET and the maximum tilt angle at which usable data can be collected. Indeed, greater ice thickness degrades image quality and contrast as it increases charging, blurring, and multi-scattering events that cause images to be noisier

(Radermacher 1988) and poses one of the most critical problems for alignment and classification of subtomograms, as discussed further below.

When the specimen is too thick for beam penetration, it can be sectioned with a diamond knife (Al-Amoudi *et al.* 2004) or milled thin with a focused ion beam (FIB) (Marko *et al.* 2007) prior to imaging. While this enables the study of thicker specimens by cryoET, cryo-sectioning is, unfortunately, an extremely challenging technique that yields sections laden with artifacts, from compression and curving to crevasses and cracks (Al-Amoudi *et al.* 2003, 2005). On the other hand, while FIB milling produces less severe artifacts, it destroys the bulk of the specimen, and it can be very difficult to focus the milling process on the region of interest. An alternative method to study thick frozen-hydrated specimens is X-ray tomography (Wang *et al.* 2000; Le Gros *et al.* 2005), albeit typically at resolutions an order of magnitude lower than cryoET.

The conductivity of holey and/or continuous amorphous carbon support films commonly used in cryoET is decreased at cryogenic temperatures, resulting in the accumulation of charge during imaging. Charging and beam-induced motion are exacerbated by specimen curvature (*i.e.*, frozen menisci)(Chen *et al.* 2008) and ice thickness (Brink *et al.* 1998), and are therefore more prominent and more likely to occur in cryoET the higher the tilt angle (Galaz-Montoya *et al.* 2016a). Although DDDs permit collecting images as short movies, which can be corrected for beam-induced motion (Brilot *et al.* 2012), different particles in the imaged specimen may move in different directions (Campbell *et al.* 2012). Strategies have emerged to correct for this in SPA cryoEM, where individual particles can be motion corrected (Scheres 2014). However, extensive dose fractionation in cryoET with DDDs typically yields movie frames at each tilt angle with considerably lower contrast and higher noise than movie frames in SPA cryoEM. Aligning whole movie frames can produce micrographs that are only locally unblurred, sometimes blurring areas of the image that were not originally blurry. This can happen both in thick, cellular tomograms (Fig. 6), as well as in tomograms of isolated macromolecules (not shown). Single particles are not always readily detectable in the individual frames of movies in cryoET images. Furthermore, cryoET images often cannot be completely broken down into discrete particles because the specimen is continuous (*e.g.*, cells). Therefore, local unblurring strategies that can correct extremely low-dose images anisotropically and without discontinuities are needed to maximize productive data usage in cryoET, as recently proposed for SPA cryoEM (Zheng *et al.* 2016). Given the extensive

dose fractionation and greater radiation damage due to total cumulative dose, accomplishing this for cryoET is foreseeably more challenging than for SPA cryoEM. To a first approximation, gold fiducials might serve the purpose of guiding local motion correction for extremely low-dose cryoET movie frames, in addition to serving their normal role in tiltseries alignment.

Tiltseries alignment

Gold fiducials are commonly used as markers in cryoET so that the precise 3D orientation of each image in a tiltseries can be determined. Unfortunately, such gold markers have been shown to undergo independent beam-induced motions and therefore their positions are not strictly consistent and predictable across a tiltseries (Comolli and Downing 2005; Noble and Stagg 2015). This effect varies considerably among specimens and can be a limiting factor in reconstruction quality. Furthermore, gold fiducials cause extremely strong artifacts ("streaks") that can obscure features in tomograms and interfere with subtomograms alignment. Strategies to regularize gold fiducial artifacts have been proposed (Song *et al.* 2012; Maiorca *et al.* 2014; Han *et al.* 2015), some of which are available in IMOD, but do not seem to be routinely applied. Alternatively, many fiducial-less alignment methods have been developed (Liu *et al.* 1995; Brandt *et al.* 2001; Renken and McEwen 2003; Castaño-Díez *et al.* 2007) but usually require that the specimen itself exhibits high-contrast features. These procedures are still prone to error since specimens can also undergo deformations with cumulative dose, and in most cases gold fiducials yield better tiltseries alignment. While subnanometer resolution can be achieved in cryoSPT with standard fiducial-based alignment (Schur *et al.* 2013), circumventing the known errors inherent in this methodology can improve resolution (Bartesaghi *et al.* 2012). Indeed, if made easily applicable, overcoming the limitations of fiducial-based alignment (Iwasaki *et al.* 2005; Bartesaghi *et al.* 2012; Zhang and Ren 2012) might facilitate achieving subtomogram averages beyond the 8–12 Å resolution range routinely for various types of specimens.

Tomographic reconstruction methodologies

A century ago, Johann Radon postulated that a function could be precisely reconstructed from an infinite set of its projections (Radon 1917). This concept was a stepping-stone in the development of computed tomography (CT), which has found applications in medical imaging, astronomy, and TEM (Van Heel 1987). Indeed, a variety of methods have been developed to

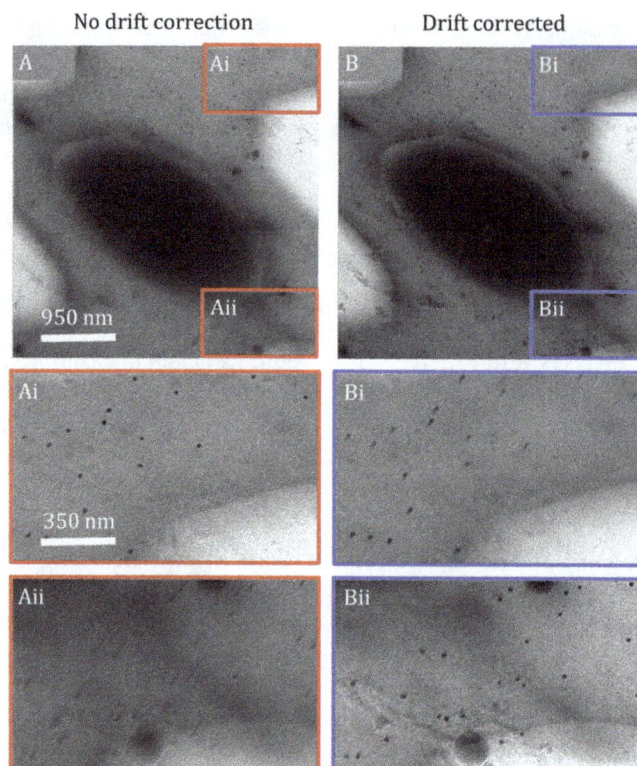

Fig. 6 Incomplete, local unblurring of cryoET images by whole-frame motion correction. Image of a mouse platelet at 57° tilt without (**A**) and with (**B**) motion correction applied to 21 frames collected with a K2 DDD on a JEM3200FSC microscope. Blurring is anisotropic both before (**Ai and Aii**) and after (**Bi and Bii**) whole-frame motion correction. While motion correction by iterative frame alignment (Galaz-Montoya et al. 2016b) improves the overall image (**B**), the extent of improvement varies in different parts of the image (**Bii**). Unexpectedly, a region that was not originally blurry (**Ai**) becomes blurry after motion correction (**Bi**) while a different region is effectively unblurred (**Aii** versus **Bii**). This suggests that different parts of the specimen are subject to divergent apparent motions, possibly due to charging effects and/or motion perpendicular to the imaging plane, and therefore local motion correction methods are needed for cryoET images, similar to those applied per particle in SPA cryoEM

reconstruct the 3D structure of biological specimens from a set of 2D TEM projection images in known orientations. Mathematically, all of these methods are based on the central section theorem (DeRosier and Klug 1968a, b; Crowther *et al.* 1970; Crowther 1971), according to which the Fourier transform (FT) of a 2D projection from a 3D specimen corresponds to a section through the center of the specimen's representation in Fourier space. Thus, one can construct the 3D FT of the specimen by combining the FTs of all the 2D projection images inserted into a 3D FT volume, and computing the inverse Fourier transform of the volume (*i.e.*, "direct Fourier inversion") (Ludtke *et al.* 1999; Belnap 2015). Although this approach is conceptually simple, interpolation in Fourier space is non-trivial, and different strategies produce different artifacts. While Fourier inversion has become the standard approach in SPA (Penczek *et al.* 2004; Penczek 2010), it is not widely used in cryoET (Heymann and Belnap 2007).

The most popular reconstruction method for cryoET due to its speed and relatively easy implementation for large volumes is weighted back-projection (WBP), a real-space equivalent of the central section theorem (Gilbert 1972; Radermacher 2007). It consists of literally "projecting back" the densities of 2D projections as rays into a 3D reconstruction volume. Appropriate weighting of the back-projected densities is needed to avoid the implicit low-pass filtering effect of WBP. However, WBP yields reconstructions with very poor contrast and strong "streaking" artifacts compared to iterative algebraic reconstruction methods such as the simultaneous iterative reconstruction technique (SIRT) (Gilbert 1972) or algebraic reconstruction techniques (ART) (Marabini *et al.* 1998). On the other hand, iterative algebraic methods are typically much slower, can diverge for some datasets or sometimes destroy high-resolution information, and determining *a priori* the optimal number of iterations to use and other parameters is not generally possible.

The problem of tomographic reconstruction is still being actively researched, with many novel methods described over the last decade (Díez *et al.* 2007;

Batenburg and Sijbers 2009; Wan *et al.* 2011; Kunz and Frangakis 2014; Turoňová *et al.* 2015; Zhou *et al.* 2015; Chen *et al.* 2016). A few methods have been proposed to recover some of the missing information in limited-angle tomography, for example, by using convex projections (Carazo and Carrascosa 1987). Recently, the reconstruction method proposed by Chen and Förster (2014) was shown to restore some of the missing information by iterative extrapolation. Total variation minimization (TMV, or "regularization") based on compressed sensing (CS) (Donoho 2006) has been applied successfully to ET in material sciences, yielding structures with minimal missing wedge artifacts and improved contrast (Saghi *et al.* 2011; Goris *et al.* 2012; Leary *et al.* 2013). Variations of this method have also been demonstrated for cryoET specimens (Aganj *et al.* 2007; Song *et al.* 2012). Recently, the iterative compressed-sensing optimized non-uniform (ICON) reconstruction (Deng *et al.* 2016) method demonstrated that CS can restore missing information in noisy cryoET data of both cells and isolated macromolecules, thereby minimizing missing wedge artifacts and yielding measurably better reconstructions than WBP.

At present, no single optimal algorithm has emerged, and the vast majority of users adopt whichever algorithm is most conveniently available or recommended by the software they have selected for tiltseries alignment. Reconstruction methods need to be carefully chosen depending on the goals of the study in question and the data collection parameters. For example, the success of TMV has been reported to depend on the tilt scheme used during data collection, while SIRT is less sensitive to variations in total dose or tilt scheme (Chen *et al.* 2014). It is important to note that certain algorithms are optimized for direct interpretation of tomograms but can destroy high-resolution information and are therefore suboptimal for averaging subtomograms if achieving higher resolution is the end goal. For example, SIRT delivers tomograms with much better contrast than WBP at the expense of introducing artifacts at high resolution. Furthermore, some algorithms rely on the individual images in a tiltseries having high contrast, such as the filtered iterative reconstruction technique (FIRT) (Chen *et al.* 2016), which does not seem to provide any advantages over WBP when applied to cryoET data. Interestingly, it has also been proposed that several reconstruction techniques can be applied sequentially to guide the choice of optimal parameters. For example, an initial SIRT reconstruction can guide the selection of the regularization parameter for TMV reconstruction, which can in turn help to choose adequate gray values to run a final reconstruction with the

discrete algebraic reconstruction technique (DART) (Goris *et al.* 2013).

Contrast transfer function determination and correction for tilted specimens

The contrast transfer function (CTF) (Erickson and Klug 1971) of the electron microscope modifies the amplitude of the signal in cryoEM micrographs (Toyoshima and Unwin 1988) in an oscillatory, resolution-dependent manner. While it is a function of several parameters, the only one that varies significantly during an imaging session is the defocus. In SPA cryoEM, CTF correction is a well-established, largely automated, and straightforward process, with various approaches achieving comparable results. Indeed, the "CTF challenge" recently compared many of the multiple algorithms available to perform CTF correction in SPA cryoEM (Marabini *et al.* 2015). On the other hand, as the defocus is directly related to the specimen height in the column with respect to the focal plane in the electron microscope, ET specimens produce a CTF that varies across the imaging plane due to the tilted geometry and, to a lesser extent, due to the thickness of the specimen. Compensating for these defocus gradients in images of cryoET tiltseries requires more complicated correction strategies than those implemented for SPA cryoEM. Without CTF correction, the resolution of subtomogram averages will typically be limited to 20–100 Å, depending on the imaging parameters. Resolving the structure of macromolecules by cryoSPT to better than 20 Å resolution is not yet a routine procedure, with the yearly average resolution being above this threshold (Fig. 4).

Several approaches have been implemented to determine the CTF and/or correct for it in cryoET data (Mindell and Grigorieff 2003; Winkler and Taylor 2003; Fernandez *et al.* 2006; Xiong *et al.* 2009; Zanetti *et al.* 2009; Voortman *et al.* 2011), with the most successful approach so far being that by (Schur *et al.* 2013). The latter approach achieved cryoSPT *in vitro* at subnanometer resolution for the first time, and has continued to yield structures at even higher resolution when combined with modern instrumentation, such as DDDs, and algorithmic improvements in image processing. The paramount achievement of Schur *et al.* 2013 using images taken with a charge-coupled device (CCD) was initially heavily dependent on stage eucentricity and stability during data collection and accurate autofocusing, as well as on a high particle density and automated data collection. Thin ice was also essential, as the specimen was assumed to be co-planar throughout the tomograms. So, while this was an effective proof of concept, these conditions might not be straightforward to achieve in a typical lab for all specimens. A year prior, a hybrid methodology combining concepts and data processing strategies from SPA cryoEM and cryoSPT resolved GroEL at subnanometer resolution as well (Bartesaghi *et al.* 2012), but this hybrid approach cannot be easily applied to cellular data. Another successful method (Eibauer *et al.* 2012) also recently achieved subnanometer resolution (Bharat *et al.* 2015), making use of a pair of additional high-contrast images collected away from the imaging area to interpolate the defocus in the region of interest. While this method was successful, it significantly increased data collection and processing complexity and assumed that the cryoEM support grid was flat (though not necessarily parallel to the imaging plane). Unfortunately, grid bending and "cryo-crinkling" of the carbon support mesh are common artifacts (Booy and Pawley 1993). Therefore, there is no guarantee that interpolation of the defocus at the imaging site by measuring the defocus in adjacent sites several microns away will always be accurate. Indeed, increased exposure near the imaging area due to lengthy focusing routines can induce deformations that compromise the accuracy of tiltseries alignment (Khoshouei *et al.* 2016).

Recently, a per-particle CTF correction method in 3D for cryoSPT was proposed (Galaz-Montoya *et al.* 2015) and demonstrated *in vitro* at near subnanometer resolution, without making any of the aforementioned assumptions (accurate defocusing during data collection, thin ice, unbent specimen, etc.), using only a few hundred icosahedrally symmetric virus particles (Galaz-Montoya *et al.* 2016a). Most importantly, the defocus gradient due to tilting was fitted by directly measuring the power spectrum in strips of constant defocus, similar to the method proposed by Fernández *et al.* (2006), except that it was done on each individual image (*i.e.*, different images in a tiltseries were never combined and the defocus gradient was linearly fit on a per-image basis, instead of relying on a single value from the central region of the image to compute the gradient). Indeed, DDDs now allow measuring the defocus directly from each image in a tiltseries, even at high tilt, which is essential under experimental settings for which the actual defocus might significantly differ from the target defocus and vary widely across the images of a tiltseries.

Performing CTF correction in 3D for tomographic reconstructions, including per-particle corrections (Bharat *et al.* 2015), has further been demonstrated to yield improvements compared to corrections considering 2D information only (Kunz and Frangakis 2016), as first theoretically proposed for virus reconstructions in SPA cryoEM nearly two decades ago when the depth of field started to become a resolution-limiting factor (Jensen and Kornberg 2000). The method proposed by Jensen and Kornberg to compensate for the depth of field was later generalized mathematically (Kazantsev *et al.* 2010).

The missing wedge

The effective thickness of ET specimens increases with the secant of the tilt angle, meaning that at 60° the specimen is twice as thick than at 0°, and at 70° it is nearly three times as thick. This effect degrades image quality rapidly at higher tilt angles and, in most cases, $\sim 60°$ is the highest tilt worth expending dose on. This means that typical ET tiltseries span only $\sim 2/3$ of complete tomographic angular sampling. The missing angular range is termed the "missing wedge," and leads to a variety of 3D reconstruction artifacts (Radermacher 1988), where features or particles are distorted in different ways, depending on their orientation with respect to the missing wedge.

Data collection methods alternative to canonical single-axis ET have been proposed to reduce the deleterious effects of the missing wedge in cryoET, such as dual-axis tomography (Penczek *et al.* 1995; Mastronarde 1997; Tong *et al.* 2006; Xu *et al.* 2007), conical tilt (Lanzavecchia *et al.* 2005), and multiple-axis tomography (Messaoudi *et al.* 2006). While these techniques reduce the missing wedge (to a missing pyramid, a missing cone, or a smaller missing region in general depending on how many tiltseries around different axes are combined), complete coverage is still not achieved. Furthermore, dose fractionation becomes increasingly problematic when two or more tiltseries are collected from the same imaging area, compromising the ability to align the images in the tiltseries accurately due to an exceedingly low SNR in individual images. In a recent proof-of-concept study, FIB milling was used to shape cellular material into a needle that could be fully rotated (*i.e.*, from −90° to 90°) (Narayan *et al.* 2012) and imaged by scanning electron microscopy (SEM). This study demonstrated that atom probe tomography (APT) (Miller *et al.* 2012) can be applied to chemically map freeze-dried cells with a thin metal coat in 3D. However, this was a unique experiment on unique equipment; it remains to be demonstrated whether an analogous approach could become widely applicable to frozen-hydrated cells using cryoET. In this direction, a recent study (Saghi *et al.* 2016) combined needle-shape FIB milling of the specimen with SEM imaging and tomographic reconstruction using CS. While these are exciting advances, added complications in data collection, storage, and processing preclude multiple-axis cryoET and APT of biological specimens from being routinely applied. A more promising proof of concept visualized bacterial cells by cryoET using a novel cylindrical holder (Palmer and Löwe 2014).

Missing wedge compensation for subtomogram classification and alignment

In cryoSPT, subtomograms need to be correctly aligned to each other or to a common reference before they can be averaged coherently. Preventing "missing wedge bias" is an essential step to accomplish this. Without correction, the missing wedge is the strongest feature in individual subtomograms and tends to bias the alignment of any two given subvolumes, favoring orientations with maximum density overlap as opposed to optimizing the overlap of matching structural features (Fig. 7).

Given the current state of SPT, accurate classification of noisy subtomograms with a missing wedge remains one of the biggest challenges. The missing wedge, as well as missing information between tilts, can make accurate classification statistically impossible in specific cases. Popular techniques used in cryoEM SPA such as multivariate statistical analysis (MSA) (Van Heel and Frank 1981; Frank *et al.* 1982; Van Heel 1984) can be tricky to apply to SPT, because the missing wedge, which is often the strongest feature, is difficult to exclude when the particles have already been rotated for alignment. A classification technique for subtomograms based on 2D reprojections has been proposed to overcome the uncertainties in classification introduced by the missing wedge (Yu *et al.* 2010, 2013). While this method would still be, in principle, subject to the degeneracy problem that arises in SPA cryoEM (Fig. 1), it may allow for fast, initial 2D classification of particles from cellular tomograms without the concern of overlapping densities.

Multiple methods have been proposed to identify and compensate for the missing wedge. Those based on normalization of cross correlation maps are the simplest and have been used successfully for template matching (Frangakis *et al.* 2002) and SPT (Nicastro *et al.* 2006; Schmid *et al.* 2006). These methods have the advantage of not requiring explicit identification of "missing voxels" in Fourier space. Other algorithms identify the missing wedge by presumed *a priori* knowledge of its exact location (Stölken *et al.* 2011) or by selection of a threshold value in Fourier space to constrain correlation (Bartesaghi *et al.* 2008; Förster *et al.* 2008; Schmid and

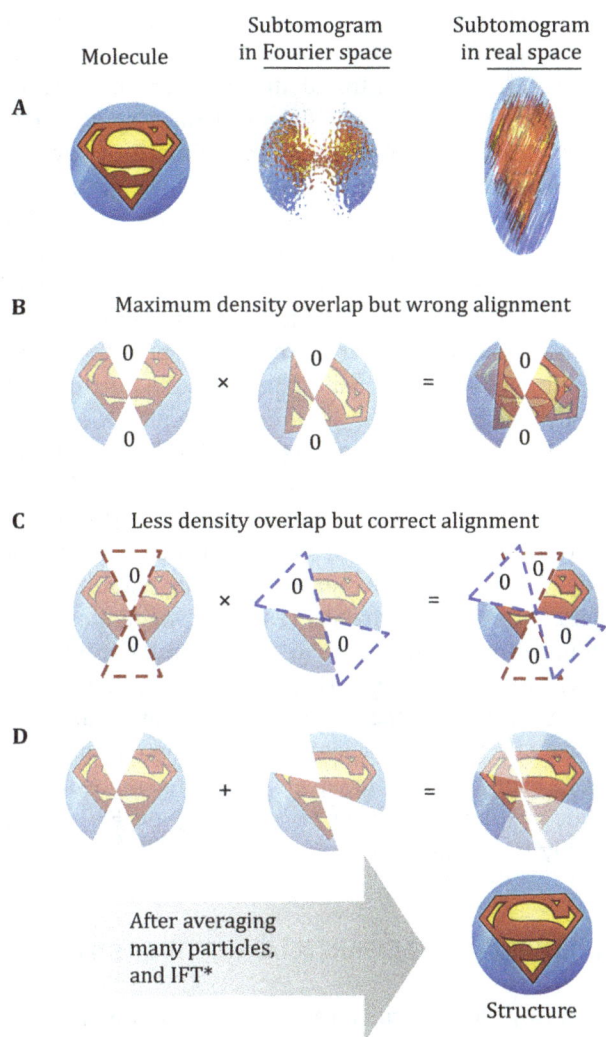

Fig. 7 Missing wedge bias in Fourier space. **A** Cartoon representation of a molecule to image (*left*), its representation as a subtomogram in Fourier space after limited-angle tomographic reconstruction (*center*), and the effects that the "missing wedge" of information due to limited-angle imaging and dose fractionation due to limited tolerable electron dose have on the subtomogram in real space (*i.e.,* individual subtomograms suffer from elongation and high levels of noise). **B, C** Since the comparison metric most commonly used in SPT is cross correlation, greater density overlap tends to increase the similarity score during alignment. This "missing wedge bias" towards favoring larger density overlaps can cause misalignment of subtomograms; however, it can be compensated for by cross correlation map normalization or other methods. **D** After averaging many correctly aligned subtomograms in different orientations, the missing wedge can be "filled in", and the associated elongation artifacts disappear while improving the SNR and the resolution of the macromolecular complex of interest. *Note:* The Fourier transform of the Superman logo would appear visually as an abstract and somewhat random pattern, in which it would be difficult to discern correct feature overlaps. For this reason, we have represented the image in Fourier space using the real-space image as a proxy to facilitate interpretation of the impact of missing wedge bias

Booth 2008). The rationale behind thresholding is that, for individual subtomograms, ∼95% of the structural information is concentrated in approximately the top 1% strongest Fourier coefficients (Amat *et al.* 2010). Other proposed methods use an estimate of the true structure (Heumann *et al.* 2011) or fill in the missing information with data derived from the population average (Bostina *et al.* 2007). However, the latter methods assume structural homogeneity among particles. On the other hand, methods relying on thresholding assume that Fourier voxels corresponding to the missing wedge have small values and thus a threshold can be easily assigned to identify them. One caveat inherent to the thresholding method is that simple operations such as filtration and masking can alter the values of voxels inside or outside the missing wedge in different ways, leading to misidentification of missing wedge voxels and therefore increasing subtomogram alignment error. A dynamic, resolution-dependent, per-volume thresholding method was recently introduced (Galaz-Montoya *et al.* 2016a), which produces much more robust identification of the missing wedge than the use of a single threshold value, and better alignment results than adaptive low-pass filtering using estimates of the resolution (Hrabe *et al.* 2012) or FSC curves directly (Galaz-Montoya *et al.* 2015).

IMPROVING SINGLE PARTICLE TOMOGRAPHY

Measuring the defocus directly from experimental images, coupled with several other methodological improvements such as an optimized tilt-image acquisition scheme (Hagen *et al.* 2017) and damage compensation (Bammes and Jin 2014), also known as exposure filtering (Grant and Grigorieff 2015), as well as the usage of cutting-edge technology, are some of the latest improvements that have permitted cryoSPT to achieve ∼4 Å resolution in the best case reported to date (Schur *et al.* 2016).

Ideally, one would like to produce a well-resolved 3D map with isotropic resolution for any specimen, be it particles extracted from cells, or macromolecules in solution. Covering the full 180° rotation about any single axis would yield a tiltseries without a missing wedge, but while this has been successfully implemented in material sciences, it is still not generally possible for biological specimens. Proof-of-concept methodologies that solve this problem have been proposed, such as combining FIB and SEM (Narayan *et al.* 2012; Saghi *et al.* 2016), or using a novel cylindrical specimen holder for cryoET (Palmer and Löwe 2014). Continued development of cylindrical holders or applying needle shaping of bulk specimens by FIB to cryoET could have a significant positive impact on cryoSPT, particularly for macromolecules *in situ*.

The ongoing development of tomographic reconstruction methods will also propel biological discoveries by cellular cryoSPT in the years to come. A current limiting factor is the reliable, automated identification of heterogeneous macromolecular complexes in crowded environments. Due to the missing wedge and the low SNR of cellular tomograms, automated particle picking is prone to bias and high false-positive rates (Yu and Frangakis 2014; Hrabe 2015; Kunz *et al.* 2015), though the use of neural networks has been proposed to reduce such issues (Yu and Frangakis 2011). Reliable automation of tiltseries reconstruction into tomograms will also increase the throughput of studies by cryoET (Mastronarde and Held 2017). Automated segmentation of features in tomograms remains poor and can be extremely subjective and time consuming when performed manually (Hecksel *et al.* 2016), particularly for thick frozen-hydrated specimens. The use of compressed sensing and other methodologies that improve the quality of tomographic reconstructions should facilitate more accurate particle picking and more objective, automated annotation of cellular tomograms.

Ongoing methodological developments and automation of algorithms on various fronts will also continue to improve image processing. For example, movie-mode data collection with relatively high contrast using DDDs now allows decoupling the decay in image quality due to radiation damage versus that stemming from beam-induced motions. Taking advantage of this, an "exposure filtering" method was recently proposed (Grant and Grigorieff 2015), which optimizes the SNR of every frame collected, thereby minimizing the deleterious effects of radiation damage. This idea was also proposed in an earlier study that filtered DDD frames based on cumulative exposure to compensate for radiation damage (Bammes and Jin 2014; Wang *et al.* 2014).

Since many subvolumes in random orientations are typically combined in SPT, there is seldom a need to use traditional small tilt steps of 1°–2° during data collection. As long as there are sufficient images to permit accurate alignment of the tiltseries, much larger tilt steps may be used in most cases (perhaps as high as 10°), depending on the size of the specimen. This can dramatically reduce data collection and processing times as well as data storage requirements, with no negative impact on final averages if enough subtomograms are averaged. Using a larger tilt step for cryoSPT also has the positive effect that each tilt image has a higher dose, permitting the application of some methods that require high contrast in each image of a tiltseries (Iwasaki *et al.* 2005; Bartesaghi *et al.* 2012; Zhang and Ren 2012). Unfortunately, to this day many cryoSPT studies continue to collect data using canonical parameters originally designed for cryoET of unique, non-repeating structures such as cells.

Several unconventional proposed improvements to general cryoEM have been reviewed in (Massover 2011) and might yield beneficial effects for cryoET and cryoSPT as well. For example, the development of specimen holders that are more stable and allow collecting data at slower exposure rates with "resting" periods in between exposures might make biological specimens more impervious to radiation damage (Chen *et al.* 2008; Karuppasamy *et al.* 2011).

Novel specimen grids (Rhinow and Kühlbrandt 2008; Pantelic *et al.* 2010; Yoshioka *et al.* 2010; Russo and Passmore 2016) made with materials that more readily dissipate charge and heat during imaging can minimize beam-induced specimen motions as well as the bubbling and distortions caused by radiation damage. The widespread adoption of such grids will also improve cryoET and cryoSPT and accelerate discoveries by these methodologies.

The problem of low-contrast at tolerable doses in cryoEM micrographs was first partially alleviated by the introduction of better illumination sources (field emission guns, opposed to metal filaments) (Tonomura *et al.* 1979) and the usage of large defocus settings for contrast enhancement (Erickson and Klug 1970). More recently, energy filters (Zhu *et al.* 1997) and phase plates (Danev and Nagayama 2001; Majorovits *et al.* 2007) have provided dramatic improvements in image contrast (Dai *et al.* 2013). Most significantly, the Volta phase plate (Danev *et al.* 2014) was recently used to demonstrate cryoSPT at subnanometer resolution (Khoshouei *et al.* 2016), averaging merely ~1400 ribosome subtomograms. Earlier phase plates caused strong ringing or "fringe" artifacts due to the cut-on frequency determined by the size of the central hole responsible for the phase shift, necessitating cumbersome computational correction (Kishchenko *et al.* 2015). Hole-less phase plates, such as the Volta phase plate, reduce such artifacts and are thus poised to revolutionize cryoET and cryoSPT. Unfortunately, a number of technical problems still remain that impede routine, straightforward use of this technology. The progressive change of phase with dose on Volta phase plates as well as the defocus gradient at high tilt both necessitate taking phase plates significantly out of focus. These factors greatly complicate CTF correction, particularly at high tilt (Sharp *et al.* 2017).

DDDs are revolutionizing the field and also provide images with better contrast compared to CCDs by virtue of their improved DQE, allowing for the correction of beam-induced blurring at the whole-frame, or at the per-particle level in SPA cryoEM. A step further, a local, anisotropic motion correction method recently proposed for SPA cryoEM (Zheng *et al.* 2016) might find applications in effectively unblurring lower-SNR images of highly tilted, non-repeating specimens, which sometimes cannot be corrected with current methods (Fig. 6).

Lastly, the application of hybrid techniques such as combining FIB milling (Wang *et al.* 2012; Villa *et al.* 2013; Mahamid *et al.* 2016) and correlative light microscopy (Zhang 2013; Hampton *et al.* 2017) with cryoET and cryoSPT has opened the window for observing dynamic cellular processes in regions of thick mammalian cells at nanometer resolution. Further developments to make these hybrid methodologies accessible and routine are likely to lead the next generation of insights into the biochemistry occurring within cells.

Abbreviations

CryoET	Cryo-electron tomography
CTF	Contrast transfer function
DDD	Direct detection device
SPT	Single particle tomography
STA	Subtomogram averaging

Acknowledgements We thank Boxue Ma from Dr. Wah Chiu's lab at the NCMI, Baylor College of Medicine (Houston, TX, USA), for collecting the platelet tiltseries, and Dr. Anil Sood's lab at MD Anderson Cancer Center (Houston, TX, USA) for providing the platelet specimen.

Compliance with Ethical Standards

Conflict of interest Jesús G. Galaz-Montoya and Steven J. Ludtke declare that they have no conflicts of interest.

Human and animal rights and informed consent This article does not contain any studies with human or animal subjects performed by any of the authors.

References

Adrian M, Dubochet J, Lepault J, McDowall AW (1984) Cryo-electron microscopy of viruses. Nature 308(5954):32–36

Aganj I, Bartesaghi A, Borgnia M, Liao HY (2007) Regularization for inverting the radon transform with wedge consideration. In: 4th IEEE International Symposium on Biomedical Imaging: From Nano to Macro. pp. 217–220

Agulleiro JI, Vazquez F, Garzon EM, Fernandez JJ (2012) Hybrid computing: CPU + GPU co-processing and its application to tomographic reconstruction. Ultramicroscopy 115:109–114

Al-Amoudi A, Dubochet J, Gnaegi H, Lüthi W, Studer D (2003) An oscillating cryo-knife reduces cutting-induced deformation of vitreous ultrathin sections. J Microsc 212(1):26–33

Al-Amoudi A, Norlen LP, Dubochet J (2004) Cryo-electron microscopy of vitreous sections of native biological cells and tissues. J Struct Biol 148(1):131–135

Al-Amoudi A, Studer D, Dubochet J (2005) Cutting artefacts and cutting process in vitreous sections for cryo-electron microscopy. J Struct Biol 150(1):109–121

Amat F, Comolli LR, Moussavi F, Smit J, Downing KH, Horowitz M (2010) Subtomogram alignment by adaptive Fourier coefficient thresholding. J Struct Biol 171(3):332–344

Asano S, Fukuda Y, Beck F, Aufderheide A, Förster F, Danev R, Baumeister W (2015) A molecular census of 26S proteasomes in intact neurons. Science 347(6220):439–442

Asano S, Engel BD, Baumeister W (2016) In situ cryo-electron tomography: a post-reductionist approach to structural biology. J Mol Biol 428(2):332–343

Baker LA, Rubinstein JL (2010) Chapter fifteen—radiation damage in electron cryomicroscopy. Methods Enzymol 481:371–388

Bammes BE, Jin L (2014) Method of electron beam imaging of a specimen by combining images of an image sequence. Direct Electron, Lp. U.S. Patent 8,809,781

Bammes BE, Jakana J, Schmid MF, Chiu W (2010) Radiation damage effects at four specimen temperatures from 4 to 100 K. J Struct Biol 169(3):331–341

Bartesaghi A, Sprechmann P, Liu J, Randall G, Sapiro G, Subramaniam S (2008) Classification and 3D averaging with missing wedge correction in biological electron tomography. J Struct Biol 162(3):436–450

Bartesaghi A, Lecumberry F, Sapiro G, Subramaniam S (2012) Protein secondary structure determination by constrained single-particle cryo-electron tomography. Structure 20(12):2003–2013

Batenburg KJ, Sijbers J (2009) Adaptive thresholding of tomograms by projection distance minimization. Pattern Recogn 42(10):2297–2305

Baumeister W (2002) Electron tomography: towards visualizing the molecular organization of the cytoplasm. Curr Opin Struct Biol 12(5):679–684

Belnap DM (2015) Electron microscopy and image processing: essential tools for structural analysis of macromolecules. Curr Protocols Protein Sci 17(17):2

Bharat TA, Russo CJ, Löwe J, Passmore LA, Scheres SH (2015) Advances in single-particle electron cryomicroscopy structure determination applied to sub-tomogram averaging. Structure 23(9):1743–1753

Booy FP, Pawley JB (1993) Cryo-crinkling: what happens to carbon films on copper grids at low temperature. Ultramicroscopy 48(3):273–280

Bostina M, Bubeck D, Schwartz C, Nicastro D, Filman DJ, Hogle JM (2007) Single particle cryoelectron tomography characterization of the structure and structural variability of poliovirus–receptor–membrane complex at 30 Å resolution. J Struct Biol 160(2):200–210

Brandt S, Heikkonen J, Engelhardt P (2001) Automatic alignment of transmission electron microscope tilt series without fiducial markers. J Struct Biol 136(3):201–213

Brandt F, Carlson L-A, Hartl FU, Baumeister W, Grünewald K (2010) The three-dimensional organization of polyribosomes in intact human cells. Mol Cell 39(4):560–569

Briegel A, Ortega DR, Mann P, Kjær A, Ringgaard S, Jensen GJ (2016) Chemotaxis cluster 1 proteins form cytoplasmic arrays in vibrio cholerae and are stabilized by a double signaling domain receptor DosM. Proc Natl Acad Sci USA 113(37):10412–10417

Brilot AF, Chen JZ, Cheng A, Pan J, Harrison SC, Potter CS, Carragher B, Henderson R, Grigorieff N (2012) Beam-induced motion of vitrified specimen on holey carbon film. J Struct Biol 177(3):630–637

Brink J, Sherman MB, Berriman J, Chiu W (1998) Evaluation of charging on macromolecules in electron cryomicroscopy. Ultramicroscopy 72(1):41–52

Campbell MG, Cheng A, Brilot AF, Moeller A, Lyumkis D, Veesler D, Pan J, Harrison SC, Potter CS, Carragher B, Grigorieff N (2012) Movies of ice-embedded particles enhance resolution in electron cryo-microscopy. Structure 20(11):1823–1828

Carazo JM, Carrascosa JL (1987) Information recovery in missing angular data cases: an approach by the convex projections method in three dimensions. J Microsc 145(1):23–43

Carbajal-González BI, Heuser T, Fu X, Lin J, Smith BW, Mitchell DR, Nicastro D (2013) Conserved structural motifs in the central pair complex of eukaryotic flagella. Cytoskeleton 70(2):101–120

Cardone G, Grünewald K, Steven AC (2005) A resolution criterion for electron tomography based on cross-validation. J Struct Biol 151(2):117–129

Cassidy CK, Himes BA, Alvarez FJ, Ma J, Zhao G, Perilla JR, Schulten K, Zhang P (2015) CryoEM and computer simulations reveal a novel kinase conformational switch in bacterial chemotaxis signaling. eLife 4:e08419

Castaño-Díez D, Al-Amoudi A, Glynn A-M, Seybert A, Frangakis AS (2007) Fiducial-less alignment of cryo-sections. J Struct Biol 159(3):413–423

Castaño-Díez D, Kudryashev M, Arheit M, Stahlberg H (2012) Dynamo: a flexible, user-friendly development tool for subtomogram averaging of cryo-EM data in high-performance computing environments. J Struct Biol 178(2):139–151

Chen Y, Förster F (2014) Iterative reconstruction of cryo-electron tomograms using nonuniform fast Fourier transforms. J Struct Biol 185(3):309–316

Chen JZ, Sachse C, Xu C, Mielke T, Spahn CM, Grigorieff N (2008) A dose-rate effect in single-particle electron microscopy. J Struct Biol 161(1):92–100

Chen D, Goris B, Bleichrodt F, Mezerji HH, Bals S, Batenburg KJ, de With G, Friedrich H (2014) The properties of SIRT, TVM, and DART for 3D imaging of tubular domains in nanocomposite thin-films and sections. Ultramicroscopy 147:137–148

Chen Y, Zhang Y, Zhang K, Deng Y, Wang S, Zhang F, Sun F (2016) FIRT: filtered iterative reconstruction technique with information restoration. J Struct Biol 195(1):49–61

Cheng N, Wu W, Steven AC, Thomas J, Black L (2012) Bubblegrams reveal the inner body of bacteriophage phiKZ. Microsc Microanal 18(S2):112–113

Comolli LR, Downing KH (2005) Dose tolerance at helium and nitrogen temperatures for whole cell electron tomography. J Struct Biol 152(3):149–156

Cosslett VE (1978) Radiation damage in the high resolution electron microscopy of biological materials: a review. J Microsc 113(2):113–129

Crowther RA (1971) Procedures for three-dimensional reconstruction of spherical viruses by Fourier synthesis from electron micrographs. Philos Trans R Soc Lond B 261(837):221–230

Crowther RA (2010) From envelopes to atoms: the remarkable progress of biological electron microscopy. Adv Protein Chem Struct Biol 81:1–32

Crowther RA, DeRosier DJ, Klug A (1970) The reconstruction of a three-dimensional structure from projections and its application to electron microscopy. Proc R Soc Lond A 317(1530): 319–340

Dai W, Fu C, Raytcheva D, Flanagan J, Khant HA, Liu X, Rochat RH, Haase-Pettingell C, Piret J, Ludtke SJ (2013) Visualizing virus assembly intermediates inside marine cyanobacteria. Nature 502(7473):707–710

Dalm D, Galaz-Montoya JG, Miller JL, Grushin K, Villalobos A, Koyfman AY, Schmid MF, Stoilova-McPhie S (2015) Dimeric organization of blood coagulation factor VIII bound to lipid nanotubes. Sci Rep 5:11212

Danev R, Nagayama K (2001) Transmission electron microscopy with Zernike phase plate. Ultramicroscopy 88(4):243–252

Danev R, Buijsse B, Khoshouei M, Plitzko JM, Baumeister W (2014) Volta potential phase plate for in-focus phase contrast transmission electron microscopy. Proc Natl Acad Sci USA 111(44):15635–15640

Darrow MC, Sergeeva OA, Isas JM, Galaz-Montoya JG, King JA, Langen R, Schmid MF, Chiu W (2015) Structural mechanisms of mutant huntingtin aggregation suppression by the synthetic chaperonin-like CCT5 complex explained by cryoelectron tomography. J Biol Chem 290(28):17451–17461

Davies KM, Strauss M, Daum B, Kief JH, Osiewacz HD, Rycovska A, Zickermann V, Kühlbrandt W (2011) Macromolecular organization of ATP synthase and complex I in whole mitochondria. Proc Natl Acad Sci USA 108(34):14121–14126

De Rosier DJ, Klug A (1968) Reconstruction of three dimensional structures from electron micrographs. Nature 217(5124): 130–134

Deng Y, Chen Y, Zhang Y, Wang S, Zhang F, Sun F (2016) ICON: 3D reconstruction with 'missing-information' restoration in biological electron tomography. J Struct Biol 195(1):100–112

Díez DC, Mueller H, Frangakis AS (2007) Implementation and performance evaluation of reconstruction algorithms on graphics processors. J Struct Biol 157(1):288–295

Donoho DL (2006) Compressed sensing. IEEE Trans Inf Theory 52(4):1289–1306

Dubochet J, Booy FP, Freeman R, Jones AV, Walter CA (1981) Low temperature electron microscopy. Annu Rev Biophys Bioeng 10(1):133–149

Dubochet J, Adrian M, Chang JJ, Homo JC, Lepault J, McDowall AW, Schultz P (1988) Cryoelectron microscopy of vitrified specimens. Q rev biophys 21(2):129–228

Eibauer M, Hoffmann C, Plitzko JM, Baumeister W, Nickell S, Engelhardt H (2012) Unraveling the structure of membrane proteins in situ by transfer function corrected cryo-electron tomography. J Struct Biol 180(3):488–496

Eibauer M, Pellanda M, Turgay Y, Dubrovsky A, Wild A, Medalia O (2015) Structure and gating of the nuclear pore complex. Nat Commun 6:7532

Eisenstein M (2016) The field that came in from the cold. Nat Methods 13(1):19–22

Erickson HP, Klug A (1970) The Fourier transform of an electron micrograph: effects of defocussing and aberrations, and implications for the use of underfocus contrast enhancement. Ber Bunsenges Phys Chem 74(11):1129–1137

Erickson HP, Klug A (1971) Measurement and compensation of defocusing and aberrations by Fourier processing of electron micrographs. Philos Trans R Soc Lond B 261(837):105–118

Fernandez J-J (2012) Computational methods for electron tomography. Micron 43(10):1010–1030

Fernández JJ, Li S, Crowther RA (2006) CTF determination and correction in electron cryotomography. Ultramicroscopy 106(7):587–596

Fontana J, Jurado KA, Cheng N, Engelman A, Steven AC (2015) Exploiting the susceptibility of HIV-1 nucleocapsid protein to radiation damage in tomo-bubblegram imaging. Microsc Microanal 21(S3):545–546

Förster F, Hegerl R (2007) Structure determination in situ by averaging of tomograms. Methods Cell Biol 79:741–767

Förster F, Medalia O, Zauberman N, Baumeister W, Fass D (2005) Retrovirus envelope protein complex structure in situ studied by cryo-electron tomography. Proc Natl Acad Sci USA 102(13): 4729–4734

Förster F, Pruggnaller S, Seybert A, Frangakis AS (2008) Classification of cryo-electron sub-tomograms using constrained correlation. J Struct Biol 161(3):276–286

Frangakis AS, Böhm J, Förster F, Nickell S, Nicastro D, Typke D, Hegerl R, Baumeister W (2002) Identification of macromolecular complexes in cryoelectron tomograms of phantom cells. Proc Natl Acad Sci USA 99(22):14153–14158

Frank J, Verschoor A, Boublik M (1982) Multivariate statistical analysis of ribosome electron micrographs: L and R lateral views of the 40 S subunit from HeLa cells. J Mol Biol 161(1): 107–133

Galaz JG, Flanagan JF, Schmid MF, Chiu W, Ludtke S (2012) Single particle tomography in EMAN2. Microsc Microanal 18(S2): 552

Galaz-Montoya JG, Flanagan J, Schmid MF, Ludtke SJ (2015) Single particle tomography in EMAN2. J Struct Biol 190(3):279–290

Galaz-Montoya JG, Hecksel CW, Baldwin PR, Wang E, Weaver SC, Schmid MF, Ludtke SJ, Chiu W (2016a) Alignment algorithms and per-particle CTF correction for single particle cryoelectron tomography. J Struct Biol 194(3):383–394

Galaz-Montoya JG, Hecksel CW, Chin J, Wang R, Lewis CW, Haemmerle M, Schmid MF, Ludtke SJ, Sood AK, Chiu W (2016b) Computational tools to improve visualization by cryo-electron tomography. Biophys J 110:159a

Gilbert P (1972) Iterative methods for the three-dimensional reconstruction of an object from projections. J Theor Biol 36(1):105–117

Glaeser RM (1971) Limitations to significant information in biological electron microscopy as a result of radiation damage. J Ultrastruct Res 36(3):466–482

Glaeser RM, Taylor KA (1978) Radiation damage relative to transmission electron microscopy of biological specimens at low temperature: a review. J Microsc 112(1):127–138

Goris B, Van den Broek W, Batenburg KJ, Mezerji HH, Bals S (2012) Electron tomography based on a total variation minimization reconstruction technique. Ultramicroscopy 113:120–130

Goris B, Roelandts T, Batenburg KJ, Mezerji HH, Bals S (2013) Advanced reconstruction algorithms for electron tomography: from comparison to combination. Ultramicroscopy 127:40–47

Grant T, Grigorieff N (2015) Measuring the optimal exposure for single particle cryo-EM using a 2.6 Å reconstruction of rotavirus VP6. Elife 4:e06980

Grange M, Vasishtan D, Grünewald K (2017) Cellular electron cryo tomography and in situ sub-volume averaging reveal the context of microtubule-based processes. J Struct Biol 197(2):181–190

Grubb DT (1974) Radiation damage and electron microscopy of organic polymers. J Mater Sci 9(10):1715–1736

Grünewald K, Medalia O, Gross A, Steven AC, Baumeister W (2002) Prospects of electron cryotomography to visualize macromolecular complexes inside cellular compartments: implications of crowding. Biophys Chem 100(1):577–591

Hagen WJ, Wan W, Briggs JA (2017) Implementation of a cryo-electron tomography tilt-scheme optimized for high resolution subtomogram averaging. J Struct Biol 197(2):191–198

Hampton CM, Strauss JD, Ke Z, Dillard RS, Hammonds JE, Alonas E, Desai TM, Marin M, Storms RE, Leon F, Melikyan GB, Santangelo PJ, Spearman PW, Wright ER (2017) Correlated fluorescence microscopy and cryo-electron tomography of virus-infected or transfected mammalian cells. Nat Protoc 12(1):150–167

Han R, Wang L, Liu Z, Sun F, Zhang F (2015) A novel fully automatic scheme for fiducial marker-based alignment in electron tomography. J Struct Biol 192(3):403–417

Harris A, Cardone G, Winkler DC, Heymann JB, Brecher M, White JM, Steven AC (2006) Influenza virus pleiomorphy characterized by cryoelectron tomography. Proc Natl Acad Sci USA 103(50):19123–19127

Harris AK, Meyerson JR, Matsuoka Y, Kuybeda O, Moran A, Bliss D, Das SR, Yewdell JW, Sapiro G, Subbarao K (2013) Structure and accessibility of HA trimers on intact 2009 H1N1 pandemic influenza virus to stem region-specific neutralizing antibodies. Proc Natl Acad Sci USA 110(12):4592–4597

Hart RG (1968) Electron microscopy of unstained biological material: the polytropic montage. Science 159(3822):1464–1467

Hawkes PW (2007) The electron microscope as a structure projector. In: Frank J (ed) Electron tomography. Springer, New York, pp 83–111

Hecksel CW, Darrow MC, Dai W, Galaz-Montoya JG, Chin JA, Mitchell PG, Chen S, Jakana J, Schmid MF, Chiu W (2016) Quantifying variability of manual annotation in cryo-electron tomograms. Microsc Microanal 22:1–10

Henderson R, Chen S, Chen JZ, Grigorieff N, Passmore LA, Ciccarelli L, Rubinstein JL, Crowther RA, Stewart PL, Rosenthal PB (2011) Tilt-pair analysis of images from a range of different specimens in single-particle electron cryomicroscopy. J Mol Biol 413(5):1028–1046

Henderson R, Sali A, Baker ML, Carragher B, Devkota B, Downing KH, Egelman EH, Feng Z, Frank J, Grigorieff N (2012) Outcome of the first electron microscopy validation task force meeting. Structure 20(2):205–214

Heumann JM, Hoenger A, Mastronarde DN (2011) Clustering and variance maps for cryo-electron tomography using wedge-masked differences. J Struct Biol 175(3):288–299

Heymann JB, Belnap DM (2007) Bsoft: image processing and molecular modeling for electron microscopy. J Struct Biol 157(1):3–18

Hoppe W, Langer R, Knesch G, Poppe C (1968) Protein-kristall-strukturanalyse mit elektronenstrahlen. Naturwissenschaften 55(7):333–336

Hoppe W, Gassmann J, Hunsmann N, Schramm HJ, Sturm M (1974) Three-dimensional reconstruction of individual negatively stained yeast fatty-acid synthetase molecules from tilt series in the electron microscope. Hoppe-Seyler's Z Phys Chem 355(11):1483–1487

Hrabe T (2015) Localize. pytom: a modern webserver for cryo-electron tomography. Nucleic Acids Res 43(W1):W231–W236

Hrabe T, Chen Y, Pfeffer S, Kuhn Cuellar L, Mangold A-V, Förster F (2012) PyTom: a python-based toolbox for localization of macromolecules in cryo-electron tomograms and subtomogram analysis. J Struct Biol 178(2):177–188

Hu B, Margolin W, Molineux IJ, Liu J (2013) The bacteriophage t7 virion undergoes extensive structural remodeling during infection. Science 339(6119):576–579

Huiskonen JT, Hepojoki J, Laurinmäki P, Vaheri A, Lankinen H, Butcher SJ, Grünewald K (2010) Electron cryotomography of Tula hantavirus suggests a unique assembly paradigm for enveloped viruses. J Virol 84(10):4889–4897

Iancu CV, Wright ER, Heymann JB, Jensen GJ (2006) A comparison of liquid nitrogen and liquid helium as cryogens for electron cryotomography. J Struct Biol 153(3):231–240

Iwasaki K, Mitsuoka K, Fujiyoshi Y, Fujisawa Y, Kikuchi M, Sekiguchi K, Yamada T (2005) Electron tomography reveals diverse conformations of integrin αIIbβ3 in the active state. J Struct Biol 150(3):259–267

Jensen GJ, Kornberg RD (2000) Defocus-gradient corrected back-projection. Ultramicroscopy 84(1):57–64

Karuppasamy M, Karimi Nejadasl F, Vulovic M, Koster AJ, Ravelli RB (2011) Radiation damage in single-particle cryo-electron microscopy: effects of dose and dose rate. J Synchrotron Radiat 18(3):398–412

Kazantsev IG, Klukowska J, Herman GT, Cernetic L (2010) Fully three-dimensional defocus-gradient corrected backprojection in cryoelectron microscopy. Ultramicroscopy 110(9):1128–1142

Khoshouei M, Pfeffer S, Baumeister W, Förster F, Danev R (2016) Subtomogram analysis using the Volta phase plate. J Struct Biol 197(2):94–101

Kishchenko GP, Danev R, Fisher R, He J, Hsieh C, Marko M, Sui H (2015) Effect of fringe-artifact correction on sub-tomogram averaging from Zernike phase-plate cryo-TEM. J Struct Biol 191(3):299–305

Knauer V, Hegerl R, Hoppe W (1983) Three-dimensional reconstruction and averaging of 30S ribosomal subunits of Escherichia coli from electron micrographs. J Mol Biol 163(3):409–430

Kosinski J, Mosalaganti S, von Appen A, Teimer R, DiGuilio AL, Wan W, Bui KH, Hagen WJ, Briggs JA, Glavy JS (2016) Molecular architecture of the inner ring scaffold of the human nuclear pore complex. Science 352(6283):363–365

Koster AJ, Grimm R, Typke D, Hegerl R, Stoschek A, Walz J, Baumeister W (1997) Perspectives of molecular and cellular electron tomography. J Struct Biol 120(3):276–308

Koyfman AY, Schmid MF, Gheiratmand L, Fu CJ, Khant HA, Huang D, He CY, Chiu W (2011) Structure of Trypanosoma brucei flagellum accounts for its bihelical motion. Proc Natl Acad Sci USA 108(27):11105–11108

Kremer JR, Mastronarde DN, McIntosh JR (1996) Computer visualization of three-dimensional image data using IMOD. J Struct Biol 116(1):71–76

Kudryashev M, Castaño-Díez D, Stahlberg H (2012) Limiting factors in single particle cryo electron tomography. Comput Struct Biotechnol J 1(2):1–6

Kudryashev M, Castaño-Díez D, Deluz C, Hassaine G, Grasso L, Graf-Meyer A, Vogel H, Stahlberg H (2016) The structure of the mouse serotonin 5-HT 3 receptor in lipid vesicles. Structure 24(1):165–170

Kunz M, Frangakis AS (2014) Super-sampling SART with ordered subsets. J Struct Biol 188(2):107–115

Kunz M, Frangakis AS (2016) Three-dimensional CTF correction improves the resolution of electron tomograms. J Struct Biol 197(2):114–122

Kunz M, Yu Z, Frangakis AS (2015) M-free: mask-independent scoring of the reference bias. J Struct Biol 192(2):307–311

Lanzavecchia S, Cantele F, Bellon PL, Zampighi L, Kreman M, Wright E, Zampighi GA (2005) Conical tomography of freeze-fracture replicas: a method for the study of integral membrane proteins inserted in phospholipid bilayers. J Struct Biol 149(1):87–98

Le Gros MA, McDermott G, Larabell CA (2005) X-ray tomography of whole cells. Curr Opin Struct Biol 15(5):593–600

Leary R, Saghi Z, Midgley PA, Holland DJ (2013) Compressed sensing electron tomography. Ultramicroscopy 131:70–91

Leschziner AE, Nogales E (2006) The orthogonal tilt reconstruction method: an approach to generating single-class volumes with no missing cone for ab initio reconstruction of asymmetric particles. J Struct Biol 153(3):284–299

Li S, Sun Z, Pryce R, Parsy ML, Fehling SK, Schlie K, Siebert CA, Garten W, Bowden TA, Strecker T, Huiskonen JT (2016) Acidic pH-induced conformations and LAMP1 binding of the Lassa virus glycoprotein spike. PLoS Pathog 12(2):e1005418

Liu Y, Penczek PA, McEwen BF, Frank J (1995) A marker-free alignment method for electron tomography. Ultramicroscopy 58(3):393–402

Ludtke SJ, Baldwin PR, Chiu W (1999) EMAN: semiautomated software for high-resolution single-particle reconstructions. J Struct Biol 128(1):82–97

Mahamid J, Pfeffer S, Schaffer M, Villa E, Danev R, Cuellar LK, Förster F, Hyman AA, Plitzko JM, Baumeister W (2016) Visualizing the molecular sociology at the HeLa cell nuclear periphery. Science 351(6276):969–972

Maimon T, Elad N, Dahan I, Medalia O (2012) The human nuclear pore complex as revealed by cryo-electron tomography. Structure 20(6):998–1006

Maiorca M, Millet C, Hanssen E, Abbey B, Kazmierczak E, Tilley L (2014) Local regularization of tilt projections reduces artifacts in electron tomography. J Struct Biol 186(1):28–37

Majorovits E, Barton B, Schultheiss K, Perez-Willard F, Gerthsen D, Schröder RR (2007) Optimizing phase contrast in transmission electron microscopy with an electrostatic (Boersch) phase plate. Ultramicroscopy 107(2):213–226

Marabini R, Herman GT, Carazo JM (1998) 3D reconstruction in electron microscopy using ART with smooth spherically symmetric volume elements (blobs). Ultramicroscopy 72(1):53–65

Marabini R, Carragher B, Chen S, Chen J, Cheng A, Downing KH, Frank J, Grassucci RA, Heymann JB, Jiang W, Jonic S (2015) CTF challenge: result summary. J Struct Biol 190(3):348–359

Marko M, Hsieh C, Schalek R, Frank J, Mannella C (2007) Focused-ion-beam thinning of frozen-hydrated biological specimens for cryo-electron microscopy. Nat Methods 4(3):215–217

Massover WH (2011) New and unconventional approaches for advancing resolution in biological transmission electron microscopy by improving macromolecular specimen preparation and preservation. Micron 42(2):141–151

Mastronarde DN (1997) Dual-axis tomography: an approach with alignment methods that preserve resolution. J Struct Biol 120(3):343–352

Mastronarde DN, Held SR (2017) Automated tilt series alignment and tomographic reconstruction in IMOD. J Struct Biol 197(2):102–113

Mattei S, Glass B, Hagen WJ, Kräusslich HG, Briggs JA (2016) The structure and flexibility of conical HIV-1 capsids determined within intact virions. Science 354(6318):1434–1437

McEwen BF, Downing KH, Glaeser RM (1995) The relevance of dose-fractionation in tomography of radiation-sensitive specimens. Ultramicroscopy 60(3):357–373

Medalia O, Typke D, Hegerl R, Angenitzki M, Sperling J, Sperling R (2002a) Cryoelectron microscopy and cryoelectron tomography of the nuclear pre-mRNA processing machine. J Struct Biol 138(1):74–84

Medalia O, Weber I, Frangakis AS, Nicastro D, Gerisch G, Baumeister W (2002b) Macromolecular architecture in eukaryotic cells visualized by cryoelectron tomography. Science 298(5596):1209–1213

Messaoudi C, Loubresse NG, Boudier T, Dupuis-Williams P, Marco S (2006) Multiple-axis tomography: applications to basal bodies from Paramecium tetraurelia. Biol Cell 98(7):415–425

Milazzo A-C, Leblanc P, Duttweiler F, Jin L, Bouwer JC, Peltier S, Ellisman M, Bieser F, Matis HS, Wieman H (2005) Active pixel sensor array as a detector for electron microscopy. Ultramicroscopy 104(2):152–159

Milazzo A-C, Moldovan G, Lanman J, Jin L, Bouwer JC, Klienfelder S, Peltier ST, Ellisman MH, Kirkland AI, Xuong N-H (2010) Characterization of a direct detection device imaging camera for transmission electron microscopy. Ultramicroscopy 110(7):741–744

Miller MK, Kelly TF, Rajan K, Ringer SP (2012) The future of atom probe tomography. Mater Today 15(4):158–165

Mindell JA, Grigorieff N (2003) Accurate determination of local defocus and specimen tilt in electron microscopy. J Struct Biol 142(3):334–347

Murata K, Zhang Q, Galaz-Montoya JG, Fu C, Coleman ML, Osburne MS, Schmid MF, Sullivan MB, Chisholm SW, Chiu W (2017) Visualizing adsorption of cyanophage P-SSP7 onto marine *Prochlorococcus*. Sci Rep 7:44176

Murray SC, Galaz-Montoya JG, Tang G, Flanagan JF, Ludtke SJ (2014) EMAN2. 1-a new generation of software for validated single particle analysis and single particle tomography. Microsc Microanal 20(S3):832–833

Nans A, Kudryashev M, Saibil HR, Hayward RD (2015) Structure of a bacterial type III secretion system in contact with a host membrane in situ. Nat Commun 6:10114

Narayan K, Prosa TJ, Fu J, Kelly TF, Subramaniam S (2012) Chemical mapping of mammalian cells by atom probe tomography. J Struct Biol 178(2):98–107

Nicastro D, Schwartz C, Pierson J, Gaudette R, Porter ME, McIntosh JR (2006) The molecular architecture of axonemes revealed by cryoelectron tomography. Science 313(5789):944–948

Noble AJ, Stagg SM (2015) Automated batch fiducial-less tilt-series alignment in Appion using Protomo. J Struct Biol 192(2): 270–278

Nogales E (2016) The development of cryo-EM into a mainstream structural biology technique. Nat Methods 13(1):24–27

Oettl H, Hegerl R, Hoppe W (1983) Three-dimensional reconstruction and averaging of 50S ribosomal subunits of *Escherichia coli* from electron micrographs. J Mol Biol 163(3):431–450

Ortiz JO, Förster F, Kürner J, Linaroudis AA, Baumeister W (2006) Mapping 70S ribosomes in intact cells by cryoelectron tomography and pattern recognition. J Struct Biol 156(2):334–341

Ortiz JO, Brandt F, Matias VR, Sennels L, Rappsilber J, Scheres SH, Eibauer M, Hartl FU, Baumeister W (2010) Structure of hibernating ribosomes studied by cryoelectron tomography in vitro and in situ. J Cell Biol 190(4):613–621

Palmer CM, Löwe J (2014) A cylindrical specimen holder for electron cryo-tomography. Ultramicroscopy 137:20–29

Pantelic RS, Meyer JC, Kaiser U, Baumeister W, Plitzko JM (2010) Graphene oxide: a substrate for optimizing preparations of frozen-hydrated samples. J Struct Biol 170(1):152–156

Pascual-Montano A, Taylor KA, Winkler H, Pascual-Marqui RD, Carazo J-M (2002) Quantitative self-organizing maps for clustering electron tomograms. J Struct Biol 138(1):114–122

Penczek PA (2010) Chapter one-fundamentals of three-dimensional reconstruction from projections. Methods Enzymol 482:1–33

Penczek P, Marko M, Buttle K, Frank J (1995) Double-tilt electron tomography. Ultramicroscopy 60(3):393–410

Penczek PA, Renka R, Schomberg H (2004) Gridding-based direct Fourier inversion of the three-dimensional ray transform. JOSA A 21(4):499–509

Peralta B, Gil-Carton D, Castaño-Díez D, Bertin A, Boulogne C, Oksanen HM, Bamford DH, Abrescia NG (2013) Mechanism of membranous tunnelling nanotube formation in viral genome delivery. PLoS Biol 11(9):e1001667

Pfeffer S, Brandt F, Hrabe T, Lang S, Eibauer M, Zimmermann R, Förster F (2012) Structure and 3D arrangement of endoplasmic reticulum membrane-associated ribosomes. Structure 20(9):1508–1518

Pfeffer S, Woellhaf MW, Herrmann JM, Förster F (2015a) Organization of the mitochondrial translation machinery studied in situ by cryoelectron tomography. Nat Commun 6:6019

Pfeffer S, Burbaum L, Unverdorben P, Pech M, Chen Y, Zimmermann R, Beckmann R, Förster F (2015b) Structure of the native Sec61 protein-conducting channel. Nat Commun 6:6019

Radermacher M (1988) Three-dimensional reconstruction of single particles from random and nonrandom tilt series. J Electron Microsc Tech 9(4):359–394

Radermacher M (2007) Electron tomography. Springer, New York

Radermacher M, Wagenknecht T, Verschoor A, Frank J (1986) A new 3-d reconstruction scheme applied to the 50 s ribosomal subunit of E. coli. J Microsc 141(1):1–2

Radon J (1917) Uber die Bestimmung von Funktionen durch ihre Integralwerte langs gewissez Mannigfaltigheiten, Ber. Verh. Sachs. Akad. Wiss. Leipzig, Math Phys Klass, 69

Renken C, McEwen B (2003) Markerless alignment: bridging the gap between theory and practice. Microsc Microanal 9(S02):1170–1171

Rhinow D, Kühlbrandt W (2008) Electron cryo-microscopy of biological specimens on conductive titanium–silicon metal glass films. Ultramicroscopy 108(7):698–705

Riedel C, Vasishtan D, Siebert CA, Whittle C, Lehmann MJ, Mothes W, Grünewald K (2017) Native structure of a retroviral envelope protein and its conformational change upon interaction with the target cell. J Struct Biol 197(2):172–180

Russo CJ, Passmore LA (2016) Ultrastable gold substrates: properties of a support for high-resolution electron cryomicroscopy of biological specimens. J Struct Biol 193(1):33–44

Saghi Z, Holland DJ, Leary R, Falqui A, Bertoni G, Sederman AJ, Gladden LF, Midgley PA (2011) Three-dimensional morphology of iron oxide nanoparticles with reactive concave surfaces. A compressed sensing-electron tomography (CS-ET) approach. Nano Lett 11(11):4666–4673

Saghi Z, Divitini G, Winter B, Leary R, Spiecker E, Ducati C, Midgley PA (2016) Compressed sensing electron tomography of needle-shaped biological specimens–Potential for improved reconstruction fidelity with reduced dose. Ultramicroscopy 160:230–238

Schaller RR (1997) Moore's law: past, present and future. IEEE Spectr 34(6):52–59

Scheres SH (2014) Beam-induced motion correction for sub-megadalton cryo-EM particles. elife 3:e03665

Schmid MF (2011) Single-particle electron cryotomography (cryoET). Adv Protein Chem Struct Biol 82:37–65

Schmid MF, Booth CR (2008) Methods for aligning and for averaging 3D volumes with missing data. J Struct Biol 161(3):243–248

Schmid MF, Paredes AM, Khant HA, Soyer F, Aldrich HC, Chiu W, Shively JM (2006) Structure of halothiobacillus neapolitanus carboxysomes by cryo-electron tomography. J Mol Biol 364(3):526–535

Schmid MF, Hecksel CW, Rochat RH, Bhella D, Chiu W, Rixon FJ (2012) A tail-like assembly at the portal vertex in intact herpes simplex type-1 virions. PLoS Pathog 8(10):e1002961

Schur FK, Hagen W, de Marco A, Briggs JA (2013) Determination of protein structure at 8.5 Å resolution using cryo-electron tomography and subtomogram averaging. J Struct Biol 184(3):394–400

Schur FK, Hagen WJ, Rumlová M, Ruml T, Müller B, Kräusslich H-G, Briggs JA (2015a) Structure of the immature HIV-1 capsid in intact virus particles at 8.8 A resolution. Nature 517(7535):505–508

Schur FK, Dick RA, Hagen WJ, Vogt VM, Briggs JA (2015b) The structure of immature virus-like Rous sarcoma virus Gag particles reveals a structural role for the p10 domain in assembly. J Virol 89(20):10294–10302

Schur FK, Obr M, Hagen WJ, Wan W, Jakobi AJ, Kirkpatrick JM, Sachse C, Kräusslich HG, Briggs JA (2016) An atomic model of HIV-1 capsid-SP1 reveals structures regulating assembly and maturation. Science 353(6298):506–508

Schwartz CL, Heumann JM, Dawson SC, Hoenger A (2012) A detailed, hierarchical study of Giardia lamblia's ventral disc reveals novel microtubule-associated protein complexes. PLoS ONE 7(9):e43783

Shahmoradian SH, Galaz-Montoya JG, Schmid MF, Cong Y, Ma B, Spiess C, Frydman J, Ludtke SJ, Chiu W (2013) TRiCs tricks inhibit huntingtin aggregation. eLife 2:e00710

Sharp T, Koster A, Gros P (2016) Heterogeneous MAC initiator and pore structures in a lipid bilayer by phase-plate cryo-electron tomography. Cell Rep 15(1):1–8

Sharp TH, Faas FG, Koster AJ, Gros P (2017) Imaging complement by phase-plate cryo-electron tomography from initiation to pore formation. J Struct Biol 197(2):155–162

Song K, Comolli LR, Horowitz M (2012) Removing high contrast artifacts via digital inpainting in cryo-electron tomography: an application of compressed sensing. J Struct Biol 178(2):108–120

Stoffler D, Feja B, Fahrenkrog B, Walz J, Typke D, Aebi U (2003) Cryo-electron tomography provides novel insights into nuclear pore architecture: implications for nucleocytoplasmic transport. J Mol Biol 328(1):119–130

Stölken M, Beck F, Haller T, Hegerl R, Gutsche I, Carazo J-M, Baumeister W, Scheres SH, Nickell S (2011) Maximum likelihood based classification of electron tomographic data. J Struct Biol 173(1):77–85

Sun L, Young LN, Zhang X, Boudko SP, Fokine A, Zbornik E, Roznowski AP, Molineux IJ, Rossmann MG, Fane BA (2014) Icosahedral bacteriophage [Phi] X174 forms a tail for DNA transport during infection. Nature 505(7483):432–435

Taylor KA, Glaeser RM (1974) Electron diffraction of frozen, hydrated protein crystals. Science 186(4168):1036–1037

Tong J, Arslan I, Midgley P (2006) A novel dual-axis iterative algorithm for electron tomography. J Struct Biol 153(1):55–63

Tonomura A, Matsuda T, Todokoro H, Komoda T (1979) Development of a field emission electron microscope. J Electron Microsc 28(1):1–11

Toyoshima C, Unwin N (1988) Contrast transfer for frozen-hydrated specimens: determination from pairs of defocused images. Ultramicroscopy 25(4):279–291

Turoňová B, Marsalek L, Davidovič T, Slusallek P (2015) Progressive stochastic reconstruction technique (PSRT) for cryo electron tomography. J Struct Biol 189(3):195–206

Van Heel M (1984) Multivariate statistical classification of noisy images (randomly oriented biological macromolecules). Ultramicroscopy 13(1):165–183

Van Heel M (1987) Angular reconstitution: a posteriori assignment of projection directions for 3D reconstruction. Ultramicroscopy 21(2):111–123

Van Heel M, Frank J (1981) Use of multivariate statistics in analysing the images of biological macromolecules. Ultramicroscopy 6(2):187–194

Villa E, Schaffer M, Plitzko JM, Baumeister W (2013) Opening windows into the cell: focused-ion-beam milling for cryo-electron tomography. Curr Opin Struct Biol 23(5):771–777

Voortman LM, Stallinga S, Schoenmakers RH, van Vliet LJ, Rieger B (2011) A fast algorithm for computing and correcting the CTF for tilted, thick specimens in TEM. Ultramicroscopy 111(8): 1029–1036

Wan W, Briggs JAG (2016) Chapter thirteen-cryo-electron tomography and subtomogram averaging. Methods Enzymol 579:329–367

Wan X, Zhang F, Chu Q, Zhang K, Sun F, Yuan B, Liu Z (2011) Three-dimensional reconstruction using an adaptive simultaneous algebraic reconstruction technique in electron tomography. J Struct Biol 175(3):277–287

Wang Y, Jacobsen C, Maser J, Osanna A (2000) Soft X-ray microscopy with a cryo scanning transmission X-ray microscope: II. Tomography. J Microsc 197(Pt 1):80–93

Wang K, Strunk K, Zhao G, Gray JL, Zhang P (2012) 3D structure determination of native mammalian cells using cryo-FIB and cryo-electron tomography. J Struct Biol 180(2):318–326

Wang Z, Hryc CF, Bammes B, Afonine PV, Jakana J, Chen DH, Liu X, Baker ML, Kao C, Ludtke SJ, Schmid MF (2014) An atomic model of brome mosaic virus using direct electron detection and real-space optimization. Nat Commun 5:4808

Winkler H (2007) 3D reconstruction and processing of volumetric data in cryo-electron tomography. J Struct Biol 157(1):126–137

Winkler H, Taylor KA (1999) Multivariate statistical analysis of three-dimensional cross-bridge motifs in insect flight muscle. Ultramicroscopy 77(3):141–152

Winkler H, Taylor KA (2003) Focus gradient correction applied to tilt series image data used in electron tomography. J Struct Biol 143(1):24–32

Wu W, Thomas JA, Cheng N, Black LW, Steven AC (2012) Bubblegrams reveal the inner body of bacteriophage φKZ. Science 335(6065):182

Xiong Q, Morphew MK, Schwartz CL, Hoenger AH, Mastronarde DN (2009) CTF determination and correction for low dose tomographic tilt series. J Struct Biol 168(3):378–387

Xu P, Donaldson LA, Gergely ZR, Staehelin LA (2007) Dual-axis electron tomography: a new approach for investigating the spatial organization of wood cellulose microfibrils. Wood Sci Technol 41(2):101–116

Yoshioka C, Carragher B, Potter CS (2010) Cryomesh: a new substrate for cryo-electron microscopy. Microsc Microanal 16(01):43–53

Yu Z, Frangakis AS (2011) Classification of electron sub-tomograms with neural networks and its application to template-matching. J Struct Biol 174(3):494–504

Yu Z, Frangakis AS (2014) M-free: Scoring the reference bias in sub-tomogram averaging and template matching. J Struct Biol 187(1):10–19

Yu L, Snapp RR, Ruiz T, Radermacher M (2010) Probabilistic principal component analysis with expectation maximization (PPCA-EM) facilitates volume classification and estimates the missing data. J Struct Biol 171(1):18–30

Yu L, Snapp RR, Ruiz T, Radermacher M (2013) Projection-based volume alignment. J Struct Biol 182(2):93–105

Zanetti G, Riches JD, Fuller SD, Briggs JA (2009) Contrast transfer function correction applied to cryo-electron tomography and sub-tomogram averaging. J Struct Biol 168(2):305–312

Zhang P (2013) Correlative cryo-electron tomography and optical microscopy of cells. Curr Opin Struct Biol 23(5):763–770

Zhang L, Ren G (2012) IPET and FETR: experimental approach for studying molecular structure dynamics by cryo-electron tomography of a single-molecule structure. PLoS ONE 7(1):e30249

Zhao X, Zhang K, Boquoi T, Hu B, Motaleb MA, Miller KA, James ME, Charon NW, Manson MD, Norris SJ (2013) Cryoelectron tomography reveals the sequential assembly of bacterial flagella in Borrelia burgdorferi. Proc Natl Acad Sci USA 110(35):14390–14395

Zheng SQ, Palovcak E, Armache JP, Cheng Y, Agard DA (2016) Anisotropic correction of beam-induced motion for improved single-particle electron cryo-microscopy. BioArxiv. doi:10.1101/061960

Zhou Z, Li Y, Zhang F, Wan X (2015) FASART: an iterative reconstruction algorithm with inter-iteration adaptive NAD filter. Bio-Med Mater Eng 26(s1):S1409–S1415

Zhu J, Penczek PA, Schröder R, Frank J (1997) Three-dimensional reconstruction with contrast transfer function correction from energy-filtered cryoelectron micrographs: procedure and application to the 70s Escherichia coli ribosome. J Struct Biol 118(3):197–219

5

Structure determination of a human virus by the combination of cryo-EM and X-ray crystallography

Zheng Liu[1,6], Tom S. Y. Guu[2], Jianhao Cao[3,5], Yinyin Li[3], Lingpeng Cheng[4], Yizhi Jane Tao[2✉], Jingqiang Zhang[3✉]

[1] Department of Biophysics, Health Science Centre, Peking University, Beijing 100191, China
[2] Department of BioSciences, Rice University, Houston, TX 77005, USA
[3] School of Life sciences, Sun Yat-sen University, Guangzhou 510275, China
[4] School of Life Science, Tsinghua University, Beijing 100084, China
[5] State Key Laboratory of Organ Failure Research, Institute of Antibody Engineering, School of Biotechnology, Southern Medical University, Guangzhou 510515, China
[6] *Present address:* Department of Biochemistry and Molecular Biophysics, Columbia University, New York 10032, USA

Abstract Virus 3D atomic structures provide insight into our understanding of viral life cycles and the development of antiviral drugs. X-ray crystallography and cryo-EM have been used to determine the atomic structure of viruses. However, limited availability of biological samples, biosafety issues due to virus infection, and sometimes inherent characteristics of viruses, pose difficulties on combining both methods in determining viral structures. These have made solving the high resolution structure of some medically important viruses very challenging. Here, we describe our recently employed protocols for determining the high-resolution structure of the virus-like particle of hepatitis E virus (HEV), a pathogen of viral hepatitis in human. These protocols include utilizing recombinant baculovirus system to generate sufficient amount of virus particles, single-particle cryo-EM to get an intermediate resolution structure as a phasing model, and X-ray crystallography for final atomic structure determination. Our protocols have solved the hepatitis E virus structure to the resolution of 3.5 Å. The combined methodology is generally applicable to other human infectious viruses.

Keywords Structure determination, Cryo-EM, X-ray crystallography, Human virus

INTRODUCTION

X-ray crystallography and cryo-EM are two main techniques for visualizing virus 3D structures. One prerequisite for both methods is sufficient amount of samples. For some viruses, samples are purified either directly from their hosts or from tissue culture. Unfortunately, for some viruses including hepatitis E virus, neither sources provide adequate biological sample. In addition, poor infectivity of the virus and biosafety problems further complicate their structural study. In such cases, an efficient in vitro system for generating its substitute, recombinant virus-like particles (VLPs) (Liu et al. 2016), became necessary. The assembled HEV VLPs were expressed without its complete viral genetic materials, therefore bypassing the biosafety issues. Hepatitis E virus (HEV), discovered by immune electron microscopy (Balayan et al. 1983), is the sole member of the genus *Herpesvirus* within the family *Hepeviridae*. It causes acute viral hepatitis in human. Neither stools of the infected patients nor cell culture were able to produce enough sample. Thus, recombinant systems were

Zheng Liu and Tom S. Y. Guu have contributed equally to this work.

✉ Correspondence: ytao@rice.edu (Y. J. Tao), lsszhjq@sysu.edu.cn (J. Zhang)

developed to provide materials for this investigation. The HEV virus-like particles were first obtained by the use of baculovirus insect system expressing the aa112–608 fragment of open reading frame 2 (ORF2) (Li et al. 1997), and then 3D structures of both $T = 1$ and $T = 3$ particles were solved at low resolution using cryo-EM (Xing et al. 2010). When using *E. Coli* to express the HEV capsid protein, only P2 particles were assembled by expressing the protein of aa394–606 (Li et al. 2009). The P2 particles and their complexes with monoclonal antibodies were subsequently determined by X-ray Crystallography (Tang et al. 2011). The high-resolution atomic structure of HEV VLP were eventually determined by several laboratories independently (Guu et al. 2009; Yamashita et al. 2009; Liu et al. 2011).

Here, we present our protocol for determining high resolution of the HEV VLP by combining EM and X-ray crystallography techniques. The protocol, utilizing recombinant baculovirus system to obtain sufficient amount of virus particles, and single-particle cryo-EM to get an intermediate resolution structure as a phasing model, and then X-ray crystallography for final structure determination, solved the hepatitis E virus structure to the resolution of 3.5 Å. The high-resolution structure reveals intricate details in HEV-VLP, including three linear domains—S, P1, and P2—arranged in a manner different from caliciviruses, which also possess the three domains. This combined approach can also be applied to structural investigation of other infectious viruses with low abundance in nature, high infectivity, and other unique structural features, which may make it challenging to study using either cryo-EM or X-crystallography alone.

OVERVIEW OF EXPERIMENTAL DESIGN

Our protocol can be grouped into three sections (Fig. 1). The first section (Step 1–9) describes the procedures to generate sufficient and suitable virus-like particles for single-particle cryo-EM study (Fig. 2). The second section (Step 10–20) details the steps of performing single-particle cryo-EM to get a 3D density map (Fig. 3). The third section (Step 21–26) encompasses the details of using X-ray method to yield the final high-resolution structure (Fig. 4).

In the first part of the protocol, we will provide an overview on how to purify the virus gene, to construct recombinant plasmid, to scale-up expression, and to carry out purification. Since making recombinant plasmids and preparing cell culture involve multiple detailed steps, it is strongly recommended to follow manufacturers' manuals.

In the second set of steps, we describe the details of how to freeze cryo-EM sample, to operate transmission electron microscope, and to perform image processing. Once the micrographs are collected, the particles of the virus are picked out. The images of particles are then assigned with a group of positional parameters including translation and orientation information by refining an initial model. The main image processing procedures are performed using programs from EMAN (Ludtke et al. 1999) and IMIRS package (Liang et al. 2002).

In the final section, the crystallization and data processing are detailed. For obtaining the phase

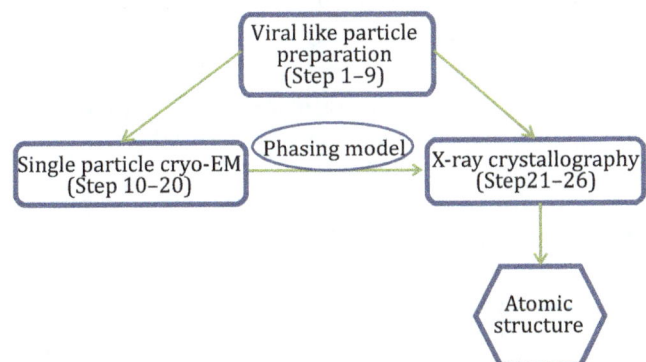

Fig. 1 The flowchart for the determination of the HEV virus-like particle structure using the baculovirus insect system, single-particle cryo-EM, and X-ray crystallography

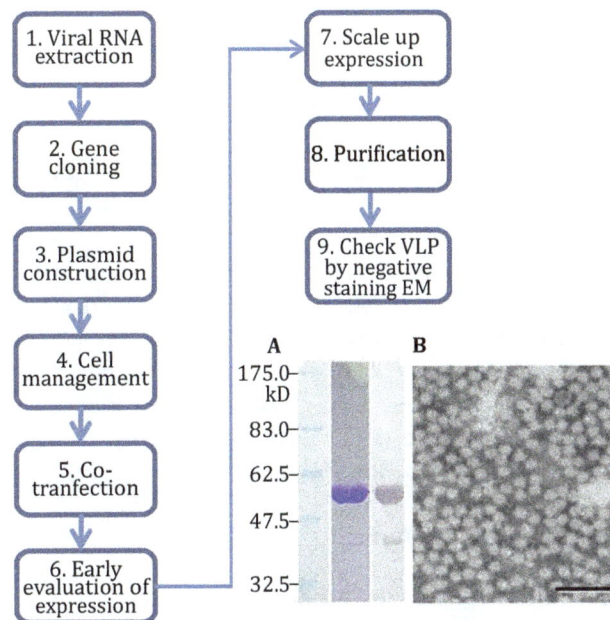

Fig. 2 The HEV viral capsid protein expression and virus-like particle assembly. An outline of the procedures used for the expression and purification of HEV capsid protein is shown. The results of protein SDS-PAGE electrophoresis, Western blotting, and particle observation by negative stain microscopy are shown on the *right*. The scale bar in **B** is 100 nm

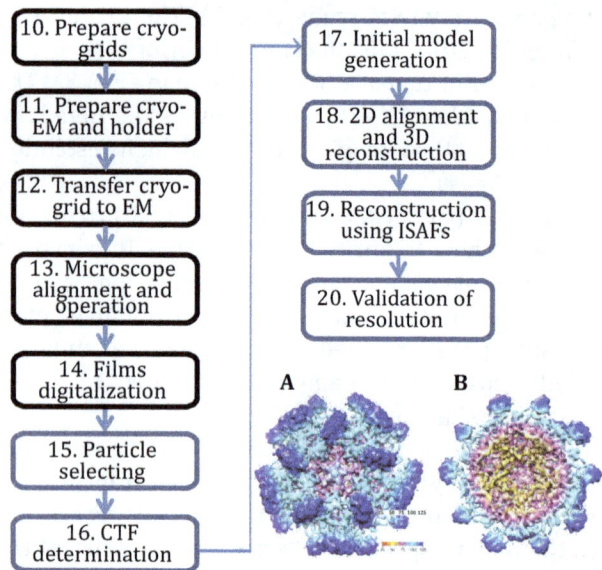

Fig. 3 The cryo-EM and image processing of 3D structure of hepatitis E virus. An outline of the procedure used for the single-particle analysis of HEV virus-like particle is shown. Steps involving the cryo-EM are highlighted in *green*, whereas steps involving the image processing are shown in *blue*. The reconstruction at 9.8-Å resolution is shown on the *right*. **A** The density map is viewed along the 5-fold axis. **B** The internal view of the virus-like particle

information, cryo-EM model from the second step is used as an initial phasing model.

MATERIALS AND EQUIPMENT

Materials and reagents for producing virus-like particle

DNA purification kits, PCR kit, *E. coli* DH5α, and TOP10 strain. Cell lines and cell culture media: sf9, sf21, high five cells; TC-100 media, EX-CELL 405 media (Sigma-Aldrich, USA), Grace Media, tissue culture flasks, BD BaculoGold linearized baculovirus DNA (BD Bioscience, USA), Cellfectin reagent (Invitrogen, USA), transfer vector (Pharmingen, San Diego, USA).

Equipment for Cryo-EM

SO163 films (Kodak, USA); D19 (for low dose film development) (Kodak, USA).

JEM-2010 electron microscope (JEOL, Japan) equipped with a Gatan 626 Cryo-holder (Gatan, USA), R2/1, 200-mesh holey grids (Quantifoil, Germany), ion coater (Eiko Engineering Co., LTD), model 655 pumping station (Gatan, USA), home-made plunge freezer, and LS-8000ED scanner (Nikon, Japan).

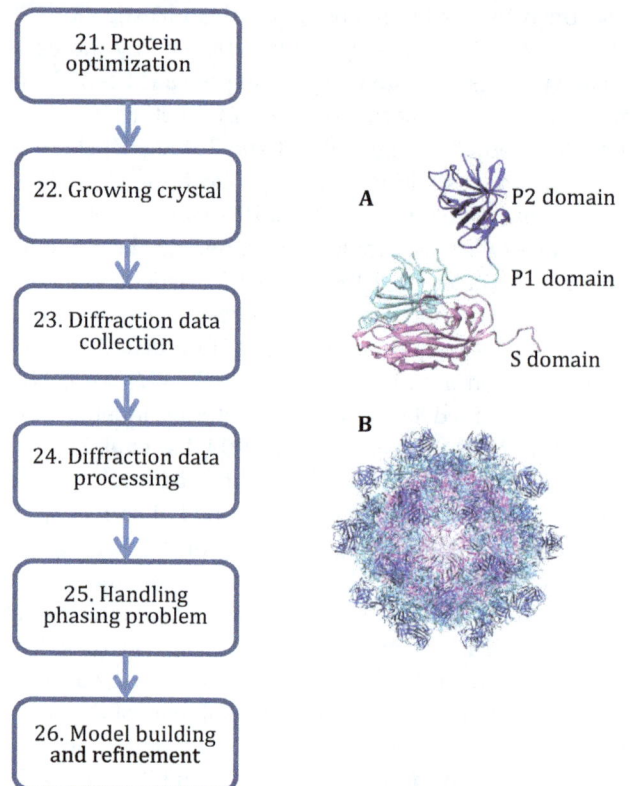

Fig. 4 The X-ray crystallography of virus-like particle of hepatitis E virus. The procedures to obtain 3.5-Å resolution structure are shown. **A** The monomer of the HEV capsid protein is shown on the *right*. **B** The whole model of the HEV virus-like particle viewed from along the 5-fold axis

Software packages for cryo-EM reconstruction

EMAN (Ludtke et al. 1999), IMIRS (Liang et al. 2002).

Software packages for X-ray crystallography

General locked rotation function (GLRF) (Tong and Rossmann 1997), MAVE (Read and Kleywegt 2001), MAPROT (Stein et al. 1994), MAPMAN (Kleywegt and Jones 1996), SFALL (Ten Eyck 1977), RSTATS (Collaborative 1994), AVE and RAVE (Jones 1992), O (Jones et al. 1991), CNS (Brunger et al. 1998).

SUMMARIZED PROCEDURE

1. Viral RNA extraction is performed on the stools from the patients of hepatitis E;
2. The ORF2 of hepatitis E virus is cloned into TA vector;
3. The target gene is inserted to the plasmid for transfection;

4. The sf9 and high five cells are cultured for transfection and expression;
5. The transfecting plasmid with interest gene is used to transfect sf9 cells;
6. Early evaluation of the HEV capsid protein expression is performed in P1 cells, which is then followed by baculovirus amplification;
7. Scale up expression using high five cells and high-titer baculoviruses;
8. The virus-like particle is purified using ultracentrifugation;
9. The VLP sample is checked by negative staining EM;
10. Prepare cryo-grids;
11. Cool down EM and holder;
12. Transfer cryo-grid to the column of the electron microscope;
13. Microscope alignment is careful operated;
14. Films digitalization using Niko scanner;
15. Viral particles are selected using EMAN boxer;
16. CTF determination is performed by CTFIT;
17. Initial model generation using minimum phase residue method;
18. 2D alignment and 3D reconstruction for individual particle images;
19. Reconstruction of all the aligned particle images using ISAFs;
20. Validation of resolution;
21. Protein preparation for crystallization;
22. Growing crystal;
23. Diffraction data collection;
24. Diffraction data processing;
25. Handle phasing problem;
26. Model building and refinement.

PROCEDURE

Capsid protein expression and VLP purification

Viral RNA extraction

The gene of target proteins of infectious viruses is usually obtained from various sources: (1) clinical samples including throat swab, stool, body fluids, etc.; (2) synthesized gene by commercial company; (3) gifted plasmid containing the gene of interest. To obtain viral gene, in this study, stool sample was first collected from patient with confirmed hepatitis E (non-A, non-B acute hepatitis) infection. The stool sample was preserved at −80 °C for less than two weeks before being subjected to RNA extraction.

The stool sample was suspended with 4×volume of phosphate-buffered saline and then the mixed specimen was homogenized by vortexing for 1 min. The resulted mixture was subjected to centrifugation steps with a speed of about 1500 RCF first and the supernatant was applied to another centrifugation with 12,000 r/min at 4 °C for 20 min to further remove the sediment. The total RNA extraction from the pretreated stool sample was performed using commercial RNA extraction kit. The extracted product was stored at −80 °C for the downstream usage including PCR.

Gene cloning

This step aims to amplify the target gene and then to transfer it into a proper vector or plasmid. Amplification of the entire HEV ORF2 by PCR was performed with the following primers:

HEV-D2 (59-TGGGTTCGCGACCATGCGCCCTCG-39),
HEV-U2 (59-CAACAGAAAGAAGGGGGGCACAAG-39).

The PCR product was first examined by DNA gel, and then was subjected to gene sequencing. After obtaining the PCR product with accurate sequence of the ORF2, we used TA cloning to transfer the PCR products using TA Cloning Kit with PCR 2.1 vector (Invitrogen, USA).

Plasmid construction

Target genes were cloned into a transfer vector which was used to manipulate the viral genome. Here, we used PVL1391 based on homologous recombination.

(A) The T vector with HEV-ORF2 gene was digested with *NruI* and *XbaI*, and the resultant 2-kb fragment was ligated with a transfer vector pVL1393 digested with *SmaI* and *XbaI*.
(B) Set up a DNA ligation to fuse the digested pVL1393 vector and the cDNA fragment. Typically 100 ng of the linear plasmid fragment was ligated with three-fold molar excess of the insert in 10 µL reaction volume.
(C) Transformed the ligation reaction into competent cells, for example *E. coli* DHα5 and plated onto LB agar plates containing 100 µg/mL ampicillin.
(D) Checked the transfer vector if it contained the gene of choice. Picked up 4–5 colonies and grew 1 mL precultures and then used the HEV-D2 and HEV-U2 primers for PCR analysis to check target gene.
(E) Precipitated 10 µg DNA with 300 mmol/L Na/Acetate pH 5.2 (final concentration) and three volumes of ethanol 100%. Resuspended DNA in 20 µL sterile ultrapure H_2O. Took an aliquot to measure the DNA concentration and stored at −20 °C.

[**CAUTION!**] Site-specific transposition and cotransfection. For generating recombinant baculoviruses, there are two different available methods. Both methods make use of the baculovirus genome engineered into a bacterial plasmid, which was subsequently ampllifed in *E. coli*. The first method is based on site-specific transposition (Tn7 transposition) of an expression cassette into the baculovirus genome in *E. coli*; the second employs a transfer vector and viral DNA that are cotransfected into insect cells and utilizes host enzyme-mediated homologous recombination.

Cell management

In general, insect cells are handled in a laminar flow hood under aseptic conditions. All cell culture experiments are carried out at 27 °C. The basic work on cells is classified into three steps: cell recovery, cell passage, and cell storage.

(A) Cell recovery.
  i. Removed vial of cells from liquid nitrogen and placed in water bath at 37 °C. Thawed them rapidly in the water bath.
  ii. Decontaminated the outside of the vial by spraying with 70% ethanol.
  iii. Transferred the 1 mL thawed cell suspension directly into the 4 mL of media, and then incubated flask at 27 °C and allowed cells to attach for 30–45 min.
  iv. After cells were attached, removed the medium, and fed cells with 5 mL of fresh medium. After 24 h, exchanged with fresh medium.

(B) Cell passage.
  v. When cells grew until 90% confluence (Fig. 5), removed the growing culture and detached cells by tapping the flask.
  vi. Took an aliquot of the cell suspension, counted cells, and determined their viability.
  vii. Added an appropriate volume of medium to the flask and aliquotted the cells into three new flasks with a starting density of 0.5×10^6 cells/mL and incubated cells at 27 °C.

(C) Cell storage.
  viii. Prepared cryo-vials, cool them on ice. Centrifuged cells at 100–150 g for 10 min at RT and removed supernatant.
  ix. Resuspended cells at the proper density. Then, transferred 1 mL to each sterile cryo-vial.
  x. Placed the cells at −20 °C for 1 h, then stored at −80 °C for 24–48 h. Transferred them to Dewars filled with liquid nitrogen for long-term storage.

Cotransection

To generate recombinant baculoviruses, the transfer vector is cotransfected with linearized viral DNA in insect cells.

(A) Seeded a 6-well plate using 1.5×10^6 sf9 cells per well in 1.6 mL insect cell culture medium (TC100 + 10% FBS) and let the cells adhere for 20 min at 27 °C (Fig. 5C). Meanwhile, mixed 4 μg of DNA transfer vector with 1 μg of linearized DNA in 100 μL of sterile 150 mmol/L NaCl.
(B) Added 200 μL DNA/cell-transfection solution drop to the cells, and incubated at 27 °C. Four hours after the cotransfection, added 2 mL of insect cell medium to the cell layer and returned to the incubator for at least 5 days.
(C) From the second day, observed cells daily by an inverted microscope and searched for infected cells. After 5 days, carefully collected the supernatant by centrifugation at 200 g for 10 min (P0) and stored at 4 °C.
  The cotransfection product is referred to as the initial virus stock (P0). It can be used for further evaluation of the target protein expression and for large-scale expression.

Early evaluation of protein expression

P0 can be used for an initial screening to provide the first indication of the expression level.

(A) Seeded a 6-well plate using 1.5×10^6 sf9 cells per well and let the cells adhere for 20 min at 27 °C.

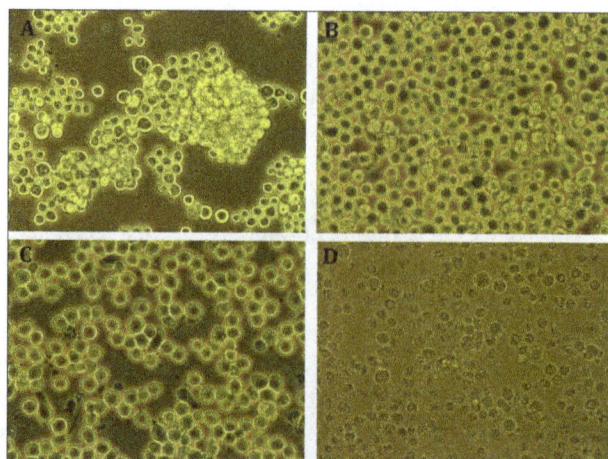

Fig. 5 Cell management of sf9 insect cell line. **A** The aggregated cells showed suboptimal status. **B** The cells achieved 90% confluence indicating a time point for passage. **C** Cells for infection of viral stock. **D** Cell debris after 4-day postinfection

Discarded the medium; added 300 μL of fresh medium and 150 μL of P0 to attached cells followed by 1 h incubation at 27 °C.

(B) Added 3 mL of insect cell medium and returned to the incubator for 48 h. Collected the infected cells by centrifuging at 200 g for 10 min at 4 °C.

(C) Resuspended cells in 1 mL of lysis buffer containing 1% Tween20 and 400 U/mL of DNase Type I with EDTA-free protease inhibitor and sonicated them for 30 s. Took 15 μL aliquots and add 5 μL of 4× SDS loading buffer.

(D) Analyzed the different samples using SDS-PAGE with coomassie staining and/or with western blotting in case of low expression levels.

Scale up expression of protein

There are two common strategies to scale up the expression of protein: one uses cells grown in suspension in the Fermentor/Bioreactor and the other uses large tissue flasks with adherent culture. We used the second method to scale up our protein expression.

To ensure the viral stock is of high infectious efficiency, one (P1) or two (P2) or even more round of amplification can produce virus stocks of high titers. The yield of the recombinant protein is affected by a number of factors. Thus, once a concentrated baculovirus stock has been made, optimization experiments are required before large scale expression. In general, the optimization is performed for two variables: the amount of infectious virus and the best time for harvesting. In our HEV expression experiment, we used the MO1 at 10 and harvested our expressing cell after 3 days of postinfection.

Purification of virus-like particle

Instead of using FPLC, we used gradient ultracentrifugation for HEV-VLP purification.

(A) Infected cells and media were pelleted at 10,000 g for 90 min and the pellet was discarded.

(B) The supernatant was pelleted again using Beckman rotor SW28 at 25,000 r/min at 4 °C for 2 h. The resulting pellet was resuspended in 4.5 mL PBS at 4 °C overnight.

(C) The resuspended pellet was mixed with 1.95 g CsCl and centrifuged by using rotor SW55Ti at 35,000 g at 4 °C for 24 h.

(D) The tube containing CsCl mixed with virus-like particles was punctured with a needle to collect fractions

of different densities. Each fraction was diluted with PBS and dialyzed against PBS to remove CsCl.

(E) After removing the CsCl, the sample was subjected to concentration check. For long-term storage, the sample was aliquoted and stored at −20 °C.

Check VLP by negative staining EM

To evaluate the overall quality of our HEV-VLP samples, negative staining was implemented.

(A) About 3 μL of purified sample was applied to a newly glow-discharged grid coated with carbon film.

(B) After about 2 min adsorption, the sample on the grid was blotted using a filter paper.

(C) A droplet of 2% phosphotungstic acid was applied to the grid, and after 1 min, the stain was wicked away again using a filter paper.

(D) The grid was dried in air and examined under TEM operated at 120 kV.

Single-particle cryo-EM

Preparation of cryo-grids

Once the sample was deemed adequate for cryo-EM, the next step was to prepare cryo-EM grids. To make an optimized grid with thin ice and good particle distribution, it was a trial-error procedure to find out optimal blotting time and blotting force. In general, preparation of cryo-EM grids was carried out in four steps.

(A) Cleaned grids. In order to remove any residual polymer on the grids, they were subjected to washing with chloroform.

(B) Treated grids. For samples that prefer to stay at the hole edge of the grid, one can evaporate carbon on the holey grid to coat a thin layer of carbon film over the surface of the grid. This step is optional.

(C) Glow discharge. The glow discharge converts the naturally hydrophobic carbon-coating layer of the grids into hydrophilic. Placed cleaned grids into the chamber of the glow-discharge equipment and closed the chamber. Pumped the bell jar and glowed discharge for 20 s at 25 mA and 7×10^{-2} mbar.

(D) Sample plunge-freezing. Cooled down freezing apparatus by filling up liquid nitrogen. Then ethane was poured slightly into the inner container of the freezer. 4 μL of purified HEV-VLP sample with a concentration of 2.5 mg/mL was applied to a quantifoil R2/1 100 mesh grid. After 1 min, the excessive sample was removed with a filter paper. The prepared grids were transferred into liquid nitrogen Dewars for storage.

Prepare EM and cryo-holder

Before loading the sample to the column of microscope, some preparation was needed for the instrument and for loading the sample.

(A) Checked the EM status. Read the logbook and checked if the previous session was normal. Ensure the cryo-cycle was finished before starting this new session.
(B) Checked the column vacuum ($<3 \times 10^{-5}$ Pa) and the gun vacuum (0.1×10^{-6} Pa). Made sure the high tension (HT) is ready.
(C) Filled up the anti-contamination device of the microscope with LN2. Made sure the stage is centered.
(D) Prepared cryo-holder. A cryo-holder is usually stored on a dry pumping station. To start a new cryo-EM session, the cryo-holder was recommended to regenerate the zeolite desiccant, and to achieve a high vacuum in the cryo-holder Dewars.

Transfer cryo samples into electron microscope

Loading a cryo-EM grid to the column of microscope is a critical step for cryo-EM. In particular, for 200 kV microscope, this step brings air in to the column and can disrupt the vacuum status. Thus, this step requires much attention when operating the microscope.

(A) Inserted the cryo-holder into the cryo-transfer workstation. Cooled down the workstation by filling with liquid nitrogen.
(B) Plugged the cold stage controller cable to the cryo-holder. Kept monitoring the temperature and added liquid nitrogen if necessary until the temperature stabilized at around -194 °C.
(C) Quickly transferred the cryo-grid box to the workstation. Loaded the grid to the sample holder slot and made sure the clip ring on top of the grid was firmly seated in place. Closed the cryo-shutter, and moved the entire workstation with holder and grid to the microscope console.
(D) Turned the goniometer airlock switch from 'AIR' to 'PUMP'. Quickly transferred the cryo-EM holder to the stage by aligning the holder airlock pin with the goniometer slot, and slightly pushed it straight in until stopped. When the airlock opened and the green light was on, turned the holder slowly clockwise about 30° until stopped, made sure to hold it firmly so that it was sucked in slowly by

vacuum. Continued to turn the holder $\sim 60°$ clockwise until stopped, and then let it in slowly.
(E) Added liquid nitrogen to the specimen holder Dewar. Waited for the microscope column vacuum to recover (about $1 \sim 3 \times 10^{-5}$ Pa). Opened the specimen holder shutter and then opened the gun valve.

Alignment and operation of electron microscope

Before starting data collection, the microscope should be carefully aligned to ensure it runs at an optimal state. The following is the detailed protocols for the alignment of our JEM-2010 for the HEV project data collection.

At the beginning, set up the high voltage to 200 kV in the HT ramp-up program, set step to 0.1 kV, time to 30 min, and then started the program.

(A) Aligned the TEM by following these steps in order:

i. Gun alignment. At MAG 50 k, focused the beam (C2), activated the node Wobbler, pressed the GUN Deflection button, adjusted the GUN DEF X and Y to minimize swipe of the illumination disk, and the GUN SHIFT to center the disk.
ii. Condenser alignment. Centered the beam with the GUN SHIFT X and Y at SPOT SIZE 1, and BEAM SHIFT X and Y at SPOT SIZE 5, repeated several times until the beam position was stable when the SPOT SIZE was varied.
iii. Inserted and selected a proper condenser aperture size (e.g., the third one, 40 mm). Centered the aperture by spreading the beam with C2 to half the screen size, centered the illumination disk by adjusting the mechanical X and Y shift of the C2 aperture; focused the beam (C2), and centered the beam with BEAM SHIFT X and Y. Repeated several times.
iv. Voltage centering. This step is to minimize the image movement when the image focus is varied. At a higher MAG of 120 k, centered and focused a recognizable sample feature. Activated the HT WOBBLER and the BRIGHT TILT; adjusted the DEF X and Y so that the feature at the center of the small fluorescent screen remained stationary.
v. Adjusted beam tilt (also called the pivot). At MAG 150 k, focused and centered the beam, activated the CONDEF ADJ TILT button, turned the white toggle switch TILT to X position. This wobbled the beam by modulating the deflection lens current. Used the SHIFT X and the DEF X knobs to form a single spot. Repeated for the TILT Y position.

vi. Corrected the astigmatism of the object lens at a magnification of 600 k.

(B) Imaging at low electron dose.

i. Made sure the temperature was stable at −170 °C.
ii. Opened the filament and checked if the grid was in the "Search Mode."
iii. Checked the grid quality under "Search Mode" to determine whether it showed evenly particle distribution and thin ice. If so, the grid could be used for data collection.
iv. Shifted to "Focus Mode," adjusted the z-height of the grid, then focused on an area near the exposure site.
v. Shifted to "Photo/Exposure Mode," loaded a film and pressed the button "Photo" to expose.

[Note] The temperature of the sample had to be under −168 °C throughout all of the procedures.

Films development and digitalization

A new type recorder, termed direct detector, is much superior to photographic film and CCD (Liu and Zhang 2014). In the following, we will describe the procedures for developing films which were used during HEV structural study.

(A) Exposed films were first put into a rack for development in batch. Made sure that the room temperature and solution were at 20 °C.
(B) Emerged the rack in Developer for exactly 12 min at 20 °C. Drained excessive solution in <10 s.
(C) Washed the rack in water for 1 min and drained excessive water as much as possible. Emerged the rack in Fixer for 4 min and then drained excessive solution. Washed fixed films with running water for 30–60 min, then air dried them.

In our experiment, we used Nikon LS-8000ED scanner. The scanning condition was set at 4000 dpi with a pixel size of 1.27 Å.

[CAUTION!] Recorders of electron microscope. Photographic film and CCD are two traditional recorders for electron microscope. CCD is superior to films considering the convenience of usage. It can output direct digital images without extra workload for development and digitalization that are necessary for films. It also allows for continuous data collection without the need to exchange it and for imaging automation on microscope. Recently, thanks to a new type recorder called electron direct detector device (DDD), cryo-EM field is experiencing revolution. DDD possesses both film and CCD advantages including high DQE and direct

outputting digital images. In addition, it allows collecting movies (multiple frames) for each exposure. With image processing, one can correct drift induced by electron interacting with sample, which is a long-standing difficult problem for high-resolution cryo-EM.

[CRITICAL STEP] The films with obvious astigmatism, contamination, shift of samples, ice crystals, and few particles were excluded.

Boxing particles

To make sure only good quality micrographs were used for further image processing, the first step was to select the particles from micrographs.

(A) The command "boxer" in EMAN software package was used to call the particle-picking module. The box size was set to 256 pixels for HEV-LP selection (Fig. 6).
(B) All particles are boxed manually. The center of the boxes should be near the center of particles.
(C) All the coordinates of these particles were stored in box files.

CTF determination

Among many of the parameters (defocus, B-factor, noises, etc.) of the contrast transfer function (CTF), the defocus value is the most critical parameter (Jiang et al. 2012). In addition to determining the CTF parameters, the Thon rings in power spectrum of micrographs can be used for evaluating the quality of micrographs. The circular Thon rings indicate good quality images regardless of other factors; non-circular Thon rings show astigmatisms (Fig. 7).

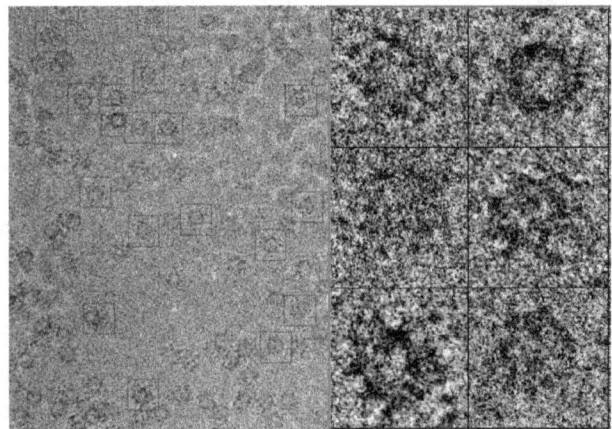

Fig. 6 Cryo-EM micrograph of hepatitis E virus particles. *Left*: an area of a micrograph with particles highlighted in *red boxes*. *Right*: six selected particles

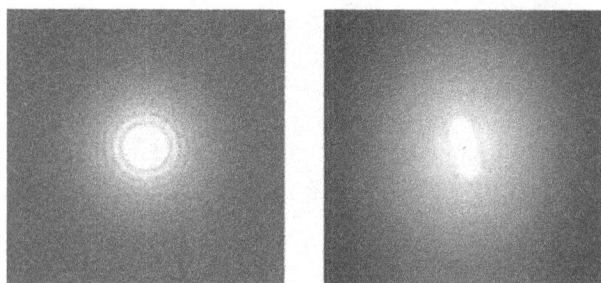

Fig. 7 Power spectrums of two representative micrographs. *Left*: power spectrum of a good micrograph displays round thon rings. *Right*: power spectrum of a suboptimal micrograph with drift and astigmatism reveals oval-shaped thon rings

This step was performed with 'ctfit' in EMAN1.9 software package. The parameters were set as following: accelerate voltage 200 kV, Cs of EM 2.0 mm, step size of images 1.27 A/pixel, contrast amplitude of images 0.1.

Reconstruct initial model

Three-dimensional (3D) reconstruction of a pool of particle images requires their positional parameters. These parameters were initially assigned by aligning the particle images against an initial model and then iteratively refined by aligning with the reconstructed map from the previous iteration.

(A) Used *ortall* to assign orientation and center of all particles. The parameters were: radius of mask 124, pixel size 1.27 Å/pixel; the maximum and minimum Fourier radius for common line search: minR = 3, maxR = 9.

(B) Selected 7–10 particles with smaller phase residues. Some rules should be followed in picking up particles to build a template. All particles should have similar defocus values and be picked up from the same photograph. The orientation should not be near the 2/3/5-fold symmetrical axes.

(C) From *template.log*, located three particles whose cross-correlation residues were as small as possible, and then used them to compute an initial model. The HEV models used in the study are shown in Fig. 8.

2D alignment and 3D reconstruction

Once the initial model has been generated, all the particles' orientation could be assigned and refined based on this model and later reconstructed maps. Here is the procedure.

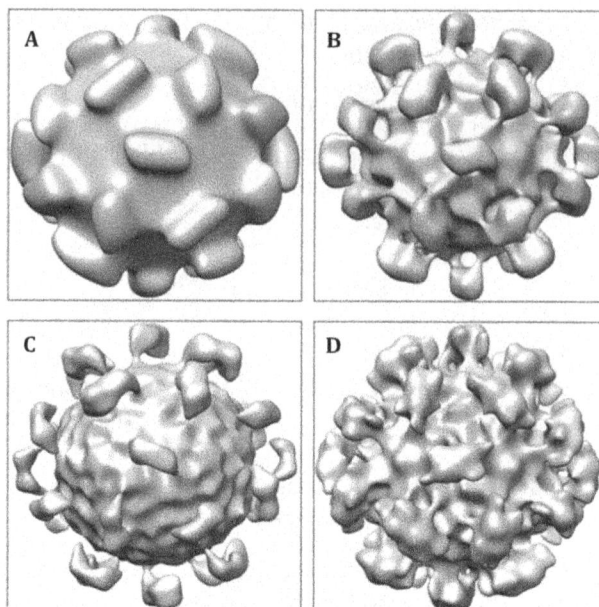

Fig. 8 Four initial models reconstructed from different groups of particles. The map showing a consistent icosahedral spike pattern was the most promising initial model

(A) Used *eliminateort* to refine orientation and center of all particles with cross-common line phase residue method. The running parameters of *eliminateort* were as follows: threshold of cross-common line phase residue was 55, image size of particles was 256, minimum and maximum value of Fourier radius were 3 and 9. The result of *eliminateort* was stored in file *definedort.dat*.

(B) Reconstructed density map (*reconstruct definedort.dat)*. Need for CTF correction: yes; diameter for reconstruction (odd): 511. maximum Fourier radius for reconstruction: 45. The *reconstruct* produces three map files: *icos3 f.map*, *icos2 f.map*, and *icos5 f.map*.

(C) Produced projections of 3D density map. Ran command line with projection size 256 and projection step *projecticos_auto icos5f.map ptemps.dat*.

(D) Used the 20 projections above as models and made FFT with the following command: *masktrans ptemps.dat mask 121 prj*. File *ptemps.dat* was a model containing 20 particles projections including orientation and center. 121 was the radius.

(E) Refined orientations and centers of all particles: *refineall ptemps.dat*. Some parameters were needed for input.
Scanning step size: 1.27; minimum and maximum Fourier radius: 3 and 9; the step size for refining orientations of particles: default value; the step size for refining centers of particles: default value; threshold for phase residue: 55.

(F) Made FFT of new orientation and center files: *masktrans newOrt.dat 129 hev.*

(G) Repeated Step C–F until the orientations and centers were stable.

Reconstruction using recISAFs

The icosahedral symmetry-adapted functions (ISAFs) reconstruction method was developed by Hongrong Liu et al., which takes advantage of the icosahedral symmetry of virus particles (Liu et al. 2008). It can reconstruct the same set of images to higher resolution when compared to conventional approach. The following was the setting for *recISAFs*: orientation file name was *x.dat*; number of particles was 891; diameter of particle (odd only) was 261. Maximum Fourier radius in reconstruction depended on the procedure; need for CTF correction: yes; memory of the computer (G): 2G. The program produced three map files: *sph3.map*, *sph2.map*, and *sph5.map*.

Validation of resolution

In general, the resolution of the 3D reconstruction is determined using the Fourier shell correlation (FSC) between two reconstructions that are obtained using the half datasets. In IMIRS, the resolution evaluation was performed using three steps as described below. From the FSC curve, we used the 0.5 criterion to calculate the reported resolution for the final 3D reconstruction.

(A) Splited the dataset into two parts (*splitdataset *.dat out1.dat out2.dat*).

(B) Reconstructed the *out1.dat* and *out2.dat* (*reconstruct *1.dat*).

(C) Calculated the correlation of the two maps with the program *frc*.

X-ray crystallography

Protein expression

The sf21 insect cell line was used to express HEV capsid proteins for crystallization using the detail procedures described below.

(A) The recombinant bacmid DNA obtained using the Bac-to-Bac® Baculovirus Expression System (Version D, Invitrogen) was used to transfect sf21 cells.

(B) The P1 viral stock was harvested by centrifuging down the cell debris at 3000 *g* for 15 min. The cell pellet was then lysed to detect the presence of recombinant proteins by SDS-PAGE gels. The P1 viral stock was then amplified twice to generate a P3 viral stock with higher titer.

(C) For recombinant protein expression, sf21 cells were mixed with 10% of the P3 viral stock and incubated for 48 h at 27 °C postinfection. The harvested insect cell pellet was sonicated in lysis buffer. The lysate was clarified by centrifugation at 25,000 *g* for 30 min.

(D) His-tagged capsid protein of HEV (HEV-CP) was purified using a Ni–NTA column, an anion-exchange (Q) column and a gel-filtration column. The gel-filtration chromatogram indicated that HEV-CP was eluted as homo-dimers.

(E) The final purified HEV-CP was at least 95% pure as indicated by SDS-PAGE. Approximately 25 mg of purified HEV-CP could be obtained from each liter of sf21 cells.

Crystallization

(A) To search for crystallization conditions, an automated robot (Hydra II plus) and 96-well screen trays were used for high throughput screening. Many solvent conditions were tested using commercial screening kits including those from Qiagen (AmSO4, Classics, Classics-Lite, pHClear I & II, MbClass I & II, MPD, Anions, Cations, PEG, and Cryos) and Hampton Research (Index HT, Crystal Screen HT, and SaltRX).

(B) The initial crystallization condition was optimized by varying several factors including pH, temperature, precipitant or protein concentration, precipitant type, volume ratio between protein and mother liquor, and the size of the hanging drop. Microseeding and the addition of the detergent *n*-tetradecyl-b-D-maltoside accelerated crystal growth and improved crystal quality.

(C) For data collection, HEV-CP crystals were flashed frozen in liquid nitrogen in cryo-protectant made of the mother liquor supplemented with 20% of glycerol (Fig. 9).

Data collection for HEV crystals

The large unit cell dimensions of the HEV-CP crystals suggest the formation of VLPs during crystallization. Our data collection was performed at the Advanced Photon Source (APS) (Argonne National Laboratory, Lemont, IL) on beamline SBC-19-ID with a wavelength of 0.97934 Å using a 3 × 3 mosaic CCD detector (ADSC Quantum 315).

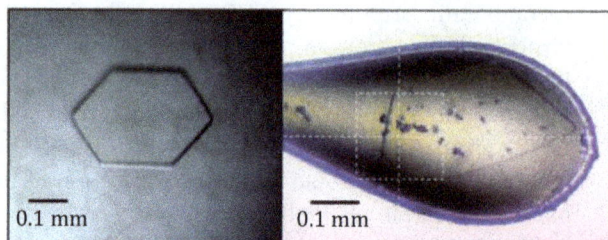

Fig. 9 Optimized HEV crystals. *Left*: an optimized crystal as the result of crush seeding and detergent addition. *Right*: a crystal in the cryo-loop taken during data collection

(A) Exposure time. With intense and focused synchrotron radiation, an exposure time of only a few seconds is usually sufficient. HEV-CP diffraction data were collected using a 10 s exposure time.

(B) Detector distance at APS. A 400-mm detector distance was sufficient to produce well-resolved diffraction spots to 3.5-Å resolution.

(C) Oscillation angle. Preliminary test shots showed our HEV crystals had a high mosaicity (0.6–1.2). While the use of large oscillation angels can help to collect whole reflections, reflection overlapping becomes a serious problem. In our case, a 0.5° oscillation angle produced balanced results.

(D) Oscillation range. In order to determine how many degrees of data collection were needed for a complete dataset, a few test images (usually with oscillation angles of 0°, 45°, and 90°) were taken to check the space group of the crystal.

With many screened crystals (60–70) in total and all randomly oriented), we were able to collect several full data sets with good mosaicity and completeness. The X-ray diffraction data was summarized in Table 1.

Table 1 Diffraction data statistics

Data collection	
Space group	P6$_3$
Unit cell dimension (Å)	$a, b = 241.1; c = 519.9$
Resolution (Å)	60–3.5
Total number of frames	239 from two crystals
Total number of reflection	5,236,044
Unique reflection	214,958
I/σ	11.5 (2.9)
Redundancy	6.9 (6.5)
Completeness (%)	93.7 (92.8)
R_{merge}	20.9 (67.6)

Diffraction data processing

The observed diffraction data were recorded as raw image files (*.img* or *.osc* files). These data must be indexed, integrated, and scaled in order to obtain their reciprocal space indices (h, k, l) and their corresponding diffraction intensity (*I*) over the estimated error in intensity (σ). HEV-CP diffraction data were processed using HKL2000 (Otwinowski and Minor 1997), which consists of XdisplayF, Denzo, and Scalepack.

To orient the HEV-VLP into the crystal unit cell, we performed a self-rotation search using the program GLRF (Tong and Rossmann 1997) as described below. The solution of the GLRF is expressed in polar angles (φ, ψ, κ), and rotation symmetries can be found when κ is equal to 180°, 120°, and 72°, corresponding to the icosahedral two-fold, three-fold, and five-fold symmetry.

(A) Two-dimensional searches (φ and $\psi = 0$–180°) were performed by fixing κ at 180°, 120°, and 72°. The solutions identifying the positions of all the NCS elements expected for an icosahedral particle (Fig. 10A–C).

(B) The exact positions of the symmetry axes were fine-tuned by carrying out a series of slow self-rotation searches using a smaller search interval (e.g., 0.2°) (Fig. 10D). Since the three-fold NCS axis of the HEV-VLP is parallel to the crystallographic 6$_3$ axis, there is only one degree of freedom—rotation about the crystal 6$_3$ axis.

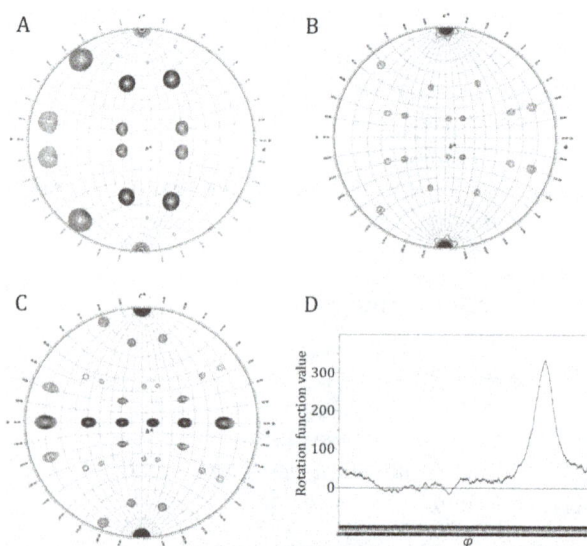

Fig. 10 The GLRF self-rotation search results for the crystallographic data. A–C Fast GLRF shows the symmetry solutions around five-, three-, and two-fold axes. **A** $\kappa = 72°$, **B** $\kappa = 120°$, **C** $\kappa = 180°$. **D** Slow GLRF is used to determine the exact value for φ angle, by fixing $\kappa = 72°$ and $\psi = 142.62°$. The result indicates $\varphi = 49.00°$

(C) The difference between our computationally determined φ angle and the $\varphi 0$ angle of an icosahedron at the standard 222 orientation, $\Delta\varphi$, defines the actual orientation of our HEV-VLP relative to that of a standard icosahedron. Because all symmetry axes are related by fixed angles in an icosahedron, φ can be accurately computed from the large numbers of symmetry axes presented in an icosahedron. The averaged angle for $\Delta\varphi$ was determined to be 49.00°. Knowing how the crystallographic VLP sits in the unit cell allowed us to relate this orientation to the EM standard orientation (Fig. 11 *Left*)—this angle, when expressed in Eulerian angles, was $\alpha = 49.000°$, $\beta = -20.905°$, $\gamma = 0.000°$ (Fig. 11 *Right*).

(D) Considering the unit cell dimensions and the size of the VLP, there should be one VLP in each unit cell. The particle should sit on the *c*-axis, but its *z* coordinate is arbitrary in the space group P63. For convenience, we fixed the particle center at the origin ($x = 0, y = 0, z = 0$).

Handling the phasing problem

Our initial phasing model is a 14-Å cryo-EM HEV-VLP reconstructed from the Step 10–20. The EM particle was first rotated from the standard EM orientation by $\alpha = 49.000°$, $\beta = -20.905°$, $\gamma = 0.000°$ using either MAVE (skew command) or MAPROT (Stein et al. 1994). A mask was then created for the rotated map. The density from the EM map inside the mask (in a P1 cell) was projected into the P63 unit cell using MAVE (expand command). The resulting map, also called the "P-cell map," showed tight packing of the EM particles. Due to the high symmetry of icosahedral viruses, NCS averaging is a very powerful option for phase improvement and phase extension (Kleywegt and Jone 1996). Our crystallographic asymmetric unit contains 1/3 of the VLP particle, which consists of 20 capsid protein subunits. Therefore, there are a total of 20 NCS operators.

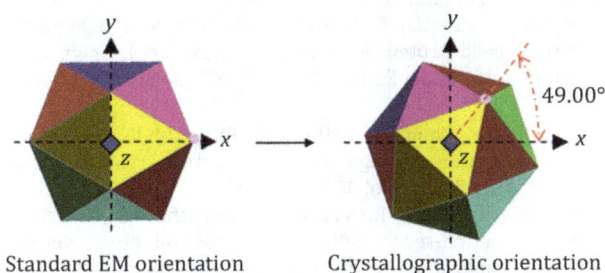

Fig. 11 Rotation of the HEV density map. *Left*: the cryo-EM density map orientation. *Right*: the crystallographic orientation as determined by GLRF

Several factors are crucial to a successful phase extension.

(A) Initial resolution. The goal was to maximize the overlap between the crystallographic data set and the EM model (Navaza 2008).

(B) Magnification. It is known that electron microscopy cannot determine the precise scale of the particle—there can be up to ~5% error in the estimation (Dodson 2001).

(C) Orientation. Although the average φ angle was determined by GLRF, a small incremental range of φ angle can be tested for precise orientation.

(D) Contour level. The importance of defining an appropriate mask around the EM density map cannot be overstated. A high contour level may exclude potential real density, resulting in missing structural information. A low contour level, however, may include excessive noise, introducing errors into phase calculations.

These four parameters can be adjusted to generate the initial phasing model. The quality of the phasing model is assessed primarily by the *R*-factor and correlation coefficient. Phase extension was executed using RAVE (Jones 1992; Kleywegt and Jones 1996) combined with several CCP4 programs (Collaborative 1994). SFALL generated Fcalc from the density map. RSTATS then scaled Fobs and Fcalc together, producing statistics to evaluate the agreement between the two sets of structure factors. With the calculated structure factors, FFT created a 2Fo-Fc map. This new calculated map was averaged by AVE using the 20 NCS operators for 10 cycles. The improved map was then back-transformed to a slightly higher resolution, typically by one-reciprocal lattice point, than at which it was originally calculated. The phase estimates at the extended resolution, though noisy, were fed back into the cyclical procedure, and subsequently refined by NCS averaging. As a result, the phase was gradually extended from low resolution to the desired high-resolution shell. In our studies, phase extension from 14 to 3.5 Å required a total of 123 steps. Our final phase extension yielded an averaging *R*-factor of 31.7 and a correlation coefficient of 79.7 at 3.5-Å resolution.

Model building

Our 3.5-Å density map showed highly continuous density with few breaks, prominent side chains for the aromatic residues, and a high percentage of β sheets, consistent with the secondary structure prediction and the characteristics of small RNA viruses. Model building of a single HEV-CP protomer was performed using program O (Jones et al. 1991). In addition, our final 3.5-Å map was sharpened using $B = -150$ Å^2 to enhance the side chain density. Refinement of the model was performed in CNS (Brunger et al. 1998). Once a monomeric CP model was built, the entire HEV-VLP can then be generated from a single protomer using XPAND (Kleywegt) and MOLEMAN (Kleywegt). Our deposited structure (3HAG) has a Rwork and Rfree of 27.67% and 28.60%.

RESULT AND DISCUSSION

This protocol contains three commonly used methodolgies for solving virus atomic structures: protein expression using baculovirus, single-particle cryo-EM, and X-ray crystallography. Each of them includes multiple intricate steps. Expressing HEV capsid proteins by baculovirus and insect cells helped to bypass the issue of safety and sample shortage. Due to the inherent HEV structural flexibility (as contributed by its flexible surface spikes) and the instrumentation limitation of electron microscope, cryo-EM alone was not able to solve the HEV VLP structure to high resolution. On the other hand, the enormous size of virus particle (typically measured in mega Dalton) makes particle crystallization, X-ray diffraction data collection and processing very challenging. Thus, solving virus structure with X-ray crystallography alone has been shown to be a difficult endeavor. Fortunately, using the low resolution cryo-EM reconstruction as the initial phase model and combining it with X-ray diffraction data by taking advantage of averaging electron density maps of the highly symmetrical virus particle, we were able to determine the HEV VLP structure to 3.5 Å resolution. These combined approaches can be implemented to solve the atomic structures of other medically significant viruses.

Our VLP structure is the first high-resolution structure of HEV viral particle and reveals a number of details different from those of calicivirus to which HEV was previously classified. First, the three domains S, P1, and P2 are arranged in a linear sequence in HEV-CP. Second, P1 domain adopts a different protein fold, which forms trimeric protrusions. These structural details provide an important framework for better understanding of the assembly of hepatitis E virus and insights into finding potential antiviral targets to combat this virus.

Acknowledgments This work was supported by a grant from The National Nature Science Foundation of China (31170691) and the Robert A. Welch Foundation (C-1565 to YJT).

Compliance with Ethical Standards

Conflict of Interest All authors declare that they have no conflict of interest.

Human and Animal Rights and Informed Consent This article does not contain any studies with human or animal subjects performed by any of the authors.

References

Balayan MS, Andjaparidze AG, Savinskaya SS, Ketiladze ES, Braginsky DM, Savinov AP, Poleschuk VF (1983) Evidence for a virus in non-A, non-B hepatitis transmitted via the fecal-oral route. Intervirology 20:23–31

Brunger AT, Adams PD, Clore GM, DeLano WL, Gros P, Grosse-Kunstleve RW, Jiang JS, Kuszewski J, Nilges M, Pannu NS, Read RJ, Rice LM, Simonson T, Warren GL (1998) Crystallography & NMR system: a new software suite for macromolecular structure determination. Acta Crystallogr D Biol Crystallogr 54:905–921

Collaborative CP (1994) The CCP4 suite: programs for protein crystallography. Acta Crystallogr Sect D, Biol Crystallogr 50:760

Dodson EJ (2001) Using electron-microscopy images as a model for molecular replacement. Acta Crystallogr D 57:1405–1409. doi:10.1107/S0907444901013415

Guu TS, Liu Z, Ye Q, Mata DA, Li K, Yin C, Zhang J, Tao YJ (2009) Structure of the hepatitis E virus-like particle suggests mechanisms for virus assembly and receptor binding. Proc Natl Acad Sci USA 106:12992–12997. doi:10.1073/pnas.0904848106

Jiang W, Guo F, Liu Z (2012) A graph theory method for determination of cryo-EM image focuses. J Struct Biol 180:343–351. doi:10.1016/j.jsb.2012.07.005

Jones T (1992) A set of averaging programs. In: Dodson E, Gover S, Wolf W (eds) Molecular replacement. SERC Daresbury Laboratory, Warrington, pp 91–105

Jones TA, Zou JY, Cowan SWT, Kjeldgaard M (1991) Improved methods for building protein models in electron density maps and the location of errors in these models. Acta Crystallogr Sect A: Found Crystallogr 47:110–119

Kleywegt GJ, Jones TA (1996) xdlMAPMAN and xdlDATAMAN-programs for reformatting, analysis and manipulation of biomacromolecular electron-density maps and reflection data sets. Acta Crystallogr D Biol Crystallogr 52:826–828

Li TC, Yamakawa Y, Suzuki K, Tatsumi M, Razak MA, Uchida T, Takeda N, Miyamura T (1997) Expression and self-assembly of empty virus-like particles of hepatitis E virus. J Virol 71:7207–7213

Li S, Tang X, Seetharaman J, Yang C, Gu Y, Zhang J, Du H, Shih JW, Hew CL, Sivaraman J, Xia N (2009) Dimerization of hepatitis E virus capsid protein E2 s domain is essential for virus-host interaction. PLoS Pathog 5:e1000537. doi:10.1371/journal.ppat.1000537

Liang Y, Ke EY, Zhou ZH (2002) IMIRS: a high-resolution 3D reconstruction package integrated with a relational image database. J Struct Biol 137:292–304

Liu Z, Zhang J-J (2014) Revolutionary breakthrough of structure determination-recent advances of electron direct detection device application in cryo-EM. Acta Biophys Sin 30:1–12

Liu H, Cheng L, Zeng S, Cai C, Zhou ZH, Yang Q (2008) Symmetry-adapted spherical harmonics method for high-resolution 3D single-particle reconstructions. J Struct Biol 161:64–73. doi:10.1016/j.jsb.2007.09.016

Liu Z, Tao YJ, Zhang J (2011) Structure and function of the hepatitis E virus capsid related to hepatitis E pathogenesis. In: Mukomolov DS (ed) Viral hepatitis—selected issues of pathogenesis and diagnostics. InTech, New York, pp 141–152

Liu Z, Guo F, Wang F, Li TC, Jiang W (2016) 2.9 Å resolution cryo-EM 3D reconstruction of close-packed virus particles. Structure 24:319–328. doi:10.1016/j.str.2015.12.006

Ludtke SJ, Baldwin PR, Chiu W (1999) EMAN: semiautomated software for high-resolution single-particle reconstructions. J Struct Biol 128:82–97

Navaza J (2008) Combining X-ray and electron-microscopy data to solve crystal structures. Acta Crystallogr Sect D, Biol Crystallogr 64:70–75. doi:10.1107/S0907444907053334

Otwinowski Z, Minor W (1997) Processing of X-ray diffraction data collected in oscillation mode. Method Enzymol 276:307–326. doi:10.1016/S0076-6879(97)76066-X

Read RJ, Kleywegt GJ (2001) Density modification: theory and practice. In: Turk D, Johnson L (eds) Methods in macromolecular crystallography. IOS Press, Amsterdam, pp 123–135

Stein PE, Boodhoo A, Armstrong GD, Cockle SA, Klein MH, Read RJ (1994) The crystal structure of pertussis toxin. Structure 2:45–57

Tang X, Yang C, Gu Y, Song C, Zhang X, Wang Y, Zhang J, Hew CL, Li S, Xia N, Sivaraman J (2011) Structural basis for the neutralization and genotype specificity of hepatitis E virus. Proc Natl Acad Sci USA 108:10266–10271. doi:10.1073/pnas.1101309108

Ten Eyck LF (1977) Efficient structure-factor calculation for large molecules by the fast Fourier transform. Acta Crystallogr Sect A 33:486–492

Tong L, Rossmann MG (1997) Rotation function calculations with GLRF program. Methods Enzymol 276:594

Xing L, Li TC, Mayazaki N, Simon MN, Wall JS, Moore M, Wang CY, Takeda N, Wakita T, Miyamura T, Cheng RH (2010) Structure of hepatitis E virion-sized particle reveals an RNA-dependent viral assembly pathway. J Biol Chem 285:33175–33183. doi:10.1074/jbc.M110.106336

Yamashita T, Mori Y, Miyazaki N, Cheng RH, Yoshimura M, Unno H, Shima R, Moriishi K, Tsukihara T, Li TC, Takeda N, Miyamura T, Matsuura Y (2009) Biological and immunological characteristics of hepatitis E virus-like particles based on the crystal structure. Proc Natl Acad Sci USA 106:12986–12991. doi:10.1073/pnas.0903699106

6

Pseudomonas sp. LZ-Q continuously degrades phenanthrene under hypersaline and hyperalkaline condition in a membrane bioreactor system

Yiming Jiang[1], Haiying Huang[1], Mengru Wu[2], Xuan Yu[1], Yong Chen[1], Pu Liu[3], Xiangkai Li[1✉]

[1] Ministry of Education Key Laboratory of Cell Activities and Stress Adaptations, School of Life Science, Lanzhou University, Lanzhou 730000, China
[2] State Key Laboratory of Microbial Resources, Institute of Microbiology, Chinese Academy of Sciences, Beijing 100101, China
[3] Department of Development Biology Sciences, School of Life Science, Lanzhou University, Lanzhou 730000, China

Graphical Abstract

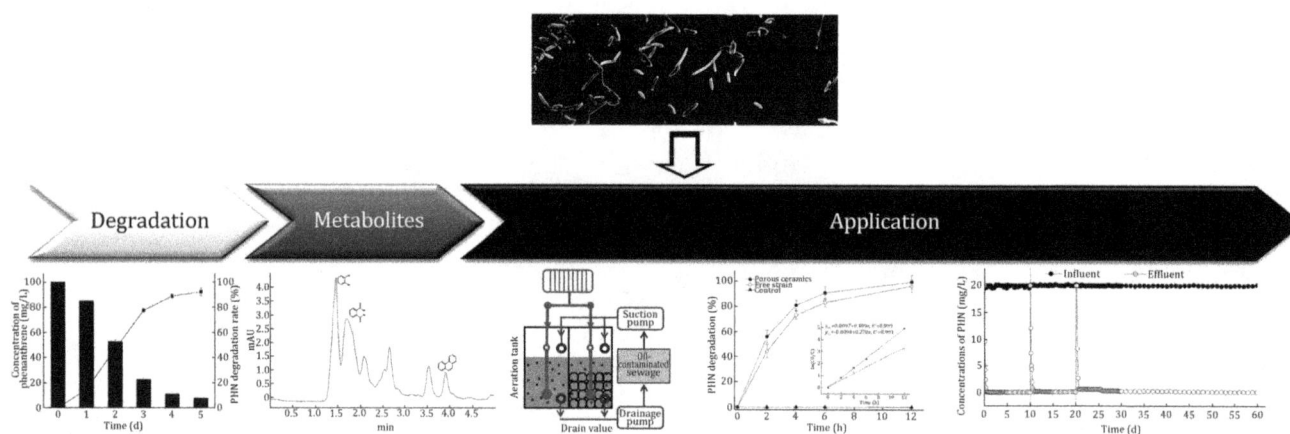

Abstract Phenanthrene is one of the most recalcitrant components of crude oil-contaminated wastewater. An efficient phenanthrene-degrading bacterium *Pseudomonas* sp. strain named LZ-Q was isolated from oil-contaminated soil near the sewage outlet of a petrochemical company. *Pseudomonas* sp. LZ-Q is able to degrade 1000 mg/L phenanthrene in Bushnell-Hass mineral salt medium. It also degrades other polycyclic aromatic hydrocarbons such as naphthalene, anthracene, pyrene, petrol, and diesel at broad ranges of salinities of 5 g/L to 75 g/L, pHs of 5.0–10.0, and temperatures of 10–42 °C. Therefore, *Pseudomonas* sp. LZ-Q could be a good candidate for remediation of polycyclic aromatic hydrocarbon (PAH)-contaminated wastewater. A membrane bioreactor (MBR) was applied to investigate the remediation ability of the strain LZ-Q. Wastewater containing phenanthrene with pH of 8, salinity of 35 g/L, and COD of 500 mg/L was continuously added to the system (HRT = 3 h). Results showed that *Pseudomonas* sp. LZ-Q is capable of degrading 96% of 20 mg/L phenanthrene and 94% of 500 mg/L COD for 60 days in a continuous mode. These results showed that the MBR system with strain LZ-Q might be a good approach for PAHs' remediation in industrial wastewaters.

Yiming Jiang, Haiying Huang and Mengru Wu contributed equally to this work.

✉ Correspondence: xkli@lzu.edu.cn (X. Li)

Keywords   *Pseudomonas* sp. LZ-Q, Phenanthrene degradation, Immobilization microorganisms, Hypersaline and hyperalkaline wastewater, Membrane bioreactor (MBR)

INTRODUCTION

Nowadays, water pollution is becoming an increasingly concerned environmental problem (Schwarzenbach et al. 2010). In the northwest of China, water pollution events occur frequently. Among the refractory organics causing water pollution events, polycyclic aromatic hydrocarbons (PAHs) are considered to be the most environmentally significant and hazardous to human health (Wang et al. 2013). PAHs are a group of organic chemicals consisting of two or more fused benzene rings that are in linear, angular, and cluster arrangements (Bamforth and Singleton 2005). Most of the PAHs are toxic, mutagenic, carcinogenic, and recalcitrant (Wu et al. 2010; Patel et al. 2012). PAHs released into the environment would cause serious risks to natural environment, fishery, agriculture and human health (Wang et al. 2013). Therefore, controlling PAHs pollution is an urgent task in water protection.

Phenanthrene (PHN) identified as one of the priority pollutants is a typical PAH and some of its derivatives are carcinogenic (Jerina et al. 2012). Microbes are able to degrade PHN and provide an ideal bioremediation approach. For example, *Pseudomonas stutzeri* ZP2 can degrade more than 90% of PHN at 1000 ppm in 6 days (Zhao et al. 2009). *Pseudomonas* sp. JM2 isolated from active sewage sludge of a chemical plant removes 50 mg/L PHN within 4 days (Ma et al. 2012). However, successful applications of using microbes to remediate PHN in industrial wastewater are still scarce (Lefebvre and Moletta 2006). Industrial wastewater with wide ranges of pHs and hypersalinities inhibits microbial respiration rates, reduces enzymes' activity, and elevates osmotic pressure of cells. Many of the known PHN-degrading bacteria cannot survive well under such condition and function properly (Kunst and Rapoport 1995; Metcalf 2003). Therefore, searching for a PHN degrading strain with survival ability in industrial wastewater is critical for applications of PHN's remediation.

Membrane bioreactor (MBR) is proven to be a good method for wastewater treatment and attracts extensive attentions, because MBR is characterized with high pollutant removal efficiency. Removal of PAHs by MBR has been studied (González et al. 2012). However, membrane fouling is one significant limitation when using MBR to treat wastewater (Tang et al. 2010; Zhang et al. 2011). Previous studies have demonstrated that microorganism immobilization technology (MIT) can improve the effectiveness of sewage treatment and enable cells to separate from aqueous solution easily and reduce the membrane fouling. MIT has been widely applied in industrial operation and works efficiently (Juang et al. 2008; Bai et al. 2009; Ting and Sun 2000; Yan and Viraraghavan 2001). Immobilizing microbiota B500 on macro-porous carriers enhances the removal efficiency for contaminants in wastewater (Park et al. 2005). Iron-oxidizing bacteria immobilized onto polyurethane foam decrease the risk of membrane fouling and increase the efficiency of pollutants' degradation (Zhou et al. 2008).

In this study, a bacterial strain LZ-Q utilizing PHN as the sole carbon source was isolated from petroleum-contaminated soil. Ceramics are used as carriers of strain LZ-Q in MBR system. Strain LZ-Q degraded PHN efficiently and this MBR with immobilized strain LZ-Q cells showed ability of degradation of PHN in artificial petrochemical wastewater in long period.

RESULTS AND DISCUSSION

Isolation and characterizations of strain LZ-Q

Four bacterial strains with the ability of phenanthrene degradation were isolated from petrochemical-contaminated soils in Lanzhou reach of the Yellow River, China. All four strains were gram-negative and aerobic. By comparison of 16S rRNA gene sequences, LZ-Q (GenBank No.: KR140091, CCTCC No.: M2015564), LZ-O (GenBank No.: KR140089), and LZ-G (GenBank No.: KR140088) were closely related to *Pseudomonas* spp. and LZ-P (GenBank No.: KR140090) was related to *Rhizobium* sp. Growths of the isolated strains using phenanthrene (1000 mg/L) as the sole carbon source were determined at pH 7, 180 r/min, and 28 °C. All isolated strains can grow under such condition and strain LZ-Q reached highest optical density (OD_{600nm}) of 0.25 after 168 h incubation (Fig. 1A). This result suggests that all four strains can degrade PHN, and strain LZ-Q was chosen for further study as it showed higher growth when using PHN as sole carbon source. The strain LZ-Q was short rod-shaped bacterium. The colonies of strain LZ-Q were mostly small, opaque, circular or irregular oval-shaped and oyster white-colored with moist and luster surface. ViTek phenotype analysis showed that strain LZ-Q

Fig. 1 A Growth curves of isolated strains and *E. coli*. **B** Phylogenetic tree based on 16S rRNA gene sequence showing the relationship between corresponding sequences of the genus *Pseudomonas* genus

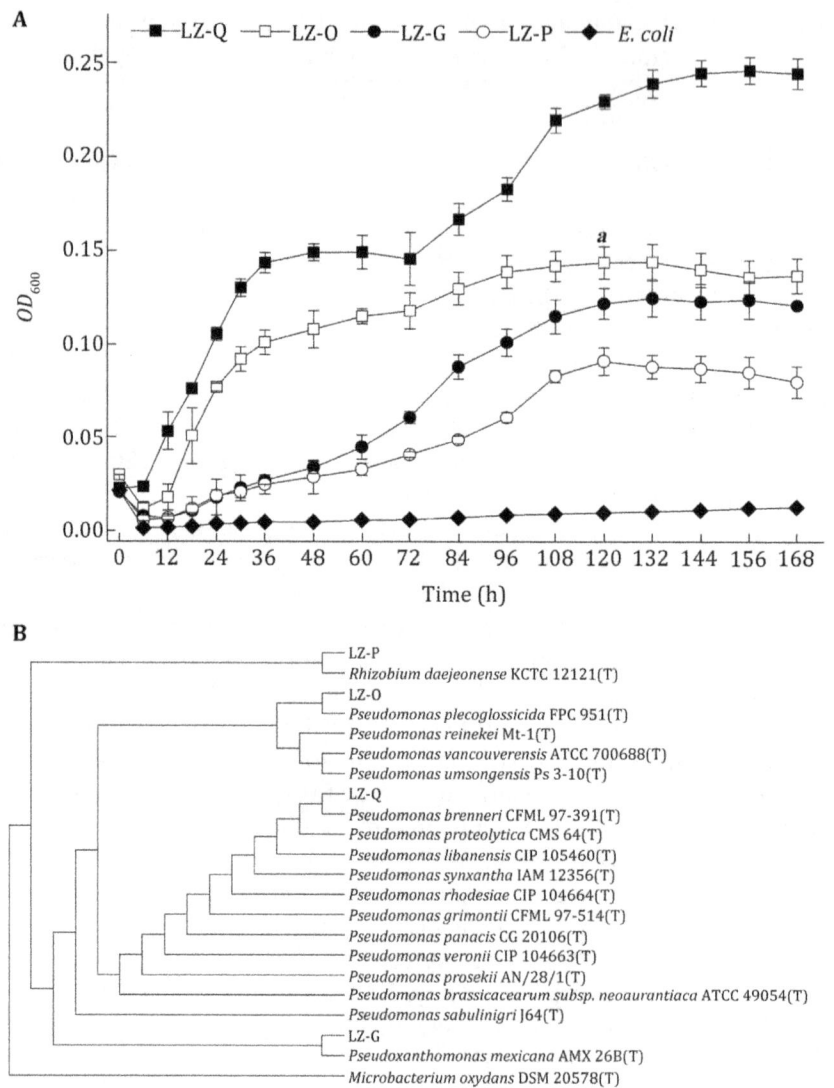

was 95% closely related to *Pseudomonas fluorescens* (Table 1). A phylogenetic tree based on neighbor-joining algorithm demonstrated that LZ-Q clustered with *P. brenneri*, which falls within the *P. fluorescens* group, at a bootstrap value of 75% (Fig. 1B). Our data also showed that strain LZ-Q degrades various PAHs, petrol, and diesel, suggesting it might be an ideal strain for PAH bioremediation (Table 1).

Table 1 ViTek report of strain LZ-Q

Biochemical details

2	APPA⁻	−	3	ADO	−	4	PyrA	+	5	IARL	−	7	dCEL	−	9	BGAL	−
10	H$_2$S	−	11	BNAG	−	12	AGLTp	−	13	dGLU	+	14	GGT	−	15	OFF	−
17	BGLU	−	18	dMAL	−	19	dMAN	−	20	dMNE	+	21	BXYL	−	22	BAlap	−
23	ProA⁻	+	26	LIP	−	27	PLE	−	29	TyrA	+	31	URE	−	32	dSOR	−
33	SAC-	−	34	dTAG	−	35	dTRE	+	36	CIT	+	37	MNT	−	39	5 KG	−
40	ILATk-	−	41	AGLU	−	42	SUCT	+	43	NAGA	−	44	AGAL	−	45	PHOS	−
46	GlyA-	−	47	ODC	−	48	LDC	−	53	IHISa	−	56	CMT	−	57	BGUR	−
58	O129R	+	59	GGAA	−	61	IMLTa	+	62	ELLM	−	64	lLATa	−			

"+" and "−" represent whether the strain can utilize the substrate or not

Characterization of phenanthrene degradation in strain LZ-Q

In order to determine the optimum degradation condition, strain LZ-Q was cultivated in BH medium with 100 mg/L PHN under pHs ranging from 5 to 10 and temperatures varying from 10 °C to 42 °C. Highest OD_{600nm} was achieved at pH 7.0 and 28 °C, which indicates that the optimum growth and PHN degrading condition for strain LZ-Q is at pH 7.0 and 28 °C (Fig. 2A, B). Under the optimum growth condition, strain LZ-Q degraded 92.27% PHN after 5 days cultivation (Fig. 2C).

Studies about *Pseudomonas* spp. which can degrade phenanthrene have been reported previously. Bacterial strain *P.* sp. Ph6-gfp isolated from clover grown in a PAH-contaminated site showed a 81.1% decrease of phenanthrene (50 mg/L) within 15 days (Sun et al. 2014), and strain *P.* stutzeri ZP2 isolated from soil in oil refinery fields in Shanghai China could reduce about 96% PHN (250 mg/L) within 6 days (Janbandhu and Fulekar 2011). In line with previous studies, strain LZ-Q that is closely related to *Pseudomonas* genus can utilize a range extension of refractory organics and degrade high concentrations of phenanthrene (Lin et al. 2014). The degradation rate is higher than *P.* sp. but lower than *P.* stutzeri ZP2 (Zhao et al. 2009; Sun et al. 2014). In addition, the results also showed that strain LZ-Q can grow at pHs of 5.0–10.0 and temperatures ranging from 10 °C to 42 °C, suggesting that it was capable of degrading PHN using phenanthrene as the sole carbon source at broad pHs and temperatures. Therefore, even though the degradation rate of PHN is not the most efficient, strain LZ-Q could degrade PHN more efficiently as the bio-mediation condition in wastewater treatment plant is variable.

Degradation of PHN is often affected by the high salinity (Haritash and Kaushik 2009). To further determine the influence of high salinity on growth and PHN degradation of strain LZ-Q, the adaptability to salinity was investigated in BH medium with the addition of different concentrations of NaCl (5, 10, 35, 40, 50, 75 and 100 g/L) after 120 h of incubation. Strain LZ-Q degraded PHN with 5 g/L to 75 g/L NaCl (Fig. 3A). These results showed that strain LZ-Q could degrade PHN efficiently under hypersaline condition. As the high salinity of industrial wastewater usually restricts microorganisms to remove pollutants, studies about bacterial strains which degraded PHN under hypersaline condition and were used in industrial wastewater plant were limited (Kunst and Rapoport 1995; Metcalf 2003). *P.* sp. BZ-3 degraded 75% phenanthrene under the 20 g/L NaCl (Lin et al. 2014). Our data showed that strain LZ-Q also remove phenanthrene contaminants

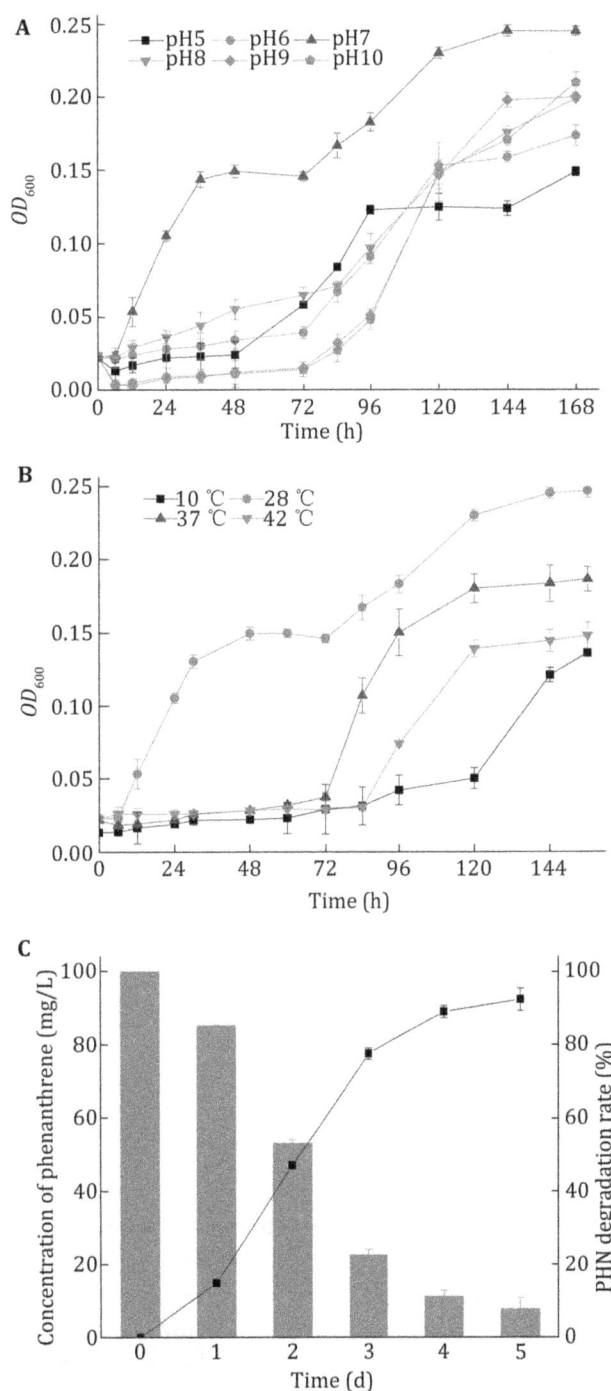

Fig. 2 **A** Optimum pH conditions of strain LZ-Q. **B** Optimum temperature conditions of strain LZ-Q. **C** Biodegradation rate of PHN under optimum conditions

under high salinity (75 g/L NaCl) and high alkalinity (pH 9) conditions efficiently. Results showed that LZ-Q's PHN degradation ability which is 3% lower than *P.* stutzeri ZP2, is not the highest at optimum conditions. But it can degrade PHN at pHs and temperatures which other strains cannot, suggesting that strain LZ-Q is a potential strain for application.

Fig. 3 A OD_{600} and PHN degradation rate of LZ-Q under saline condition. **B** The antibiotic resistance of strain LZ-Q

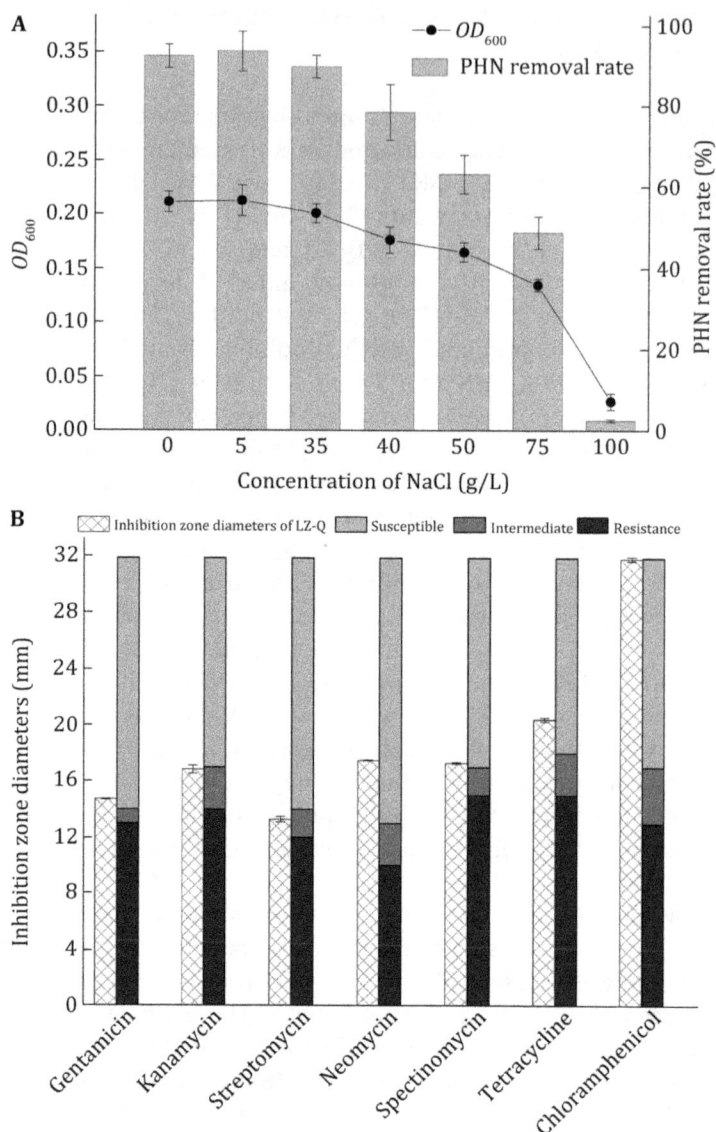

Previous studies proposed that multi-drug-resistant strains have a higher risk of spreading antibiotic resistance genes to indigenous flora (Masakorala et al. 2013). The antibiotic sensitive tests showed strain LZ-Q was sensitive to gentamicin, neomycin, spectinomycin, tetracycline, chloramphenicol, and intermediately susceptible to streptomycin and kanamycin (Fig. 3B). These results demonstrate that strain LZ-Q displays a profile of low resistance to antibiotics. Thus, *P.* sp. LZ-Q could be a suitable bioremediation additive in in situ wastewater treatment.

All the results provide the evidence that *P.* sp. LZ-Q might be a good potential candidate for the bioremediation of phenanthrene-contaminated industrial sewage under hypersaline and hyperalkaline conditions.

Metabolic pathways of phenanthrene degradation by strain LZ-Q

Two metabolic pathways including salicylic acid pathway and phthalic acid pathway were reported to degrade phenanthrene by bacteria (Prabhu and Phale 2003). Salicylic acid and catechol are intermediates of the salicylic acid pathway, phthalic acid is generated in the phthalic acid pathway (Peng et al. 2008; Haritash and Kaushik 2009). Strain LZ-Q utilizes salicylic acid, catechol, and phthalic acid as carbon sources (Table 2), suggesting that strain LZ-Q degrades phenanthrene through both salicylic acid pathway and phthalic acid pathway. The possible pathways for PHN degradation by strain LZ-Q are elucidated via HPLC method in this study. HPLC analysis data showed that peaks of

Table 2 The diversity of degradable substrates by strain LZ-Q

Substrates	Growth situations	Substrates	Growth situations
Phenanthrene	++	Diesel	++
Naphthalene	++	Salicylic acid	+
Anthracene	++	Phthalic acid	++
Pyrene	++	Diphenylamine	+
Petrol	+		

+: Moderate, ++: good

phthalic acid, catechol, and phenanthrene appeared at a retention time of 1.65, 1.47, and 3.91 min, respectively (Fig. 4A, 4B, and 4C). After three-day cultivation of strain LZ-Q in BH/PHN, the peak at 3.91 min decreased and two main peaks occurred at 1.65 and 1.47 min (Fig. 4D), suggesting that phthalic acid and catechol

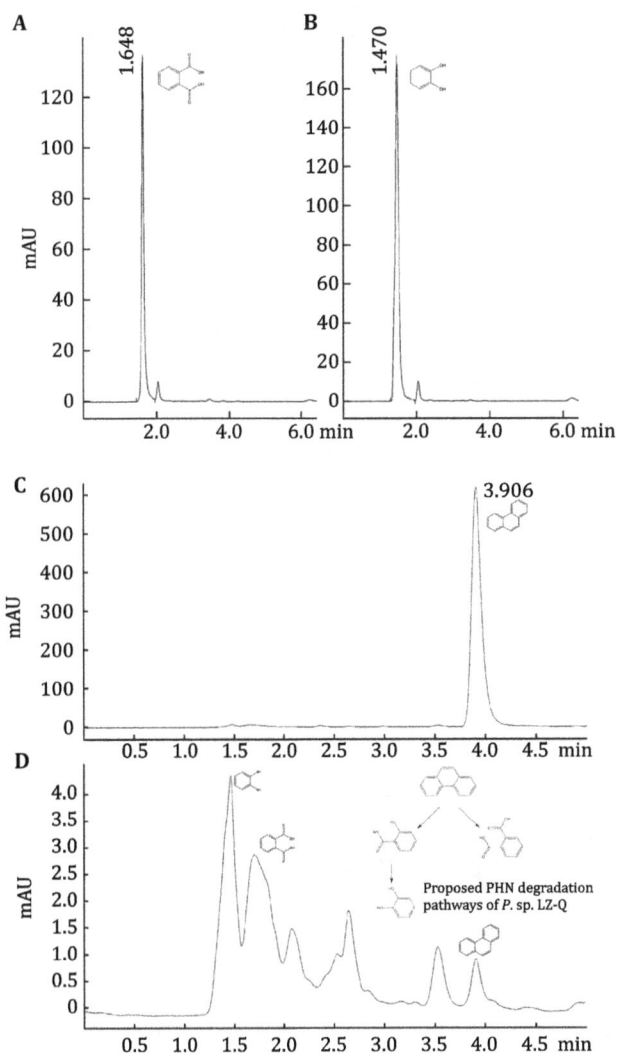

Fig. 4 Metabolite analysis using HPLC. Standard substances of phthalic acid (**A**), catechol (**B**) and phenanthrene (**C**); **D** Metabolite analysis

were generated which is consistent with the previous result that strain LZ-Q could degrade salicylic acid, catechol, and phthalic acid. These results revealed that there are two PHN possible degradation pathways for LZ-Q (inset of Fig. 4D).

Studies of metabolic variations of phenanthrene biodegradation have been proved by HPLC method, such as studies of *Pseudomonas putida* NCIB 9816, *Mycobacterium* sp., and *Burkholderia* sp. strain BC1 (Boldrin et al. 1993; Yang et al. 1994; Chowdhury et al. 2014). Among *Pseudomonas* genus, studies about metabolic pathways of phenanthrene degradation showed a variation. *Pseudomonas* strain BZ-3 degraded PHN through salicylic acid pathway (Lin et al. 2014). According to the HPLC result, dimethylphthalate was detected as the intermediate product during PHN degradation. Therefore, *Pseudomonas* sp. USTB-RU biodegrade PHN via the phthalic acid pathway (Masakorala et al. 2013). In this study, strain LZ-Q degraded PHN both via salicylic acid pathway and phthalic acid pathway. This result is in agreement with the previous work revealing that *Pseudomonas* sp. N7 can degrade PHN via both pathways (Jia et al. 2008).

Immobilized strain LZ-Q in MBR degrades PHN continuously and efficiently under hypersaline and hyperalkaline conditions

Our previous results suggest that LZ-Q is suitable for application in hypersaline and hyperalkaline wastewater treatment. An MBR system was setup to test LZ-Q's ability to degrade PHN in wastewater. In our study, ceramics are used as adsorbing carriers. Results show that strain LZ-Q grew well on the surface and in the ostioles of ceramics. In order to determine the degradation ability of strain LZ-Q in MBR, microorganisms were treated with synthetic wastewater with 20 mg/L phenanthrene (pH 8 and 35 g/L NaCl). The PHN degradation and COD removal were detected in free-bacteria reactors (FBR) and immobilized-microorganisms reactors (IMR). Decomposition rates of COD and PHN were faster in IMRs than in FBRs. The PHN degradation rate reached 90.68% in IMRs at 6 h, while it was only 82.04% in FBRs with no significant differences ($P > 0.05$). A COD removal rate of 90.4% was achieved when using ceramics as carriers, whereas it was only 78.6% by free strain (Fig. 5). COD removal rate in IMRs had significant difference as compared with that in FBRs ($P < 0.05$). These results reveal that ceramic carriers with strain LZ-Q spur COD removal, suggesting that immobilized technique is suitable for wastewater treatment and similar to report about immobilizing bacteria onto ceramic carriers could remove COD more efficiently during operation (Kariminiaae-Hamedaani et al. 2003). Parameswarappa reported that ceramic material was a good choice for wastewater treatment and immobilization technology enhanced the degrading efficiency of ethylbenzene by *Pseudomonas fluorescens*-CS2 (Parameswarappa et al. 2008). In the batch and semi-continuous treatments, ceramics with immobilized consortia could remove COD, phosphate, nitrate, and H_2S effectively with removal

rates of 89%, 77%, 99%, and 99.8% for 1 month (Nagadomi et al. 2000). In a packed bed bioreactor, 82% of the influent COD was removed within 160 days of operation (Kariminiaae-Hamedaani et al. 2003).

In addition, degradation of PHN in both FBRs and IMRs fitted to the exponential function and followed a pseudo-first-order kinetic model with rate constants of 0.372/h ($R^2 = 0.999$) and 0.290/h ($R^2 = 0.991$), respectively (inset of Fig. 5A), and agrees with previous studies on PAHs biodegradation, such as *Pseudomonas aeruginosa* strain PAH-1 whose PHN degradation characteristics fitted to pseudo-first-order kinetic model (Ma et al. 2011).

The MBRs were operated at 19–21 °C with a HRT of 3 h, and influent COD and PHN concentrations were maintained to 500 and 20 mg/L, respectively. Based on different pHs (8, 9 and 10) of influent, process was divided into three phases. In the first phase, the effluent concentrations of COD and PHN were reduced to 26 and

Fig. 5 PHN degradation curve (**A**) and COD removal rate curve (**B**) of strain LZ-Q using in MBR. Reactor without strains was used as a control

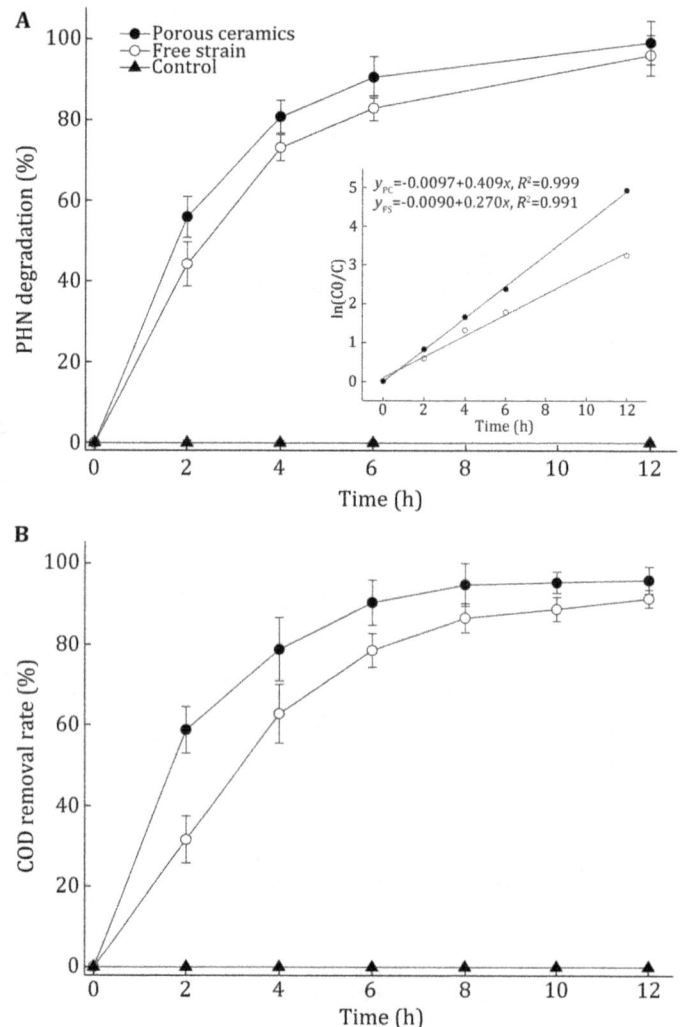

0.9 mg/L in 8 h. At the 12th hour of the second phase, the COD removal rate of 94.4% and the PHN degradation rate of 95.4% were achieved. In the final phase, the COD removal rate reached 94.3% at the 12th hour, and concentration of PHN dropped to 0.7 mg/L with the degradation rate of 96.5% at the 16th hour. Effluent concentrations of COD and PHN were maintained around 30 and 0.8 mg/L in all three phases (Fig. 6). The MBRs used in this study operated stably for 60 days and degradation ability showed no signs of decreasing.

Membrane bioreactor is a good system to treat industrial wastewater (Melin et al. 2006). It has been reported that MBR could remove more than 94%–98% COD and TOC (Scholz and Fuchs 2000). Cirja reported removal of organic micropollutants by microorganism in MBR (Cirja et al. 2008). *Streptomyces* sp. QWE-35 degraded 200 mg/L naphthalene in a MBR and could be a potential candidate for coal gasification wastewater treatment (Xu et al. 2014). However, in industrial operations of MBR, using free strain to treat wastewater is not efficient, as free strains are hard to separate from aqueous solution and cause membrane fouling (Zhang et al. 2011). Membrane fouling still limits the widespread application of MBR (Tang et al. 2010). Immobilized microorganism technology is a solution for these disadvantages (Xu et al. 2012). Ceramics is a kind of carrier which can decrease the membrane fouling, enhance the removal rate, harbor the long-term usability in wastewater treatment (Kariminiaae-Hamedaani et al. 2003; Bai et al. 2009). In our study, the system

Fig. 6 Changes in COD and PHN concentrations of the influent and effluent in the MBR with different pH

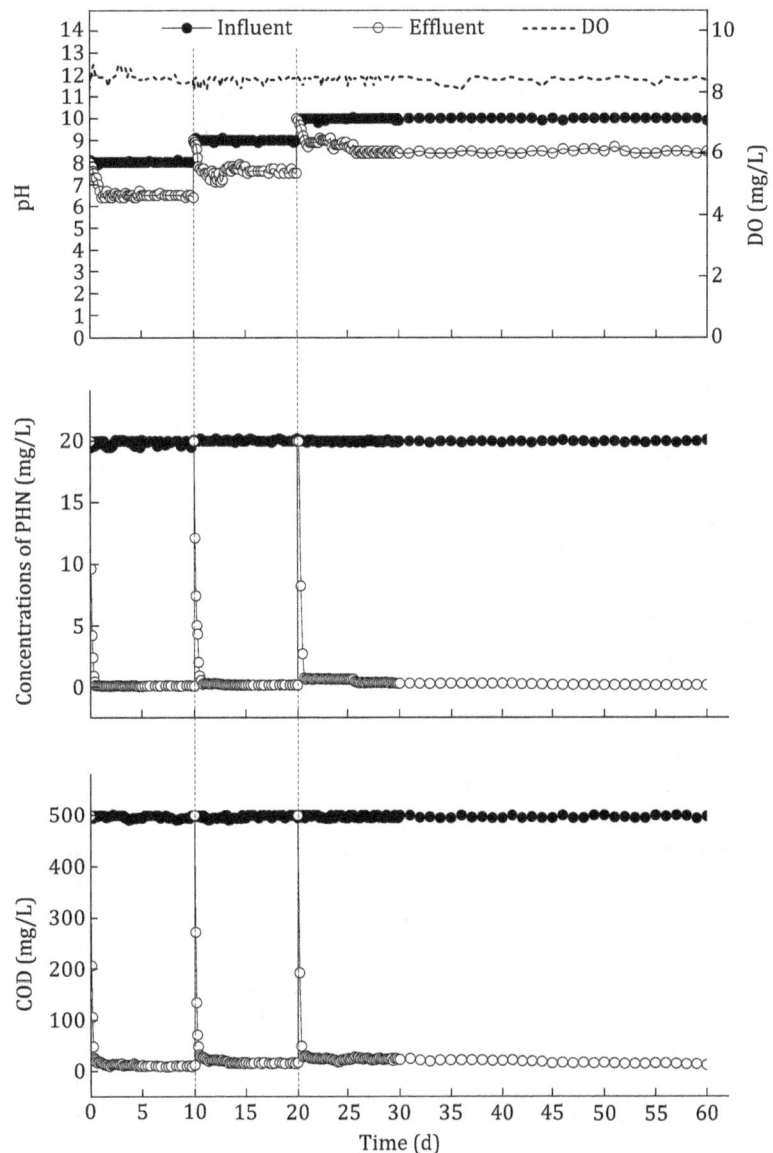

immobilized strain LZ-Q onto ceramic carriers in membrane-bioreactor was able to remove more than 96% of the influent phenanthrene and 94% of the influent COD during 60 days of operation when the HRT was 3 h, and degradation ability showed no signs of decreasing.

Hypersaline and hyperalkaline conditions are also limitations of efficiency of wastewater treatment in MBR as biological treatment is strongly inhibited by salts (mainly NaCl) (Lefebvre and Moletta 2006). Soltani reported that microorganisms only degraded 50% PHN in MBR system under high salinity conditions (Soltani et al. 2010). When the salt concentration was up to 84 mg/L, MBR system requires 73 operating days adaptation to recover to the normal COD removal efficiency (Artiga et al. 2008). However, our data indicated that strain LZ-Q could degrade PAHs including phenanthrene efficiently under the hyperhaline and hyperalkaline conditions. The MBR system with immobilized strain LZ-Q operated stably for 60 days without decreasing. All these results suggesting immobilized strain LZ-Q in MBR is a candidate system for degrading phenanthrene in saline-alkali wastewater.

CONCLUSIONS

An efficient phenanthrene degradation strain LZ-Q, which was isolated from petroleum-contaminated soil in Lanzhou reach of the Yellow River, was identified as *P.* sp. by ViTek 2 and 16S rRNA gene sequencing. The antibiotic-sensitive strain LZ-Q grew in BH medium containing 1 g/L phenanthrene as the carbon source and had the ability to degrade phenanthrene with a broad range of salinities (5–75 g/L), pHs (5–10), and temperatures (10–42 °C). Strain LZ-Q degraded phenanthrene efficiently through salicylic acid pathway and phthalic acid pathway. In the BH/PHN (100 mg/L) medium, the PHN degradation rate of strain LZ-Q was 92.27% in 5 days. In addition, strain LZ-Q could use other organic compounds as carbon sources, such as naphthalene, anthracene, pyrene, petrol, and diesel. Compared with free strain, adsorptive ceramic carrier of strain LZ-Q spurs removal rates of COD and PHN from 78.6% to 90.4% and 82.04% to 90.68% at 6 h. Continuous treatment performance of MBR with immobilized strain LZ-Q operated powerfully and stably in this study. All the observations indicated that MBR with immobilized *P.* sp. LZ-Q could be a candidate to remediate phenanthrene-contaminated saline-alkali sewage.

MATERIALS AND METHODS

Sample collection

Soil sampling was carried out near the sewage outlet of a petrochemical company (36°02′N, 103°61′E, Lanzhou, China) in Gansu province, which is in upper reach of the Yellow River in April, 2013. The climate of sampling site is a typical temperate and monsoonal continental climate, with an annual mean air temperature of 9.3 °C and precipitation of 360 mm (Wu et al. 2011). Samples were collected at depth of 15 cm at 22 °C and pH 6.2. After collection, samples were immediately transferred to the laboratory in sterilized aluminum boxes and stored at −80 °C.

Reagents and culture media

Phenanthrene, salicylic acid, catechol, phthalic acid, and methyl alcohol were HLC grade, and all the other chemicals were of analytical grade.

Bushnell-Hass mineral salt medium (BH) was composed of (g/L) NaCl 5, KH_2PO_4 1, K_2HPO_4 1, NH_4NO_3 1, $MgSO_4 \cdot 7H_2O$ 0.2, $CaCl_2 \cdot 2H_2O$ 0.02, and $FeCl_3$ 0.05. Luria–Bertani (LB) medium was composed of (g/L) peptone 10, NaCl 5, and yeast extract 10. Per liter solid LB contained 15 g agar powder. One liter of synthetic wastewater (SW) consisted of NaCl 35 g, glucose 450 mg, $FeCl_2 \cdot 4H_2O$ 200 mg, $CaCl_2 \cdot 2H_2O$ 200 mg, $MgCl_2 \cdot 6H_2O$ 300 mg, $CuCl_2 \cdot 2H_2O$ 0.6 mg, $ZnCl_2$ 1 mg, H_3BO_3 1 mg, $MnCl_2 \cdot 4H_2O$ 10 mg, $CoCl_2 \cdot 6H_2O$ 1 mg, $NaMoO_4 \cdot 2H_2O$ 0.2 mg, $NiCl_2 \cdot 6H_2O$ 1 mg. NH_4Cl, and K_2HPO_4 were added into the SW to maintain COD:N:P ratio of 200:5:1. The pH of the SW was adjusted with 25 g/L $NaHCO_3$. All prepared media were autoclaved at 121 °C for 20 min.

Enrichment and isolation of phenanthrene-degrading strain

Phenanthrene-degrading strains were isolated from the mixed soil samples taken from Lanzhou reach of Yellow River. 1 g soil was added into 100 mL BH supplied with 50 mg phenanthrene as sole carbon and energy source. Enrichments were incubated at 28 °C and 180 r/min. After 5 days, 1 mL of enriched aqueous culture was transferred to another BH/phenanthrene (500 mg/L) medium. Consecutive enrichment processes were repeated three times for every 5 days until microbial consortium was developed in the medium. Then, 100 μL dilute aqueous culture was spread on the solid BH plate. After 2 days, developed colonies on the plates were isolated and inoculated into the BH/phenanthrene medium to confirm their potential of degrading phenanthrene. Repeated plate streaking was employed to ensure purity of the isolates. A mixture of the bacterial solution and 50% glycerol with a ratio of 1:1 (*v/v*) were stored at −20 °C.

Characterization and identification of microorganism

The isolated bacteria were examined using observation of morphological features. Gram staining was performed before biochemical tests were done with the Vitek 2 System (bioMerieux Industry, Marcyl'Etoile, France) according to the manufacturer's instructions.

Molecular identification was carried out by phylogenetic analysis following the 16S rRNA sequencing. The strain was first activated in 5 mL LB on a shaker at 30 °C and 180 r/min for 12 h. Amplification of gene fragments encoding 16S rRNA was performed using the universal 16S rRNA primers (*E. coli* 27F and 1492R). Fragments sequencing was done by Shanghai Majorbio Bio-pharm Technology Co. Ltd (Shanghai, China) and sequences were analyzed at EzTaxon (www.eztaxon.org/) database. Phylogenetic tree was generated by MEGA (Tamura et al. 2007).

Determination of growth characterization and degradation ability

Growth on phenanthrene

The optimum temperature test was carried out under the temperatures of 10, 28, 37, and 42 °C. Optimum pH was tested at pH 5, 6, 7, 8, 9, and 10. Growth curves were determined by measuring OD_{600nm} in the medium using a spectrophotometer. The isolated strain was separately inoculated into the 100 mL sterilized BH medium with 100 mg/L phenanthrene as sole carbon and cultivated at pH 7, 180 r/min, and 28 °C. *Escherichia coli* was used as negative control.

Growth on other carbon sources

Growth of the pure culture on other carbon substrates, including naphthalene, anthracene, pyrene, salicylic acid, phthalic acid, diphenylamine, petrol, and diesel was tested to find out the potential degradable substrate as sole carbon and energy source. All experiments were in triplicate.

Degradation of phenanthrene

1 mL aliquots of isolated strain were added into 100 mL sterilized BH medium containing 100 mg/L phenanthrene for 5 days. Quantity of phenanthrene concentrations and varieties of catabolic intermediates were determined using HPLC. Bacterial solutions were filtered through disposable filters (0.45 μm). Phenanthrene and putative metabolites were separated with a silica C18 column (4.6 × 150 mm). The mobile phase consisted of 80% methanol and 20% water at a flow rate of 1 mL/min and room temperature. Eluants were monitored by UV-Vis light detection at a wavelength of 254 nm and qualified using an external standard calibration curve.

Tolerance of salinity and antibiotics

Salinity tolerance of the isolate was assayed on BH/PHN(100 mg/L) liquid containing 5, 35, 40, 50, 75, and 100 g/L NaCl. The OD_{600nm} was reported in 120 h of incubation. Antibiotic sensitivity tests were carried out by the Kirby-Bauer antibiotic susceptibility disk diffusion method (Kirby et al. 1966). All experiments were in triplicate.

Bacteria immobilization and reactor setup

Ceramic-microorganism adsorptive carriers were spherical in shape with the diameter of 2–3 mm, and were prepared as follows: washing with water for three times, and then immersed in 5% HCl, neutral water, and 5% NaOH for 2 h successively. 100 mL modified ceramics was placed into 500-mL flasks containing 200 mL bacterial culture for 3 days with changing LB liquid medium every 24 h. The obtained carriers were washed with normal saline and stored in physiological saline at 4 °C before using.

This research was carried out in continuous aerating MBRs with 2 L (3 L of total volume) (Fig. 7). The MBRs operated with micro-filtration (MF) flat-sheet (FS) membrane module with nominal porosity of 0.4 μm. Free-bacteria or immobilized microorganism carriers were added into reactors. All the other operating conditions for reactors were kept the same. MBRs were operated at 19–21 °C and were run at 8–12 mg/L DO with the help of an aeration pump. Hydraulic retention time (HRT) was 3 h. The experiment lasted for 60 days and each reactor was set up in triplicate.

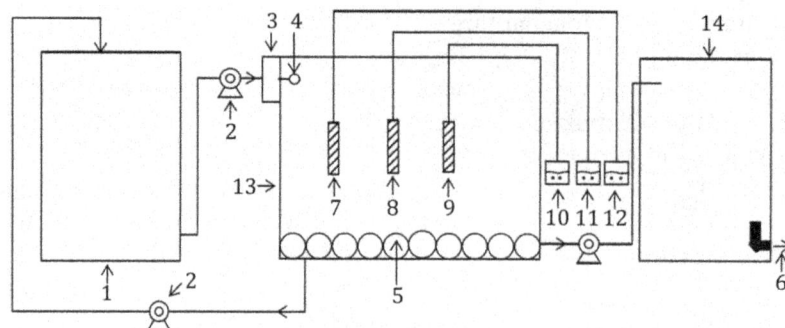

Fig. 7 Simplified scheme of the immobilized microorganisms reactor. *1*. Sewage storage tank; *2*. Suction pump; *3*. Rotameter; *4*. Inlet; *5*. Aerator; *6*. Outlet; *7*. DO sensor; *8*. Temperature sensor; *9*. pH sensor; *10*. pH monitor; *11*. Temperature monitor; *12*. DO monitor; *13*. Aeration tank; *14*. Membrane reactor

Morphological observations

Morphological observation of scanning electron microscopy (SEM) was done with Hitachi S-3400 N Scanning Electron Microscope (Hitachi High-Technologies Corporation, Tokyo, Japan). Sample preparation for SEM was carried out according to the methods reported before (Zhou et al. 2009).

Chemical analysis

The chemical oxygen demand (COD_{Cr}) was determined by the standard method based on potassium dichromate ($K_2Cr_2O_7$) oxidization method (Zhou et al. 2009).

Acknowledgments This study was supported by National Natural Science Foundation of China (31470224 and 31200085); MOST international cooperation grant (2014DFA91340) and Gansu Provincial International Cooperation (134WCGA176).

Compliance with Ethical Standards

Conflict of Interest Yiming Jiang, Haiying Huang, Mengru Wu, Xuan Yu, Yong Chen, Pu Liu, and Xiangkai Li declare that they have no conflict of interest.

Human and Animal Rights and Informed Consent This article does not contain any studies with human or animal subjects performed by any of the authors.

References

Artiga P, García-Toriello G, Méndez R, Garrido J (2008) Use of a hybrid membrane bioreactor for the treatment of saline wastewater from a fish canning factory. Desalination 221:518–525

Bai X, Ye Z, Qu Y, Li Y, Wang Z (2009) Immobilization of nanoscale Fe-0 in and on PVA microspheres for nitrobenzene reduction. J Hazard Mater 172:1357–1364

Bamforth SM, Singleton I (2005) Bioremediation of polycyclic aromatic hydrocarbons: current knowledge and future directions. J Chem Technol Biotechnol 80:723–736

Boldrin B, Tiehm A, Fritzsche C (1993) Degradation of phenanthrene, fluorene, fluoranthene, and pyrene by a *Mycobacterium* sp. Appl Environ Microbiol 59:1927–1930

Chowdhury PP, Sarkar J, Basu S, Dutta TK (2014) Metabolism of 2-hydroxy-1-naphthoic acid and naphthalene via gentisic acid by distinctly different sets of enzymes in *Burkholderia* sp. strain BC1. Microbiology 160:892–902

Cirja M, Ivashechkin P, Schäffer A, Corvini PF (2008) Factors affecting the removal of organic micropollutants from wastewater in conventional treatment plants (CTP) and membrane bioreactors (MBR). Rev Environ Sci Bio/Technol 7:61–78

González D, Ruiz LM, Garralón G, Plaza F, Arévalo J, Parada J, Pérez J, Moreno B, Gómez MÁ (2012) Wastewater polycyclic aromatic hydrocarbons removal by membrane bioreactor. Desalination Water Treat 42:94–99

Haritash A, Kaushik C (2009) Biodegradation aspects of polycyclic aromatic hydrocarbons (PAHs): a review. J Hazard Mater 169:1–15

Janbandhu A, Fulekar M (2011) Biodegradation of phenanthrene using adapted microbial consortium isolated from petrochemical contaminated environment. J Hazard Mater 187:333–340

Jerina DM, Yagi H, Lehr RE, Thakker DR, Schaefer-Ridder M, Karle JM, Levin W, Wood AW, Chang RL, Conney AH (2012) The bay-region theory of carcinogenesis by polycyclic aromatic hydrocarbons. Polycycl Hydrocarb Cancer 1:173–188

Jia Y, Yin H, Ye JS, Peng H, He BY, Qin HM, Zhang N, Qiang J (2008) Characteristics and pathway of naphthalene degradation by *Pseudomonas* sp. N7. Environ Sci 29:756–762

Juang R, Chung T, Wang M, Lee D (2008) Experimental observations on the effect of added dispersing agent on phenol biodegradation in a microporous membrane bioreactor. J Hazard Mater 151:746–752

Kariminiaae-Hamedaani H-R, Kanda K, Kato F (2003) Wastewater treatment with bacteria immobilized onto a ceramic carrier in an aerated system. J Biosci Bioeng 95:128–132

Kirby WM, Baner AW, Sherris KC, Truck M (1966) Antibiotic susceptibility testing by a standard single disc method. Am J Clin Pathol 45:493–502

Kunst F, Rapoport G (1995) Salt stress is an environmental signal affecting degradative enzyme synthesis in *Bacillus subtilis*. J Bacteriol 177:2403–2407

Lefebvre O, Moletta R (2006) Treatment of organic pollution in industrial saline wastewater: a literature review. Water Res 40:3671–3682

Lin M, Hu X, Chen W, Wang H, Wang C (2014) Biodegradation of phenanthrene by *Pseudomonas* sp. BZ-3, isolated from crude oil contaminated soil. Int Biodeterior Biodegrad 94:176–181

Ma C, Wang Y, Zhuang L, Huang D, Zhou S, Li F (2011) Anaerobic degradation of phenanthrene by a newly isolated humus-reducing bacterium, *Pseudomonas aeruginosa* strain PAH-1. J Soils Sediments 11:923–929

Ma J, Xu L, Jia L (2012) Degradation of polycyclic aromatic hydrocarbons by *Pseudomonas* sp. JM2 isolated from active sewage sludge of chemical plant. J Environ Sci 24:2141–2148

Masakorala K, Yao J, Cai M, Chandankere R, Yuan H, Chen H (2013) Isolation and characterization of a novel phenanthrene (PHE) degrading strain *Psuedomonas* sp. USTB-RU from petroleum contaminated soil. J Hazard Mater 263:493–500

Melin T, Jefferson B, Bixio D, Thoeye C, De Wilde W, De Koning J, van der Graaf J, Wintgens T (2006) Membrane bioreactor technology for wastewater treatment and reuse. Desalination 187:271–282

Metcalf E (2003) Wastewater engineering, treatment and reuse. McGraw-Hill, New York

Nagadomi H, Kitamura T, Watanabe M, Sasaki K (2000) Simultaneous removal of chemical oxygen demand (COD), phosphate, nitrate and H_2S in the synthetic sewage wastewater using porous ceramic immobilized photosynthetic bacteria. Biotechnol Lett 22:1369–1374

Parameswarappa S, Karigar C, Nagenahalli M (2008) Degradation of ethylbenzene by free and immobilized *Pseudomonas fluorescens*-CS2. Biodegradation 19:137–144

Park D, Lee DS, Joung JY, Park JM (2005) Comparison of different bioreactor systems for indirect H_2S removal using iron-oxidizing bacteria. Process Biochem 40:1461–1467

Patel V, Cheturvedula S, Madamwar D (2012) Phenanthrene degradation by *Pseudoxanthomonas* sp. DMVP2 isolated from hydrocarbon contaminated sediment of Amlakhadi canal, Gujarat, India. J Hazard Mater 201:43–51

Peng RH, Xiong AS, Xue Y, Fu XY, Gao F, Zhao W, Tian YS, Yao QH (2008) Microbial biodegradation of polyaromatic hydrocarbons. FEMS Microbiol Rev 32:927–955

Prabhu Y, Phale P (2003) Biodegradation of phenanthrene by *Pseudomonas* sp. strain PP2: novel metabolic pathway, role of biosurfactant and cell surface hydrophobicity in hydrocarbon assimilation. Appl Microbiol Biotechnol 61:342–351

Scholz W, Fuchs W (2000) Treatment of oil contaminated wastewater in a membrane bioreactor. Water Res 34:3621–3629

Schwarzenbach RP, Egli T, Hofstetter TB, Von Gunten U, Wehrli B (2010) Global water pollution and human health. Annu Rev Environ Resour 35:109–136

Soltani S, Mowla D, Vossoughi M, Hesampour M (2010) Experimental investigation of oily water treatment by membrane bioreactor. Desalination 250:598–600

Sun K, Liu J, Gao Y, Jin L, Gu Y, Wang W (2014) Isolation, plant colonization potential, and phenanthrene degradation performance of the endophytic bacterium *Pseudomonas* sp. Ph6-gfp Scientific reports 4

Tamura K, Dudley J, Nei M, Kumar S (2007) MEGA4: molecular evolutionary genetics analysis (MEGA) software version 4.0. Mol Biol Evol 24:1596–1599

Tang S, Wang Z, Wu Z, Zhou Q (2010) Role of dissolved organic matters (DOM) in membrane fouling of membrane bioreactors for municipal wastewater treatment. J Hazard Mater 178:377–384

Ting YP, Sun G (2000) Use of polyvinyl alcohol as a cell immobilization matrix for copper biosorption by yeast cells. J Chem Technol Biotechnol 75:541–546

Wang X-T, Miao Y, Zhang Y, Li Y-C, Wu M-H, Yu G (2013) Polycyclic aromatic hydrocarbons (PAHs) in urban soils of the megacity Shanghai: occurrence, source apportionment and potential human health risk. Sci Total Environ 447:80–89

Wu M, Nie M, Wang X, Su J, Cao W (2010) Analysis of phenanthrene biodegradation by using FTIR, UV and GC-MS. Spectrochim Acta, Part A 75:1047–1050

Wu SM, Huang SY, Fu BQ, Liu GY, Chen JX, Chen MX, Yuan ZG, Zhou DH, Weng YB, Zhu XQ, Ye DH (2011) Seroprevalence of *Toxoplasma gondii* infection in pet dogs in Lanzhou, Northwest China. Parasit Vectors 4:64

Xu P, Zeng GM, Huang DL, Feng CL, Hu S, Zhao MH, Lai C, Wei Z, Huang C, Xie GX, Liu ZF (2012) Use of iron oxide nanomaterials in wastewater treatment: a review. Sci Total Environ 424:1–10

Xu P, Ma W, Han H, Jia S, Hou B (2014) Isolation of a naphthalene-degrading strain from activated sludge and bioaugmentation with it in a MBR treating coal gasification wastewater. Bull Environ Contam Toxicol 94:1–7

Yan G, Viraraghavan T (2001) Heavy metal removal in a biosorption column by immobilized *M. rouxii* biomass. Bioresour Technol 78:243–249

Yang Y, Chen RF, Shiaris MP (1994) Metabolism of naphthalene, fluorene, and phenanthrene: preliminary characterization of a cloned gene cluster from *Pseudomonas putida* NCIB 9816. J Bacteriol 176:2158–2164

Zhang D, Zeng X, Li W, He H, Ma P, Falandysz J (2011) Selection of optimum formulation for biosorbing lead and cadmium from aquatic solution by using PVA-SA's immobilizing *Lentinus edodes* residue. Desalination Water Treat 31:107–114

Zhao H-P, Wu Q-S, Wang L, Zhao X-T, Gao H-W (2009) Degradation of phenanthrene by bacterial strain isolated from soil in oil refinery fields in Shanghai China. J Hazard Mater 164:863–869

Zhou L, Bai X, Li Y, Ma P (2008) Immobilization of micro-organism on macroporous polyurethane carriers. Environ Eng Sci 25:1235–1242

Zhou L, Li Y, Bai X, Zhao G (2009) Use of microorganisms immobilized on composite polyurethane foam to remove Cu (II) from aqueous solution. J Hazard Mater 167:1106–1113

7

Combining biophysical methods to analyze the disulfide bond in SH2 domain of C-terminal Src kinase

Dongsheng Liu[1,2], David Cowburn[2 ✉]

[1] iHuman Institute, ShanghaiTech University, Shanghai 201203, China
[2] Department of Biochemistry, Albert Einstein College of Medicine, Bronx, NY 10461, USA

Abstract The Src Homology 2 (SH2) domain is a structurally conserved protein domain that typically binds to a phosphorylated tyrosine in a peptide motif from the target protein. The SH2 domain of C-terminal Src kinase (Csk) contains a single disulfide bond, which is unusual for most SH2 domains. Although the global motion of SH2 domain regulates Csk function, little is known about the relationship between the disulfide bond and binding of the ligand. In this study, we combined X-ray crystallography, solution NMR, and other biophysical methods to reveal the interaction network in Csk. Denaturation studies have shown that disulfide bond contributes significantly to the stability of SH2 domain, and crystal structures of the oxidized and C122S mutant showed minor conformational changes. We further investigated the binding of SH2 domain to a phosphorylated peptide from Csk-binding protein upon reduction and oxidation using both NMR and fluorescence approaches. This work employed NMR, X-ray cryptography, and other biophysical methods to study a disulfide bond in Csk SH2 domain. In addition, this work provides in-depth understanding of the structural dynamics of Csk SH2 domain.

Keywords C-terminal Src kinase, Src homology 2, Disulfide bond, Nuclear magnetic resonance

INTRODUCTION

C-terminal Src kinase (Csk) and Csk-homologous kinase (Chk) are members of the CSK family of protein tyrosine kinases. These proteins suppress the activity of Src family kinases (SFKs) by selectively phosphorylating the conserved C-terminal tail regulatory tyrosine (Nada et al. 1991, 1993; Chong et al. 2005, 2006). Csk and Chk both contain SH3, SH2, and kinase domains, which are separated by the SH3–SH2 and SH2-kinase linkers (Fig. 1A). The Csk SH2 domain is crucial in stabilizing the kinase domain in the active conformation (Shekhtman et al. 2001; Mikkola and Gahmberg 2010; Grebien et al. 2011). A disulfide bond in the SH2 is suggested to regulate Csk kinase activity (Mills et al. 2007), although the extent is possibly highly assay-specific (Kemble and Sun 2009). The subcellular localization and activity of Csk are also regulated by its SH2 domain (Chong et al. 2005).

Interactions between SH2 domain and tyrosine kinase domain regulate tyrosine kinase signaling networks (Wong et al. 2005; Ia et al. 2010; Mikkola and Gahmberg 2010). With regard to this, Wojcik et al. (2010) for instance described a potent and highly specific FN3 monobody binding to the Abl SH2 domain, which inhibits the kinase (Grebien et al. 2011). The results showed that intramolecular interaction between the SH2 and kinase domains in Bcr-Abl is both necessary and sufficient for the high catalytic activity of the enzyme. Disruption of this interface inhibits the downstream events critical for chronic myelogenous leukemia signaling.

Disulfide bonds are mostly found in secretory proteins and in extracellular domains of membrane proteins. Cytosolic proteins, which contain cysteine residues that are in close proximity to each other, may

✉ Correspondence: david.cowburn@einstein.yu.edu (D. Cowburn)

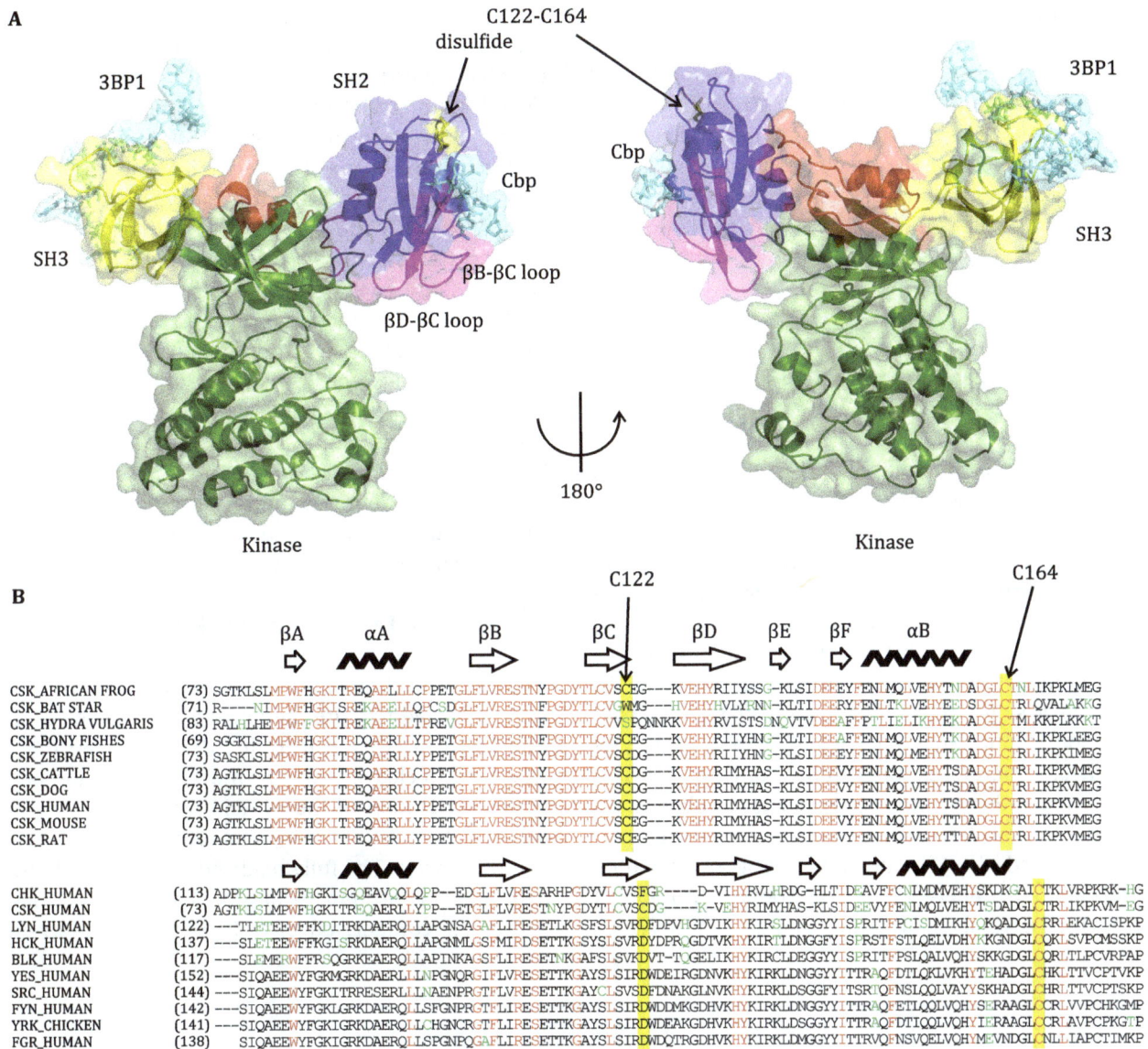

Fig. 1 Structure of Csk and sequence alignment of Csk and Src family kinase. **A** Monomeric structure of Csk with bound 3BP1 (to SH3) and CBP peptide (to SH2). The overall structure is plotted with the active Csk1K9A-A. The positions of PEP-3BP1 and CBP peptides are modeled from the structure of 1JEG and 1SPS, respectively. **B** Sequence alignments of Csk family (*upper*) and Csk, Chk and SFKs (*lower*)

function as oxidation sensors; when the reductive potential of the cell fails, they oxidize and trigger cellular response mechanisms (Sevier and Kaiser 2002). Mills et al. studied the unique disulfide bond of Csk SH2 that is absent in other known SH2 domains. The kinase activity of full-length Csk is apparently reduced by an order of magnitude upon formation of the disulfide bond in the SH2 domain (Mills et al. 2007). Disulfide bond formation is speculated to exert considerable effects on residues within the kinase domain, most notably within the active-site cleft. Given that most cellular compartments exhibit a reducing environment, disulfide bonds are possibly reduced in the cytosol

(Sevier and Kaiser 2002). SH2 sequence alignments from different Csks, Chks, and SFKs show that C122 and C162 are found in most Csks (Fig. 1B).

In this work, we used multiple biophysical methods to investigate the SH2 disulfide bond and the interaction of SH2 with a phosphotyrosine ligand. Comparison of the NMR chemical shift of the oxidized and reduced SH2 domain reveals the major difference that appeared around residue C122 compared with small chemical shift perturbation around residue C164. Denaturation studies have suggested that disulfide bond contributes significantly to the stability of the SH2 domain. Binding affinity of SH2 toward Csk-binding protein (Cbp)-phosphorylated peptide was studied. The reduced and oxidized forms of Csk can both bind with a Cbp peptide, with the reduced SH2 showing a slightly stronger affinity. Crystal structures of both oxidized SH2 and C122S SH2 were solved and refined. Comparison of the crystal structure of the different forms of SH2 suggests that only minor structural changes resulted from the disulfide bond. Analysis of biophysical data of the unusual disulfide bond provided insights into the role of the disulfide bond in SH2 domain.

RESULTS

SH3–SH2 linker contributes to the stability of SH2 domain

According to Pfam (Punta et al. 2012), the SH2 domain of Csk begins at residue W82. We constructed our first SH2 domain using the residues M80–A178. Two set of peaks were observed in ^{1}H–^{15}N HSQC spectrum of purified, uniform ^{15}N-labeled reduced sample, such as the side chain peak of W82 (Fig. 2A). We postulated that the conformational heterogeneity results from *cis-trans* isomerization of the nearby proline P81. We introduced a single point mutation in P81A–SH2 construct and found that the isomerization was not diminished. Therefore, the isomerization probably comes from other residues. Previous studies have suggested that an interaction exists between the SH3–SH2 linker and the SH2 domain (Wong et al. 2005; Mikkola and Gahmberg 2010), so a longer version of SH2 (referred to as L-SH2 and contains A73–A178)

was constructed. With the addition of this part of the SH3–SH2 linker into the SH2 domain, isomerization was essentially abolished. This phenomenon made the detailed investigation of the SH2 disulfide bond and ligand binding via NMR practical. The X-ray structure of the oxidized form of L-SH2 (PDB ID: 3EAC) showed that structure of the N-terminal linker area (A73–P81) is well defined (Fig. 2B) with traceable electron density compared with the flexible C-terminus (E174–A178). The crystal structure also revealed that the linker region fold back on the SH2 domain (Fig. 2C). Hydrophobic interactions between M80 and W82 fix the linker position. Additionally, the hydrophobic interactions between L77 in the linker and L149, F83, and Y116 make the linker fold back to the SH2 and thus stabilizing the structure (Fig. 2B). Contact between H84 and M173 was also observed, and this interaction made the N- and C-terminals spatially close to each other. Thus, compared with previous studies, the present work used this longer version of Csk SH2.

Oxidation and reduction of Csk L-SH2

During purification of the Csk L-SH2 domain, 50 mmol/L DTT was used to elute the protein from the chitin resin. The L-SH2 was used in reduced form, given that DTT was used to cleave L-SH2 from its intein fusion partner. Following elution of the protein from the ion exchange column, the L-SH2 became a mixture of the reduced and oxidized forms. The fully oxidized form was obtained by exposing the protein to air for several days. Figure 3 shows the comparison of NMR spectra of the reduced

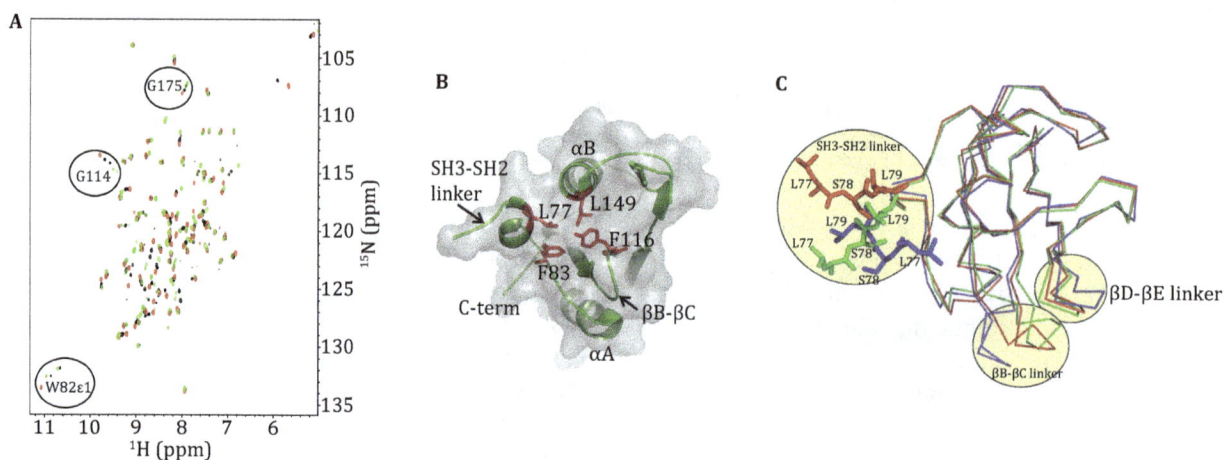

Fig. 2 ^{1}H–^{15}N HSQC spectra of different Csk SH2 domains and SH3–SH2 linker region in the crystal structure. **A** SH2 containing M80–A178 (*blue*), P81A SH2 (*green*), and L(Linker)-SH2 containing A73–A178 (*red*). **B** Crystal structure of L-SH2 showing the hydrophobic interaction between SH3–SH2 linker and L149, F83, and Y116 in the SH2 domain. **C** Comparison of SH2 structures in active (*green*), inactive (*red*) full length (PDB ID: 1K9A), and isolated form (*blue*) (PDB ID: 3EAC)

and oxidized forms of L-SH2. The oxidized form was obtained from the purified protein dissolved in water at approximately 1 mg/mL and then exposed to air for approximately a week at room temperature and pH 7.2. ^{1}H–^{15}N HSQC spectra show that the peaks associated with the reduced form disappeared after oxidation, whereas the oxidized peaks appeared. We found that 10 mmol/L DTT is sufficient to reduce the SH2 domain within 10 h, whereas a higher DTT concentration (150 mmol/L) was used in other studies (Mills et al. 2007). Figure 3C shows the oxidized form of SH2 with 10 mmol/L DTT, and the $t_{1/2}$ of the reduction reaction was 2.6 ± 0.1 h.

The differences in chemical shifts were measured after assigning both the spectra of the oxidized and reduced forms (Fig. 3B). Most of the changes in chemical shift took place in the residues close to the disulfide bond area, especially near the residue C122. This finding suggested that the C122 area was likely to undergo greater structural changes compared with the C164 residue area.

The NMR chemical shifts of ^{13}C$^\alpha$ and ^{13}C$^\beta$ of cysteine residue can be discriminated between cysteine in its reduced and oxidized states (Sharma and Rajarathnam 2000). The observed C$^\alpha$ shifts for oxidized and reduced forms cysteine were 55.5 ± 2.5 and 59.3 ± 3.2 ppm, respectively. The C$^\beta$ chemical shifts of reduced and oxidized cysteine spanned a wider range. The observed C$^\beta$ shifts for the oxidized and reduced cysteine were 40.7 ± 3.8 and 28.4 ± 2.4 ppm, respectively. All of the ^{13}C$^\alpha$ and ^{13}C$^\beta$ chemical shifts of cysteine residues are listed in Table 1 with redox status of the residue indicated in brackets. Residue C119 cannot form a disulfide bond in all of the listed conditions and thus the chemical shift of C119 remained unchanged under either the reduced or oxidized condition. The chemical shift of the ^{13}C$^\alpha$ of C122 and C164 changed into −5.7 and −0.5 ppm upon the formation of the disulfide bond, respectively. For ^{13}C$^\beta$ of C122 and C164, the changes in chemical shift were 16.6 and 13.9 ppm upon the formation of the disulfide bond, respectively. Considerably

Table 1 ^{13}C$^\alpha$ and ^{13}C$^\beta$ chemical shift of all cysteine and cysteine residues in L-SH2 domain

	C119	C122	C164
C$^\alpha$(ox)	57.1(S–H)	55.1(S–S)	58.1(S–S)
C$^\alpha$(red)	57.1(S–H)	60.8(S–H)	58.6(S–H)
C$^\alpha$(red)–Cbp	56.6(S–H)	60.8(S–H)	58.6(S–H)
C$^\beta$(ox)	29.3(S–H)	45.7(S–S)	42.0(S–S)
C$^\beta$(red)	29.4(S–H)	29.1(S–H)	28.1(S–H)
C$^\beta$(red)–Cbp	29.3(S–H)	29.2(S–H)	28.3(S–H)

"Red" or "ox" status of the residue is indicated in bracket

slight changes in ^{13}C chemical shift of these three residues were associated with Cbp peptide binding, suggesting no direct interaction exists between Cbp and the C122–C164 disulfide bond region.

Disulfide bond formation can significantly enhance the thermal stability of SH2 domain

Intrinsic fluorescence of oxidized and reduced form of L-SH2 domain, as well as that of the cysteine mutants C122S and C164S L-SH2, was measured to examine the role of the disulfide bond in the stability of the Csk SH2 domain. The fluorescence spectra of all the mutants were similar to the spectrum of the wild-type domain, which displays a prominent tryptophan maximum emission and excitation at 320 and 288 nm, respectively. Equilibrium denaturation curves exhibit a single, cooperative transition indicative of two-state unfolding. The thermodynamic stability of these oxidized and reduced L-SH2 domains, as well as that of the mutants at equilibrium, were characterized by determining the free energy of unfolding by using guanidine hydrochloride-induced unfolding experiments monitored by tryptophan fluorescence (Fig. 4A). The curves were fitted using the linear extrapolation model to obtain the ΔG for unfolding in H$_2$O (Santoro and Bolen 1988). The free energies $\Delta G^0_{U,W}$ of the reduced, oxidized, C122S and C164S L-SH2 are 27.5 ± 0.7, 42.2 ± 1.6, 29.4 ± 0.9, and 26.3 ± 0.9 kJ/mol, respectively. All of the three forms of L-SH2 without disulfide bond were greatly destabilized compared with L-SH2 containing disulfide bonds with free energy value reductions ranging from 12.8 to 15.9 kJ/mol. Interestingly, substitutions of the C122 and C164 with Ser residues caused a different destabilization. The free energy measurements were also consistent with conservation of the C122 and C164 (Fig. 1B). Formation of the disulfide bond assisted in the packing of αB to the hydrophobic region of the protein consisting of βC and βD (Fig. 4B). The S–S distance of the disulfide bond was 2.0 Å, whereas the O–S distance in the C122S mutant was 3.7 Å (Fig. 4B). As the structure refinement for the oxidized form of L-SH2 progressed, the maps indicated dual conformations for the residue C122 (Fig. 4C), which were successfully modeled as such. Therefore, the high-resolution structures of oxidized L-SH2 reveal dual conformations of disulfide bond that were not observed in the lower-resolution structure of full-length Csk. X-ray crystallography data were also consistent with the NMR observation, in which most of the changes in chemical shift took place in the residues close to C122, and this area was likely to undergo greater structural change than in the C164 residue area (Fig. 3B). Although the side chain

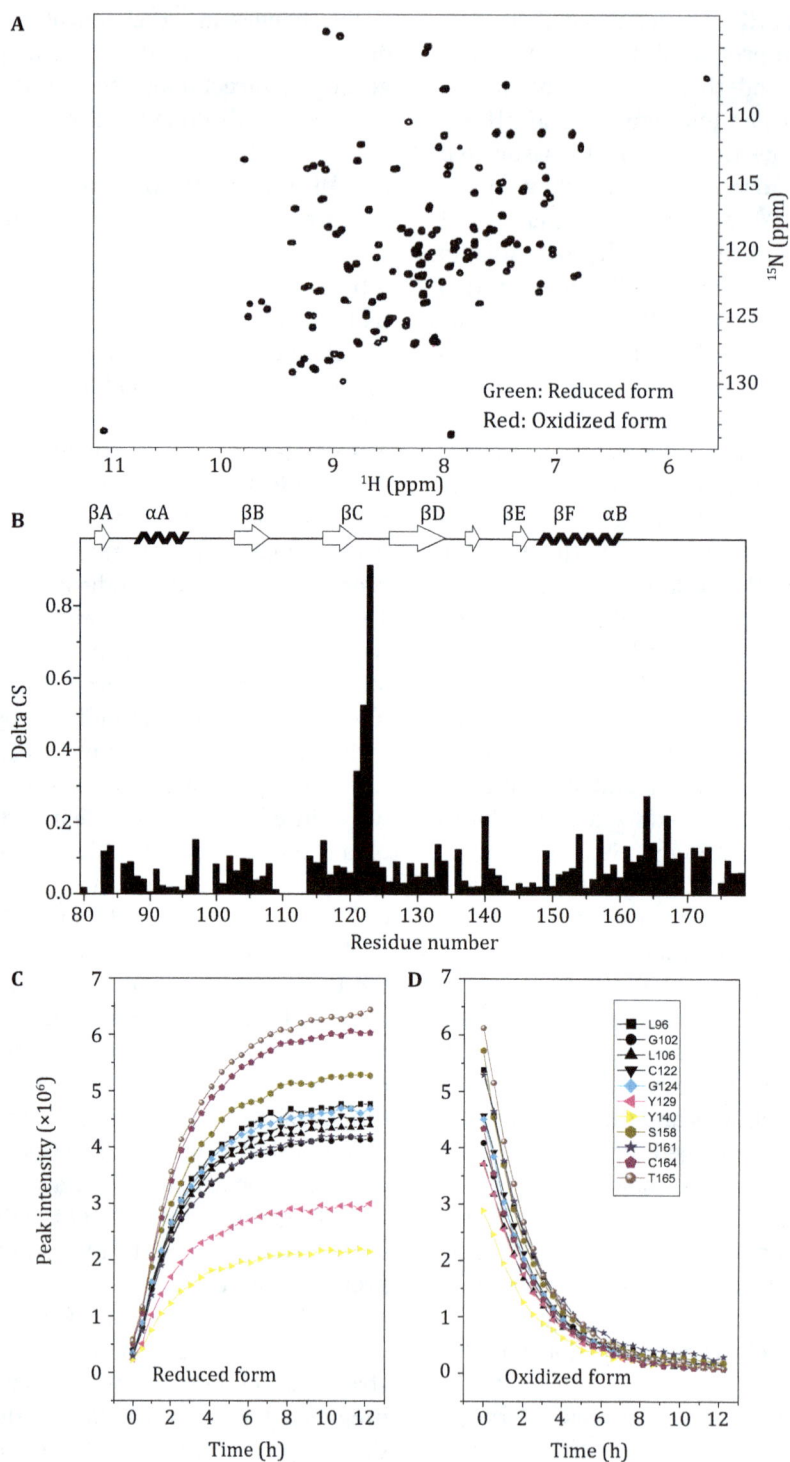

Fig. 3 Comparison of the oxidized and reduced form of SH2. **A** Comparison of ^{1}H–^{15}N HSQC of oxidized (*red*) and reduced (*green*) SH2. **B** Combined chemical shift change of oxidized and reduced form SH2. The difference was calculated as $\Delta\delta_{\text{tot}} = [(\Delta\delta_{\text{H}})^2 + (0.154\Delta\delta_{\text{N}})^2]^{1/2}$. **C** Increase in peak intensity of the reduced form of SH2 upon addition of 10 mmol/L DTT. **D** Reduction in peak intensity of oxidized form of SH2 upon addition of 10 mmol/L DTT. The buffer contains 50 mmol/L Tris–HCl (pH 7.5) with and without 10 mmol/L DTT, 1.0 mmol/L EDTA, 0.01% (*w*/*v*) NaN$_3$, 5% D$_2$O, and 0.1 mmol/L DSS

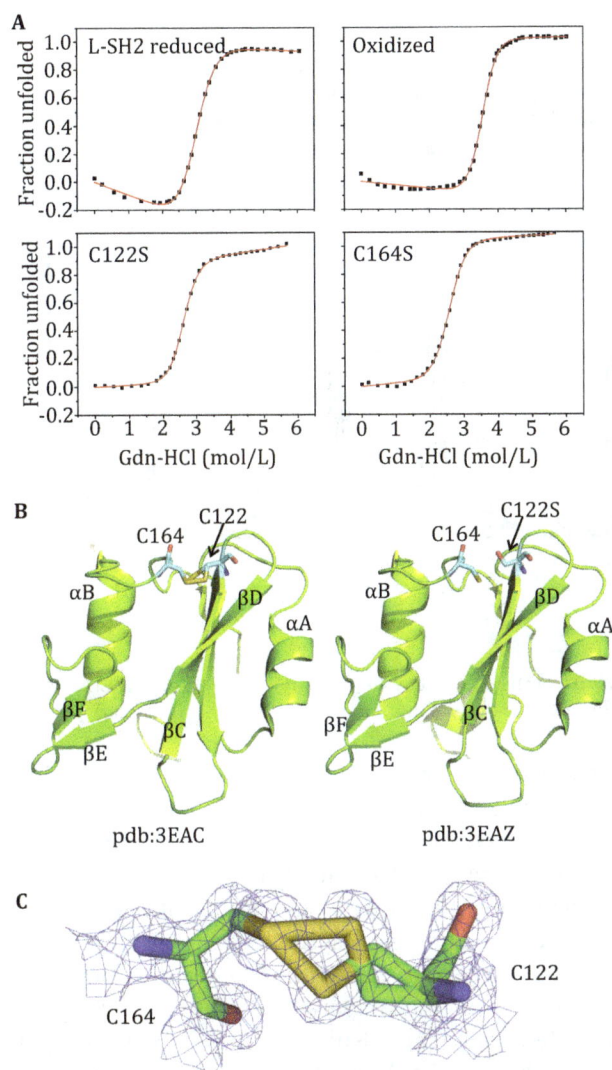

Fig. 4 Guanidine hydrochloride denaturation of reduced, oxidized, C122S, and C164S SH2 domain of Csk. **A** The free energies $\Delta G_{U,W}^0$ of reduced, oxidized, C122S, and C164S L-SH2 were 27.5 ± 0.7, 42.2 ± 1.6, 29.4 ± 0.9, and 26.3 ± 0.9 kJ/mol, respectively. **B** Comparison of crystal structures of oxidized (*left*) and C122S mutant (*right*) of L-SH2. **C** Electron density map for disulfide bond C122–C164. The density is contoured at the 1 σ level

shows the binding of Cbp phosphopeptide with Csk L-SH2. Figure 5A shows the overlay of ^{1}H–^{15}N HSQC spectra of the Csk L-SH2 with (red) and without (green) Cbp peptide. Figure 5B shows the differences in the chemical shift of the two spectra plotted against the residue number. The perturbed residues on the Csk L-SH2 suggested that βB-βC loop, βD, and βD-βE loop underwent significant conformational changes following Cbp peptide binding. N111 and Y112 were absent from ^{1}H–^{15}N HSQC spectrum in the absence of ligand; therefore, data on chemical shift perturbation were not available for these two residues. In addition, βA–αA loop residues at the N-terminal underwent smaller chemical shift perturbation upon binding of the phosphopeptide. Comparison of the SH2 structures in putative active full length, inactive full length (Ogawa et al. 2002), and isolated form (PDBs 1K9A-A, 1K9A-C, and 3EAC) also showed that the position of βD-βE and βB-βC loops significantly changed among the active form, inactive form, and isolated SH2. These results suggested that Cbp binding caused a conformational change in the βB-βC, βD, and βD-βE loops and may possibly adjust the conformation in the kinase domain via the contact proposed in a previous study (Ogawa et al. 2002).

Reduced L-SH2 binds to the phosphorylated tyrosine ligand with slightly stronger affinity than oxidized L-SH2

To study the interaction between L-SH2 and Cbp, we used a 10-residue peptide with the sequence ISAM-pYSSVNK derived from human Cbp protein (Wong et al. 2005). Figure 6A shows the binding of the 10-amino acid peptide ligand with L-SH2. The dissociation constants K_d extracted from each titration curve were 0.52 ± 0.05 and 0.99 ± 0.08 μmol/L for the reduced and oxidized SH2, respectively, suggesting that the reduced form can bind slightly more efficiently than the oxidized form. To compare the relative binding constant of the reduced and oxidized forms in a single experiment, we used the ^{1}H–^{15}N HSQC spectrum of the mixture of the oxidized and reduced form of L-SH2 (200 μmol/L). Cbp was titrated with L-SH2 at a final concentration of 100 μmol/L. Well-resolved Y129 and G162 peaks were used to calculate $\frac{Kd_{ox}}{Kd_{red}}$, obtaining 1.5 ± 0.2 and 2.1 ± 0.1, respectively (Fig. 6B, C). The NMR result confirms that the reduced form of L-SH2 binds to Cbp slightly stronger than the oxidized form. The relative ratio of the different forms of peaks observed did not change with time, suggesting that Cbp binding did not alter the dynamics of the formation and breakage of the disulfide bond. Cbp peptide binding of

direction of C122S residue was slightly different, the overall backbone position was nearly identical to that of the oxidized form and superimposes on the structures of Cα atoms that display a root-mean-square deviation of 0.19 Å.

Csk SH2 is perturbed upon Cbp peptide binding

To study the interaction of Csk L-SH2 with Cbp peptide, we performed 3D triple-resonance NMR experiments on L-SH2 samples with and without Cbp peptide. Figure 5

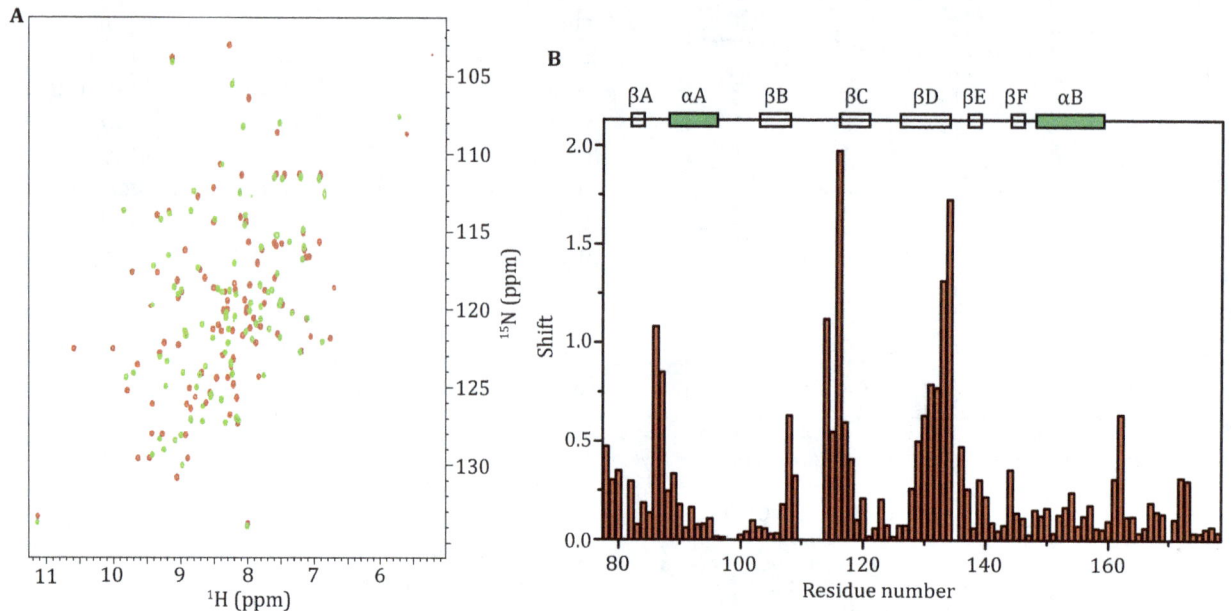

Fig. 5 Binding of Cbp peptide with Csk SH2. **A** Overlay of ^{1}H–^{15}N-HSQC spectra of the Csk SH2 with (*red*) and without (*green*) Cbp peptide. **B** Combined chemical shift change of SH2 with and without Cbp. The difference was calculated as $\Delta_{tot} = [(\Delta_H)^2 + (0.154\Delta_N)^2]^{1/2}$

the activation status of kinase domain may be modulated by selecting the reduced form over the oxidized form rather than directly changing the redox status of the Csk-SH2.

DISCUSSION

Intramolecular interactions between the SH2 domain and tyrosine kinase domain are critical in regulation of catalytic activity. Destabilizing mutations in the SH2 domain and linker region often cause diseases (Filippakopoulos et al. 2009). Studies have suggested that the SH3–SH2 linker interacts with the SH2 domain (Shekhtman et al. 2001; Wong et al. 2005; Mikkola and Gahmberg 2010), and this phenomenon was confirmed by our study. In the crystal structure of the L-SH2, this linker region folds back on the SH2 domain during the hydrophobic interactions, suggesting that the SH3–SH2 domain linker is necessary to stabilize the SH2 domain.

Studies have suggested that the kinase activity of full-length Csk decreases by an order of magnitude upon the formation of the disulfide bond in the distal SH2 domain. Prevention of the reduction of disulfide bond leads to a tenfold reduction in kinase activity, which can be restored upon re-introduction of the reducing agent into the reaction (Mills et al. 2007). Direct and specific inactivation of protein tyrosine kinases in the Src and FGFR families through reversible cysteine oxidation is also observed in alternative procedures, where Csk is

mildly activated by DTT (Kemble and Sun 2009). These results suggest that direct redox regulation possibly exists in specific PTKs. In this study, the thermodynamic stability of the oxidized and reduced L-SH2 domains, as well as that of the mutants at equilibrium, was characterized by determining the free energy of unfolding. All of the three forms of SH2 without a disulfide bond were greatly destabilized compared with the SH2 containing a disulfide bond. The disulfide bond formation significantly enhances the thermal stability of SH2 domain and assists in the packing of αB into the hydrophobic region of the protein consisting of βC and βD. The overall backbone position of C122S and the oxidized form in the crystal structure is nearly identical, suggesting that the disulfide stability results from reduction of entropy of the unfolded state.

Csk interacts with Cbp/PAG, which is localized in membrane microdomains enriched with cholesterol, glycosphingolipids, and lipid rafts and is subsequently recruited into the reaction space (Kawabuchi et al. 2000). The interaction occurs between the SH2 domain of Csk and the SFK-phosphorylated Tyr-314 of Cbp (Kawabuchi et al. 2000). In our study, a 10-residue Cbp peptide bound with Csk SH2 with affinity at the μmol/L level, and the chemical shift perturbation pattern was similar to that of the long Cbp peptide (Tanaka et al. 2013), suggesting that the pY314 area was dominant during the interaction between SH2 and Cbp. Compared with the large chemical shift perturbation in L-SH2 caused by Cbp peptide binding, formation of the

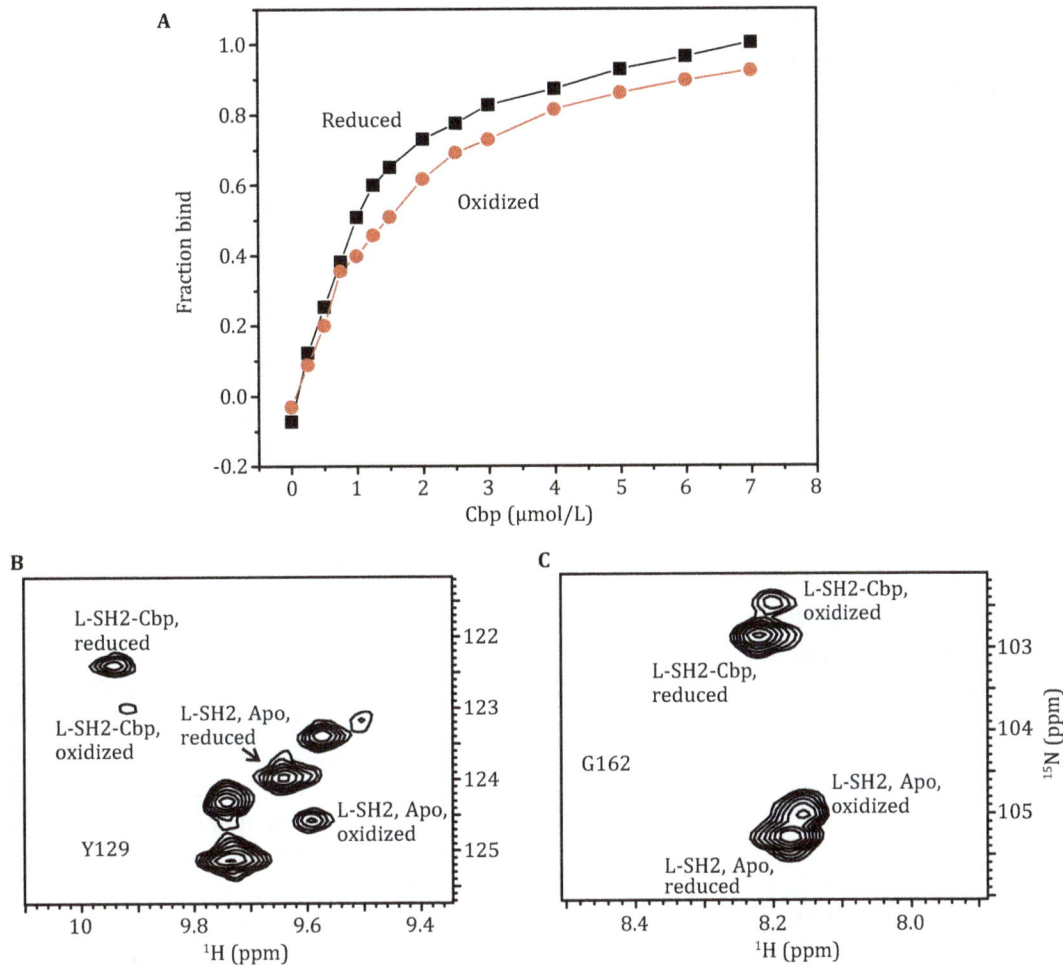

Fig. 6 Cbp peptide binding to Csk L-SH2 monitored by fluorescence and NMR. **A** Titration of reduced (*green*) and oxidized (*red*) Csk L-SH2 with Cbp peptide. **B** and **C** HSQC of the mixture of oxidized and reduced form of SH2 (200 μmol/L). Cbp was added up to a final concentration of 100 μmol/L. The residues showing well-resolved four peaks (Y129 and G162 were chosen) can be used in the calculation of relative affinity

disulfide bond caused only subtle changes in the L-SH2 domain. In addition, affinity and NMR measurement suggested that the reduced form of L-SH2 can bind with Cbp peptide slightly more efficiently than the oxidized form. Our results also show that Cbp binding did not alter the dynamics of formation and breakage of the disulfide bond; therefore, Cbp peptide binding did not change the redox status of the Csk-SH2. The present work illustrates the combination of NMR, X-ray cryptography, and other biophysical methods in investigating the disulfide bond in SH2 domain. Moreover, this work provides a more complete understanding of the structure and dynamics of SH2 domain. However, mechanistic details linking the disulfide bond to the regulation of Csk kinase activity remain elusive until further structural study of the disulfide bond in the context of full-length Csk.

MATERIALS AND METHODS

Protein expression and purification

The human Csk SH2 domain genes containing the residues A73–A178 were amplified and cloned into the expression vector pTWIN1 (New England Biolabs) as described previously (Liu et al. 2009). The plasmid was transformed into a BL21 (DE3) RIL component cell (Stratagene, 230245). The cells were grown to mid-log phase and induced with 0.5 mmol/L IPTG at 25 °C overnight. After centrifugation, the cells were resuspended in 35 mL of Buffer A (50 mmol/L Tris–HCl, pH 7.5, 200 mmol/L NaCl) and passed through a French pressure cell twice. The cell lysate was centrifuged at 12,000*g* for 20 min at 4 °C. The clarified cell extract was loaded into 6 mL of chitin beads. The cleavage of the intein-tag was induced by equilibrating the chitin beads with 50 mmol/L DTT, 50 mmol/L KH_2PO_4–K_2HPO_4 at pH 7.2 for 24 h. The target protein was eluted and further purified by a Mono Q column on Äkta system (GE Healthcare). The Cbp peptide (ISAMpYSSVNK) was synthesized by GenScript (Piscataway, NJ) using standard solid-phase peptide synthesis methods and then resuspended in purified water before use.

NMR spectroscopy

All NMR experiments were performed at 298 K on 800 MHz spectrometers equipped with triple-resonance cryoprobes. Protein solutions were prepared under the following buffer conditions: 0.8 mmol/L protein in 50 mmol/L Tris–HCl (pH 7.5), 1.0 mmol/L EDTA, 0.01% (w/v) NaN_3, 5% D_2O, and 0.1 mmol/L DSS (4,4-dimethyl-4-silapentane-1-sulfonate). DTT (10 mmol/L) was added whenever necessary. In the 3D triple-resonance experiments, HNCA, HNCO, HNCACB, and CBCA(CO)NH were collected for backbone resonance assignment. The ^1H chemical shifts were referenced to internal DSS. The ^{13}C and ^{15}N chemical shifts were referenced indirectly using the $^1H/^{13}C$ or $^1H/^{15}N$ frequency ratios of the zero point: 0.101329118 (^{15}N) and 0.251449530 (^{13}C) (Live et al. 1984; Wishart et al. 1995). The combined change in chemical shift of a particular residue upon ligation with the kinase domain was calculated as $\Delta\delta = [(\Delta\delta_H)^2 + (0.154\Delta\delta_N)^2]^{1/2}$, where $\Delta\delta_H$ and $\Delta\delta_N$ correspond to the changes in amide proton and nitrogen chemical shift, respectively. The weight factor for ^1H and ^{15}N is determined from the ratio of the average variances of the amide nitrogen and proton chemical shifts observed for the 20 common amino acid residues in proteins deposited in the BioMagResBank (Mulder et al. 1999). For those residues showing well-resolved four peaks (reduced-bound, reduced-apo, oxidized-bound, and oxidized-apo), the relative binding affinity was calculated as follows:

$$\frac{Kd_{ox}}{Kd_{red}} = \frac{\frac{[SH2_{ox}][Cbp]}{[SH2_{ox}\cdot Cbp]}}{\frac{[SH2_{red}][Cbp]}{[SH2_{red}\cdot Cbp]}} = \frac{[SH2_{ox}][SH2_{red}\cdot Cbp]}{[SH2_{red}][SH2_{ox}\cdot Cbp]}.$$

Peak intensity represents the relative concentration of each fraction.

Crystallization, data collection, and structural analysis

The oxidized form of the Csk L-SH2 domain or the C122S mutant was crystallized using the hanging drop vapor diffusion method. Protein solution (1 μL, 15 mg/mL) was mixed with reservoir solution [1 μL, 100 mmol/L Bis-Tris (pH 7.3), 22% PEG 4000] and incubated in reservoir solution (1 mL) at 25 °C. Within one day, crystals typically grew as rod-shaped structures with dimensions 300 μm × 20 μm × 20 μm. X-ray diffraction data were collected, integrated, and scaled using HKL2000 (Otwinowski and Minor 1997). Structure-factor amplitudes were calculated using TRUNCATE (Bailey 1994). Data on diffraction were consistent with the orthorhombic space group P_{212121}, with unit cell dimensions as follows: $a = 37.1$ Å, $b = 48.0$ Å, $c = 50.0$ Å and $a = 37.5$ Å, $b = 48.1$ Å, $c = 49.9$ Å for the oxidized and C122S mutant, respectively. Unless otherwise stated, all programs used for structural and crystallographic analyses were located within the CCP4 interface of the CCP4 suite (Bailey 1994). The structure of Csk L-SH2 or C122S mutant domain was solved using the molecular replacement method. Initial phases were obtained using MOLREP, and the co-ordinates of the SH2 domain with Protein Data Bank (PDB) entry code 1K9A were used as search models (Ogawa et al. 2002). Manual model rebuilding was performed in Coot (Emsley and Cowtan 2004), and maximum likelihood refinement was performed using REFMAC5 (Murshudov et al. 1997). Ordered water molecules were initially added into the model using Coot and eventually added manually. The program PROCHECK (Laskowski et al. 1993) was used to assess the quality of the final structures. Data collection and refinement statistics are shown in Table 2. Co-ordinates were deposited in PDB under accession codes 3EAC (oxidized) and 3EAZ (C122S mutant).

Fluorescence measurement

Fluorescence titrations were performed at 25 °C in 3 mL of 50 mmol/L Tris-HCl (pH 7.5) using excitation and emission wavelengths of 288 nm and 320 nm, respectively. The 10-residue synthetic phosphopeptide corresponding to the specific SH2-binding site in Cbp was titrated with the L-SH2 sample. The typically adopted protein concentration is 5 μmol/L. For the denaturation experiments, ultrapure guanidine hydrochloride (Sigma S0933) was used. Proteins were exposed to guanidine hydrochloride concentrations ranging from 0 to 6 mol/L at 0.1–0.2 mol/L steps. The concentrations of guanidine hydrochloride of the solutions were determined by measuring the index of refraction and by using the following equation: $d/d_0 = 1 + 0.2710\, W + 0.0330\, W^2$ (Tanford et al. 1966), where W is the weight fraction of guanidine hydrochloride in the solution, d is the density of the solution, and d_0 is the density of water. Reversibility of the guanidine hydrochloride-induced unfolding reaction and the time necessary to re-establish equilibrium were

Table 2 Data collection and refinement statistics

	Csk L-SH2 oxidized	Csk L-SH2 C122S
Cell constant [a, b, c (Å)]	37.1, 48.0, 50.0	37.5, 48.2, 49.9
Resolution from map calculation (Å)	29.45–1.37	25.45–1.31
Space group	P212121	P212121
Number of reflections	18,298	22,354
Completeness (%)	99.7	99.8
Refinement [R (%)/R_{free} (%)]	0.209/0.217	0.196/0.226
Water molecules	106	118
Ramachandran plot (favor/allowed/disallowed)	81.8/18.1/0	87.6/12.4/0

determined by diluting a concentrated protein solution containing 6 mol/L guanidine hydrochloride into the buffer and by monitoring the fluorescence spectrum of the protein as a function of time. The signals were normalized to the fraction of unfolded species using the standard relation $F_{\text{unf}} = (I - I_{\text{N}})/(I_{\text{U}} - I_{\text{N}})$, where N and U stand for the fluorescence intensity of the native and fully unfolded species, respectively. F_{unf} values were calculated from the linear extrapolation of the pre- and post-unfolding baselines. The unfolding free energy and m values in the relationship were obtained from the fitting of the denaturation data into a two-state model using the standard equation for all four L-SH2 protein samples. The apparent free energy of unfolding was determined according to the linear equation $\Delta G_{\text{U,W}}^{0} = \Delta G_{\text{U,D}}^{0} + m[\text{D}]$, where $\Delta G_{\text{U,W}}^{0}$ is the apparent free energy of unfolding in water, $\Delta G_{\text{U,D}}^{0}$ is the apparent free energy of unfolding at a denaturant concentration [D], and m is the coefficient expressing the dependence of the free energy of unfolding on the denaturant concentration.

Accession numbers

PDB number: 3EAC (oxidized) and 3EAZ (C122S mutant). The backbone chemical shifts of SH2 with and without Cbp peptide were deposited in BMRB with accession numbers 7141 and 7140, respectively.

Abbreviations

Csk	C-terminal Src kinase
SH2	Src homology 2
Cbp	Csk-binding protein
SFK	Src family kinase
HSQC	Heteronuclear single quantum correlation

Acknowledgments Dongsheng Liu thanks ShanghaiTech University and Shanghai Municipal Government for their financial support. David Cowburn thanks the NIH GM (66354 and 47021) for their support. We are grateful to New York Structural Biology Center for the NMR resources. We express our gratitude to Ronald D. Seidel, James Love, Wanhui Hu, Gaojie Song, and Ya Yuan for their help and insightful discussions. We thank John A. Schwan of X4C for the assistance in using the National Synchrotron Light Source, Brookhaven National Laboratory.

Compliance with ethical standards

Conflict of interest Dongsheng Liu and David Cowburn declare that they have no conflict of interest.

Human and animal rights and informed consent This article does not contain any studies with human or animal subjects performed by any of the authors.

References

Bailey S (1994) The ccp4 suite: programs for protein crystallography. Acta Crystallogr D 50:760–763

Chong YP, Mulhern TD, Cheng HC (2005) C-terminal Src kinase (CSK) and CSK-homologous kinase (CHK)—endogenous negative regulators of Src-family protein kinases. Growth Factors 23:233–244

Chong YP, Chan AS, Chan KC, Williamson NA, Lerner EC, Smithgall TE, Bjorge JD, Fujita DJ, Purcell AW, Scholz G, Mulhern TD, Cheng HC (2006) C-terminal Src kinase-homologous kinase (CHK), a unique inhibitor inactivating multiple active conformations of Src family tyrosine kinases. J Biol Chem 281:32988–32999

Emsley P, Cowtan K (2004) Coot: model-building tools for molecular graphics. Acta Crystallogr D Biol Crystallogr 60:2126–2132

Filippakopoulos P, Muller S, Knapp S (2009) SH2 domains: modulators of nonreceptor tyrosine kinase activity. Curr Opin Struct Biol 19:643–649

Grebien F, Hantschel O, Wojcik J, Kaupe I, Kovacic B, Wyrzucki AM, Gish GD, Cerny-Reiterer S, Koide A, Beug H, Pawson T, Valent P, Koide S, Superti-Furga G (2011) Targeting the SH2-kinase interface in Bcr-Abl inhibits leukemogenesis. Cell 147:306–319

Ia KK, Mills RD, Hossain MI, Chan K-C, Jarasrassamee B, Jorissen RN, Cheng H-C (2010) Structural elements and allosteric mechanisms governing regulation and catalysis of CSK-family kinases and their inhibition of Src-family kinases. Growth Factors 28:329–350

Kawabuchi M, Satomi Y, Takao T, Shimonishi Y, Nada S, Nagai K, Tarakhovsky A, Okada M (2000) Transmembrane phosphoprotein Cbp regulates the activities of Src-family tyrosine kinases. Nature 404:999–1003

Kemble DJ, Sun G (2009) Direct and specific inactivation of protein tyrosine kinases in the Src and FGFR families by reversible cysteine oxidation. Proc Natl Acad Sci USA 106:5070–5075

Laskowski RA, Macarthur MW, Moss DS, Thornton JM (1993) PROCHECK: a program to check the stereochemical quality of protein structures. J Appl Crystallogr 26:283–291

Liu D, Xu R, Cowburn D (2009) Segmental isotopic labeling of proteins for nuclear magnetic resonance. Methods Enzymol 462:151–175

Live DH, Davis DG, Agosta WC, Cowburn D (1984) Long range hydrogen bond mediated effects in peptides: nitrogen-15 NMR study of gramicidin S in water and organic solvents. J Am Chem Soc 106:1939–1941

Mikkola ET, Gahmberg CG (2010) Hydrophobic interaction between the SH2 domain and the kinase domain is required for the activation of Csk. J Mol Biol 399:618–627

Mills JE, Whitford PC, Shaffer J, Onuchic JN, Adams JA, Jennings PA (2007) A novel disulfide bond in the SH2 domain of the C-terminal Src kinase controls catalytic activity. J Mol Biol 365:1460–1468

Mulder FA, Schipper D, Bott R, Boelens R (1999) Altered flexibility in the substrate-binding site of related native and engineered high-alkaline Bacillus subtilisins. J Mol Biol 292:111–123

Murshudov GN, Vagin AA, Dodson EJ (1997) Refinement of macromolecular structures by the maximum-likelihood method. Acta Crystallogr D Biol Crystallogr 53:240–255

Nada S, Okada M, MacAuley A, Cooper JA, Nakagawa H (1991) Cloning of a complementary DNA for a protein-tyrosine kinase that specifically phosphorylates a negative regulatory site of p60c-src. Nature 351:69–72

Nada S, Yagi T, Takeda H, Tokunaga T, Nakagawa H, Ikawa Y, Okada M, Aizawa S (1993) Constitutive activation of Src family kinases in mouse embryos that lack Csk. Cell 73:1125–1135

Ogawa A, Takayama Y, Sakai H, Chong KT, Takeuchi S, Nakagawa A, Nada S, Okada M, Tsukihara T (2002) Structure of the carboxyl-terminal Src kinase, Csk. J Biol Chem 277:14351–14354

Otwinowski Z, Minor W (1997) Processing of X-ray diffraction data collected in oscillation mode. Method Enzymol 276:307–326

Punta M, Coggill PC, Eberhardt RY, Mistry J, Tate J, Boursnell C, Pang N, Forslund K, Ceric G, Clements J, Heger A, Holm L, Sonnhammer EL, Eddy SR, Bateman A, Finn RD (2012) The Pfam protein families database. Nucleic Acids Res 40:D290–D301

Santoro MM, Bolen DW (1988) Unfolding free energy changes determined by the linear extrapolation method. 1. Unfolding of phenylmethanesulfonyl alpha-chymotrypsin using different denaturants. Biochemistry 27:8063–8068

Sevier CS, Kaiser CA (2002) Formation and transfer of disulphide bonds in living cells. Nat Rev Mol Cell Biol 3:836–847

Sharma D, Rajarathnam K (2000) C-13 NMR chemical shifts can predict disulfide bond formation. J Biomol NMR 18:165–171

Shekhtman A, Ghose R, Wang D, Cole PA, Cowburn D (2001) Novel mechanism of regulation of the non-receptor protein tyrosine kinase Csk: insights from NMR mapping studies and site-directed mutagenesis. J Mol Biol 314:129–138

Stevens R, Stevens L, Price NC (1983) The stabilities of various thiol compounds used in protein purifications. Biochem Educ 11:70

Tanaka H, Akagi K, Oneyama C, Tanaka M, Sasaki Y, Kanou T, Lee YH, Yokogawa D, Dobenecker MW, Nakagawa A, Okada M, Ikegami T (2013) Identification of a new interaction mode between the Src homology 2 (SH2) domain of C-terminal Src kinase (Csk) and Csk-binding protein (Cbp)/phosphoprotein associated with glycosphingolipid microdomains. J Biol Chem 288(21):15240–15254

Tanford C, Kawahara K, Lapanje S (1966) Proteins in 6-M guanidine hydrochloride. Demonstration of random coil behavior. J Biol Chem 241:1921–1923

Wishart DS, Bigam CG, Holm A, Hodges RS, Sykes BD (1995) H-1, C-13 and N-15 random coil NMR chemical-shifts of the common amino-acids.1. investigations of nearest-neighbor effects. J Biomol NMR 5:332

Wojcik J, Hantschel O, Grebien F, Kaupe I, Bennett KL, Barkinge J, Jones RB, Koide A, Superti-Furga G, Koide S (2010) A potent and highly specific FN3 monobody inhibitor of the Abl SH2 domain. Nat Struct Mol Biol 17:519

Wong L, Lieser SA, Miyashita O, Miller M, Tasken K, Onuchic JN, Adams JA, Woods VL Jr, Jennings PA (2005) Coupled motions in the SH2 and kinase domains of Csk control Src phosphorylation. J Mol Biol 351:131–143

8

Simulated microgravity potentiates generation of reactive oxygen species in cells

Fanlei Ran[1], Lili An[2], Yingjun Fan[2], Haiying Hang[2✉], Shihua Wang[1✉]

[1] Key Laboratory of Pathogenic Fungi and Mycotoxins of Fujian Province, Key Laboratory of Biopesticide and Chemical Biology of Education Ministry, School of Life Sciences, Fujian Agriculture and Forestry University, Fuzhou 350002, China
[2] Key Laboratory for Protein and Peptide Pharmaceuticals, Institute of Biophysics, Chinese Academy of Sciences, Beijing 100101, China

Abstract   Microgravity (MG) and space radiation are two major environmental factors of space environment. Ionizing radiation generates reactive oxygen species (ROS) which plays a key role in radiation-induced DNA damage. Interestingly, simulated microgravity (SMG) also increases ROS production in various cell types. Thus, it is important to detect whether SMG could potentiate ROS production induced by genotoxins including radiation, especially at a minimal level not sufficient to induce detectable ROS. In this study, we treated mouse embryonic stem (MES) cells with H_2O_2 and SMG for 24 h. The concentration of H_2O_2 used was within 30 µmol/L at which intracellular ROS was the same as that in untreated cells. Exposure of cells to SMG for 24 h did not induce significantly higher levels of intracellular ROS than that of control cells either. Simultaneous exposure of cells to both SMG- and H_2O_2-induced ROS and apoptosis in MES cells. Although incubation in medium containing 5 or 30 µmol/L H_2O_2 induced a small enhancement of DNA double-strand breaks (DSBs), the addition of SMG treatment dramatically increased DSB levels. Taken together, SMG can significantly potentiate the effects of H_2O_2 at a low concentration that induce a small or negligible change in cells on ROS, apoptosis, and DNA damage. The results were discussed in relation to the combined effects of space radiation and MG on human body in this study.

Keywords H_2O_2, SMG, ROS production, DNA damage, Apoptosis

INTRODUCTION

For manned space exploration, it is urgent to investigate the effects of the space environment on human health. Of all the known space environmental factors, microgravity (MG) and space radiation have been recognized as the two major environmental factors. Because of the cost effectiveness and limited access to space flight, simulated microgravity (SMG) on Earth has been widely used in space life research. The integrity of genomic DNA is important for normal physiological functions of cells and DNA damage is related to many diseases such as cancer and aging among others (Lombard et al. 2005; Hoeijmakers 2009). Thus, it is important to investigate the effects of space environment on cellular DNA damage.

It is well known that ionizing radiation (IR) generates reactive oxygen species (ROS) which plays important roles in DNA damage induced by radiation (Tominaga et al. 2004). Interestingly, several lines of evidence showed that SMG increased ROS production in some cell types, such as the PC12 cells, SH-SY5Y cells, and MEF

Fanlei Ran and Lili An have contributed equally to this work.

✉ Correspondence: hh91@ibp.ac.cn (H. Hang), wshyyl@sina.com (S. Wang)

cells (Wang et al. 2009; Qu et al. 2010; Li et al. 2015). It has been reported that SMG delayed the rejoining of double-strand breaks (DSBs) induced by IR and increased the genotoxic effects of IR (Mognato et al. 2009). Mognato et al. also reported that SMG treatment decreased the surviving fraction and increased the *HPRT* mutant frequency in human peripheral blood lymphocytes (Mognato and Celotti 2005). We asked whether SMG could potentiate ROS production and DNA damage induced by space radiation. In the real space environment, space radiation and microgravity act continuously on the body together. Owing to the limitation of the experimental conditions, ionizing radiation and SMG treatment have to be separated into two processes. Thus, in this study, we used H_2O_2 instead of radiation and SMG at the same time and investigated whether simulated microgravity could potentiate ROS generation, DNA damage, and apoptosis. Since radiation level inside a space shuttle or a satellite may be too low to induce ROS, we are particularly interested in the following question: when SMG itself cannot induce ROS in a model cell, and the concentration of H_2O_2 is kept low so ROS cannot be induced by H_2O_2 under 1G, whether SMG can induce ROS in the model cell treated with the low concentration of H_2O_2. So far, there have been no reports on the combined effects of SMG and low concentration of H_2O_2 on ROS production and DNA damage. In this study, we found that SMG exposure for 24 h or H_2O_2 treatment at a concentration below 30 μmol/L for 24 h under 1G could not enhance ROS above untreated mouse embryonic stem (MES) cells, but the combination of these two treatments significantly induced ROS in MES cells. SMG also potentiated the effects of H_2O_2 on DNA damage and apoptosis. The results were discussed in relation to the combined effect of space radiation and MG on human body in this study.

RESULTS

Combined effects of SMG and H_2O_2 on ROS production in wild-type MES cells

To investigate the combined effects of SMG and H_2O_2 in ROS production in wild-type MES cells, H_2O_2 at the indicated concentrations was added to the media of the cells under 1G and SMG, respectively, and the intracellular ROS level was analyzed by 2′,7′2 dichlorodihydrofluorescein diacetate (DCF-DA) staining. As shown in Fig. 1, the relative DCF fluorescence was slightly higher in the cells cultured under SMG than that in the cells cultured under 1G. However, the difference was not

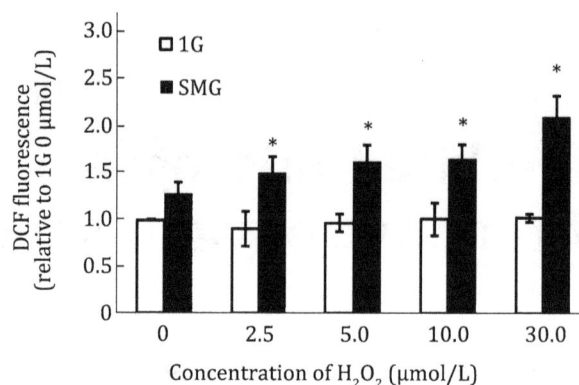

Fig. 1 Effects of SMG and H_2O_2 treatment on ROS production in wild-type MES cells. Wild-type MES cells were cultured under 1G or SMG for 24 h and treated with H_2O_2 at the indicated concentrations at the same time. Then the ROS activity was analyzed with flow cytometry. The experiments were performed thrice independently. The data are shown as mean ± SD. Student's t test, *$p < 0.05$

statistically significant. This was consistent with our previous report (Li et al. 2015). In the cells cultured under 1G, treatment of the cells with low concentrations of H_2O_2 (from 2.5 to 30 μmol/L) did not alter the intracellular ROS production significantly either. Interestingly, at each indicated concentration of H_2O_2, we observed significantly increased intracellular ROS production in the cells cultured under SMG than that in the cells cultured under 1G. These results indicate that SMG triggers ROS production in MES cells incubated in medium containing H_2O_2 at the concentration of 30 μmol/L or lower.

Potentiation of SMG to the effect of H_2O_2 on DNA damage

ROS can inflict DNA lesions (Schieber and Chandel 2014). To investigate the combined effect of SMG and H_2O_2 on DNA damage in MES cells, H_2O_2 was added to the media of the cells cultured under SMG and 1G, respectively, at the indicated concentrations, and the DNA damage was analyzed by comet assay. The comet assay is a sensitive method for measuring DNA lesions in single cells. The amount of DNA migration under electric potential indicates the amount of DNA damage in the cell. As shown in Fig. 2, there was no significant difference in DNA damage between the cells cultured under SMG and those cultured under 1G, which was consistent with our previous report (Li et al. 2015). Although 5 or 30 μmol/L H_2O_2 did not enhance intracellular ROS levels, it was able to cause higher levels of DNA damage under 1G (Fig. 2B). This elevated level of DNA lesions was small but statistically significant (data not shown). In contrast, when treated with 5 or

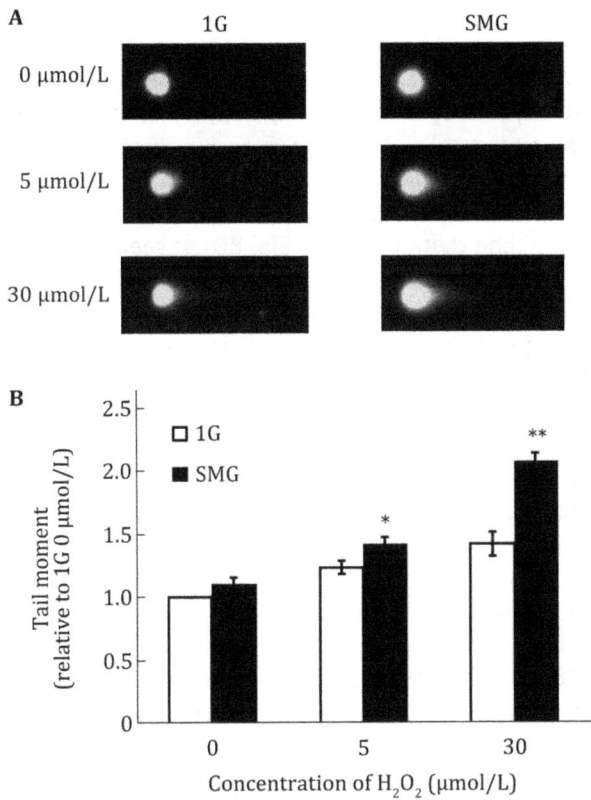

Fig. 2 Effects of SMG and H_2O_2 treatment on DNA damage in wild-type MES cells. Wild-type MES cells were cultured under 1G or SMG for 24 h and treated with H_2O_2 at the indicated concentrations at the same time. Then DNA damage was evaluated using neutral comet assay. The representative results of comet assay are shown in **A** and the quantitative comparison of comet tail moments are shown in **B**. At least 50 cells were scored for analysis of the comet tail moment. The experiments were performed thrice independently. The data are shown as mean $\pm$ SD. Student's t test, $*p < 0.05$, $**p < 0.01$

30 μmol/L H_2O_2, the relative tail moment of the cells cultured under SMG was significantly higher than that cultured under 1G. These results indicate that SMG potentiates the effect of H_2O_2 on DNA damage.

NAC significantly suppresses DNA damage in MES cells treated with both SMG and H_2O_2

N-acetylcysteine (NAC) is a widely used ROS scavenger (Dhouib et al. 2016). As shown in Fig. 3, 1 mmol/L NAC effectively suppressed ROS induced by the combined treatments of H_2O_2 and SMG. 1 mmol/L NAC also effectively suppressed DNA damage induced by the combined treatments of H_2O_2 and SMG (Fig. 4), suggesting that the DNA lesions inflicted by the combined treatments of H_2O_2 and SMG are mediated by ROS production in cells.

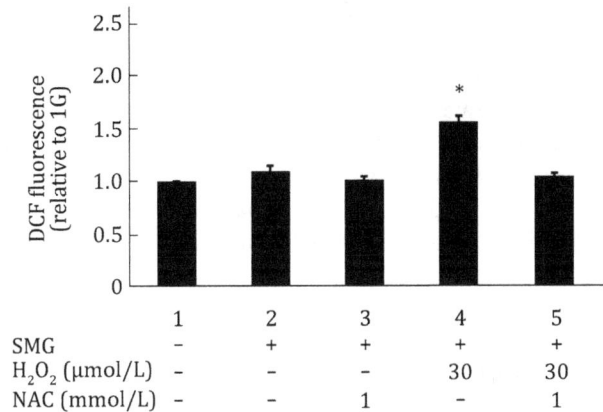

Fig. 3 NAC significantly reduces ROS production in MES cells treated with both SMG and H_2O_2. Flow cytometric analysis of ROS activity in wild-type MES cells. *Column 1* wild-type MES cells cultured under 1G without any treatment; *Column 2* wild-type MES cells cultured under SMG without any treatment; *Column 3* wild-type MES cells maintained under SMG and treated with 1 mmol/L NAC; *Column 4* wild-type MES cells cultured under SMG and treated with 30 μmol/L H_2O_2; *Column 5* wild-type MES cells cultured under SMG and treated with 30 μmol/L H_2O_2 as well as 1 mmol/L NAC. The experiments were performed thrice independently. The data are shown as mean $\pm$ SD. Student's t test, $*p < 0.05$ compared with *Column 1*

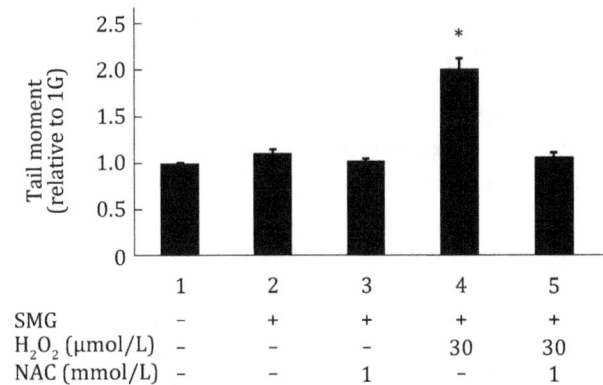

Fig. 4 NAC significantly reduces DNA damage in MES cells treated with both SMG and H_2O_2. DNA damage was assayed by neutral comet assay in wild-type MES cells. *Column 1* wild-type MES cells cultured under 1G without any treatment; *Column 2* wild-type MES cells cultured under SMG without any treatment; *Column 3* wild-type MES cells cultured under SMG and treated with 1 mmol/L NAC; *Column 4* wild-type MES cells cultured under SMG and treated with 30 μmol/L H_2O_2; *Column 5* wild-type MES cells cultured under SMG and treated with 30 μmol/L H_2O_2 as well as 1 mmol/L NAC. The experiments were performed thrice independently. The data are shown as mean $\pm$ SD. Student's t test, $*p < 0.05$ compared with *Column 1*

NAC significantly reduces apoptosis in MES cells treated with both SMG and H_2O_2

As shown above, SMG potentiated the effect of H_2O_2 on DNA damage in MES cells, and DNA damage leads to apoptosis (Zhang et al. 2016). Previously we observed that SMG itself was unable to induce apoptosis in wild-type MES cells (Li et al. 2015).We asked whether SMG could potentiate the effect of H_2O_2 on apoptosis in MES cells. As shown in Fig. 5, SMG triggered apoptosis in MES cells treated with 30 mmol/L H_2O_2 (Fig. 5). NAC treatment effectively reversed the increased apoptosis. Our results indicate that enhanced ROS mediates apoptosis induced by the combined treatments of H_2O_2 and SMG.

DISCUSSION

In this study, we found that SMG exposure alone or H_2O_2 treatment at a low concentration alone could not enhance ROS production in MES cells. However, the combination of these two treatments significantly induced ROS production (Fig. 2). SMG also potentiated the effects of H_2O_2 on DNA damage and apoptosis. Furthermore, ROS scavenger NAC could inverse these effects in MES cells treated with both SMG and H_2O_2 (Figs. 4 and 5).

In mammalian cells, there is a balance of ROS production and scavenging (Aon et al. 2010). A small increase in ROS levels only activates signaling pathways to initiate biological processes, but high levels of ROS also result in damage to DNA, protein, or lipids (Schieber and Chandel 2014). In this study, treatment of the MES cells cultured under 1G with 30 μmol/L H_2O_2 did not significantly alter the intracellular ROS production.

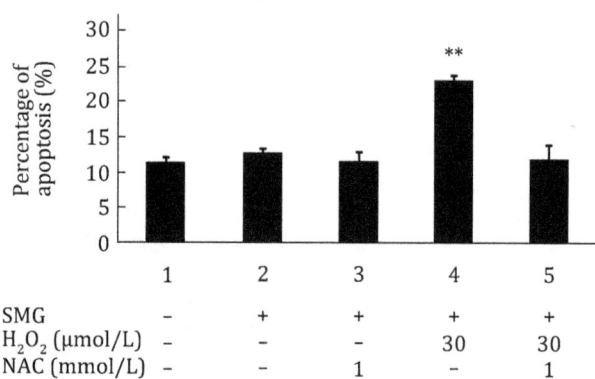

	1	2	3	4	5
SMG	–	+	+	+	+
H_2O_2 (μmol/L)	–	–	–	30	30
NAC (mmol/L)	–	–	1	–	1

Fig. 5 NAC significantly reduces apoptosis in MES cells treated with both SMG and H_2O_2. Flow cytometric analysis of apoptosis in wild-type MES cells. *Column 1* wild-type MES cells cultured under 1G without any treatment; *Column 2* wild-type MES cells cultured under SMG without any treatment; *Column 3* wild-type MES cells cultured under SMG and treated with 1 mmol/L NAC; *Column 4* wild-type MES cells cultured under SMG and treated with 30 μmol/L H_2O_2; *Column 5* wild-type MES cells cultured under SMG and treated with 30 μmol/L H_2O_2 as well as 1 mmol/L NAC. The experiments were performed thrice independently. The data are shown as mean $\pm$ SD. Student's *t* test, **$p < 0.01$ compared with *Column 1*

Consistently, we did not observe increased apoptosis in these cells. It seems that cultured under 1G, MES cells could effectively scavenge the increased ROS induced by 30 μmol/L H_2O_2, avoiding the damage of the oxidative stress to the cells. Although 5 or 30 μmol/L H_2O_2 did not enhance intracellular ROS levels, it was able to cause slightly but also significantly higher levels of DNA damage than that under 1G (Fig. 2B). It seems that H_2O_2 as low as 5 μmol/L could induce increased DNA damage before it was scavenged by the cells.

Wang et al. reported that SMG increased the amount of ROS in rat PC12 cells (Wang et al. 2009). Previously, we observed significant SMG-induced ROS production and DNA damage in $Rad9^{-/-}$ MES but not in wild-type MES cells (Li et al. 2015). In this study, SMG treatment potentiated 30 μmol/L H_2O_2-induced ROS production, as well as DNA damage and apoptosis in wild-type MES cells, indicating the synergistic effects of SMG and H_2O_2. Altogether, these results indicate that SMG is a weak genotoxic stress and could break the balance of ROS production and scavenging under the stress of low dose of H_2O_2. However, the precise mechanisms need further investigation.

In addition to microgravity, space radiation is another key detrimental factor in space environment. It has been reported that the superoxide increased by 16.5% after 15 min of 5 cGy radiation in A549 cells (Chen et al. 2015). Manganese superoxide dismutase (SOD_2) could be induced by IR with the dose as low as 2 cGy (Veeraraghavan et al. 2011). To our knowledge, there is no report on ROS production induced by IR at the doses lower than 2 cGy. It should be noted that at orbital altitudes near that of the International Space Station, the dose-equivalent to the astronauts is about 0.3 Sv per year (an absorbed dose of 1 Gy by alpha particles will lead to an equivalent dose of 20 Sv) (Thirsk et al. 2009), which means the effects of space radiation alone on ROS production may be not significant. However, we have shown that SMG potentiated ROS production induced by low concentrations of H_2O_2. Thus in the real space environment, microgravity may also potentiate space radiation-induced ROS production and DNA damage, which should be tested in the real space experiments. Furthermore, during manned space travel, the astronauts experience stressful conditions such as loneliness, tension, and lack of exercise. These may also lead to increased ROS levels in the astronauts. Thus whether SMG potentiates ROS production induced by these conditions deserves further investigation.

NAC is widely used in ROS scavenging (Dhouib et al. 2016). Here, we found that NAC could effectively suppress SMG and H_2O_2-induced ROS production, DNA damage as well as apoptosis in MES cells. Wang et al.

also reported the protective effects of NAC on ROS production and cell senescence under SMG treatment (Wang et al. 2009). Qu et al. showed that antioxidants, isorhamnetin and luteolin, could protect neuroblastoma SH-SY5Y cells against microgravity-induced oxidative stress (Qu et al. 2010). These results indicate that antioxidants such as NAC might be used in the protection of ROS stress induced by the combined effects of SMG and other factors of space environment, which may provide valuable strategy for health protection in manned space exploration.

MATERIALS AND METHODS

3D-clinostat

The 3D-clinostat which was used for SMG treatment was provided by Center for Space Science and Applied Research, Chinese Academy of Sciences (Jiang et al. 2008). By employing simultaneous rotations on two axes, the 3D-clinostat can produce an environment with an average of 10^{-3} G, thus simulating microgravity conditions.

Cell culture

MES cells, originally derived from Joyner's laboratory (Auerbach et al. 2000), were cultured on gelatin-coated flasks in standard ES cell medium with leukemia inhibitory factor (LIF) according to Joyner AL without a feeder layer (Matise et al. 2000). Cells were seeded in culture flasks (Becton–Dickinson) and were maintained under 1G for 18 h so that the cells could adhere to the flasks. Then the flasks were filled with fresh medium and the air bubbles were eliminated in order to diminish turbulence and shear forces. The 3D-clinostat was placed in an incubator with an atmosphere of 95% air/5% CO_2 at 37 °C. The day on which the flasks were placed on the clinostat was designated as Day 0. We did not change the culture medium during the experimental period.

The MES cells maintained under 1G or SMG were treated with H_2O_2 at the concentrations of 0, 2.5, 5, 10, and 30 μmol/L for 24 h. For antioxidant treatment, the cells were also treated with ROS scavenger NAC (1 mmol/L) for 24 h.

Apoptosis assays

MES cells were seeded at 5×10^5 cells per 25 cm^2 culture flask. After treatment, the cells were trypsinized with 0.1% trypsin at 37 °C (Sigma), then washed twice with cold PBS, and resuspended in $1\times$ binding buffer at 1×10^6 cells/mL. After that, the cells were stained with Alexa Fluor® 488 annexin V and PI (Invitrogen) at room temperature for 15 min for flow cytometric analysis.

Comet assay

Comet assay was performed according to the protocol of Singh et al. (1988) with minor modifications. Firstly, we pre-coated the slides with a thin layer of 1% normal melting agarose. Secondly, the cells were harvested and resuspended at a concentration of 5×10^5 cells/mL. 20 μL of each suspension was added to 80 μL of pre-melted 0.75% low melting agarose and the contents were pipetted onto the pre-coated slide. Thirdly, the slides were immersed in neutral lysis solution in the dark at 4 °C for 2 h. For unwinding of the DNA, the slides were immersed in 1 × TBE buffer in the dark at 4 °C for 30 min. After that, the slides were exposed to ~0.74 V/cm for 25 min in the horizontal electrophoresis chamber. Following electrophoresis, we stained the slides with propidium iodine (PI). Fluorescence images were viewed with a microscope and analyzed by CASP-1.2.2 software (University of Wroclaw).

ROS activity assays

The cells were stained with 20 μmol/L 2′,7′dichlorodihydrofluorescein diacetate (DCF-DA) (Sigma, USA), and intracellular ROS activity was examined (Shen et al. 2001). The fluorogenic probe DCF-DA is cell-permeable. It diffuses into cells and is deacetylated into the non-fluorescent DCFH by cellular esterases. While, in the presence of ROS, DCFH is rapidly oxidized to highly fluorescent DCF. The fluorescence intensity was measured by flow cytometry (FACSCalibur, Becton–Dickinson, USA) with excitation settings of 488 nm and emission settings of 530 nm, respectively.

Statistical analysis

The data are shown as mean ± SD. The statistical significance of the difference was analyzed by the Student's t test. $p < 0.05$ was considered statistically significant.

Abbreviations

MG	Microgravity
IR	Ionizing radiation
ROS	Radical oxygen species
SMG	Simulated microgravity
MES	Mouse embryonic stem
DSB	Double-strand breaks

DCF-DA	2′,7′2 dichlorodihydrofluorescein diacetate
NAC	N-acetylcysteine
SOD_2	Manganese superoxide dismutase
LIF	Leukemia inhibitory factor
PI	Propidium iodine

Acknowledgments This work was supported by the "Strategic Priority Research Program" of the Chinese Academy of Sciences (XDA04020202-13 and XDA04020413) and the National Natural Science Foundation of China (31370792 and 11179040).

Compliance with ethical standards

Conflict of interest Fanlei Ran, Lili An, Yingjun Fan, Haiying Hang, and Shihua Wang declare that they have no conflict of interest.

Human and animal rights and informed consent This article does not contain any studies with human or animal subjects performed by any of the authors.

References

Aon MA, Cortassa S, O'Rourke B (2010) Redox-optimized ROS balance: a unifying hypothesis. Biochim Biophys Acta 1797:865–877

Auerbach W, Dunmore JH, Fairchild-Huntress V, Fang Q, Auerbach AB, Huszar D, Joyner AL (2000) Establishment and chimera analysis of 129/SvEv- and C57BL/6-derived mouse embryonic stem cell lines. BioTechniques, vol 29, pp 1024–1028, 1030, 1032

Chen N, Wu L, Yuan H, Wang J (2015) ROS/Autophagy/Nrf2 pathway mediated low-dose radiation induced radio-resistance in human lung adenocarcinoma A549 cell. Int J Biol Sci 11:833–844

Dhouib IE, Annabi A, Jallouli M, Elfazaa S, Lasram MM (2016) A minireview on N-acetylcysteine: an old drug with new approaches. Life Sci 151:359–363

Hoeijmakers JH (2009) DNA damage, aging, and cancer. N Engl J Med 361:1475–1485

Jiang Y, Li W, Wang L, Zhang Z, Zhang B, Wu H (2008) Several new type of clinostat. Space Med Med Eng 21:368–371

Li N, An L, Hang H (2015) Increased sensitivity of DNA damage response-deficient cells to stimulated microgravity-induced DNA lesions. PLoS One 10:e0125236

Lombard DB, Chua KF, Mostoslavsky R, Franco S, Gostissa M, Alt FW (2005) DNA repair, genome stability, and aging. Cell 120:497–512

Matise MP, Auerbach AB, Joyner AL (2000) Production of targeted embryonic stem cell clones. In: Joyner AL (ed) Gene targeting: a practical approach, 2nd edn. Oxford University Press, New York, pp 101–132

Mognato M, Celotti L (2005) Modeled microgravity affects cell survival and HPRT mutant frequency, but not the expression of DNA repair genes in human lymphocytes irradiated with ionising radiation. Mutat Res 578:417–429

Mognato M, Girardi C, Fabris S, Celotti L (2009) DNA repair in modeled microgravity: double strand break rejoining activity in human lymphocytes irradiated with gamma-rays. Mutat Res 663:32–39

Qu L, Chen H, Liu X, Bi L, Xiong J, Mao Z, Li Y (2010) Protective effects of flavonoids against oxidative stress induced by simulated microgravity in SH-SY5Y cells. Neurochem Res 35:1445–1454

Schieber M, Chandel NS (2014) ROS function in redox signaling and oxidative stress. Curr Biol 24:R453–R462

Shen YC, Chou CJ, Chiou WF, Chen CF (2001) Anti-inflammatory effects of the partially purified extract of radix *Stephaniae tetrandrae*: comparative studies of its active principles tetrandrine and fangchinoline on human polymorphonuclear leukocyte functions. Mol Pharmacol 60:1083–1090

Singh NP, McCoy MT, Tice RR, Schneider EL (1988) A simple technique for quantitation of low levels of DNA damage in individual cells. Exp Cell Res 175:184–191

Thirsk R, Kuipers A, Mukai C, Williams D (2009) The space-flight environment: the international space station and beyond. CMAJ 180:1216–1220

Tominaga H, Kodama S, Matsuda N, Suzuki K, Watanabe M (2004) Involvement of reactive oxygen species (ROS) in the induction of genetic instability by radiation. J Radiat Res 45:181–188

Veeraraghavan J, Natarajan M, Herman TS, Aravindan N (2011) Low-dose gamma-radiation-induced oxidative stress response in mouse brain and gut: regulation by NFkappaB-MnSOD cross-signaling. Mutat Res 718:44–55

Wang J, Zhang J, Bai S, Wang G, Mu L, Sun B, Wang D, Kong Q, Liu Y, Yao X, Xu Y, Li H (2009) Simulated microgravity promotes cellular senescence via oxidant stress in rat PC12 cells. Neurochem Int 55:710–716

Zhang L, Cheng X, Gao Y, Bao J, Guan H, Lu R, Yu H, Xu Q, Sun Y (2016) Induction of ROS-independent DNA damage by curcumin leads to G2/M cell cycle arrest and apoptosis in human papillary thyroid carcinoma BCPAP cells. Food Funct 7:315–325

9

Single-molecule fluorescence studies on the conformational change of the ABC transporter MsbA

Yanqing Liu[1,2], Yue Liu[1], Lingli He[1], Yongfang Zhao[1], Xuejun C. Zhang[1,2]✉

[1] National Laboratory of Biomacromolecules, National CAS Center for Excellence in Biomacromolecules, Institute of Biophysics, Chinese Academy of Sciences, Beijing 100101, China
[2] College of Life Science, University of Chinese Academy of Sciences, Beijing 100049, China

Abstract   ATP-binding cassette (ABC) transporters are found in all forms of life from microbes to humans, and transport a wide variety of substrates across the cell membrane using the energy released from ATP hydrolysis and an alternating-access mechanism. MsbA is a homodimeric ABC exporter from Gram-negative bacteria, and transports amphipathic substrates including precursors of lipopolysaccharides from the inner leaflet to the outer leaflet of the cytoplasmic membrane. Despite extensive structural and functional studies, controversies remain regarding the dynamic properties of the conformational changes of MsbA during its transport cycle in the lipid environment. Here, we used single-molecule fluorescence resonance energy transfer (smFRET) to explore the dynamic behaviors of MsbA in detergent micelles, nanodiscs, and proteoliposomes. MsbA reconstituted into liposomes showed higher transition frequency between different states on the cytoplasmic side, whereas detergent-solubilized MsbA showed higher transition frequency on the periplasmic side. Three major states were identified from this smFRET study in the functional cycle of MsbA, including an intermediate conformation between the fully opened and fully closed cytoplasmic conformations, associated with both ATP binding and hydrolysis.

Keywords   ABC transporter, Conformational change, Single-molecule fluorescence resonance energy transfer (smFRET)

INTRODUCTION

Adenosine triphosphate (ATP)-binding cassette (ABC) transporters are a large superfamily of membrane proteins that actively translocate a wide variety of substrates across the cellular membrane, driven by energy released from ATP hydrolysis (Dassa 2011; Davidson and Chen 2004; Higgins 2007; Locher 2016; Zhang et al. 2016). All ABC transporters share a common molecular architecture containing two transmembrane domains (TMDs) and two cytoplasmic nucleotide-binding domains (NBDs). The TMDs, which show high sequence diversity, form the translocation pathway and assume the role of substrate selectivity, whereas the NBDs are highly conserved and contain the ABC-characteristic motifs that bind ATP and catalyze its hydrolysis (Ward et al. 2007). ABC transporters are classified into importers and exporters depending on the direction of substrate translocation (Zhang et al. 2016).

MsbA from Gram-negative bacteria is an extensively studied ABC exporter (Doerrler and Raetz 2002). This 128-kDa homodimeric transporter is located in the cytoplasmic (inner) membrane and plays a critical role in flipping Lipid A, a precursor of lipopolysaccharides, a

✉ Correspondence: zhangc@ibp.ac.cn (X. C. Zhang)

major component of the bacterial outer membrane, from the intracellular side to periplasmic side of the inner membrane (Doerrler et al. 2001, 2004; Ruiz et al. 2009; Zhou et al. 1998). In addition, numerous hydrophobic/amphipathic small molecules can be transported by MsbA, and thus MsbA functions as a multidrug-resistance transporter. In its functional cycle, MsbA alternates between intracellular (inward)-facing and periplasmic (outward)-facing conformations, coupling ATP hydrolysis with large conformational rearrangements of the NBD dimer as well as the TMDs (Dong et al. 2005; Higgins and Linton 2004; Ward et al. 2007; Zou et al. 2009). Crystal structures and cryo-electron microscopy (cryo-EM) structures of MsbA have been reported in multiple conformational states, including an intracellular-opening apo state, an intracellular semiclosed apo state, an intracellular-closed state bound with a non-hydrolyzable ATP analog, and an intracellular-closed state in complex with both ADP (from ATP hydrolysis) and inorganic vanadate (V_i) (Mi et al. 2017; Ward et al. 2007).

In the last decade or so, a variety of techniques have been used to study the mechanism of the transport cycle of MsbA, such as EM (Moeller et al. 2015), chemical crosslinking (Doshi et al. 2010), double electron–electron resonance (Borbat et al. 2007; Zou et al. 2009; Zou and McHaourab 2010), fluorescence homotransfer (Borbat et al. 2007), and luminescence resonance energy transfer (Cooper and Altenberg 2013). However, different results were obtained, and the mechanism remains to be firmly consolidated. Moreover, membrane reconstitution (Rigaud et al. 1995), which in principle provides the target proteins with a microenvironment closer to the native membrane, has accelerated mechanistic studies of membrane proteins, including MsbA. However, there is still a lack of direct comparison of the effects of different reconstitution methods on the dynamics of transporters in general, and of MsbA in particular.

In the current study, we used the single-molecule fluorescence resonance energy transfer (smFRET) technique to study the conformational changes of MsbA in the cycle of ATP hydrolysis under three different conditions, namely in detergent micelles, nanodiscs, and liposomes. smFRET is a powerful tool for probing conformational changes in proteins with fairly high spatiotemporal resolution (Lerner et al. 2018), and it has been applied to the study of conformational dynamics of membrane proteins (Zhao et al. 2010a, b, 2011). Our smFRET data, monitored from the cytoplasmic side, showed that the population distribution of MsbA conformations is more restricted in nanodiscs than in either liposomes or micelles of the detergent n-dodecyl-ᴅ-maltopyranoside (DDM). In addition, the transition

frequency between distinct conformations of MsbA NBDs in liposomes is higher than that in either nanodiscs or DDM, suggesting that the NBD dimer is more mobile in the more native-like membrane environment than in the other two model systems. In contrast, the DDM micelles favor conformational changes of the TMD dimer of MsbA on the periplasmic side. Moreover, the ensemble of MsbA in the presence of ATP and V_i, which is believed to mimic the transition state of ATP hydrolysis, in fact represents a terminal-state that cannot be reached by simultaneous addition of ADP and V_i. These results may shed new light on mechanisms of the ABC transporter MsbA.

RESULTS

Experimental design for smFRET analysis

smFRET experiments were performed on variants of *Escherichia coli* MsbA containing a single cysteine (Cys) residue in each subunit linked to fluorophores by maleimide chemistry. Two wild-type (WT) Cys residues were first substituted with Ala, and the resulting Cys-less MsbA (denoted WT*) was used as the template to construct single-Cys mutants. To probe conformational changes of MsbA, we individually mutated several sites to Cys based on available structural information. From these constructs, variants D277C (in the periplasmic region of the TMD) and T561C (in the cytoplasmic NBD) (Fig. 1A) were selected on the basis of their high labeling efficiency (>90%), low nonspecific labeling (Supplemental Fig. S1A), and low anisotropy (Supplemental Table S1).

For the detergent-micelle assays, smFRET experiments were performed with recombinant MsbA variants which were N-terminally His-tagged and immobilized onto biotin-NTA-coated microfluidic channels in 0.05% DDM (Fig. 1B). To better mimic the membrane environment, we further performed smFRET experiments with MsbA proteins reconstituted into either liposomes or nanodiscs, both of which were made from *E. coli* total lipids, and the reconstituted proteins were surface immobilized through biotin-labeled lipids to streptavidin-decorated channels (Fig. 1B). The average diameter of the nanodiscs was estimated by EM imaging to be ∼12 nm, and the liposome diameter was estimated to be ∼100 nm. All the three preparations showed specific surface attachment (Supplemental Fig. S1B). ATPase activities of recombinant MsbA variants were measured in the three conditions. Results from the activity assays revealed that all the MsbA variants in nanodiscs showed significantly higher ATPase activity than those in either liposomes or DDM (Fig. 1C).

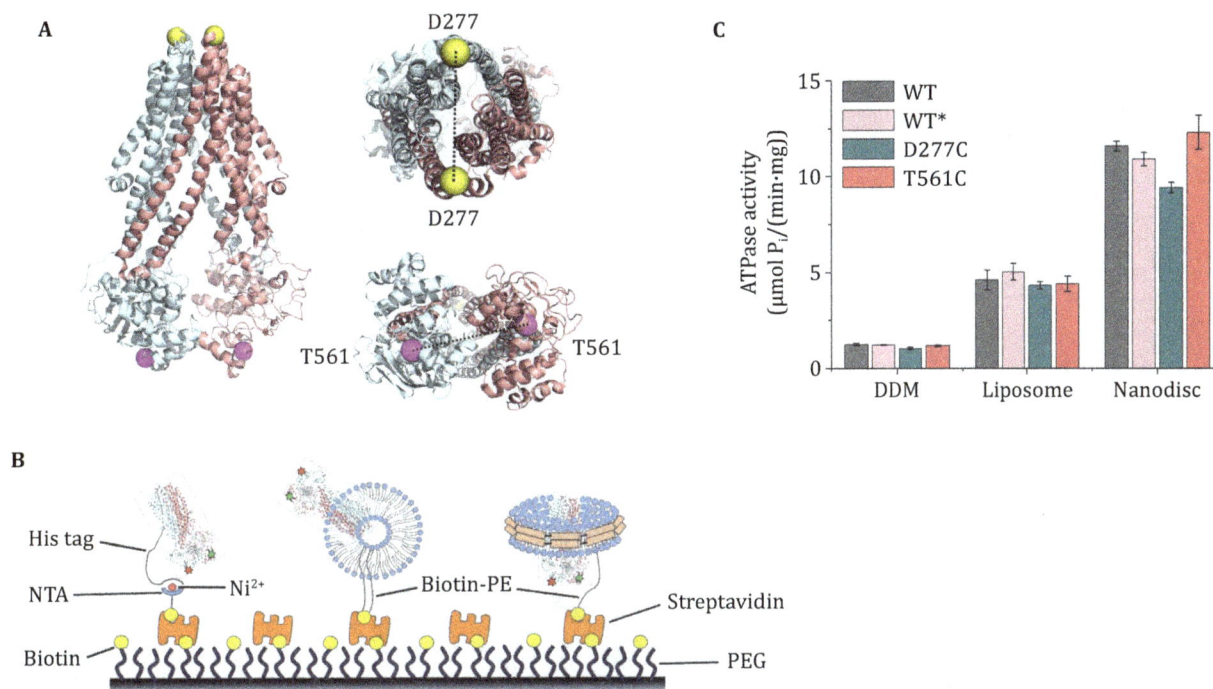

Fig. 1 Design of the smFRET experiments and ATPase activities of MsbA variants. **A** Cartoon representation of the cryo-EM structure of apo-MsbA (PDB ID: 5TV4). Positions of residues D277 and T561 are shown as spheres. **B** Schematic diagram of strategies to immobilize MsbA. **C** ATPase activities of wild-type (WT), Cys-less (WT*), D277C, and T561C variants of MsbA in DDM micelles, liposomes, and nanodiscs, respectively. In liposomes, the activities have been corrected for the fraction of protein (53%) with an outward-facing orientation of the nucleotide-binding domains (NBDs). Error bars indicate standard deviations of results from three independent experiments

For instance, the ATP-hydrolysis rate of WT MsbA in nanodiscs (24.9 ± 2.6 s^{-1} per MsbA molecule) was higher than that in DDM (2.6 ± 0.2 s^{-1}) or in liposomes (10.0 ± 0.8 s^{-1}), probably due to the nanodisc restricting MsbA to a conformation(s) favoring ATP hydrolysis. As expected, vanadate, which is an analog of inorganic phosphate (P_i) and is believed to mimic the γ-phosphate of ATP in the transition state of ATP hydrolysis (Davidson 2002; Smith and Rayment 1996; Urbatsch *et al.* 2003), inhibited the ATPase activities of all of the MsbA variants used in this study (Supplemental Fig. S2B). The variation in activity in different solubilization conditions is consistent with previous studies. For example, in the previously reported nanodisc-embedded structure of MsbA, the two NBDs of MsbA assume a perfect dimer orientating the Walker A and signature motifs to form a NBD–ATP–NBD sandwich for ATP hydrolysis, and the ATPase activity of the corresponding recombinant protein was estimated to be 14 s^{-1} (Mi *et al.* 2017). In contrast, detergent-solubilized MsbA molecules assumed a cytoplasmic-opened conformation in the crystal structures (Ward *et al.* 2007) and had an ATPase activity of 2 s^{-1} (Eckford and Sharom 2008).

Conformational changes of NBDs

Population distribution in different states

To study conformational changes of the MsbA NBDs during the ATP-hydrolysis cycle, the T561C variant was chosen to probe the smFRET signals with different ligand combinations. The mutation site, T561C, is located on the membrane-distal surface of the NBD. In the nanodiscs, the nucleotide-free (apo) state mainly produced a unimodal distribution, peaking at ~ 0.6 FRET efficiency (E_{FRET}, the value of which ranges between 0.0 and 1.0) (Fig. 2A), consistent with the cryo-EM structure of apo-MsbA in which the C_α–C_α distance between the two T561 residues is ~ 35 Å (Mi *et al.* 2017). In both ADP (5 mmol/L) and ADP (5 mmol/L)–V_i (1 mmol/L) ensembles, the MsbA population also showed a similar E_{FRET} distribution (Fig. 2A), indicating that binding of ADP either alone or with additional V_i did not promote conformational change in MsbA. In contrast, in the presence of either ATP (5 mmol/L) or the non-hydrolyzable ATP analog β,γ-imido-adenosine 5′-triphosphate (AMPPNP, 5 mmol/L), MsbA displayed an E_{FRET} peak at ~ 0.65 (Fig. 2A), revealing a slightly closer

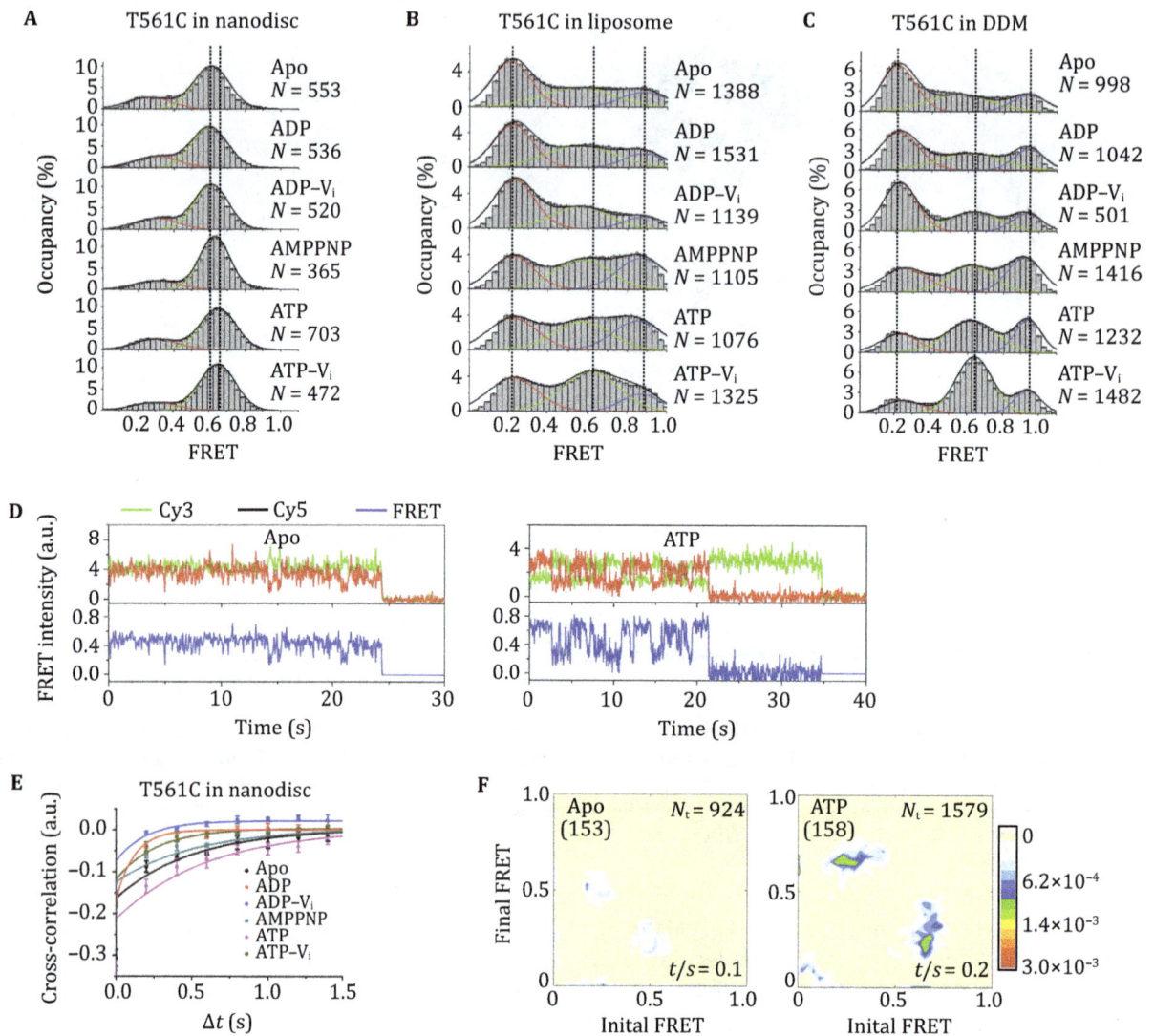

Fig. 2 Conformational dynamics of MsbA NBDs. **A–C** Fluorescence resonance energy transfer (FRET) histograms for MsbA–T561C reconstituted into nanodiscs (**A**), liposomes (**B**), and in n-dodecyl-ᴅ-maltopyranoside (DDM) (**C**), in different ligand-binding ensembles. N, the number of MsbA molecules included in the analysis. Error bars indicate standard deviations of 1000 bootstrap samples of the FRET traces. **D** Representative single-molecule fluorescence trajectories from the apo and ATP ensembles of MsbA-T561C in nanodiscs. **E** Cross-correlation analysis of fluorescence trajectories between donors and acceptors in the nanodiscs. The ATP ensemble showed a stronger negative correlation than other ensembles, indicating that MsbA molecules are more dynamic in this ensemble. Trajectories from all experimental donor–acceptor pairs were used in the calculation. Error bars indicate standard deviations of results from five independent experiments. **F** Transition-density plots of the apo and ATP ensembles of MsbA–T561C in nanodiscs. MsbA molecules undergoing recognizable transitions were manually selected to make the transition-density plots (the number of selected molecules is shown in parentheses), and three-state transitions were assumed. N_t is the number of total transitions included in the analysis; t/s is the average transition frequency per MsbA molecule. The color scale of the transition rate is from light-yellow (background, zero) to red (most populated)

arrangement between the two NBDs than in the apo state. To trap MsbA molecules in their ATP-hydrolysis transition state, we added vanadate together with ATP to the smFRET assay. Like ATP and AMPPNP, addition of ATP (5 mmol/L)–V_i (1 mmol/L) promoted a population shift toward higher E_{FRET} for MsbA–T561C (Fig. 2A).

Sharply distinct from nanodiscs, the proteoliposome of apo-MsbA displayed an intracellular-opening conformation with a unimodal distribution at $E_{FRET} \sim 0.2$ (Fig. 2B). In liposomes, both the ADP and ADP-V_i ensembles showed an E_{FRET} distribution similar to the apo state (Fig. 2B), again demonstrating that binding of either ADP or ADP-V_i does not stimulate NBD dimerization. Similar results were observed in DDM micelles (Fig. 2C), in agreement with previous spectroscopic studies and the crystal structure of detergent-solubilized apo-MsbA (Moeller *et al.* 2015; Ward *et al.* 2007; Zou *et al.* 2009). More importantly, under both liposome and DDM conditions, the E_{FRET} distribution of the AMPPNP ensemble revealed a broad profile with one peak at about 0.9, corresponding to an intracellular-closed conformation, and a similar FRET distribution was observed for the ATP ensemble (Fig. 2B, C). Interestingly, the E_{FRET} distribution showed a peak at ~ 0.6 for the ATP–V$_i$ ensemble, especially in DDM (Fig. 2B, C). These results demonstrated that, in liposomes and detergent micelles, binding of either AMPPNP or ATP thermodynamically favors NBD dimerization, resulting in two extra clusters of states of high (0.9) and intermediate (0.6) E_{FRET} values. In addition, in the presence of V$_i$, binding of ATP (which is subsequently converted to ADP) seems to particularly stabilize the intermediate-E_{FRET} state(s) that likely has a less compact NBD-dimer structure than the high-E_{FRET} state(s). An identical profile of E_{FRET}-population distributions in both liposomes and DDM, however, does not necessarily indicate the same underlying kinetics in state transitions, as shown below.

Dynamics of conformational changes

Using cross-correlation analysis (Vafabakhsh *et al.* 2015) on the donor- and acceptor-fluorescence data obtained from MsbA–T561C in nanodiscs, we found that the amplitude of negative correlation between donor- and acceptor-fluorescence trajectories in the ATP ensemble was larger than that in other ensembles (Fig. 2E). Since only transitions between drastically different FRET states contribute consistently to the negative correlation, the larger amplitude from the ATP ensemble likely reflects more frequent conformational transitions, presumably being stimulated by ATP hydrolysis. Furthermore, manual inspection of the FRET trajectories of different ligand-binding ensembles revealed that only a fraction (10%–30%) of MsbA molecules showed recognizable transitions before being photobleached. These data suggest the possibility that a given cluster of MsbA molecules of similar E_{FRET} occupy multiple states, and only some of these subpopulations are involved in frequent transitions with other clusters with distinct E_{FRET} values. These MsbA molecules capable of transition were further analyzed using hidden Markov modeling (HMM) (McKinney *et al.* 2006), resulting in estimates of the lower-limit of the dwell times (τ_{dwell}) of different clusters and the upper-limit of transition rates (κ) between clusters. Indeed, peaks in the transition-density plot from the HMM

analysis often showed minor yet recognizable shifts relative to the FRET-population distribution (*e.g.* Fig. 2F). Such shifts are likely to be related to insensitivity of the HMM method to transitions between states with only small differences in FRET values (McKinney *et al.* 2006). In the ATP ensemble, transitions between low- and intermediate-E_{FRET} states (with E_{FRET} ~ 0.3 and ~ 0.65, respectively) occurred at an average rate of ~ 0.2 s^{-1}; the apo ensemble showed an average rate of ~ 0.1 s^{-1} (Fig. 2D, F). Since a complete cycle of conformational changes requires at least two transitions, the average rate of such cycles in the ATP ensemble is estimated to be 0.1 s^{-1} or lower. Consistent with the pattern of transition rates, the dwell times in both the low- and intermediate-E_{FRET} states of apo-MsbA were ~ 5-times longer than those in the presence of ATP alone (Supplemental Fig. S2C).

Unlike in nanodiscs, in DDM micelles, MsbA NBDs showed a slightly stronger negative correlation in the apo ensemble than in the other ensembles (Supplemental Fig. S3B). Both inspection of the FRET trajectories of individual apo-MsbA molecules and the transition-density plot showed that, in DDM, conformational transitions of the apo ensemble mostly occurred between states with low (0.2) and intermediate (0.6) E_{FRET}, with an average transition frequency of ~ 0.3 s^{-1} (Supplemental Fig. S3A, C). These results suggested that, while it is mainly in the cytoplasmic-open state (E_{FRET} ~ 0.2), the NBD dimer of apo-MsbA may transiently visit the cytoplasmic semiclosed state (E_{FRET} ~ 0.6).

In liposomes, the MsbA NBDs showed fairly strong negative correlation between the donor- and acceptor-fluorescence trajectories in all apo and ligand-binding ensembles (Fig. 3B), suggesting that MsbA molecules are more dynamic in liposomes than in nanodiscs. The trajectories from individual MsbA molecules and the transition-density plot showed frequent transitions between states of low-, intermediate-, and high-FRET, centering at E_{FRET} of about 0.2, 0.6, and 0.8, respectively (Fig. 3A, C). For instance, in the apo ensemble, transitions mainly occurred between states with E_{FRET} ~ 0.2 and ~ 0.6, and only a small fraction of proteins underwent transition between 0.6 and 0.8. The transitions (mainly between the states with E_{FRET} ~ 0.2 and ~ 0.6) occurred at an average rate of ~ 0.3 s^{-1}, roughly three-times more frequent than in the nanodiscs. The ADP and ADP–V$_i$ ensembles showed dynamics similar to that of the apo ensemble. Therefore, apo-MsbA molecules mostly populate the intracellular-open conformations and 'occasionally' assume the intracellular semiclosed conformation; neither ADP- nor ADP–V$_i$-binding stimulates NBD dimerization. Furthermore, the transition-density plot of the AMPPNP ensemble showed frequent transitions between

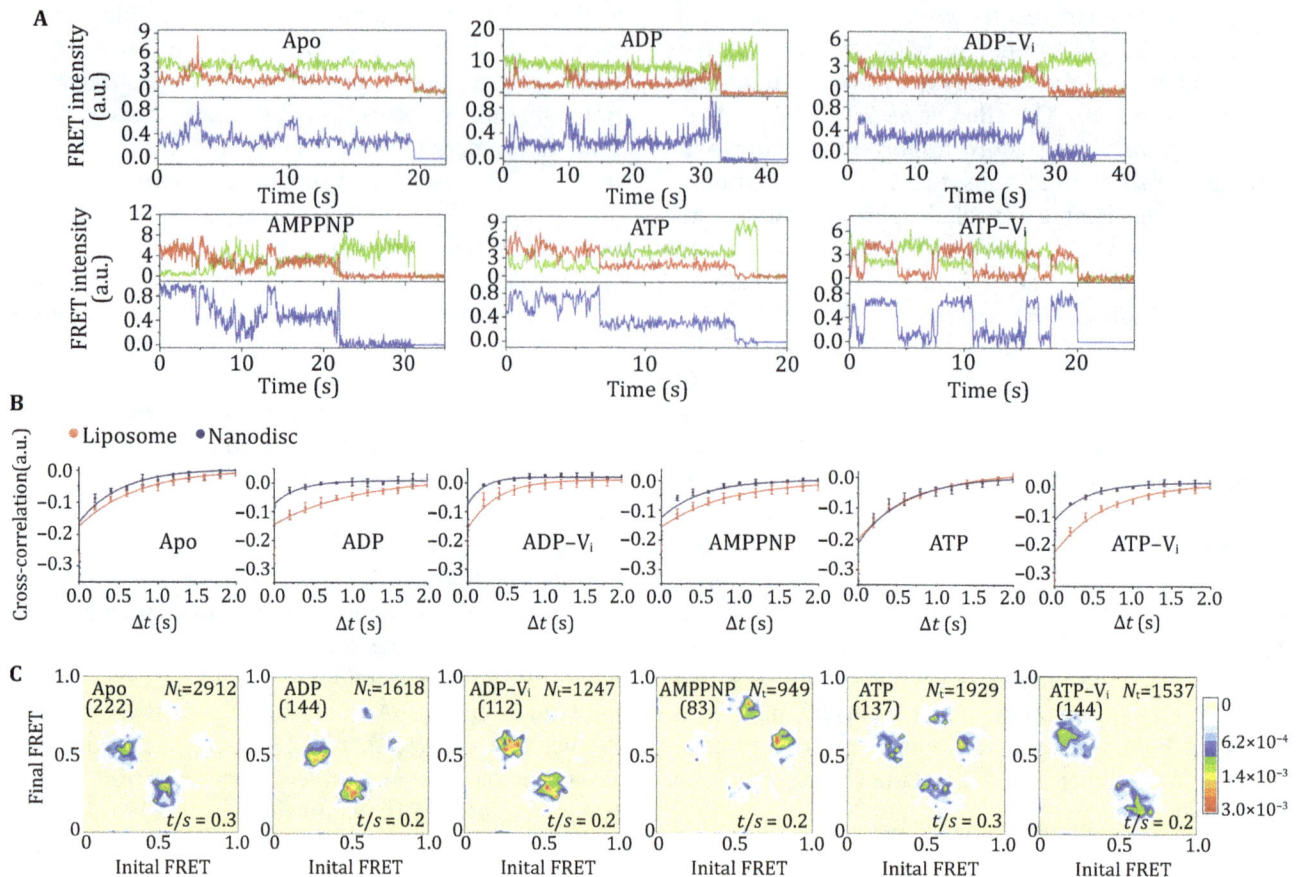

Fig. 3 Conformational dynamics of NBDs of MsbA in liposomes. **A** Representative single-molecule trajectories from the apo and different nucleotide ensembles of MsbA-T561C in liposomes. **B** Cross-correlation analysis of fluorescence trajectories between donors and acceptors in liposomes compared with nanodiscs. The results indicate that MsbA molecules in liposomes (*orange curves*) are more dynamic than those in nanodiscs (*blue curves*) in nearly all ligand-binding ensembles. Error bars indicate standard deviations of results from five independent experiments. **C** Transition-density plots of all ligand-binding ensembles of MsbA-T561C in liposomes. A three-state transition was assumed (See Fig. 2F for legend)

states with $E_{FRET} \sim 0.6$ and ~ 0.8 (Fig. 3C), and the dwell time was longer in the high-FRET state than in the other two states (Supplemental Fig. S2D). Thus, AMPPNP-binding induces dimerization of the NBDs and promotes the intracellular-closed conformation. The ATP ensemble showed a transition-density plot similar to that of the AMPPNP ensemble, with an extra transition between states with $E_{FRET} \sim 0.2$ and ~ 0.6 (Fig. 3C); the latter is probably associated with ATP hydrolysis, as suggested by E_{FRET} data in nanodiscs. Importantly, the transition-density plot of the ATP-V_i ensemble showed frequent transitions, mainly between states with $E_{FRET} \sim 0.2$ and ~ 0.6 (Fig. 3C).

Conformational changes of the TMDs

We examined the conformational dynamics of the TMDs of the MsbA-D277C variant. Each TMD contains six

transmembrane helices (TM1–TM6). Residue D277 is in a short periplasmic loop that connects TM5 and TM6. Patterns of conformational changes probed by a donor–acceptor pair at the D277C sites were expected to be reciprocal to those of T561C if the movements of the cytoplasmic and periplasmic regions of MsbA were fully coupled.

In the nanodiscs, the apo ensemble produced a uni-modal distribution peaking at $E_{FRET} \sim 0.9$, and so did the ensembles with ADP and ADP-V_i (Fig. 4A). These data indicated that MsbA adopts a periplasmic-closed conformation in the apo ensemble, and neither binding of ADP nor ADP-V_i affects this distribution. In the AMPPNP and ATP-V_i ensembles, the E_{FRET}-population peak shifted toward lower E_{FRET} of ~ 0.8 (Fig. 4A), suggesting that both ensembles had a less-compact conformation on the periplasmic side. Moreover, cross-correlation analysis showed weak negative correlations

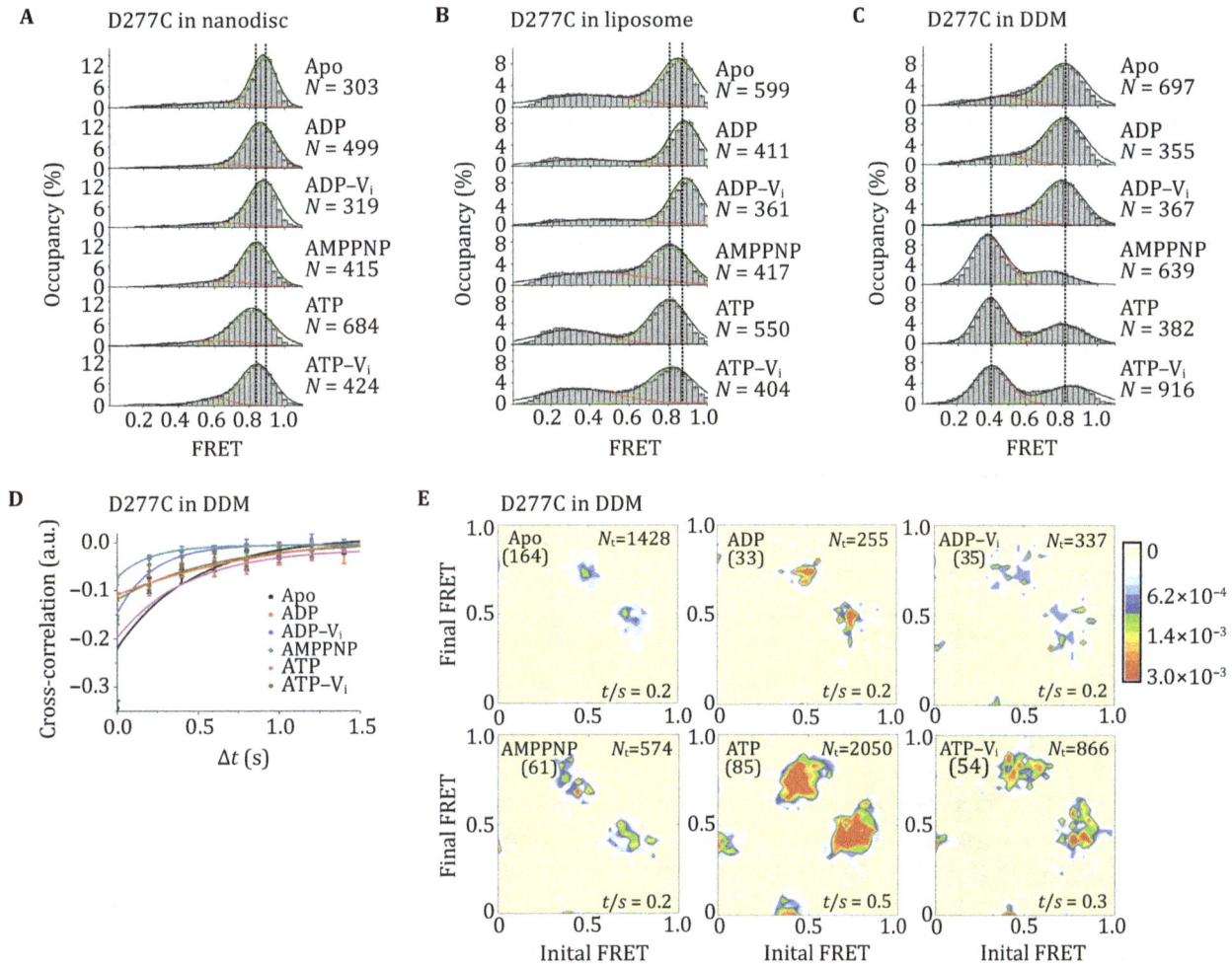

Fig. 4 Conformational dynamics of transmembrane domains (TMDs) of MsbA. **A–C** FRET histograms for MsbA–D277C reconstituted into nanodiscs (**A**), liposomes (**B**), and DDM (**C**), in apo and different ligand-binding ensembles. Error bars indicate standard deviations of 1000 bootstrap samples of the FRET traces. **D** Cross-correlation plots show slightly enhanced dynamics for all ensembles of MsbA–D277C in DDM compared with nanodiscs. Error bars indicate standard deviations of results from five independent experiments. **E** Transition-density plots of MsbA–D277C of apo and different nucleotide ensembles in DDM. A two-state transition was assumed (See Fig. 2F for legend)

between the donor- and acceptor-fluorescence trajectories in all ensembles (Supplemental Fig. S4). The transition-density plot provided additional evidence of conformational rigidity of the TMDs.

In liposomes, the effects of different ligands on the distribution of FRET states resembled that in nanodiscs (Fig. 4B). Specifically, the E_{FRET}-distribution peak centered at ~0.9 in the apo, ADP, and ADP-V_i ensembles indicated a periplasmic-closed conformation. Similar to the case in nanodiscs, addition of AMPPNP, ATP, and ATP-V_i to MsbA–D277C proteoliposomes induced a population shift toward lower FRET values, peaking at ~0.8. Notably, cross-correlation analysis showed that the amplitudes of negative correlations in the liposomes were slightly higher in all apo- and ligand-binding

ensembles than those in nanodiscs (Supplemental Fig. S5A), yet they were lower than those obtained for the T561C variant in liposomes. This suggested that conformational transitions on the periplasmic side are less frequent than those on the cytoplasmic side. This notion is visualized in the transition-density plots (Supplemental Fig. S5B). Collectively, these results suggest that the TMDs of MsbA in liposomes are more mobile than those in nanodiscs.

In detergent micelles, unimodal FRET distributions peaking at ~0.8 were obtained in the apo, ADP, and ADP-V_i ensembles of MsbA–D277C (Fig. 4C), indicating a periplasmic-closed conformation. Addition of AMPPNP, ATP, or ATP-V_i to MsbA–D277C induced a shift to a low E_{FRET} (~0.4), and presumably resulted in a periplasmic-opened conformation(s). These data clearly demonstrated that the FRET-probes at D277C undergo a large environmental change that can be reported by smFRET. Together, these observations indicated that, in the presence of ATP or an ATP analog, an open conformation(s) on the periplasmic side of MsbA is populated. Cross-correlation analysis showed that, in DDM, the negative correlation amplitudes of the ATP and apo ensembles were higher than those of the other ensembles (Fig. 4D). Transition-density plots further indicated that frequent transitions occurred between states with $E_{FRET} \sim 0.8$ and ~ 0.4 (Fig. 4E).

Together, comparison between data from the D277C and T561C variants showed that the dynamics on the periplasmic side are drastically different from those on the cytoplasmic side, and that in the experimental conditions used, movements on the two sides of MsbA are only qualitatively coupled.

DISCUSSION

Understanding the population distribution and transition dynamics between distinct states is critical to delineate the mechanisms of any ABC transporter. Usually, crystallographic studies provide snapshots of the structures of different states of the target protein one at a time, and cryo-EM studies may reveal static structural information of a few major states from one experiment. The smFRET technique complements such structural studies by adding information on both thermodynamic distributions and transition dynamics between states. According to the Boltzmann theorem, the population distribution of the target protein obtained from smFRET provides information about the relative free energy differences (ΔG) between states under either equilibrium (*e.g.* in the apo ensemble) or steady-state (*e.g.* in the ATP ensemble) conditions (see Supplemental Table S2); such states may or may not be accessible in structural studies. In addition, fluorescence trajectories from individual molecules reveal the transition dynamics (*e.g.*, the transition rates) between states (Zhu *et al.* 2017), thus drawing connections between multiple states of a chemical–kinetics network of the target protein (*e.g.*, that represented by the King–Altman plot).

In this study, we used smFRET to analyze the dynamics of MsbA during its ATP-hydrolysis cycle in three environmental conditions (*i.e.*, in DDM micelles, liposomes, and nanodiscs). Two MsbA residues, T561 and D277, were individually mutated to cysteine and labeled with fluorescent dyes, and we successfully demonstrated that they could be used to probe the

conformational changes of the NBDs and TMDs, respectively, of MsbA. In particular, the T561C variant revealed three major clusters of conformational states (with $^{T561C}E_{FRET} \sim 0.2$, ~ 0.6, and ~ 0.9) in the cytoplasmic region (Fig. 2), whereas the D277C variant revealed two such clusters ($^{D277C}E_{FRET} \sim 0.4$ and ~ 0.8) in the periplasmic region (Fig. 4). Based on available structural information, it is reasonable to assume that the states with $^{T561C}E_{FRET} \sim 0.2$, ~ 0.6, and ~ 0.9 correspond to inward-facing (cytoplasmic-open), intermediate (cytoplasmic semiclosed), and outward-facing (cytoplasmic-closed) states, respectively (Fig. 5). Importantly, transition-density plots (*e.g.*, Fig 3c) reveal that the transition from the inward-facing state to the outward-facing state must go through the intermediate state. Whereas the periplasmic-closed state ($^{D277C}E_{FRET} \sim 0.8$) seems to correspond to the cytoplasmic-open state ($^{T561C}E_{FRET} \sim 0.2$), the periplasmic-open state ($^{D277C}E_{FRET} \sim 0.4$) is likely to correspond to the cytoplasmic-closed state ($^{T561C}E_{FRET} \sim 0.9$) and probably the cytoplasmic semiclosed state ($^{T561C}E_{FRET} \sim 0.6$) as well. Thus, FRET-probes at the periplasmic D277C site seem unable to decisively distinguish the intermediate state from the outward-facing state.

By analyzing smFRET data for the T561C variant (mutated in the NBD), we conclude that, in nanodiscs, MsbA is restricted to conformations in which the two NBDs remain close to each other (with $^{T561C}E_{FRET} \sim 0.6$), in agreement with the cryo-EM structures of nanodisc-embedded MsbA (Mi *et al.* 2017), as well as the results of our ATPase-activity assays (Fig. 1C). Upon addition of AMPPNP, ATP, or ATP-V_i, the population peak shifted slightly toward higher FRET values (from 0.6 to 0.65 in nanodiscs) (Fig. 2A). Moreover, in the ATP ensemble, conformational transitions mainly occurred between states with $^{T561C}E_{FRET} \sim 0.3$ and ~ 0.65 (Fig. 2F). The average transition frequency of the conformational cycle (~ 0.1 s^{-1}) further suggests that the transporter is only "occasionally" pushed from the intermediate state ($^{T561C}E_{FRET} \sim 0.65$) into the higher-energy inward-facing state ($^{T561C}E_{FRET} \sim 0.3$) by a power stroke of ATP hydrolysis. However, comparing with the 25 s^{-1} rate of ATP hydrolysis in nanodiscs, the transition rate of MsbA in the presence of ATP is ~ 200-fold slower. This rate difference suggests the possibility that the *in vitro* ATP hydrolysis cycle does not require a large conformational change (*e.g.*, between $^{T561C}E_{FRET} \sim 0.3$ and ~ 0.65); nevertheless, the energy release from ATP hydrolysis may stimulate conformational changes from time to time.

The restriction from the nanodisc construction appears to be alleviated in both proteoliposomes and detergent micelles. In contrast to nanodiscs, in both the

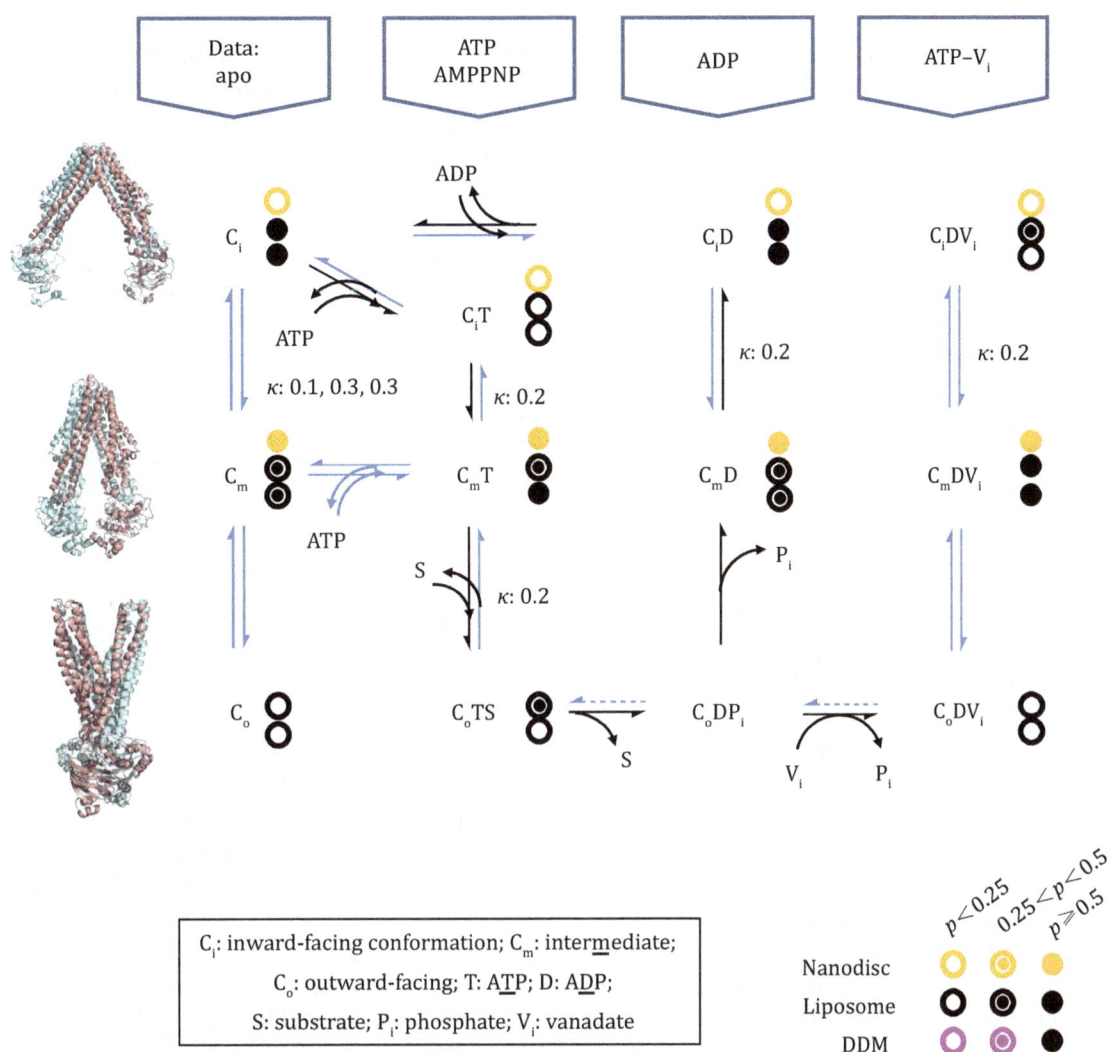

Fig. 5 King–Altman plot of the conformational transitions of MsbA. Transitions are labeled with arrows. The "canonical" transport cycle is shown with red arrows. Structures of the inward-facing (PDB ID: 3B5 W), intermediate (5TV4), and outward-facing (3B60) states are shown in ribbon presentation. Data on population probabilities (p) is taken from Supplemental Table S2, and represented by circles: open, $p < 0.25$; open with a dot, $0.25 < p < 0.5$; filled, $p \geq 0.5$. Data obtained in nanodiscs, liposomes, and DDM are colored *orange*, *green*, and *magenta*, respectively. Average rate (κ) of one-way transitions is copied from the transition-density plots

liposomes and DDM, a wide-opening conformation of the NBD dimer with a unimodal distribution peaking at $^{T561C}E_{FRET} \sim 0.2$ was observed in the apo ensemble. Thus, *in vivo*, MsbA is likely to assume an inward-facing conformation before binding ATP. In addition to the intermediate state ($^{T561C}E_{FRET} \sim 0.6$), a more compact conformation ($^{T561C}E_{FRET} \sim 0.9$) was also captured in the AMPPNP and ATP ensembles. This compact state, which can be formed by the NBD dimers of MsbA in both the liposome and DDM conditions, seems to become inaccessible in the nanodiscs. In agreement with previous reports (Borbat *et al.* 2007; Cooper and Altenberg 2013; Zou *et al.* 2009; Zou and McHaourab 2010), these observations indicate that large-amplitude

conformational dynamics can be adopted by MsbA in liposomes and DDM. Despite the apparently similar population-distributions of MsbA–T561C in the liposomes and DDM (Fig. 2), the NBD dimer of MsbA seems more dynamic in the liposomes (Fig. 3C), which is presumably closer to the native membrane environment, than in DDM or nanodiscs. In addition, under all the three conditions (*i.e.*, nanodiscs, liposomes, and DDM), upon adding ADP, no change of conformation distribution was observed relative to the apo ensemble, in agreement with previous reports (Doshi *et al.* 2010; Mi *et al.* 2017; Weng *et al.* 2010). Curiously, the outward-facing ($^{T561C}E_{FRET} \sim 0.9$) as well as the intermediate ($^{T561C}E_{FRET} \sim 0.6$) states observed in the ATP–V_i

ensemble in both the liposomes and DDM become less accessible in the ADP–V_i ensemble, indicating that in the presence of V_i, ATP hydrolysis traps MsbA in an "end"-state that is not in equilibrium with the ADP-bound state(s) (which may even not bind V_i).

By means of the D277C variant, we also found that the transmembrane region of MsbA is less dynamic than the NBDs, especially in the liposomes and nanodiscs. Under all the three conditions (i.e., nanodiscs, liposomes, and DDM), MsbA adopts a periplasmic-closed conformation ($^{D277C}E_{FRET} \sim 0.9$) in the apo, ADP, and ADP–V_i ensembles (Fig. 4). However, in the AMPPNP, ATP, and ATP–V_i ensembles, MsbA assumed a periplasmic-opened conformation ($^{D277C}E_{FRET} \sim 0.4$) in DDM. This trend was also observed in the liposomes, with the nanodiscs being the most restrictive condition. In addition, in DDM, the apo ensemble showed frequent transitions both between $^{D277C}E_{FRET} \sim 0.8$ and ~ 0.4 and between $^{T561C}E_{FRET} \sim 0.2$ and ~ 0.6 (Fig. 4E and Supplemental Fig. S3C), supporting the notion that whereas the periplasmic-closed conformation corresponds to the cytoplasmic-opened (inward-facing) state, the periplasmic-opened conformation corresponds to the intermediate state. Moreover, in DDM, the average rate of the conformational cycles of the periplasmic region of the TMD dimer in the presence of ATP was <0.25 s^{-1}, comparable with the ATP-hydrolysis rate of 3 s^{-1} per MsbA molecule. Again, the rate difference indicates that ATP hydrolysis and the cycle of conformational changes of the TMD dimer are (partially) decoupled in the experimental conditions used here. More generally, our current study strongly suggests that the choice of solubilization method markedly affects the thermodynamic distribution of states and kinetic behaviors of membrane proteins. The nanodisc construction strongly restricts the dynamics of MsbA on both sides of the lipid bilayer, whereas DDM micelles allow for wider opening of the periplasmic side of MsbA than liposomes and nanodiscs. Therefore, potential effects of the solubilization method(s) should be considered when interpreting results from in vitro studies of membrane proteins, especially concerning their dynamic behaviors.

The functional cycle of MsbA follows the general alternating-access model of transporters (Jardetzky 1966). A putative chemical–kinetics network (i.e., a King–Altman plot) of MsbA is shown in Fig. 5, in which the "canonical" functional cycle is embedded (as red arrows). On the one hand, our current study sheds new light on the function of MsbA by providing thermodynamic and kinetic data to describe this network in detail. On the other hand, an ideal ABC transporter sitting in the cellular membrane would not continuously consume ATP in the absence of substrates. Therefore, a missing piece of the picture of the MsbA functional cycle from the current study seems to be how the binding of substrate triggers ATP hydrolysis. Such a mechanism may be related to factors that have not been included in the currently used in vitro model systems. For example, the electrochemical potential of protons has been proposed to play certain roles in MsbA (Singh et al. 2016). Lacking a proper membrane potential—which may result in a shift of the equilibrium position and orientation of electrically charged membrane proteins relative to the membrane—may explain the apparent decoupling of ATP hydrolysis from substrate transport in MsbA in the current study. Future investigations, for example, of smFRET in the presence of a membrane potential, will verify such mechanistic hypotheses.

MATERIALS AND METHODS

Protein expression and purification

The gene of Cys-less MsbA (C88A/C315A, WT*) (Cooper and Altenberg 2013; Moeller et al. 2015) was synthesized commercially with NdeI/BamHI restriction sites. It was cloned into vector pET-19b containing an N-terminal His$_{10}$-tag. The WT MsbA and single point Cys mutants D277C and T561C were (back-) generated with WT* MsbA as the template using the QuikChange Mutagenesis Kit (Stratagene, CA).

The pET-19b plasmids with different MsbA-coding DNAs were transformed into E. coli strain C43. The cells were cultured in Terrific Broth to OD$_{600}$ 0.8, and induced using 0.5 mmol/L isopropyl β-D-thiogalactoside at 16 °C for 16 h. Cells were collected and resuspended in buffer A (20 mmol/L HEPES (pH 7.5), 300 mmol/L NaCl, 5 mmol/L β-mercaptoethanol, and 10% (v/v) glycerol) and disrupted at 10,000–15,000 p.s.i. using a JN-R2C homogenizer (JNBio, China). Whole cells and cell debris were removed by centrifugation at 17,000 g for 10 min, and the supernatant was ultracentrifuged at 100,000 g for 1 h. The membrane fraction collected from the pellet was solubilized in buffer A supplemented with 1% (w/v) DDM (Anatrace, OH) for 1 h at 4 °C. After a second ultracentrifugation at 100,000 g for 30 min, the supernatant was loaded onto the Ni^{2+}–nitrilotriacetate affinity column (Ni–NTA; Qiagen, IL) and washed with buffer A containing 60 mmol/L imidazole and 0.05% (w/v) DDM. The protein sample was eluted with buffer B (20 mmol/L HEPES (pH 7.5), 100 mmol/L NaCl, and 10% (v/v) glycerol) containing 300 mmol/L imidazole and 0.05% (w/v) DDM. The concentrated sample was then loaded onto a Superdex-200 10/30 column (GE

Healthcare) pre-equilibrated with buffer B containing 0.05% (*w/v*) DDM.

MsbA labeling

Protein samples (20 μmol/L) were labeled in buffer B containing 0.05% (*w/v*) DDM, 200 μmol/L Cy3 dye, and 200 μmol/L Cy5 maleimide (GE Healthcare) for 30 min at 4 °C. Free dyes were removed by loading the samples onto a Zeba™ Spin desalting column (Thermo Scientific). Labeling efficiency measurement was carried out with protein labeled by Cy5 maleimide. The extent of the labeling was estimated from absorption spectra of labeled protein by measuring peak maxima at 649 nm using a NanoPhotometer P330 (IMPLEN, Germany). The labeled-protein concentration was measured using a Bicinchoninic Acid Protein Assay Kit (CWBIO, China). The labeling efficiency at all selected mutation sites was >90%.

Reconstitution of MsbA into liposomes

E. coli total lipid extract (Avanti, AL) was dissolved at 20 mg/mL in buffer C (20 mmol/L HEPES (pH 7.5), 100 mmol/L NaCl). Before reconstitution, the lipid was extruded through a 0.4 μm filter > 11 times. The lipid bilayers were destabilized by the addition of 1% (*w/v*) DDM for 1 h at room temperature. For reconstitution for smFRET experiments, a labeled protein sample was added to lipids at a final protein:lipid ratio of 1:1000 (*w/w*) and incubated for 30 min at room temperature. Prewashed and equilibrated Bio-Beads (Bio-Rad, Inc.) were added to the proteoliposome mixtures at 100 mg/mL and incubated at 4 °C for 1 h. Then, the beads were replaced twice and incubated with the proteoliposome suspensions overnight at 4 °C. The multilayer vesicles were harvested by centrifugation at 320,000 *g* for 20 min (Georgieva *et al.* 2013). Before imaging, liposomes were extruded through a 0.1 μm filter > 21 times to obtain homogenously sized proteoliposomes. Then, the liposomes were loaded onto Ni–NTA, and the proteoliposomes oriented with NBDs facing outward were eluted and obtained.

For activity assays, liposomes were prepared by the same method, except the ratio of protein to lipid was 1:5 (*w/w*). Liposomes were extruded through a 0.4 μm filter > 11 times to obtain homogenously sized liposomes.

Reconstitution of MsbA into nanodiscs

Membrane scaffold protein MSP1E3D1 was purified as previously described (Ritchie *et al.* 2009). Purified and labeled MsbA was prepared as described above. For reconstitution, 68 μl *E. coli* total lipid (20 mg/mL) and 34 μl biotin-phosphatidylethanolamines (PE) (1 mg/mL) were added to 500 μl buffer C, and then MSP1E3D1 and MsbA were added with molar ratio MSP1E3D1:lipid:MsbA of 2:120:1. The reconstitution mixtures were rotated at 4 °C for 1 h, then prewashed and equilibrated Bio-Beads were added to the mixtures at 300 mg/mL. The beads were incubated overnight with the mixture suspensions at 4 °C. The aggregation of MsbA and empty nanodiscs was removed by loading the sample onto a Superdex-200 10/30 column.

ATPase-activity measurements

The ATPase activity of MsbA was measured by the release of P_i from ATP using the malachite green colorimetric assay (Venter *et al.* 2008). Standards containing 0–7 nmol P_i were prepared by dilution of 1 mol/L potassium phosphate buffer (pH 7.5) with concentrations from 0 to 140 μmol/L in 20 μmol/L increments in a final volume of 50 μl. Purified protein (1 μg) and reconstituted liposomes and nanodiscs (0.1 μg) were added to 50 μl of buffer B containing 5 mmol/L $MgSO_4$ and 10 mmol/L ATP with or without 0.05% (*w/v*) DDM respectively, followed by incubation for 3 min at 37 °C. The ATPase reaction was terminated by adding 300 μl fresh malachite green solution, then samples were incubated for 5 min at 20 °C. Finally, 150 μl of 34% (*w/v*) citric acid was added to each sample, followed by incubation for 30 min at 30 °C in the dark to allow color development. The samples were measured at 600 nm using a NanoPhotometer P330.

Determination of MsbA orientation in liposomes in activity assay

The functional mutant MsbA–T561C was reconstituted into proteoliposomes and labeled with Cy5 maleimide. Non-disrupted and disrupted proteoliposome (1% DDM) samples were analyzed by SDS–PAGE and using a fluorescence gel scanner (Tanon, China). The ratio of MsbA with NBDs facing outside was calculated based on comparison of the band fluorescence intensities using ImageJ software. The analysis showed that about 53% ± 1% of the MsbA molecules were oriented with the NBDs facing outside.

Single-molecule FRET imaging experiments

smFRET imaging experiments were carried out as previously described (Zhao *et al.* 2010b). Imaging chambers passivated with a mixture of polyethylene glycol (PEG) and biotin-PEG were incubated with 100 μg/mL streptavidin. Two methods of imaging were used. (1) For MsbA in detergent, His_{10}-tagged dye-labeled samples were immobilized onto the streptavidin-treated chamber surface via biotin-NTA–Ni^{2+}. They were imaged in buffer B containing 0.05% (w/v) DDM. (2) For MsbA reconstituted into nanodiscs and liposomes, samples were immobilized onto the streptavidin-treated chamber surface using biotin-PE, and the imaging experiments were carried out in buffer B. All ensembles with nucleotides were assayed in the presence of additional 5 mmol/L $MgCl_2$. To prolong the lifetime of the fluorescence, an oxygen-scavenging system (0.1% glucose, 5 mmol/L β-mercaptoethanol, 1 unit/mL glucose oxidase, 1 unit/mL catalase, and 1 mmol/L cyclo-octatetraene) was added into the imaging buffer. Images were taken at 50 ms/frame.

Fluorescence signals were collected by means of an objective-based total internal reflection fluorescent microscope, and data were acquired using the software Metamorph (Universal Imaging Corporation) (Zhao *et al.* 2010b). Data analysis was performed as described in previous studies (Juette *et al.* 2016). All assays were repeated at least three times, and the results were reproducible.

Steady-state fluorescence anisotropy measurements

To verify low anisotropy of the labeled protein samples, steady-state anisotropy measurements of Cy5-labeled MsbA (10 nmol/L) were carried out using a Hitachi F-7000 spectrofluorometer in buffer B containing 0.05% (w/v) DDM. The excitation and emission wavelengths were 646 and 662 nm, respectively. The results confirmed that dyes attached to MsbA variants with different nucleotides bound had low anisotropy (Supplemental Table S1).

Abbreviations

ABC	ATP-binding cassette (transporters)
ADP	Adenosine diphosphate
AMPPNP	β,γ-imido-adenosine 5′-triphosphate
ATP	Adenosine triphosphate
DDM	n-dodecyl-β-D-maltopyranoside
NBD	Nucleotide-binding domain
TMD	Transmembrane domain

Acknowledgements This study was supported by the National Basic Research Program of China (973 Program) (2014CB910400) and the National Natural Science Foundation of China (31522016 and 31470745). We thank James Allen, DPhil, from Liwen Bianji, Edanz Group China, for editing the English text of a draft of this manuscript.

Compliance with Ethical Standards

Conflict of interest Yanqing Liu, Yue Liu, Lingli He, Yongfang Zhao, and Xuejun C. Zhang declare that they have no conflict of interest.

Human and animal rights and informed consent This article does not contain any studies with human or animal subjects performed by any of the authors.

References

Borbat PP, Surendhran K, Bortolus M, Zou P, Freed JH, Mchaourab HS (2007) Conformational motion of the ABC transporter MsbA induced by ATP hydrolysis. PLoS Biol 5:e271

Cooper RS, Altenberg GA (2013) Association/dissociation of the nucleotide-binding domains of the ATP-binding cassette protein MsbA measured during continuous hydrolysis. J Biol Chem 288:20785–20796

Dassa E (2011) Natural history of ABC systems: not only transporters. Essays Biochem 50:19–42

Davidson AL (2002) Mechanism of coupling of transport to hydrolysis in bacterial ATP-binding cassette transporters. J Bacteriol 184:1225–1233

Davidson AL, Chen J (2004) ATP-binding cassette transporters in bacteria. Annu Rev Biochem 73:241–268

Doerrler WT, Raetz CR (2002) ATPase activity of the MsbA lipid flippase of *Escherichia coli*. J Biol Chem 277:36697–36705

Doerrler WT, Reedy MC, Raetz CR (2001) An *Escherichia coli* mutant defective in lipid export. J Biol Chem 276:11461–11464

Doerrler WT, Gibbons HS, Raetz CR (2004) MsbA-dependent translocation of lipids across the inner membrane of *Escherichia coli*. J Biol Chem 279:45102–45109

Dong J, Yang G, McHaourab HS (2005) Structural basis of energy transduction in the transport cycle of MsbA. Science 308:1023–1028

Doshi R, Woebking B, van Veen HW (2010) Dissection of the conformational cycle of the multidrug/lipidA ABC exporter MsbA. Proteins 78:2867–2872

Eckford PD, Sharom FJ (2008) Functional characterization of *Escherichia coli* MsbA: interaction with nucleotides and substrates. J Biol Chem 283:12840–12850

Georgieva ER, Borbat PP, Ginter C, Freed JH, Boudker O (2013) Conformational ensemble of the sodium-coupled aspartate transporter. Nat Struct Mol Biol 20:215–221

Higgins CF (2007) Multiple molecular mechanisms for multidrug resistance transporters. Nature 446:749–757

Higgins CF, Linton KJ (2004) The ATP switch model for ABC transporters. Nat Struct Mol Biol 11:918–926

Jardetzky O (1966) Simple allosteric model for membrane pumps. Nature 211:969–970

Juette MF, Terry DS, Wasserman MR, Altman RB, Zhou Z, Zhao H, Blanchard SC (2016) Single-molecule imaging of non-equilibrium molecular ensembles on the millisecond time-scale. Nat Methods 13:341–344

Lerner E, Cordes T, Ingargiola A, Alhadid Y, Chung SY, Michalet X, Weiss S (2018) Toward dynamic structural biology: two decades of single-molecule Forster resonance energy transfer. Science. https://doi.org/10.1126/science.aan1133

Locher KP (2016) Mechanistic diversity in ATP-binding cassette (ABC) transporters. Nat Struct Mol Biol 23:487–493

McKinney SA, Joo C, Ha T (2006) Analysis of single-molecule FRET trajectories using hidden markov modeling. Biophys J 91:1941–1951

Mi W, Li Y, Yoon SH, Ernst RK, Walz T, Liao M (2017) Structural basis of MsbA-mediated lipopolysaccharide transport. Nature 549:233–237

Moeller A, Lee SC, Tao H, Speir JA, Chang G, Urbatsch IL, Potter CS, Carragher B, Zhang Q (2015) Distinct conformational spectrum of homologous multidrug ABC transporters. Structure 23:450–460

Rigaud J-L, Pitard B, Levy D (1995) Reconstitution of membrane proteins into liposomes application to energy-transducing membrane proteins. Biochim Biophys Acta 1231:223–246

Ritchie TK, Grinkova YV, Bayburt TH, Denisov IG, Zolnerciks JK, Atkins WM, Sligar SG (2009) Reconstitution of membrane proteins in phospholipid bilayer nanodiscs. Methods Enzymol 464:211–231

Ruiz N, Kahne D, Silhavy TJ (2009) Transport of lipopolysaccharide across the cell envelope: the long road of discovery. Nat Rev Microbiol 7:677–683

Singh H, Velamakanni S, Deery MJ, Howard J, Wei SL, van Veen HW (2016) ATP-dependent substrate transport by the ABC transporter MsbA is proton-coupled. Nat Commun 7:12387

Smith CA, Rayment I (1996) X-ray structure of the magnesium(II)ADP-vanadate complex of the dictyostelium discoideum Myosin Motor Domain to 1.9 Å resolution. Biochemistry 35:5404–5417

Urbatsch IL, Tyndall GA, Tombline G, Senior AE (2003) P-glycoprotein catalytic mechanism: studies of the ADP-vanadate inhibited state. J Biol Chem 278:23171–23179

Vafabakhsh R, Levitz J, Isacoff EY (2015) Conformational dynamics of a class C G-protein-coupled receptor. Nature 524:497–501

Venter H, Velamakanni S, Balakrishnan L, van Veen HW (2008) On the energy-dependence of Hoechst 33342 transport by the ABC transporter LmrA. Biochem Pharmacol 75:866–874

Ward A, Reyes CL, Yu J, Roth CB, Chang G (2007) Flexibility in the ABC transporter MsbA: alternating access with a twist. Proc Natl Acad Sci USA 104:19005–19010

Weng JW, Fan KN, Wang WN (2010) The conformational transition pathway of ATP binding cassette transporter MsbA revealed by atomistic simulations. J Biol Chem 285:3053–3063

Zhang XC, Han L, Zhao Y (2016) Thermodynamics of ABC transporters. Protein cell 7:17–27

Zhao Y, Quick M, Shi L, Mehler EL, Weinstein H, Javitch JA (2010a) Substrate-dependent proton antiport in neurotransmitter: sodium symporters. Nat Chem Biol 6:109–116

Zhao Y, Terry D, Shi L, Weinstein H, Blanchard SC, Javitch JA (2010b) Single-molecule dynamics of gating in a neurotransmitter transporter homologue. Nature 465:188–193

Zhao Y, Terry DS, Shi L, Quick M, Weinstein H, Blanchard SC, Javitch JA (2011) Substrate-modulated gating dynamics in a Na$^+$-coupled neurotransmitter transporter homologue. Nature 474:109–113

Zhou Z, White KA, Polissi A, Georgopoulos C, Raetz CR (1998) Function of *Escherichia coli* MsbA, an essential ABC family transporter, in lipid A and phospholipid biosynthesis. J Biol Chem 273:12466–12475

Zhu Y, Zhang L, Zhang XC, Zhao Y (2017) Structural dynamics of Giα protein revealed by single molecule FRET. Biochem Biophys Res Commun 491:603–608

Zou P, Mchaourab HS (2010) Increased sensitivity and extended range of distance measurements in spin-labeled membrane proteins: Q-band double electron-electron resonance and nanoscale bilayers. Biophys J 98:L18–L20

Zou P, Bortolus M, McHaourab HS (2009) Conformational cycle of the ABC transporter MsbA in liposomes: detailed analysis using double electron-electron resonance spectroscopy. J Mol Biol 393:586–597

10

Radiolabeled cyclic RGD peptides as radiotracers for tumor imaging

Jiyun Shi[1,2], Fan Wang[1,2], Shuang Liu[3✉]

[1] Interdisciplinary Laboratory, Institute of Biophysics, Chinese Academy of Sciences, Beijing 100101, China
[2] Medical Isotopes Research Center, Peking University, Beijing 100191, China
[3] School of Health Sciences, Purdue University, West Lafayette, IN 47907, USA

Abstract   The integrin family comprises 24 transmembrane receptors, each a heterodimeric combination of one of 18α and one of 8β subunits. Their main function is to integrate the cell adhesion and interaction with the extracellular microenvironment with the intracellular signaling and cytoskeletal rearrangement through transmitting signals across the cell membrane upon ligand binding. Integrin $\alpha_v\beta_3$ is a receptor for the extracellular matrix proteins containing arginine–glycine–aspartic (RGD) tripeptide sequence. The $\alpha_v\beta_3$ is generally expressed in low levels on the epithelial cells and mature endothelial cells, but it is highly expressed in many solid tumors. The $\alpha_v\beta_3$ levels correlate well with the potential for tumor metastasis and aggressiveness, which make it an important biological target for development of antiangiogenic drugs, and molecular imaging probes for early tumor diagnosis. Over the last decade, many radiolabeled cyclic RGD peptides have been evaluated as radiotracers for imaging tumors by SPECT or PET. Even though they are called "$\alpha_v\beta_3$-targeted" radiotracers, the radiolabeled cyclic RGD peptides are also able to bind $\alpha_v\beta_5$, $\alpha_5\beta_1$, $\alpha_6\beta_4$, $\alpha_4\beta_1$, and $\alpha_v\beta_6$ integrins, which may help enhance their tumor uptake due to the "increased receptor population." This article will use the multimeric cyclic RGD peptides as examples to illustrate basic principles for development of integrin-targeted radiotracers and focus on different approaches to maximize their tumor uptake and T/B ratios. It will also discuss important assays for pre-clinical evaluations of the integrin-targeted radiotracers, and their potential applications as molecular imaging tools for noninvasive monitoring of tumor metastasis and early detection of the tumor response to antiangiogenic therapy.

Keywords   Integrin $\alpha_v\beta_3$, PET and SPECT radiotracers, Tumor imaging

INTRODUCTION

Cancer is the second leading cause of death worldwide (Siegel et al. 2015). Most patients will survive if the cancer can be detected at the early stage. Accurate and rapid detection of rapidly growing and metastatic tumors is of great importance before they become widely spread. There are several imaging modalities available for the diagnosis of cancer, including X-ray computed tomography (CT), ultrasound (US), nuclear magnetic resonance imaging (MRI), and nuclear medicine procedures. While CT, US and MRI are better suited for anatomic analysis of solid tumors, molecular imaging with positron emission tomography (PET) and single-photon emission computed tomography (SPECT) offers significant advantages with respect to sensitivity and specificity because they are able to provide the detailed information related to biochemical changes in tumor tissues at the cellular and molecular levels

✉ Correspondence: liu100@purdue.edu (S. Liu)

(Mankoff et al. 2008; Shokeen and Anderson 2009; Tweedle 2009; Correia et al. 2011; Fani and Maecke 2012; Fani et al. 2012; Gaertner et al. 2012; Laverman et al. 2012b; Jamous et al. 2013). The most sensitive molecular imaging modalities are SPECT ($\sim 10^{-10}$ mol/L) and PET (10^{-10}–10^{-12} mol/L) using radiotracers (Fani and Maecke 2012; Fani et al. 2012; Gaertner et al. 2012). According to their biodistribution properties, radiotracers are classified as those whose biodistribution is determined by their chemical and physical properties, and those whose ultimate distribution is determined by their receptor or enzyme binding. The latter class is called target-specific radiotracers. Peptides are often used as targeting biomolecules (BM) for receptor binding in order to achieve high tumor specificity. Many radiotracers have been developed to target the receptors overexpressed on tumor cells and/or tumor vasculature (Mankoff et al. 2008; Shokeen and Anderson 2009; Tweedle 2009; Correia et al. 2011; Fani and Maecke 2012; Fani et al. 2012; Gaertner et al. 2012; Laverman et al. 2012b; Jamous et al. 2013).

A large number of radiolabeled cyclic RGD (arginine–glycine–aspartic) peptides have been evaluated as SPECT or PET radiotracers for tumor imaging (Liu et al. 2005; Wu et al. 2005; Jia et al. 2006; Liu et al. 2006; Zhang et al. 2006; Alves et al. 2007; Dijkgraaf et al. 2007a, b; Liu et al. 2007; Wu et al. 2007; Jia et al. 2008; Li et al. 2008b; Liu et al. 2008a; Shi et al. 2008; Wang et al. 2008a, b; Liu et al. 2009a, b; Shi et al. 2009a, b, c; Chakraborty et al. 2010; Kubas et al. 2010; Dumont et al. 2011; Jia et al. 2011; Shi et al. 2011a, b; Zhou et al. 2011b; Nwe et al. 2012; Pohle et al. 2012; Zhou et al. 2012; Ji et al. 2013a, b; Li et al. 2013; Simecek et al. 2013; Tsiapa et al. 2013; Maschauer et al. 2014; Yang et al. 2014; Zheng et al. 2015). Many excellent review articles have appeared to cover their nuclear medicine applications (D'Andrea et al. 2006; Liu 2006; Meyer et al. 2006; Beer and Schwaiger 2008; Cai and Chen 2008; Liu et al. 2008b; Liu 2009; Stollman et al. 2009; Beer and Chen 2010; Chakraborty and Liu 2010; Dijkgraaf and Boerman 2010; Haubner et al. 2010; Beer et al. 2011; Michalski and Chen 2011; Zhou et al. 2011a; Danhier et al. 2012; Tateishi et al. 2012. This article is not intended to be an exhaustive review of current literature on radiolabeled cyclic RGD peptides. Instead, it will use the multimeric cyclic RGD peptides to illustrate some basic principles for new radiotracer development and to address some important issues associated with integrin-targeted radiotracers. It will focus on different approaches to maximize the tumor uptake and T/B ratios. Authors would apologize to those whose work has not been cited in this article.

RADIOTRACER DESIGN

Integrin-targeted radiotracer

Figure 1 shows the schematic construction of an integrin-targeted radiotracer (Liu 2006, 2009). The cyclic RGD peptide serves as a "vehicle" to carry the isotope to integrins expressed on both tumor cells and activated endothelial cells of tumor neovasculature. BFC is a bifunctional coupling agent to attach the appropriate radionuclide to a cyclic RGD peptide (Liu and Edwards 2001; Liu 2004, 2008; Liu and Chakraborty 2011). The pharmacokinetic modifying (PKM) linker is often used to improve excretion kinetics of radiotracers (Liu and Edwards 2001; Liu 2004, 2008). For a new radiotracer to be successful in clinics, it must show clinical indications for several of high-incidence tumor types (namely breast, lung, and prostate cancers). Renal excretion is necessary in order to maximize the tumor-to-background (T/B) ratios. The main objective of tumor imaging is to achieve the following goals: (1) to detect the presence of tumor at early stage, (2) to distinguish between benign and malignant tumors, (3) to follow the tumor growth and tumor response to a specific therapy (chemotherapy, radiation therapy, or combination thereof), (4) to predict success or failure of a specific therapeutic regimen, and (5) to access the prognosis of a particular tumor.

Radionuclide

The choice of radionuclide depends largely on the modality for tumor imaging. More than 80% of radiotracers for SPECT in nuclear medicine departments are ^{99m}Tc compounds due to optimal nuclear properties of ^{99m}Tc and its easy availability at low cost (Liu and Edwards 2001; Liu 2004, 2008; Liu and Chakraborty 2011). The 6-h half-life is long enough to allow radiopharmacists to carry out radiosynthesis and for physicians to collect clinically useful images. At the same

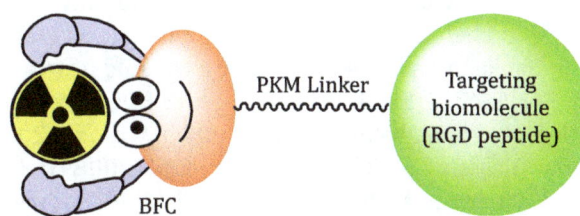

Fig. 1 Schematic presentation of the integrin $\alpha_v\beta_3$-targeted radiotracers. The cyclic RGD peptide is used as the targeting biomolecule (BM) to carry the isotope into the tumor tissue. BFC is used to attach the isotope to the targeting biomolecule. PKM linker is utilized to modify its pharmacokinetics

time, it is short enough to permit administration of 20–30 mCi of ^{99m}Tc without imposing a significant radiation dose to the patients. ^{18}F is a cyclotron-produced isotope suitable for PET. It has a half-life of 110 min. Despite its short half-life, the availability of preparative modules makes ^{18}F radiotracers more accessible to clinicians (Anderson et al. 2003). ^{64}Cu is another PET isotope to develop target-specific radiotracers. It has a half-life of 12.7 h and a β^+ emission (18%, $E_{max} = 0.655$ MeV). Despite poor nuclear properties, ^{64}Cu is a viable alternative to ^{18}F for research programs that wish to incorporate high sensitivity and spatial resolution of PET, but cannot afford to maintain the expensive isotope production infrastructure (Anderson et al. 2003). ^{68}Ga is generator-produced PET isotope with the half-life of 68 min. The ^{68}Ge–^{68}Ga generator can be used for more than a year. ^{68}Ga could become as useful for PET as ^{99m}Tc for SPECT (Maecke et al. 2005). The ^{68}Ga-labeled somatostatin analogs have been studied for PET imaging of somatostatin-positive tumors in both pre-clinical animal models and cancer patients (Henze et al. 2005; Koukouraki et al. 2006a, b). Gallium chemistry and related nuclear medicine applications have been reviewed recently (Maecke et al. 2005).

Bifunctional coupling agent (BFC)

The choice of BFC depends on the radionuclide (Liu 2004, 2008; Liu and Chakraborty 2011). Among various BFCs for ^{99m}Tc-labeling, 6-hydazinonicotinic acid (Fig. 2: HYNIC) is of great interest due to its high efficiency (rapid radiolabeling and high radiolabeling yield), the high solution stability of its ^{99m}Tc complexes, and the easy use of co-ligands for modification of biodistribution properties of ^{99m}Tc radiotracers (Liu 2004, 2005, 2008; Liu and Chakraborty 2011). In contrast, DOTA (1,4,7,10-tetraazacyclododecane-1,4,7,10-tetraacetic acid), NOTA (1,4,7-triazacyclononane-1,4,7-triacetic acid), and their derivatives (Fig. 2) have been widely used for ^{68}Ga/^{64}Cu-labeling of biomolecules due to the high hydrophilicity and in vivo stability of its ^{68}Ga/^{64}Cu chelates (Anderson et al. 2003; Maecke et al. 2005; Shokeen and Anderson 2009). Organic prosthetic groups (Fig. 2: 4-FB, 4-FBz, 2-FP, and 2-FDG) are often needed for ^{18}F-labeling (Dolle 2005; Li et al. 2007, 2008a; Glaser et al. 2008; Hausner et al. 2008; Hohne et al. 2008; Mu et al. 2008; Becaud et al. 2009; Namavari et al. 2009; Vaidyanathan et al. 2009; Jacobson and Chen 2010; Liu et al. 2010; Wangler et al. 2010; Jacobson et al. 2011; Schirrmacher et al. 2013). However, recent results indicate that the Al(NOTA) chelates is more efficient for routine radiosynthesis of ^{18}F radiotracers

using the kit formulation (McBride et al. 2009, 2010, 2012; D'Souza et al. 2011; Lang et al. 2011; Liu et al. 2011; Laverman et al. 2010, 2012a).

Integrins as molecular targets for tumor imaging

Angiogenesis is a requirement for tumor growth and metastasis (Hwang and Varner 2004; Weigelt et al. 2005). The angiogenic process depends on the vascular endothelial cell migration and invasion, and is regulated by cell adhesion receptors. Integrins are such a family of receptors that facilitate the cellular adhesion to and the migration on extracellular matrix proteins, and regulate the cellular entry and withdraw from the cell cycle (Albelda et al. 1990; Falcioni et al. 1994; Carreiras et al. 1996; Bello et al. 2001; Sengupta et al. 2001; Cooper et al. 2002; Zitzmann et al. 2002; Hwang and Varner 2004; Jin and Varner 2004; Weigelt et al. 2005; Sloan et al. 2006; Zhao et al. 2007; Hodivala-Dilke 2008; Barczyk et al. 2010; Taherian et al. 2011; Gupta et al. 2012; Sheldrake and Patterson 2014). The integrin family comprises 24 transmembrane receptors (Table 1) (Sheldrake and Patterson 2014). Their main function is to integrate the cell adhesion and interaction with the extracellular microenvironment with the intracellular signaling and cytoskeletal rearrangement through transmitting signals across the cell membrane on ligand binding. Many integrins are crucial to the tumor initiation, progression, and metastasis. Among the 24 members, the $\alpha_v\beta_3$ is studied most extensively for its role in tumor angiogenesis and metastasis (Albelda et al. 1990; Falcioni et al. 1994; Carreiras et al. 1996; Bello et al. 2001; Sengupta et al. 2001; Cooper et al. 2002; Zitzmann et al. 2002; Hwang and Varner 2004; Jin and Varner 2004; Weigelt et al. 2005; Sloan et al. 2006; Zhao et al. 2007; Hodivala-Dilke 2008; Barczyk et al. 2010; Taherian et al. 2011; Gupta et al. 2012). It is not surprising that radiolabeled cyclic RGD peptides are often called "$\alpha_v\beta_3$–targeted" radiotracers in majority of the literature (D'Andrea et al. 2006; Liu 2006; Meyer et al. 2006; Beer and Schwaiger 2008; Cai and Chen 2008; Liu et al. 2008b; Liu 2009; Stollman et al. 2009; Beer and Chen 2010; Chakraborty and Liu 2010; Dijkgraaf and Boerman 2010; Haubner et al. 2010; Beer et al. 2011; Michalski and Chen 2011; Zhou et al. 2011a; Danhier et al. 2012; Tateishi et al. 2012).

The changes in the $\alpha_v\beta_3$ expression levels and activation state have been well documented during tumor growth and metastasis (Hwang and Varner 2004; Weigelt et al. 2005; Sloan et al. 2006; Zhao et al. 2007; Hodivala-Dilke 2008; Barczyk et al. 2010; Gupta et al. 2012). The $\alpha_v\beta_3$ is expressed in low levels on epithelial cells and mature endothelial cells, but it is highly

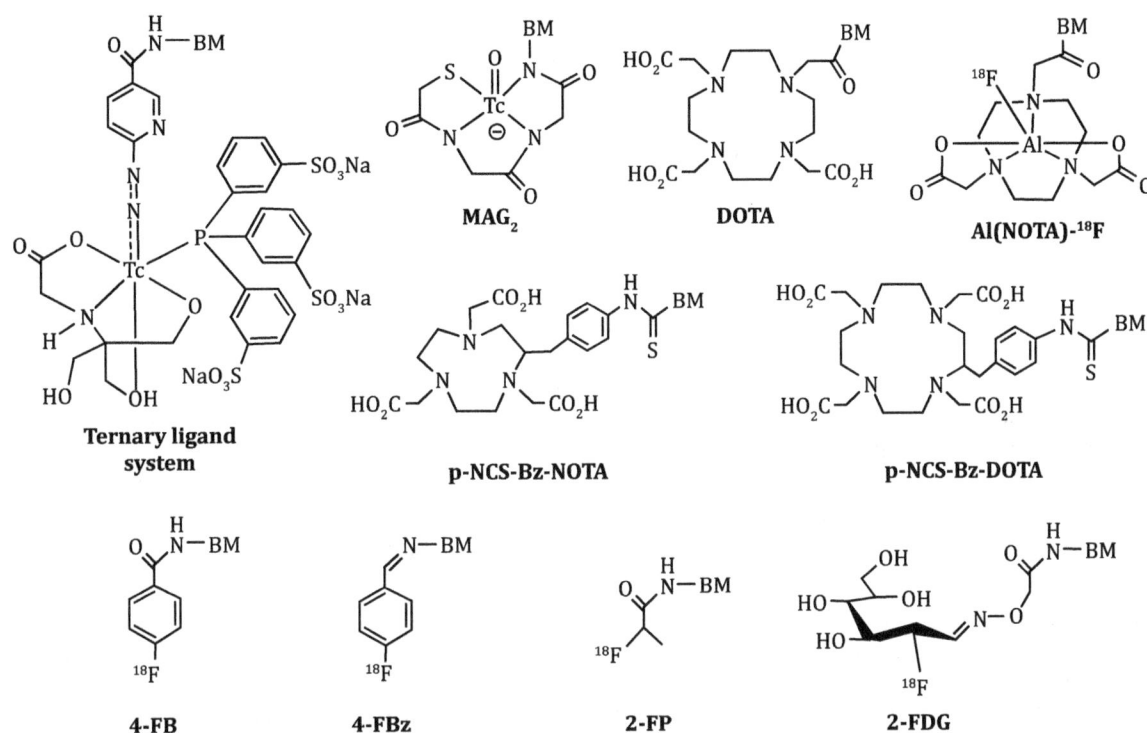

Fig. 2 Examples of BFCs useful for radiolabeling of cyclic RGD peptides. HYNIC and MAG$_2$ are useful for ^{99m}Tc-labeling while DOTA, NOTA, and their derivatives are better suited for chelation of ^{64}Cu and ^{68}Ga. For ^{18}F-labeling, 4-FB, 4-FBz, 2-FP, and 2-FDG are often used as prosthetic groups. The Al(NOTA) chelate is highly efficient for radiosynthesis of ^{18}F radiotracers using a kit formulation

Table 1 Natural integrin ligands and their corresponding recognition peptide sequences

Integrins	Recognition sequence	Natural ligands
$\alpha_v\beta_1$, $\alpha_v\beta_3$, $\alpha_v\beta_5$, $\alpha_v\beta_6$, $\alpha_v\beta_8$, $\alpha_5\beta_1$, $\alpha_8\beta_1$, $\alpha_{IIb}\beta_3$	RGD	Vitronectin, fibronectin, osteopontin, fibrinogen
$\alpha_4\beta_1$, $\alpha_9\beta_1$, $\alpha_4\beta_7$, $\alpha_E\beta_7$, $\alpha_L\beta_2$, $\alpha_M\beta_2$, $\alpha_X\beta_2$, $\alpha_D\beta_2$	LDV and related sequences	Fibronectin, vascular cell adhesion molecule 1, mucosal addressin cell adhesion molecule 1, intercellular cell adhesion molecule 1
$\alpha_1\beta_1$, $\alpha_2\beta_1$, $\alpha_{10}\beta_1$, $\alpha_{11}\beta_1$	GFOGER	Collagen, laminin
$\alpha_3\beta_1$, $\alpha_6\beta_1$, $\alpha_7\beta_1$, $\alpha_6\beta_4$	Other	Laminin

Table were adapted from Sheldrake and Patterson (2014)

expressed in many solid tumors, which include osteosarcomas, glioblastoma, melanomas, and carcinomas of lung and breast (Albelda et al. 1990; Falcioni et al. 1994; Carreiras et al. 1996; Bello et al. 2001; Sengupta et al. 2001; Cooper et al. 2002; Zitzmann et al. 2002; Hwang and Varner 2004; Jin and Varner 2004; Weigelt et al. 2005; Sloan et al. 2006; Zhao et al. 2007; Hodivala-Dilke 2008; Barczyk et al. 2010; Taherian et al. 2011; Gupta et al. 2012). Studies show that $\alpha_v\beta_3$ is overexpressed on tumor cells and activated endothelial cells of tumor neovasculature (Pilch et al. 2002; Taherian et al. 2011). It is believed that the $\alpha_v\beta_3$ expressed on endothelial cells modulate cell adhesion and

migration during angiogenesis, while the $\alpha_v\beta_3$ overexpressed on carcinoma cells potentiate metastasis by facilitating invasion and movement of tumor cells across blood vessels (Sloan and Anderson 2002; Minn et al. 2005; Dittmar et al. 2008; Lorger et al. 2009; Omar et al. 2010). It has been shown that the $\alpha_v\beta_3$ expression levels correlate with the potential for metastasis and aggressiveness of tumors, including glioma, melanoma, and carcinomas of the breast and lungs (Zhao et al. 2007; Hodivala-Dilke 2008). The $\alpha_v\beta_3$ is considered as an important biological target to develop antiangiogenic drugs (Gottschalk and Kessler 2002; Kumar 2003; Jin and Varner 2004; D'Andrea et al. 2006) and molecular imaging probes for diagnosis of tumors (D'Andrea et al. 2006; Meyer et al. 2006; Liu 2006, 2009; Beer and Schwaiger 2008; Cai and Chen 2008; Liu et al. 2008b; Stollman et al. 2009; Beer and Chen 2010; Chakraborty and Liu 2010; Dijkgraaf and Boerman 2010; Haubner et al. 2010; Beer et al. 2011; Michalski and Chen 2011; Zhou et al. 2011a; Danhier et al. 2012; Tateishi et al. 2012).

Cyclic RGD peptides as targeting biomolecules

The $\alpha_v\beta_3$ is a receptor for the extracellular matrix proteins with the exposed RGD tripeptide sequence. Theoretically, both linear and cyclic RGD peptides can be used as targeting biomolecules. A major drawback of linear RGD peptides are their low binding affinity ($IC_{50} > 100$ nmol/L), lack of specificity ($\alpha_v\beta_3$ vs. $\alpha_{IIB}\beta_3$), and rapid degradation by proteases in serum. Cyclization of RGD peptides via the linkers, such as S-S disulfide, thioether, and rigid aromatic rings, leads to the increased receptor binding affinity and selectivity (Aumailley et al. 1991; Gurrath et al. 1992; Müller et al. 1992; Haubner et al. 1996). It seems that the $\alpha_{IIB}\beta_3$ is less sensitive to variations in the RGD peptide backbone and can accommodate a larger distance or spacer than $\alpha_{IIB}\beta_3$ and $\alpha_v\beta_5$ (Pfaff et al. 1994). Incorporation of the RGD sequence into a cyclic pentapeptide framework (Fig. 3: c(RGDfV) and EMD121974) could significantly increase the binding affinity and selectivity of $\alpha_v\beta_3/\alpha_v\beta_5$ over $\alpha_{IIb}\beta_3$ (Aumailley et al. 1991; Gurrath et al. 1992; Müller et al. 1992; Pfaff et al. 1994; Haubner et al. 1996). The addition of a rigid aromatic ring into the cyclic hexapeptide structure (Fig. 3: DMP728 and DMP757) enhances the binding affinity of $\alpha_{IIB}\beta_3$ (Liu et al. 2010; Jacobson et al. 2011; Danhier et al. 2012). The structure–activity studies indicated that the amino acid residue in position 5 has little impact on $\alpha_v\beta_3/\alpha_v\beta_5$ binding affinity (Aumailley et al. 1991; Gurrath et al. 1992; Müller et al. 1992; Haubner et al. 1996). The valine (V) residue in c(RGDfV) can be replaced by lysine (K) or glutamic acid (E) to afford c(RGDfK) and c(RGDfE), respectively, without changing their $\alpha_v\beta_3/\alpha_v\beta_5$ binding affinity.

Figure 4 shows several examples of monomeric cyclic RGD peptides that have high affinity for $\alpha_v\beta_3$ and $\alpha_v\beta_5$. Among the radiotracers evaluated in pre-clinical tumor-bearing models, [18F]Galacto-RGD (Fig. 4: 2-[18F]fluoropropanamide c(RGDfK(SAA); SAA = 7-amino-L-glyero-L-galacto-2,6-anhydro-7-deoxyheptanamide) was the first one under clinical investigation for visualization of $\alpha_v\beta_3$ expression in cancer patients (Beer et al. 2005; 2007, 2008; Haubner et al. 2005). The results from imaging studies in cancer patients showed that there was sufficient $\alpha_v\beta_3$ for PET imaging. The tumor uptake of [18F]Galacto-RGD correlates with the $\alpha_v\beta_3$ levels in cancer patients (Haubner et al. 2005; Beer et al. 2007, 2008). However, the radiotracers derived from monomeric cyclic RGD peptides all had low tumor uptake with T/B ratios because of their relatively low $\alpha_v\beta_3$ binding affinity.

It must be noted that cyclic RGD peptides bind not only $\alpha_v\beta_3$ but also other integrins. While the $\alpha_v\beta_3$ plays pivotal role in the tumor growth and progression, $\alpha_{IIB}\beta_3$ is critical for the platelet aggregation during thrombosis formation. The interaction between $\alpha_v\beta_3$ and $\alpha_{IIb}\beta_3$ facilitates the adhesion of tumor cells to the vasculature and often leads to metastasis (Felding-Habermann et al. 1996; Bakewell et al. 2003). The $\alpha_v\beta_5$ is very similar to $\alpha_v\beta_3$ in the ligand binding site region and has a similar expression pattern and function to those of $\alpha_v\beta_3$. Both $\alpha_v\beta_5$ and $\alpha_v\beta_3$ are highly expressed on the activated endothelial cells and have similar roles in angiogenesis, promoting angiogenic response to different growth factors (Bakewell et al. 2003; Goodman et al. 2012). The $\alpha_v\beta_5$ has been shown to overexpress on a wide range of tumor types (Goodman et al. 2012; Boger et al. 2014). A number of tumors co-express $\alpha_v\beta_3$ and $\alpha_v\beta_5$ (Sung et al. 1998; Erdreich-Epstein et al. 2000; Graf et al. 2003; Humphries et al. 2006; Monferran et al. 2008; Bianchi-Smiraglia et al. 2013; Roth et al. 2013; Vogetseder et al. 2013; Boger et al. 2014; Navarro-Gonzalez et al. 2015), because both engage the same ECM ligands and activate complementary cell signaling pathways in order to promote tumor progression (Sung et al. 1998; Bianchi-Smiraglia et al. 2013). It was also reported that the tumor cell expression of $\alpha_v\beta_3$, $\alpha_v\beta_5$, $\alpha_5\beta_1$, $\alpha_6\beta_4$, $\alpha_4\beta_1$, and $\alpha_v\beta_6$ is correlated with the progression of various tumors (Vogetseder et al. 2013; Boger et al. 2014). The structures of other RGD-binding integrins ($\alpha_v\beta_6$, $\alpha_v\beta_8$, $\alpha_v\beta_1$ and $\alpha_8\beta_1$) have not yet been studied in details (Sheldrake and Patterson 2014).

MAXIMIZING BINDING AFFINITY VIA MULTIMERIZATION

The multivalent concept has been used to develop radiotracers with the increased tumor-targeting capability. For example, E[c(RGDfK)]$_2$ (RGD$_2$) was the first cyclic RGD dimer for development of diagnostic (99mTc) and therapeutic (90Y and 64Cu) radiotracers (Liu et al. 2001a; b; 2005, 2006, 2007, 2008a, 2015; Jia et al. 2006, 2008). RGD tetramers RGD$_4$ was also used to develop SPECT and PET radiotracers (Wu et al. 2005; Liu et al. 2007, 2008a). Both the in vitro assays and biodistribution data showed that the radiolabeled

$\alpha_v\beta_3$-targeted cyclic pentapeptides

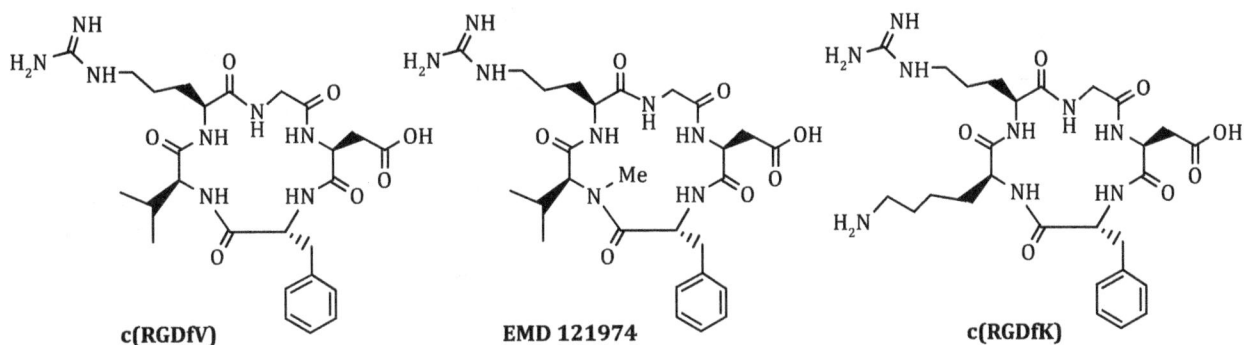

c(RGDfV)   EMD 121974   c(RGDfK)

$\alpha_{IIb}\beta_3$-targeted cyclic hexapeptides

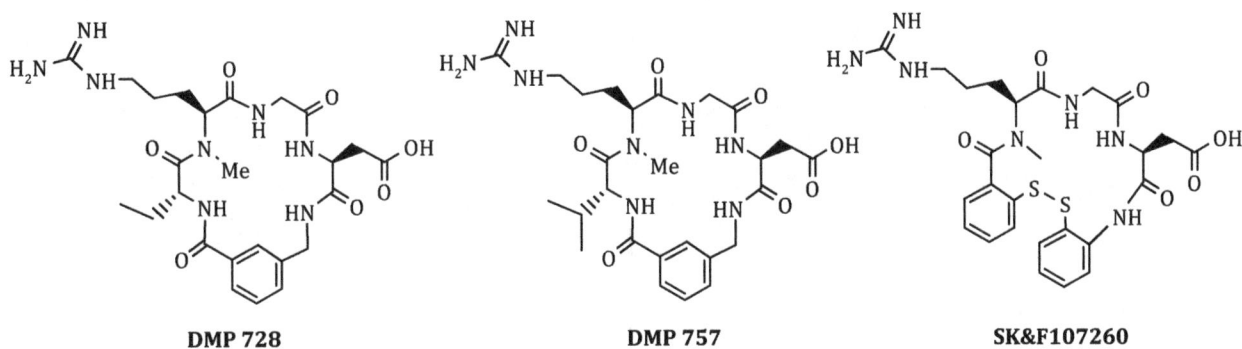

DMP 728   DMP 757   SK&F107260

Fig. 3 Examples of monomeric cyclic RGD peptides as targeting biomolecules for the development of $\alpha_v\beta_3$-targeted radiotracers. EMD121974 has been under clinical investigations as an "orphan drug" for treatment of glioblastoma either stand-alone or in combination with radiation therapy. DMP728 and DMP757 were originally developed as anti-thrombotic agents

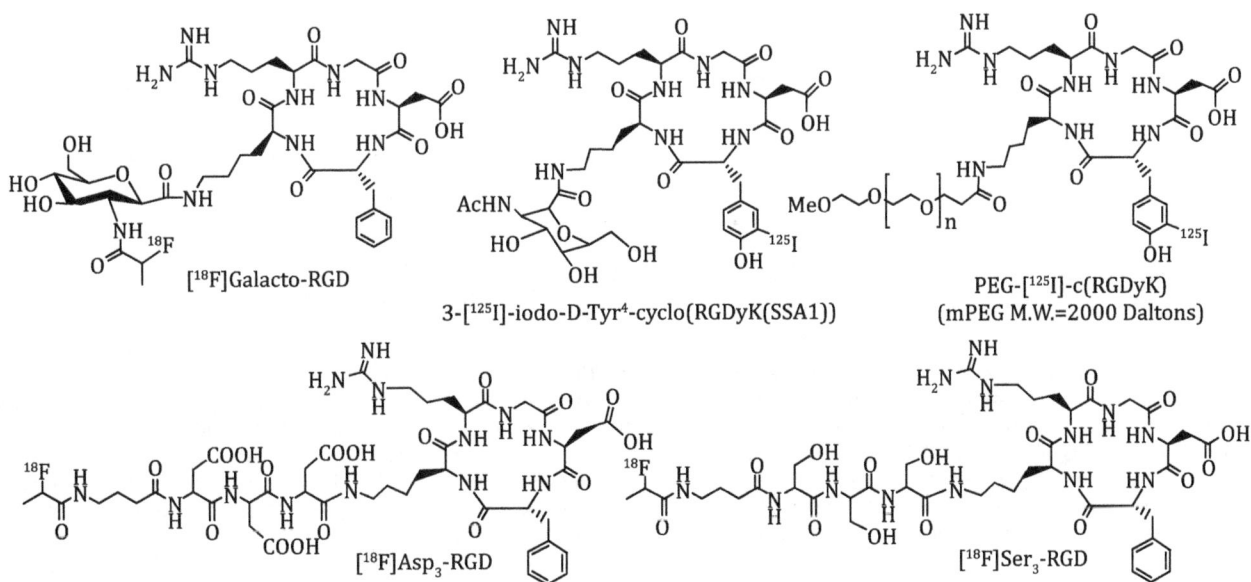

[18F]Galacto-RGD   3-[125I]-iodo-D-Tyr4-cyclo(RGDyK(SSA1))   PEG-[125I]-c(RGDyK) (mPEG M.W.=2000 Daltons)

[18F]Asp3-RGD   [18F]Ser3-RGD

Fig. 4 Examples of the radiolabeled monomeric cyclic RGD peptides as radiotracers

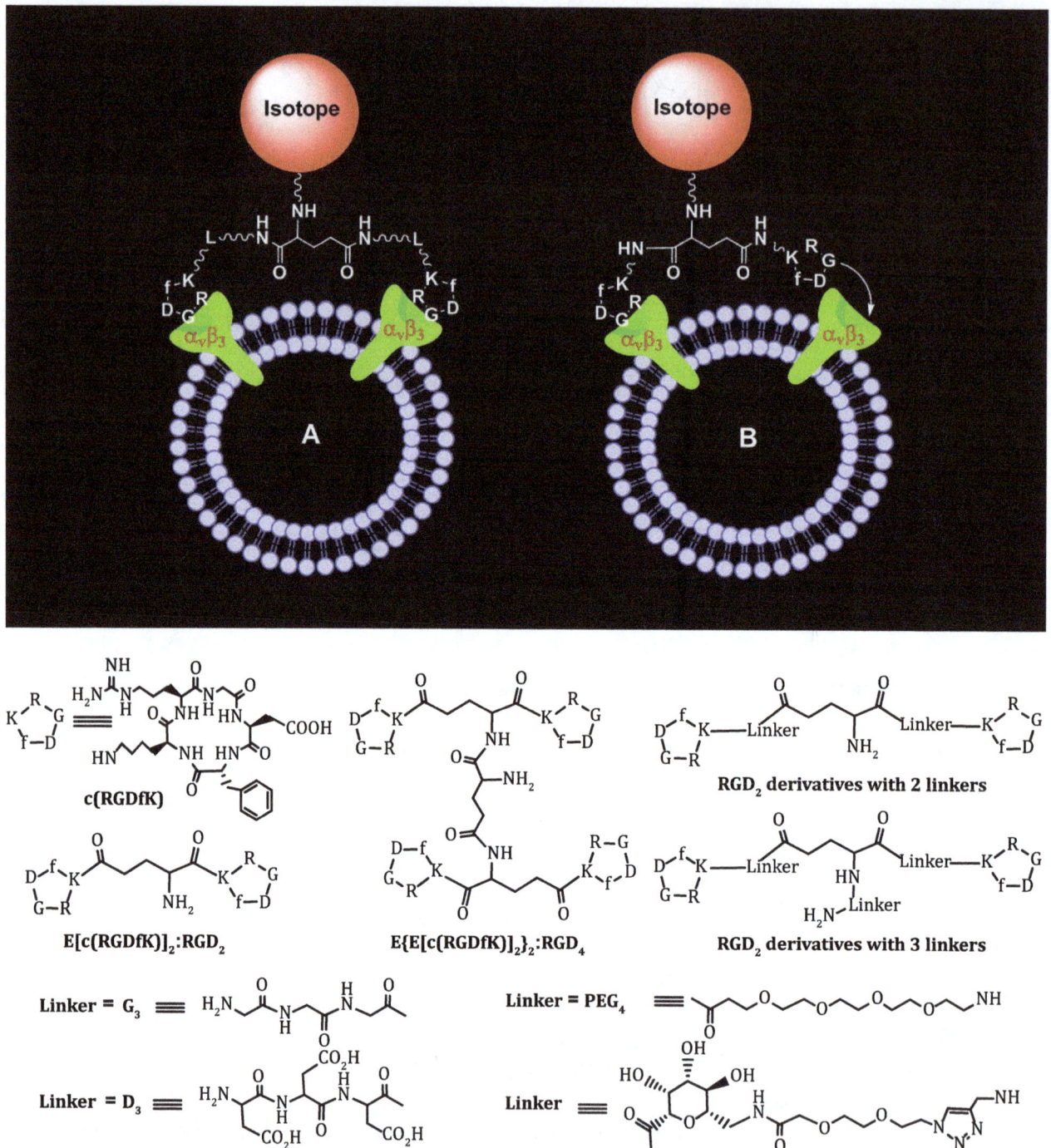

Fig. 5 Top: Schematic illustration of the interactions between cyclic RGD peptide dimers and $\alpha_v\beta_3$. **A** The distance between two RGD motifs is not long enough for simultaneous integrin $\alpha_v\beta_3$ binding. However, the RGD concentration is "locally enriched" in the vicinity of neighboring integrin $\alpha_v\beta_3$ once the first RGD motif is bound. **B** The distance between two RGD motifs is long due to the presence of two linkers (L). As a result, the cyclic RGD dimer is able to bind integrin $\alpha_v\beta_3$ in a "bivalent" fashion. In both cases, the end-result would be higher integrin $\alpha_v\beta_3$ binding affinity for the multimeric cyclic RGD peptides. **Bottom**: Selected cyclic RGD peptide dimers and tetramers useful for development of $\alpha_v\beta_3$-targeted radiotracers. The D_3, G_3, PEG_4, and sugar linkers are used to increase the distance between two RGD motifs and to improve radiotracer excretion kinetics from non-cancerous organs

(^{99m}Tc, ^{18}F, and ^{64}Cu) multimeric cyclic RGD peptides have higher $\alpha_v\beta_3$ binding affinity and better tumor uptake than their monomeric analogs (Liu et al. 2008b; Liu 2009). It is important to note that multimeric RGD peptides are not necessarily multivalent (Liu et al. 2008b; Chakraborty et al. 2010). Two factors (Fig. 5:

bivalency and enhanced local RGD concentration) contribute to the high $\alpha_v\beta_3$ binding affinity of cyclic RGD peptides (Liu et al. 2008b; Chakraborty et al. 2010). The concentration factor exists in all multimeric RGD peptides regardless of the linker length. Given the short distance (6 bonds excluding side-arms of K-residues) between two RGD motifs in E[c(RGDfK)]$_2$ and E[c(RGDyK)]$_2$, it is unlikely that they would bind to two adjacent $\alpha_v\beta_3$ sites simultaneously. However, the binding of one RGD motif to $\alpha_v\beta_3$ will increase the "local concentration" of second RGD motif in the vicinity of $\alpha_v\beta_3$ sites (Fig. 5B). The concentration factor may explain the higher tumor uptake of radiolabeled (^{99m}Tc, ^{111}In, ^{90}Y, ^{18}F, and ^{64}Cu) E[c(RGDfK)]$_2$ and E[c(RGDyK)]$_2$ than their monomeric derivatives (Beer and Chen 2010; Chakraborty and Liu 2010; Dijkgraaf and Boerman 2010; Beer et al. 2011; Michalski and Chen 2011; Zhou et al. 2011a). The key for bivalency is the distance between two cyclic RGD motifs. For example, this distance is 38 bonds in PEG$_4$-E[c(RGDfK(PEG$_4$))]$_2$ (3P-RGD$_2$: PEG$_4$ = 15-amino-4,7,10,13-tetraoxapentadecanoic acid), and 26 bonds G$_3$-E[c(RGDfK(G$_3$))]$_2$ (3G-RGD$_2$: G$_3$ = Gly-Gly-Gly), which are long enough for them to achieve the bivalency. As a result, HYNIC-3P-RGD$_2$ (IC$_{50}$ = 60 $\pm$ 3 nmol/L) and HYNIC-3G-RGD$_2$ (IC$_{50}$ = 59 $\pm$ 3 nmol/L) have much higher $\alpha_v\beta_3$ binding affinity than HYNIC-P-RGD$_2$ (P-RGD$_2$ = PEG$_4$-E[c(RGDfK)]$_2$: (IC$_{50}$ = 89 $\pm$ 7 nmol/L)) (Shi et al. 2008; Wang et al. 2008b). ^{99m}Tc-3P-RGD$_2$ and ^{99m}Tc-3G-RGD$_2$ had higher breast tumor uptake than ^{99m}Tc-P-RGD$_2$ (Fig. 6) (Shi et al. 2008; Wang et al. 2008b). Since the tumor uptake of ^{99m}Tc-3P-RGD$_2$ and ^{99m}Tc-3P-RGD$_2$ is comparable to that of ^{99m}Tc-RGD$_4$ suggests that the contribution from "concentration factor" may not be as significant as that from the "bivalency."

MAXIMIZING RADIOTRACER UPTAKE BY TARGETING MULTIPLE RECEPTORS

Two most important factors affecting the radiotracer tumor uptake are receptor binding affinity and receptor population. The receptor binding affinity is critically important for selective tumor localization and tumor uptake of radiolabeled cyclic RGD peptides (Liu et al. 2008b). The receptor population is equally important for the receptor-based molecular imaging. It will not be possible to image the tumor if that it has very limited or no receptor expression even if the receptor ligand has high receptor binding affinity. There are two approaches to maximize the target population. The first approach (Fig. 7A) involves the

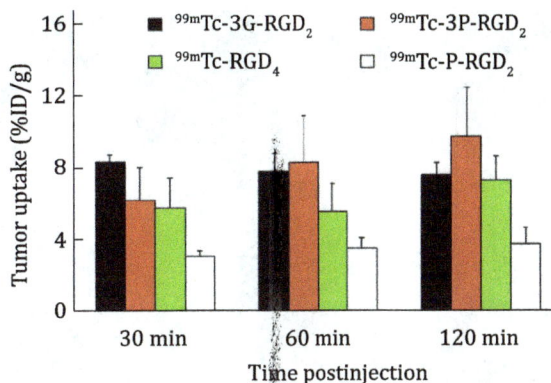

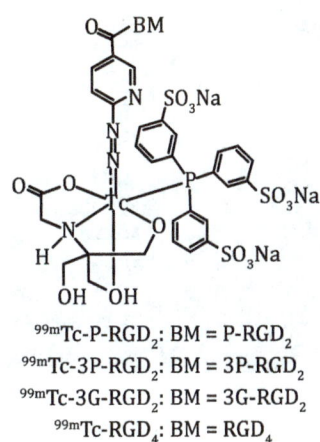

Fig. 6 Direct comparison of tumor uptake for ^{99m}Tc-P-RGD$_2$, ^{99m}Tc-3G-RGD$_2$, ^{99m}Tc-3P-RGD$_2$, and ^{99m}Tc-RGD$_4$ in athymic nude mice bearing MDA-MB-435 breast cancer xenografts. The biodistribution data were adapted from Shi et al. (2008) and Wang et al. (2008b)

use of the same cyclic RGD peptide to target two or more integrins (such as $\alpha_v\beta_3$, $\alpha_v\beta_5$, $\alpha_5\beta_1$, $\alpha_6\beta_4$, $\alpha_4\beta_1$, and $\alpha_v\beta_6$). Another approach (Fig. 7B) involves the use of a bifunctional peptide that is able to target two different receptors, such as $\alpha_v\beta_3$ and bombesin (BBN) receptor. By targeting two different receptors, the radiotracer will have more opportunities to localize in the tumor due to the larger populations of two receptors than that of a single receptor. The so-called "bivalent heterodimers" (Fig. 7) has been used to target the $\alpha_v\beta_3$ and BBN receptors (Li et al. 2008c; Liu et al. 2009c, d). The xenografted PC-3 and MDA-MB-435 tumor-bearing models were used to evaluate their tumor-targeting capability and biodistribution properties. It is well-established that the xenografted PC-3 tumors have low $\alpha_v\beta_3$ expression (Zhou et al. 2011b; Ji et al. 2013c). It was also shown that the xenografted MDA-MB-435 tumor has little BBN receptor expression (Liu et al. 2009c, d). Therefore, both PC-3 and MDA-MB-435 tumor-bearing models are not appropriate to

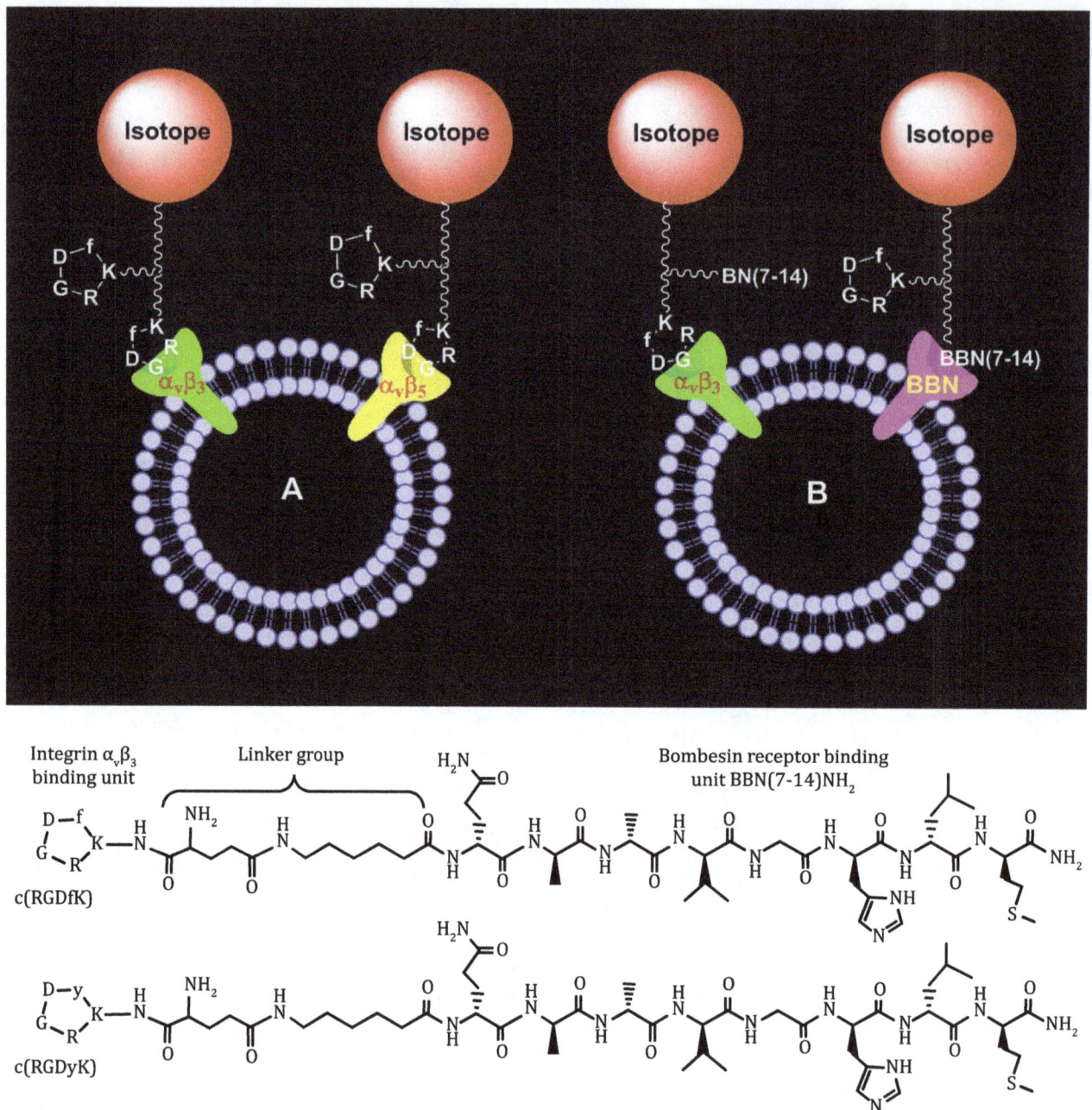

Fig. 7 Top: Schematic presentation of the interactions between the dimeric cyclic RGD peptide to target two or more integrins (such as $\alpha_v\beta_3$, $\alpha_v\beta_5$, $\alpha_5\beta_1$, $\alpha_6\beta_4$, $\alpha_4\beta_1$, and $\alpha_v\beta_6$). **B** Schematic illustration of the interactions between the bifunctional peptide and two different receptors ($\alpha_v\beta_3$ and BBN receptor). By targeting two different receptors, the radiotracer will have more opportunities to localize in the tumor because of the increased receptor population. The two targeted receptors (e.g., $\alpha_v\beta_3/\alpha_v\beta_5$ or $\alpha_v\beta_3$/BBN) must be co-localized and the distance between them must be short for the bifunctional radiotracer to achieve "simultaneous receptor binding." **Bottom**: Selected examples of bifunctional peptides containing c(RGDfK)/c(RGDyK) and Aca-BBN(7–14)NH$_2$ (ε-aminocaproic acid-Gln-Trp-Ala-Val-Gly-His-Leu-Met-NH$_2$)

prove the concept of "bivalent heterodimers." For the bifunctional radiotracers to achieve the bivalency, the $\alpha_v\beta_3$ and BBN receptors must be co-localized and the distance between them must be short. Otherwise, it would not be advantageous even if they might be able to target both individual receptors. Unfortunately, there is lack of concrete experimental data to demonstrate if the c(RGDfK)-BBN(7–14) and c(RGDyK)-BBN(7–14) conjugates are "bivalent" for tumor targeting, and whether there is indeed a "synergetic effect" between the cyclic RGD and BBN(7–14) peptides. Another challenge associated with the "bifunctional heterodimer concept" is which binding unit actually contributes to the radiotracer tumor uptake.

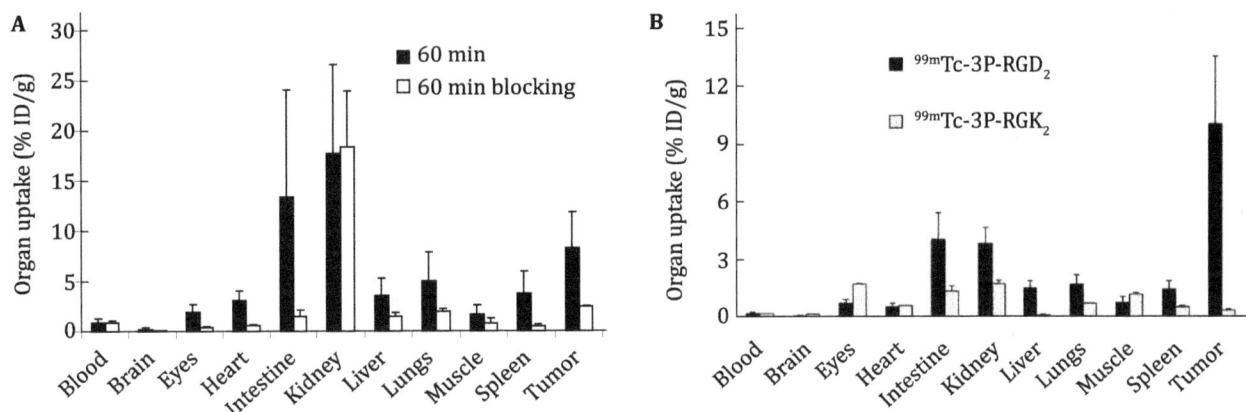

Fig. 8 **A** Comparison of organ uptake (%ID/g) for ^{99m}Tc-2P-RGD$_2$ in athymic nude mice bearing U87MG glioma xenografts in the absence or presence of excess RGD$_2$ at 60 min p.i. Co-injection of excess RGD$_2$ resulted in significant reduction in the uptake of ^{99m}Tc-2P-RGD$_2$ in the tumor and normal organs. **B** Comparison of the 60-min biodistribution data of ^{111}In-3P-RGD$_2$ and ^{111}In-3P-RGK$_2$ in athymic nude mice bearing U87MG glioma xenografts. The low tumor uptake for ^{111}In-3P-RGK$_2$ indicates that the radiolabeled cyclic RGD dimers are RGD-specific. The experimental data were adapted from Shi et al. (2011a)

INTEGRIN AND RGD SPECIFICITY

Integrin specificity

Blocking experiment (Fig. 8A) has been used to demonstrate the $\alpha_v\beta_3$ specificity of radiolabeled RGD peptides with a known $\alpha_v\beta_3$ antagonist (e.g., c(RGDfK) or RGD$_2$) as the blocking agent. This experiment is often performed by biodistribution or imaging (PET or SPECT). The blocking agent is pre- or co-injected with the radiotracer. Co-injection or pre-injection of excess blocking agents (such as RGD$_2$) will result in partial or complete blockage of the radiotracer tumor uptake (Fig. 8B). It is important to note that there is also a significant reduction in radiotracer uptake in the $\alpha_v\beta_3$-positive organs (e.g., eyes, intestine, kidneys, lungs, liver, muscle, and spleen). The normal organ uptake is consistent with the β_3 and CD31 staining data for the liver, kidneys, and lungs from the tumor-bearing athymic nude mice.

RGD specificity

There are several ways to determine the RGD specificity of radiolabeled cyclic RGD peptides, including: (1) the in vitro binding assay using ^{125}I-echistatin as the integrin-specific radioligand (Zhang et al. 2006; Wu et al. 2007; Wang et al. 2008b; Shi et al. 2009c), (2) the in vitro tissue or cellular immunohistochemical (IHC) staining assay using fluorescent probes (Zheng et al. 2014), (3) the in vivo imaging experiment (PET or SPECT) (Zhang et al. 2006; Wu et al. 2007; Wang et al. 2008b; Shi et al. 2009c), and (4) the biodistribution study (Shi et al. 2009a, 2011a, b; Chakraborty et al.

2010). In all cases, a nonsense peptide with the "scrambled sequence" will be used to prepare the corresponding radiotracer or fluorescent probe. For example, 3P-RGK$_2$ is the nonsense peptide with the composition identical to that of 3P-RGD$_2$. The $\alpha_v\beta_3$ binding affinity of DOTA-3P-RGK$_2$ (IC$_{50}$ = 596 ± 48 nmol/L) was >20× lower than that of DOTA-3P-RGD$_2$ (IC$_{50}$ = 29 ± 4 nmol/L). Similar results were also seen with FITC-3P-RGK$_2$ (IC$_{50}$ = 589 ± 73 nmol/L) and FITC-3P-RGD$_2$ (IC$_{50}$ = 32 ± 7 nmol/L). Because of the low $\alpha_v\beta_3$ affinity of DOTA-3P-RGK$_2$ (Chakraborty et al. 2010; Shi et al. 2011a, b), ^{111}In(DOTA-3P-RGK$_2$) had significantly lower ($p < 0.01$) uptake than ^{111}In(DOTA-3P-RGD$_2$) in the xenografted breast tumors and the $\alpha_v\beta_3$-positive normal organs, such as eyes, intestine, liver, lungs, and spleen (Fig. 8B) (Shi et al. 2011a). These results clearly show that the uptake of radiolabeled cyclic RGD peptides in tumors and some normal organs is indeed $\alpha_v\beta_3$-specific.

LINEAR RELATIONSHIP BETWEEN RADIOTRACER TUMOR UPTAKE AND A$_V$B$_3$ EXPRESSION

It has been shown that the radiolabeled cyclic RGD peptides are useful for non-invasive imaging of tumors in cancer patients (Beer et al. 2005, 2007, 2008; Haubner et al. 2005). It is the total $\alpha_v\beta_3$ level that will contribute the tumor uptake of a $\alpha_v\beta_3$-targeted radiotracer. The capability to visualize the $\alpha_v\beta_3$ expression provides new opportunities to characterize the tumor angiogenesis noninvasively, to select appropriate patients for antiangiogenic treatment, and to monitor the tumor response to antiangiogenic drugs. However,

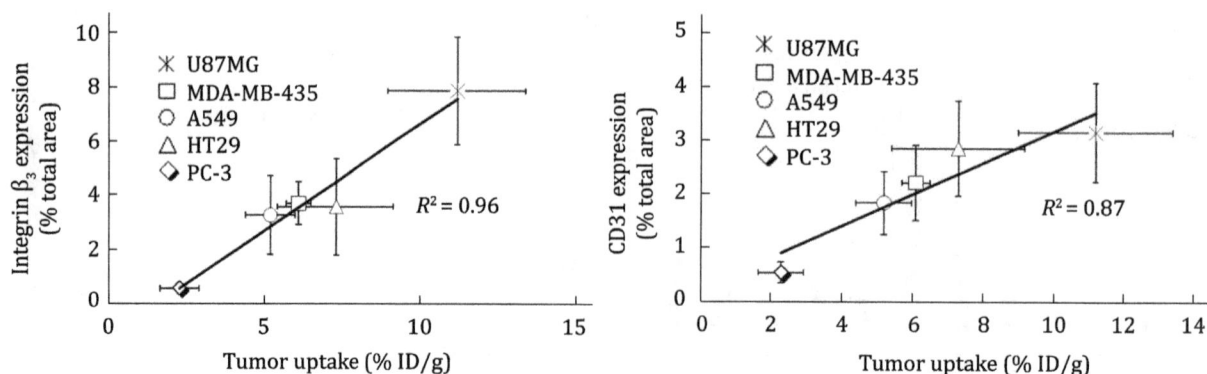

Fig. 9 Relationship between the tumor uptake (%ID/g: radioactivity density) and relative β_3 or CD31 levels in five xenografted tumors (U87MG, MDA-MB-435, A549, HT29, and PC-3). The total β_3 expression was represented by the percentage of red area over total area in each slice of tumor tissue. Each data point was derived from at least 15 different areas of same tissue ($\times 100$ magnification). Experiments were repeated three times independently with similar results. The experimental data were adapted from Zhou et al. (2011b)

there were only a few reports on the correlation between the $\alpha_v\beta_3$ expression levels and radiotracer tumor uptake (Beer et al. 2005, 2007, 2008; Haubner et al. 2005; Zhang et al. 2006).

^{99m}Tc-3P-RGD$_2$ was studied for its capability to monitor the $\alpha_v\beta_3$ expression in five different tumor-bearing animal models (U87MG, MDA-MB-435, A549, HT29, and PC-3). IHC staining was performed to determine the $\alpha_v\beta_3$ and CD31 (a biomarker for tumor vasculature) expression levels in xenografted U87MG, MDA-MB-435, A549, HT29, and PC-3 tumor tissues (Zhou et al. 2011b). It was found that the total $\alpha_v\beta_3$ expression levels on the tumor cells and tumor neo-vasculature follow the general ranking trend: U87MG > MDA-MB-435 = A549 = HT29 > PC-3. In contrast, the CD31 expression levels follow the general ranking order of U87MG = HT29 > MDA-MB-435 = A549 > PC-3 (Fig. 9). More importantly, there is an excellent relationship between the tumor uptake and the $\alpha_v\beta_3$ expression levels (Zhou et al. 2011b). The linear relationship between the tumor uptake (%ID/g) and $\alpha_v\beta_3$ density suggests that ^{99m}Tc-3P-RGD$_2$ is useful for non-invasive monitoring of the $\alpha_v\beta_3$ expression levels in cancer patients.

MONITORING TUMOR RESPONSE TO ANTIANGIOGENIC THERAPY

^{99m}Tc-3P-RGD$_2$ has been used to monitor the tumor response to antiangiogenesis treatment with linifanib (ABT-869) (Ji et al. 2013b, d), a multi-targeted receptor tyrosine kinase inhibitor targeting vascular endothelial growth factor (VEGF) and platelet-derived growth factor (PDGF) receptors (Albert et al. 2006; Shankar et al. 2007; Wong et al. 2009; Zhou et al. 2009;

Hernandez-Davies et al. 2011; Jiang et al. 2011; Tannir et al. 2011; Luo et al. 2012). We found that there was a significant decrease in tumor uptake (%ID/cm^3) and T/M ratios of ^{99m}Tc-3P-RGD$_2$ in the xenografted U87MG model, while no significant changes in tumor uptake of ^{99m}Tc-3P-RGD$_2$ were seen in the PC-3 model after linifanib treatment (Ji et al. 2013d). The uptake changes in MDA-MB-435 tumors were between those observed in the U87MG and PC-3 models (Ji et al. 2013b). This is consistent with the tumor $\alpha_v\beta_3$ expression levels (Zhou et al. 2011b). Highly vascularized tumors (e.g., U87MG) with higher level of $\alpha_v\beta_3$ and CD31 have better tumor response to linifanib therapy than poorly vascularized tumors (e.g., PC-3) with low levels of $\alpha_v\beta_3$ and CD31 (Fig. 10). Thus, ^{99m}Tc-3P-RGD$_2$ might be a screening tool to select appropriate patients who will benefits most antiangiogenic treatment. If the tumor has a high $\alpha_v\beta_3$ expression, as indicated by high tumor uptake of ^{99m}Tc-3P-RGD$_2$ at the time of diagnosis, antiangiogenic therapy would more likely be effective. If the tumor has little $\alpha_v\beta_3$ expression (low uptake of ^{99m}Tc-3P-RGD$_2$), antiangiogenic therapy would not be effective regardless the amount of antiangiogenic drug administered into the patient.

MONITORING TUMOR METASTASIS

^{99m}Tc-3P-RGD$_2$ SPECT/CT has been used as a noninvasive imaging tool to monitor the tumor growth and progression of breast cancer lung metastasis (Albert et al. 2006; Ji et al. 2013d). Figure 11 shows the SPECT/CT images of athymic nude mice ($n = 8$) with breast cancer lung metastasis. As expected, the SPECT/CT images showed no detectable metastatic breast tumor lesions in the lungs at week 4 (Fig. 11:

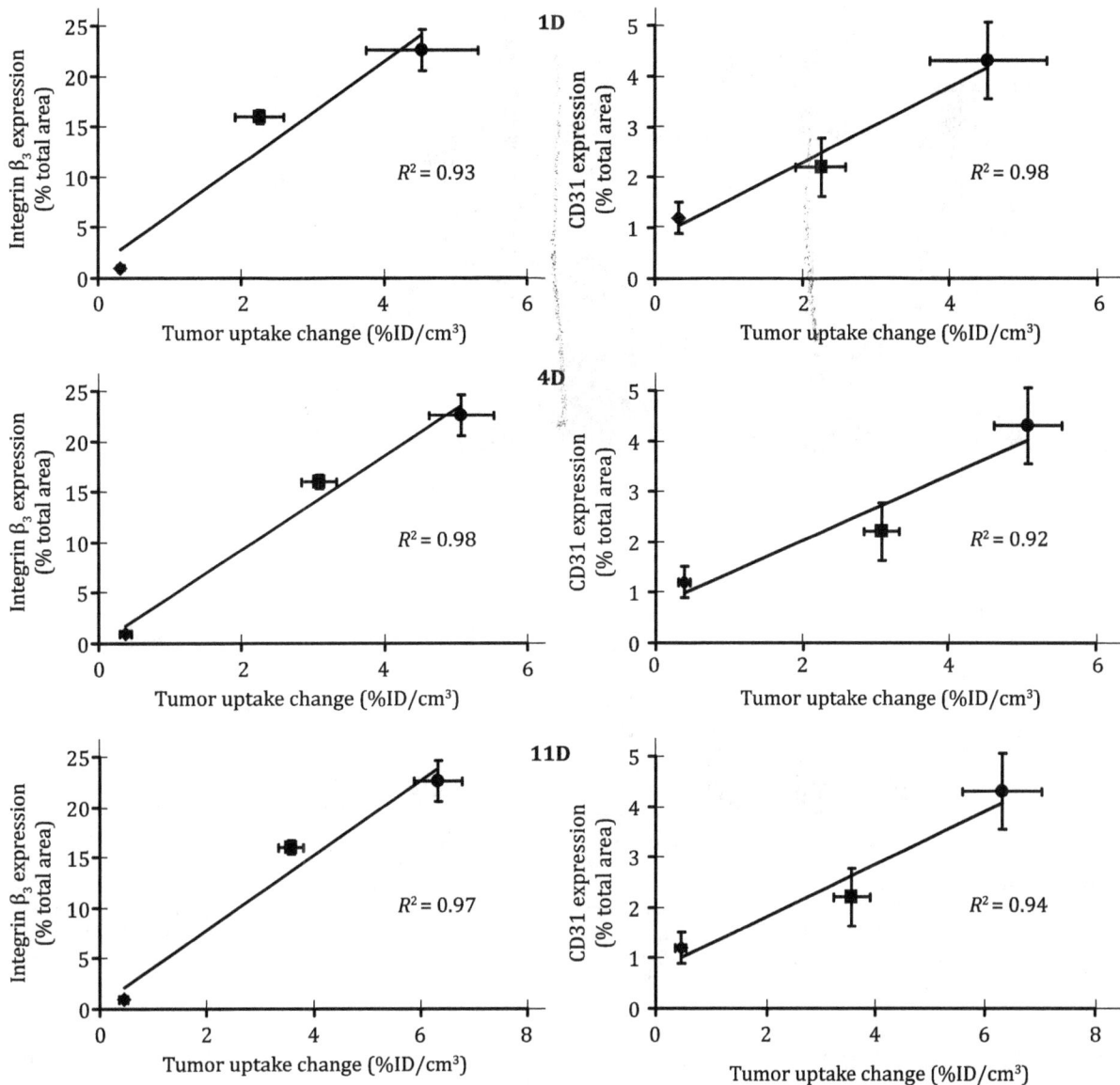

Fig. 10 Linear relationship between the %ID/cm^3 tumor uptake change at days 1 (*top*), 4 (*middle*) and 11 (*bottom*) after linifanib therapy and the expression levels of the $\alpha_v\beta_3$ (*left*) and CD31 (*right*) in three tumor-bearing animal models. The %ID/cm^3 tumor uptake values of ^{99m}Tc-3P-RGD$_2$ were calculated from SPECT/CT quantification and reported as the mean plus/minus standard error of the mean based on results from five animals ($n = 5$). The %ID/cm^3 tumor uptake change was calculated by deducting the %ID/cm^3 tumor uptake of ^{99m}Tc-3P-RGD$_2$ on days 1, 4, and 11 from its original value on -1 day (before linifanib therapy) in the same animal. The average %ID/cm^3 tumor uptake change is used as the indicator of tumor response to linifanib treatment. The experimental data were adapted from Zheng et al. (2014)

top). By week 6, small breast cancer lesions started to appear in the mediastinum and lungs. At week 8, SPECT/CT images revealed many metastatic cancer lesions in both lungs (Albert et al. 2006). Figure 11 (bottom) compares the %ID (left) and %ID/cm^3 (right) uptake values of ^{99m}Tc-3P-RGD$_2$ in the lungs. Even though the lung uptake of ^{99m}Tc-3P-RGD$_2$ (0.41 ± 0.05 %ID) at week 4 seemed to be higher than that in the control animals (0.36 ± 0.06 %ID),

this difference was not significant ($p > 0.05$) within the experimental errors. At week 6, the tumor burden in the lungs became significant. The lung uptake of ^{99m}Tc-3P-RGD$_2$ was much higher (0.89 ± 0.12 %ID, $p < 0.01$) than that in the control group. By week 8, the uptake of ^{99m}Tc-3P-RGD$_2$ in the lungs was increase to 1.40 ± 0.42 %ID. In all cases, the lung size remained relatively unchanged (1.21–1.32 cm^3) during the 8-week study period.

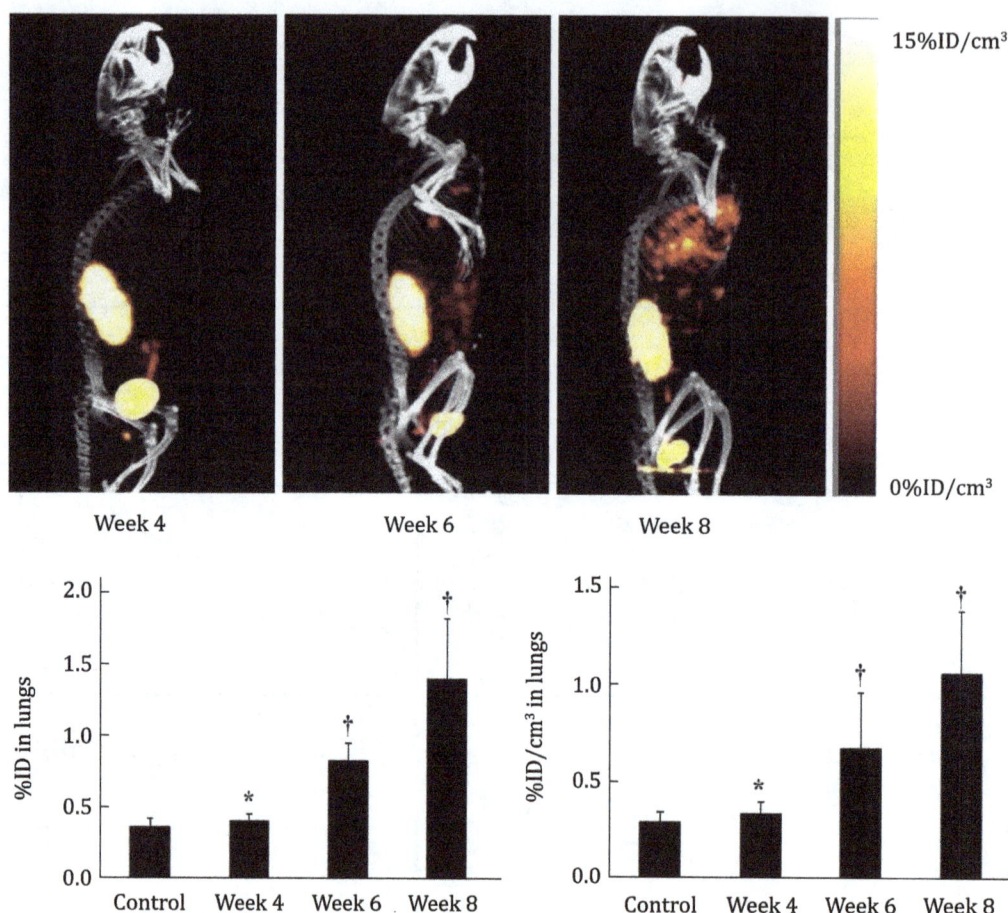

Fig. 11 Top: The 3D views of SPECT/CT images of an athymic nude mouse at week 4, 6, and 8 after tail-vein injection of 1.0×10^6 MDA-MB-231 cells. **Bottom**: The %ID (*left*) and %ID/cm^3 (*right*) uptake values of ^{99m}Tc-3P-RGD$_2$ in the lungs obtained from SPECT/CT quantification in the athymic nude mice ($n = 8$) at week 4, 6, and 8 after tail-vein injection of 1.0×10^6 MDA-MB-231 cells. Normal animals ($n = 4$) were used in the control group. $^{\dagger}p < 0.05$, significantly different from the control group; $^*p > 0.05$, no significant difference from the control group. The imaging and SPECT quantification data were from Ji et al. (2013b)

CLINICAL EXPERIENCES WITH ^{99m}TC-3P-RGD$_2$

The excellent in vivo tumor-targeting efficacy of ^{99m}Tc-3P-RGD$_2$ in animal models guaranteed its further clinical application. In a first-in-human study, ^{99m}Tc-3P-RGD$_2$ was investigated for its capability to noninvasively differentiate solitary pulmonary nodules (SPNs) (Ma et al. 2011). Among the 21 patients with SPNs, 15 (71%) were diagnosed as malignant while 6 (29%) were benign. The sensitivities for CT interpretation and ^{99m}Tc-3P-RGD$_2$ SPECT visual were 80% and 100%, respectively. All SPNs classified as indeterminate via CT can be sensitively diagnosed by ^{99m}Tc-3P-RGD$_2$ scintigraphy. ^{99m}Tc-3P-RGD$_2$ uptake in the malignant and benign nodules was well confirmed by ex vivo IHC staining of $\alpha_v\beta_3$. These results demonstrated the feasibility of using ^{99m}Tc-3P-RGD$_2$ scintigraphy in differentiating SPNs (Ma et al. 2011). A multicenter study was performed in 70 patients with suspected lung lesions

(Zhu et al. 2012). The results clearly demonstrated that ^{99m}Tc-3P-RGD$_2$ SPECT effectively detects lung malignancies, but with relatively low specificity. Whole-body planar scanning and chest SPECT are complementary for the evaluation of primary tumor and metastasis (Zhu et al. 2012). In a recently study, the potential of ^{99m}Tc-3P-RGD$_2$ SPECT in the detection of RAIR DTC lesions was conducted (Zhao et al. 2012). ^{99m}Tc-3P-RGD$_2$ SPECT identified all the target RAIR metastatic lesions, and there was a significant correlation between the mean tumor-to-background ratios and mean growth rates of target lesions. It is concluded that ^{99m}Tc-3P-RGD$_2$ imaging can be used for the localization and growth evaluation of RAIR lesions, thus providing a promising imaging strategy to monitor the efficacy of antiangiogenic therapy (Zhao et al. 2012). ^{99m}Tc-3P-RGD$_2$ SPECT was also evaluated and compared to ^{99m}Tc-MIBI for the capability to assess the breast cancer lessons (Ma et al. 2014). It was found that the mean T/NT ratio of ^{99m}Tc-3P-RGD$_2$ in malignant lesions was significantly higher than that in benign lesions (3.54 ± 1.51 vs. 1.83 ± 0.98, $p < 0.001$). The sensitivity, specificity, and accuracy of ^{99m}Tc-3P-RGD$_2$ SMM were 89.3%, 90.9%, and 89.7%, respectively, with a T/NT cut-off value of 2.40. The mean T/NT ratio of ^{99m}Tc-MIBI in malignant lesions was also significantly higher than that in benign lesions (2.86 ± 0.99 vs. 1.51 ± 0.61, $p < 0.001$). The sensitivity, specificity, and accuracy of ^{99m}Tc-MIBI SMM were 87.5%, 72.7%, and 82.1%, respectively, with a T/NT cut-off value of 1.45. According to the ROC analysis, the area under the curve for ^{99m}Tc-3P-RGD$_2$ SMM (area = 0.851) was higher than that for ^{99m}Tc-MIBI SMM (area = 0.781), but the statistical difference was not significant.

CLINICAL EXPERIENCES WITH ^{18}F-ALFATIDE AND ^{18}F-ALFATIDE II

^{18}F-labeled RGD compounds suffer from multistep and time-consuming synthetic procedures, which will limit their clinic availability. To overcome this shortcoming, the Al(NOTA) chelate has been used for ^{18}F-labeling of P-RGD$_2$ (Lang et al. 2011). The application of NOTA-AlF chelation chemistry and kit formulation allows one-step ^{18}F-labeling. Under the optimal conditions, the radiotracer [^{18}F]AlF(NOTA-P-RGD$_2$) (denoted as ^{18}F-Alfatide) was prepared in relatively high yield (42.1 ± 0.02) with more than 95% radiochemical purity. The whole radiosynthesis including post-labeling chromatographic purification was accomplished within 20 min. Nine patients with a primary diagnosis of lung cancer were examined by both static and dynamic PET imaging with ^{18}F-alfatide, and one tuberculosis patient was investigated using both ^{18}F-alfatide and ^{18}F-FDG imaging. It was found that ^{18}F-alfatide PET identified all tumors, with mean standardized uptake values of 2.90 ± 0.10. Tumor-to-muscle and tumor-to-blood ratios were 5.87 ± 2.02 and 2.71 ± 0.92, respectively. It was concluded that PET scanning with ^{18}F-alfatide allows specific imaging of avb3 expression with good contrast in lung cancer patients.

CONCLUSIONS

Over the last several years, many multimeric cyclic RGD peptides have been used to increase the radiotracer tumor-targeting capability. The fact that radiolabeled (^{18}F, ^{99m}Tc, ^{111}In, ^{64}Cu, and ^{68}Ga) cyclic RGD peptides to target multiple integrins ($\alpha_v\beta_3$, $\alpha_v\beta_5$, $\alpha_5\beta_1$, $\alpha_6\beta_4$, $\alpha_4\beta_1$, and $\alpha_v\beta_6$) will help to improve their tumor uptake due

to the "increased receptor population." In order to achieve bivalency, the distance between two cyclic RGD motifs must be long enough so that they will be able to bind the two adjacent $\alpha_v\beta_3$ sites simultaneously. Multimerization increases the uptake of radiolabeled multimeric cyclic RGD peptides in both the tumor and normal organs, and also their tumor retention times. Among the radiotracers evaluated in tumor-bearing models, the radiolabeled cyclic RGD dimers (e.g., 2P-RGD$_2$, 3P-RGD$_2$, 2G-RGD$_2$, 3G-RGD$_2$, and Galacto-RGD$_2$) show the most promising results with respect to their tumor uptake and T/B ratios. ^{99m}Tc-3P-RGD$_2$, ^{18}F-Alfatide, and ^{18}F-Alfatide II are currently under clinical investigation for tumor imaging by SPECT or PET. ^{99m}Tc-3P-RGD$_2$ offers significant advantages over both ^{18}F-Alfatide and ^{18}F-Alfatide\II because it could be routinely prepared in high yield and radiochemical purity (>95%) without post-labeling chromatographic purification and clinical availability of ^{99}Mo-^{99m}Tc generators. However, SPECT has limitations in quantification of radiotracer uptake, the speed of dynamic imaging, spatial resolution, and tissue attenuation.

Abbreviations

General terms

DCE-MRI	Dynamic contrast-enhanced magnetic resonance imaging
FITC	Fluorescein isothiocyanate isomer I
^{18}F-FDG	2-Deoxy-2-(^{18}F)fluoro-D-glucose
IHC	Immunohistochemistry
MRI	Magnetic resonance imaging
PET	Positron emission tomography
PDGFR	Platelet-derived growth factor receptors
SPECT	Single-photon emission computed tomography
VEGFR	Vascular endothelial growth factor receptors

Chelators

DOTA	1,4,7,10-Tetraazacyclododecane-1,4,7,10-tetracetic acid
HYNIC	6-Hydazinonicotinic acid
NOTA	1,4,7-triazacyclononane-1,4,7-triacetic acid

Cyclic peptides

Galacto-RGD$_2$	Glu[cyclo[Arg-Gly-Asp-D-Phe-Lys(SAA-PEG$_2$-(1,2,3-triazole)-1-yl-4-methylamide)]]$_2$ (SAA = 7-amino-L-glycero-L-galacto-2,6-anhydro-7-deoxyheptanamide, and PEG$_2$ = 3,6-dioxaoctanoic acid)
P-RGD	PEG$_4$-c(RGKfD) = cyclo(Arg-Gly-Asp-D-Phe-Lys(PEG$_4$)) (PEG$_4$ = 15-amino-4,7,10,13-tetraoxapentadecanoic acid)

RGD$_2$	E[c(RGDfK)]$_2$ = Glu[cyclo(Arg-Gly-Asp-D-Phe-Lys)]$_2$
P-RGD$_2$	PEG$_4$-E[c(RGDfK)]$_2$ = PEG$_4$-Glu[cyclo(Arg-Gly-Asp-D-Phe-Lys)]$_2$
2G-RGD$_2$	E[G$_3$-c(RGDfK)]$_2$ = Glu[cyclo[Arg-Asp-D-Phe-Lys(G$_3$)]]$_2$ (G$_3$ = Gly-Gly-Gly)
2P-RGD$_2$	E[PEG$_4$-c(RGDfK)]$_2$ = Glu[cyclo[Arg-Gly-Asp-D-Phe-Lys(PEG$_4$)]]$_2$
3G-RGD$_2$	G$_3$-E[G$_3$-c(RGDfK)]$_2$ = G$_3$-Glu[cyclo[Arg-Gly-Asp-D-Phe-Lys(G$_3$)]]$_2$
3P-RGD$_2$	PEG$_4$-E[PEG$_4$-c(RGDfK)]$_2$ = PEG$_4$-Glu[cyclo[Arg-Gly-Asp-D-Phe-Lys(PEG$_4$)]]$_2$
3P-RGK$_2$	PEG$_4$-E[PEG$_4$-c(RGDfK)]$_2$ = PEG$_4$-Glu[cyclo[Arg-Gly-Lys(PEG$_4$)-D-Phe-Asp]]$_2$)
RGD$_4$	E{E[c(RGDfK)]$_2$}$_2$ = Glu{Glu[cyclo(Arg-Gly-Asp-D-Phe-Lys)]$_2$}$_2$
6G-RGD$_4$	E{G$_3$-E[G$_3$-c(RGDfK)]$_2$}$_2$ = Glu{G$_3$-Glu[cyclo(Lys(G$_3$)-Arg-Gly-Asp-D-Phe)]-cyclo(Lys(G$_3$)-Arg-Gly-Asp-D-Phe)}-{PEG$_4$-Glu[cyclo(Lys(G$_3$)-Arg-Gly-Asp-D-Phe)]-cyclo(Lys(G$_3$)-Arg-Gly-Asp-D-Phe)}
6P-RGD$_4$	E{PEG$_4$-E[PEG$_4$-c(RGDfK)]$_2$}$_2$ = Glu{PEG$_4$-Glu[cyclo(Lys(PEG$_4$)-Arg-Gly-Asp-D-Phe)]-cyclo(Lys(PEG$_4$)-Arg-Gly-Asp-D-Phe)}-{PEG$_4$-Glu[cyclo(Lys(PEG$_4$)-Arg-Gly-Asp-D-Phe)]-cyclo(Lys(PEG$_4$)-Arg-Gly-Asp-D-Phe)}

Bioconjugates of cyclic peptides

DOTA-RGD	DOTA-c(RGDfK)
DOTA-P-RGD	DOTA-PEG$_4$-c(RGDfK)
DOTA-RGD$_2$	DOTA-E[c(RGDfK)]$_2$
DOTA-P-RGD$_2$	DOTA-PEG$_4$-E[c(RGDfK)]$_2$
DOTA-2G-RGD$_2$	DOTA-E[G$_3$-c(RGDfK)]$_2$
DOTA-2P-RGD$_2$	DOTA-E[PEG$_4$-c(RGDfK)]$_2$
DOTA-3G-RGD$_2$	DOTA-G$_3$-E[G$_3$-c(RGDfK)]$_2$
DOTA-3P-RGD$_2$	DOTA-PEG$_4$-E[PEG$_4$-c(RGDfK)]$_2$
DOTA-3P-RGK$_2$	DOTA-PEG$_4$-E[PEG$_4$-c(RGDfK)]$_2$
DOTA-Galacto-RGD$_2$	DOTA-Glu[cyclo[Arg-Gly-Asp-D-Phe-Lys(SAA-PEG$_2$-(1,2,3-triazole)-1-yl-4-methylamide)]]$_2$
DOTA-RGD$_4$	DOTA-E{E[c(RGDfK)]$_2$}$_2$
DOTA-6G-RGD$_4$	DOTA-E{G$_3$-E[G$_3$-c(RGDfK)]$_2$}$_2$
DOTA-6G-RGD$_4$	E{PEG$_4$-E[PEG$_4$-c(RGDfK)]$_2$}$_2$
FITC-3P-RGD$_2$	FITC-PEG$_4$-E[PEG$_4$-c(RGDfK)]$_2$
FITC-3P-RGK$_2$	FITC-PEG$_4$-E[PEG$_4$-c(RGDfK)]$_2$
HYNIC-RGD$_2$	HYNIC-E[c(RGDfK)]$_2$
HYNIC-P-RGD$_2$	HYNIC-PEG$_4$-E[c(RGDfK)]$_2$
HYNIC-2G-RGD$_2$	HYNIC-E[G$_3$-c(RGDfK)]$_2$
HYNIC-2P-RGD$_2$	HYNIC-E[PEG$_4$-c(RGDfK)]$_2$
HYNIC-3G-RGD$_2$	HYNIC-G$_3$-E[G$_3$-c(RGDfK)]$_2$
HYNIC-3P-RGD$_2$	HYNIC-PEG$_4$-E[PEG$_4$-c(RGDfK)]$_2$
HYNIC-Galacto-RGD$_2$	HYNIC-Glu[cyclo[Arg-Gly-Asp-D-Phe-Lys(SAA-PEG$_2$-(1,2,3-triazole)-1-yl-4-methylamide)]]$_2$
HYNIC-RGD$_4$	HYNIC-E{E[c(RGDfK)]$_2$}$_2$
NOTA-P-RGD$_2$	NOTA-PEG$_4$-E[c(RGDfK)]$_2$
NOTA-2G-RGD$_2$	NOTA-E[G$_3$-c(RGDfK)]$_2$
NOTA-2P-RGD$_2$	NOTA-E[PEG$_4$-c(RGDfK)]$_2$
NOTA-3G-RGD$_2$	NOTA-G$_3$-E[G$_3$-c(RGDfK)]$_2$
NOTA-3P-RGD$_2$	NOTA-PEG$_4$-E[PEG$_4$-c(RGDfK)]$_2$

Radiolabeled cyclic RGD peptides

^{18}F-Alfatide	[^{18}F]AlF(NOTA-P-RGD$_2$)
^{18}F-Alfatide II	[^{18}F]AlF(NOTA-2P-RGD$_2$)
^{18}F-Galacto-RGD	2-[^{18}F]fluoropropanamide c(RGDfK(SAA), SAA = 7-amino-L-glyero-L-galacto-2,6-anhydro-7-deoxyheptanamide)
^{64}Cu-P-RGD$_2$	^{64}Cu(DOTA-P-RGD$_2$)
^{64}Cu-2G-RGD$_2$	^{64}Cu(DOTA-2G-RGD$_2$)
^{64}Cu-2P-RGD$_2$	^{64}Cu(DOTA-2P-RGD$_2$)
^{64}Cu-3G-RGD$_2$	^{64}Cu(DOTA-3G-RGD$_2$)
^{64}Cu-3P-RGD$_2$	^{64}Cu(DOTA-3P-RGD$_2$)
^{68}Ga-3G-RGD$_2$	^{68}Ga(DOTA-3G-RGD$_2$)
^{68}Ga-3P-RGD$_2$	^{68}Ga(DOTA-3P-RGD$_2$)
^{111}In-P-RGD	^{111}In(DOTA-P-RGD)
^{111}In-P-RGD$_2$	^{111}In(DOTA-P-RGD$_2$)
^{111}In-2G-RGD$_2$	^{111}In(DOTA-2G-RGD$_2$)
^{111}In-2P-RGD$_2$	^{111}In(DOTA-2P-RGD$_2$)
^{111}In-3G-RGD$_2$	^{111}In(DOTA-3G-RGD$_2$)
^{111}In-3P-RGD$_2$	^{111}In(DOTA-3P-RGD$_2$)
^{111}In-Galacto-RGD$_2$	^{111}In(DOTA-Galacto-RGD$_2$)
^{111}In-6G-RGD$_4$	^{111}In(DOTA-6G-RGD$_4$)
^{111}In-6P-RGD$_4$	^{111}In(DOTA-6P-RGD$_4$)
^{99m}Tc-Galacto-RGD$_2$	[^{99m}Tc(HYNIC-Galacto-RGD$_2$)(tricine)(TPPTS)])
^{99m}Tc-RGD$_2$	[^{99m}Tc(HYNIC-RGD$_2$)(tricine)(TPPTS)])
^{99m}Tc-P-RGD$_2$	[^{99m}Tc(HYNIC-P-RGD$_2$)(tricine)(TPPTS)]
^{99m}Tc-2G-RGD$_2$	[^{99m}Tc(HYNIC-2G-RGD$_2$)(tricine)(TPPTS)]
^{99m}Tc-2P-RGD$_2$	[^{99m}Tc(HYNIC-2P-RGD$_2$)(tricine)(TPPTS)]
^{99m}Tc-3G-RGD$_2$	[^{99m}Tc(HYNIC-3G-RGD$_2$)(tricine)(TPPTS)]
^{99m}Tc-3P-RGD$_2$	[^{99m}Tc(HYNIC-3P-RGD$_2$)(tricine)(TPPTS)]
^{99m}Tc-RGD$_4$	[^{99m}Tc(HYNIC-RGD$_4$)(tricine)(TPPTS)])

Acknowledgments This work was supported, in part, by Purdue University and R21 EB017237-01 (S. L.) from the National Institute of Biomedical Imaging and Bioengineering (NIBIB).

Compliance with Ethical Standards

Conflict of Interest Jiyun Shi, Fan Wang, and Shuang Liu declare that they have no conflict of interest.

Human and Animal Rights and Informed Consent This article does not contain any studies with human or animal subjects performed by any of the authors.

References

Albelda SM, Mette SA, Elder DE, Stewart R, Damjanovich L, Herlyn M, Buck CA (1990) Integrin distribution in malignant melanoma: association of the β_3 subunit with tumor progression. Cancer Res 50:6757–6764

Albert DH, Tapang P, Magoc TJ, Pease LJ, Reuter DR, Wei RQ, Li J, Guo J, Bousquet PF, Ghoreishi-Haack NS, Wang B, Bukofzer GT, Wang YC, Stavropoulos JA, Hartandi K, Niquette AL, Soni N, Johnson EF, McCall JO, Bouska JJ et al (2006) Preclinical activity of ABT-869, a multitargeted receptor tyrosine kinase inhibitor. Mol Cancer Ther 5:995–1006

Alves S, Correia JD, Gano L, Rold TL, Prasanphanich A, Haubner R, Rupprich M, Alberto R, Decristoforo C, Santos I, Smith CJ (2007) In vitro and in vivo evaluation of a novel ^{99m}Tc(CO)3-pyrazolyl conjugate of cyclo-(Arg-Gly-Asp-D-Tyr-Lys). Bioconjug Chem 18:530–537

Anderson CJ, Green MA, Fujibayashi Y (2003) Chemistry of copper radionuclides and radiopharmaceutical products. Handb Radiopharm Radiochem Appl 401–422

Aumailley M, Gurrath M, Muller G, Calvete J, Timpl R, Kessler H (1991) Arg-Gly-Asp constrained within cyclic pentapeptides. Strong and selective inhibitors of cell adhesion to vitronectin and laminin fragment P1. FEBS Lett 291:50–54

Bakewell SJ, Nestor P, Prasad S, Tomasson MH, Dowland N, Mehrotra M, Scarborough R, Kanter J, Abe K, Phillips D, Weilbaecher KN (2003) Platelet and osteoclast β_3 integrins are critical for bone metastasis. Proc Natl Acad Sci USA 100:14205–14210

Barczyk M, Carracedo S, Gullberg D (2010) Integrins. Cell Tissue Res 339:269–280

Becaud J, Mu L, Karramkam M, Schubiger PA, Ametamey SM, Graham K, Stellfeld T, Lehmann L, Borkowski S, Berndorff D, Dinkelborg L, Srinivasan A, Smits R, Koksch B (2009) Direct one-step ^{18}F-labeling of peptides via nucleophilic aromatic substitution. Bioconjug Chem 20:2254–2261

Beer AJ, Chen X (2010) Imaging of angiogenesis: from morphology to molecules and from bench to bedside. Eur J Nucl Med Mol Imaging 37(Suppl 1):S1–S3

Beer AJ, Schwaiger M (2008) Imaging of integrin $\alpha_v\beta_3$ expression. Cancer Metastasis Rev 27:631–644

Beer AJ, Haubner R, Goebel M, Luderschmidt S, Spilker ME, Wester HJ, Weber WA, Schwaiger M (2005) Biodistribution and

pharmacokinetics of the $\alpha_v\beta_3$-selective tracer ^{18}F-galacto-RGD in cancer patients. J Nucl Med 46:1333–1341

Beer AJ, Grosu AL, Carlsen J, Kolk A, Sarbia M, Stangier I, Watzlowik P, Wester HJ, Haubner R, Schwaiger M (2007) [^{18}F]galacto-RGD positron emission tomography for imaging of $\alpha_v\beta_3$ expression on the neovasculature in patients with squamous cell carcinoma of the head and neck. Clin Cancer Res 13:6610–6616

Beer AJ, Niemeyer M, Carlsen J, Sarbia M, Nahrig J, Watzlowik P, Wester HJ, Harbeck N, Schwaiger M (2008) Patterns of $\alpha_v\beta_3$ expression in primary and metastatic human breast cancer as shown by ^{18}F-Galacto-RGD PET. J Nucl Med 49:255–259

Beer AJ, Kessler H, Wester HJ, Schwaiger M (2011) PET imaging of integrin $\alpha_v\beta_3$ expression. Theranostics 1:48–57

Bello L, Francolini M, Marthyn P, Zhang J, Carroll RS, Nikas DC, Strasser JF, Villani R, Cheresh DA, Black PM (2001) $\alpha_v\beta_3$ and $\alpha_v\beta_5$ integrin expression in glioma periphery. Neurosurgery 49:380–389 (discussion 390)

Bianchi-Smiraglia A, Paesante S, Bakin AV (2013) Integrin β_5 contributes to the tumorigenic potential of breast cancer cells through the Src-FAK and MEK-ERK signaling pathways. Oncogene 32:3049–3058

Boger C, Kalthoff H, Goodman SL, Behrens HM, Rocken C (2014) Integrins and their ligands are expressed in non-small cell lung cancer but not correlated with parameters of disease progression. Virchows Archiv 464:69–78

Cai W, Chen X (2008) Multimodality molecular imaging of tumor angiogenesis. J Nucl Med 49(Suppl 2):113S–128S

Carreiras F, Denoux Y, Staedel C, Lehmann M, Sichel F, Gauduchon P (1996) Expression and localization of αv integrins and their ligand vitronectin in normal ovarian epithelium and in ovarian carcinoma. Gynecol Oncol 62:260–267

Chakraborty S, Liu S (2010) (99m)Tc and (111)In-labeling of small biomolecules: bifunctional chelators and related coordination chemistry. Curr Top Med Chem 10:1113–1134

Chakraborty S, Shi J, Kim YS, Zhou Y, Jia B, Wang F, Liu S (2010) Evaluation of ^{111}In-labeled cyclic RGD peptides: tetrameric not tetravalent. Bioconjug Chem 21:969–978

Cooper CR, Chay CH, Pienta KJ (2002) The role of $\alpha_v\beta_3$ in prostate cancer progression. Neoplasia 4:191–194

Correia JDG, Paulo A, Raposinho PD, Santos I (2011) Radiometallated peptides for molecular imaging and targeted therapy. Dalton Trans 40:6144–6167

D'Andrea LD, Del Gatto A, Pedone C, Benedetti E (2006) Peptide-based molecules in angiogenesis. Chem Biol Drug Des 67:115–126

Danhier F, Le Breton A, Preat V (2012) RGD-based strategies to target $\alpha_v\beta_3$ integrin in cancer therapy and diagnosis. Mol Pharm 9:2961–2973

Dijkgraaf I, Boerman OC (2010) Molecular imaging of angiogenesis with SPECT. Eur J Nucl Med Mol Imaging 37(Suppl 1):S104–S113

Dijkgraaf I, Kruijtzer JA, Liu S, Soede AC, Oyen WJ, Corstens FH, Liskamp RM, Boerman OC (2007a) Improved targeting of the $\alpha_v\beta_3$ integrin by multimerisation of RGD peptides. Eur J Nucl Med Mol Imaging 34:267–273

Dijkgraaf I, Liu S, Kruijtzer JA, Soede AC, Oyen WJ, Liskamp RM, Corstens FH, Boerman OC (2007b) Effects of linker variation on the in vitro and in vivo characteristics of an ^{111}In-labeled RGD peptide. Nucl Med Biol 34:29–35

Dittmar T, Heyder C, Gloria-Maercker E, Hatzmann W, Zanker KS (2008) Adhesion molecules and chemokines: the navigation system for circulating tumor (stem) cells to metastasize in an organ-specific manner. Clin Exp Metastasis 25:11–32

Dolle F (2005) Fluorine-18-labelled fluoropyridines: advances in radiopharmaceutical design. Curr Pharm Des 11:3221–3235

D'Souza CA, McBride WJ, Sharkey RM, Todaro LJ, Goldenberg DM (2011) High-yielding aqueous ^{18}F-labeling of peptides via Al^{18}F chelation. Bioconjug Chem 22:1793–1803

Dumont RA, Deininger F, Haubner R, Maecke HR, Weber WA, Fani M (2011) Novel (64)Cu- and (68)Ga-labeled RGD conjugates show improved PET imaging of $\alpha_v\beta_3$ integrin expression and facile radiosynthesis. J Nucl Med 52:1276–1284

Erdreich-Epstein A, Shimada H, Groshen S, Liu M, Metelitsa LS, Kim KS, Stins MF, Seeger RC, Durden DL (2000) Integrins $\alpha_v\beta_3$ and $\alpha_v\beta_5$ are expressed by endothelium of high-risk neuroblastoma and their inhibition is associated with increased endogenous ceramide. Cancer Res 60:712–721

Falcioni R, Cimino L, Gentileschi MP, D'Agnano I, Zupi G, Kennel SJ, Sacchi A (1994) Expression of β_1, β_3, β_4, and β_5 integrins by human lung carcinoma cells of different histotypes. Exp Cell Res 210:113–122

Fani M, Maecke HR (2012) Radiopharmaceutical development of radiolabelled peptides. Eur J Nucl Med Mol Imaging 39:11–30

Fani M, Maecke HR, Okarvi SM (2012) Radiolabeled peptides: valuable tools for the detection and treatment of cancer. Theranostics 2:481–501

Felding-Habermann B, Habermann R, Saldivar E, Ruggeri ZM (1996) Role of β_3 integrins in melanoma cell adhesion to activated platelets under flow. J Biol Chem 271:5892–5900

Gaertner FC, Kessler H, Wester HJ, Schwaiger M, Beer AJ (2012) Radiolabelled RGD peptides for imaging and therapy. Eur J Nucl Med Mol Imaging 39(Suppl 1):S126–S138

Glaser M, Morrison M, Solbakken M, Arukwe J, Karlsen H, Wiggen U, Champion S, Kindberg GM, Cuthbertson A (2008) Radiosynthesis and biodistribution of cyclic RGD peptides conjugated with novel [^{18}F]fluorinated aldehyde-containing prosthetic groups. Bioconjug Chem 19:951–957

Goodman SL, Grote HJ, Wilm C (2012) Matched rabbit monoclonal antibodies against αv-series integrins reveal a novel $\alpha_v\beta_3$-LIBS epitope, and permit routine staining of archival paraffin samples of human tumors. Biol Open 1:329–340

Gottschalk KE, Kessler H (2002) The structures of integrins and integrin-ligand complexes: implications for drug design and signal transduction. Angew Chem 41:3767–3774

Graf MR, Prins RM, Poulsen GA, Merchant RE (2003) Contrasting effects of interleukin-2 secretion by rat glioma cells contingent upon anatomical location: accelerated tumorigenesis in the central nervous system and complete rejection in the periphery. J Neuroimmunol 140:49–60

Gupta A, Cao W, Chellaiah MA (2012) Integrin $\alpha_v\beta_3$ and CD44 pathways in metastatic prostate cancer cells support osteoclastogenesis via a Runx2/Smad 5/receptor activator of NF-κB ligand signaling axis. Mol Cancer 11:66

Gurrath M, Muller G, Kessler H, Aumailley M, Timpl R (1992) Conformation/activity studies of rationally designed potent anti-adhesive RGD peptides. Eur J Biochem/FEBS 210:911–921

Haubner R, Gratias R, Diefenbach B, Goodman SL, Jonczyk A, Kessler H (1996) Structural and functional aspects of RGD-containing cyclic pentapeptides as highly potent and selective integrin $\alpha_v\beta_3$ antagonists. J Am Chem Soc 118:7461–7472

Haubner R, Weber WA, Beer AJ, Vabuliene E, Reim D, Sarbia M, Becker KF, Goebel M, Hein R, Wester HJ, Kessler H, Schwaiger M (2005) Noninvasive visualization of the activated $\alpha_v\beta_3$ integrin in cancer patients by positron emission tomography and [^{18}F]Galacto-RGD. PLoS Med 2:e70

Haubner R, Beer AJ, Wang H, Chen X (2010) Positron emission tomography tracers for imaging angiogenesis. Eur J Nucl Med Mol Imaging 37(Suppl 1):S86–S103

Hausner SH, Marik J, Gagnon MK, Sutcliffe JL (2008) In vivo positron emission tomography (PET) imaging with an αvβ6

specific peptide radiolabeled using ^{18}F-"click" chemistry: evaluation and comparison with the corresponding 4-[^{18}F]fluorobenzoyl- and 2-[^{18}F]fluoropropionyl-peptides. J Med Chem 51:5901–5904

Henze M, Dimitrakopoulou-Strauss A, Milker-Zabel S, Schuhmacher J, Strauss LG, Doll J, Macke HR, Eisenhut M, Debus J, Haberkorn U (2005) Characterization of ^{68}Ga-DOTA-D-Phe1-Tyr3-octreotide kinetics in patients with meningiomas. J Nucl Med 46:763–769

Hernandez-Davies JE, Zape JP, Landaw EM, Tan X, Presnell A, Griffith D, Heinrich MC, Glaser KB, Sakamoto KM (2011) The multitargeted receptor tyrosine kinase inhibitor linifanib (ABT-869) induces apoptosis through an Akt and glycogen synthase kinase 3β-dependent pathway. Mol Cancer Ther 10:949–959

Hodivala-Dilke K (2008) $\alpha_v\beta_3$ integrin and angiogenesis: a moody integrin in a changing environment. Curr Opin Cell Biol 20:514–519

Hohne A, Mu L, Honer M, Schubiger PA, Ametamey SM, Graham K, Stellfeld T, Borkowski S, Berndorff D, Klar U, Voigtmann U, Cyr JE, Friebe M, Dinkelborg L, Srinivasan A (2008) Synthesis, ^{18}F-labeling, and in vitro and in vivo studies of bombesin peptides modified with silicon-based building blocks. Bioconjug Chem 19:1871–1879

Humphries JD, Byron A, Humphries MJ (2006) Integrin ligands at a glance. J Cell Sci 119:3901–3903

Hwang R, Varner J (2004) The role of integrins in tumor angiogenesis. Hematol/Oncol Clin N Am 18:991–1006

Jacobson O, Chen X (2010) PET designated flouride-18 production and chemistry. Curr Top Med Chem 10:1048–1059

Jacobson O, Zhu L, Ma Y, Weiss ID, Sun X, Niu G, Kiesewetter DO, Chen X (2011) Rapid and simple one-step F-18 labeling of peptides. Bioconjug Chem 22:422–428

Jamous M, Haberkorn U, Mier W (2013) Synthesis of peptide radiopharmaceuticals for the therapy and diagnosis of tumor diseases. Molecules 18:3379–3409

Ji S, Czerwinski A, Zhou Y, Shao G, Valenzuela F, Sowinski P, Chauhan S, Pennington M, Liu S (2013a) (99m)Tc-Galacto-RGD2: a novel ^{99m}Tc-labeled cyclic RGD peptide dimer useful for tumor imaging. Mol Pharm 10:3304–3314

Ji S, Zheng Y, Shao G, Zhou Y, Liu S (2013b) Integrin $\alpha_v\beta_3$-targeted radiotracer ^{99m}Tc-3P-RGD(2) useful for noninvasive monitoring of breast tumor response to antiangiogenic linifanib therapy but not anti-integrin $\alpha_v\beta_3$ RGD(2) therapy. Theranostics 3:816–830

Ji S, Zhou Y, Shao G, Liu S (2013c) Evaluation of K(HYNIC)(2) as a bifunctional chelator for ^{99m}Tc-labeling of small biomolecules. Bioconjug Chem 24:701–711

Ji S, Zhou Y, Voorbach MJ, Shao G, Zhang Y, Fox GB, Albert DH, Luo Y, Liu S, Mudd SR (2013d) Monitoring tumor response to linifanib therapy with SPECT/CT using the integrin $\alpha_v\beta_3$-targeted radiotracer ^{99m}Tc-3P-RGD2. J Pharmacol Exp Ther 346:251–258

Jia B, Shi J, Yang Z, Xu B, Liu Z, Zhao H, Liu S, Wang F (2006) ^{99m}Tc-labeled cyclic RGDfK dimer: initial evaluation for SPECT imaging of glioma integrin $\alpha_v\beta_3$ expression. Bioconjug Chem 17:1069–1076

Jia B, Liu Z, Shi J, Yu Z, Yang Z, Zhao H, He Z, Liu S, Wang F (2008) Linker effects on biological properties of ^{111}In-labeled DTPA conjugates of a cyclic RGDfK dimer. Bioconjug Chem 19:201–210

Jia B, Liu Z, Zhu Z, Shi J, Jin X, Zhao H, Li F, Liu S, Wang F (2011) Blood clearance kinetics, biodistribution, and radiation dosimetry of a kit-formulated integrin $\alpha_v\beta_3$-selective radiotracer ^{99m}Tc-3PRGD 2 in non-human primates. Mol Imaging Biol 13:730–736

Jiang F, Albert DH, Luo Y, Tapang P, Zhang K, Davidsen SK, Fox GB, Lesniewski R, McKeegan EM (2011) ABT-869, a multitargeted receptor tyrosine kinase inhibitor, reduces tumor microvascularity and improves vascular wall integrity in preclinical tumor models. J Pharmacol Exp Ther 338:134–142

Jin H, Varner J (2004) Integrins: roles in cancer development and as treatment targets. Br J Cancer 90:561–565

Koukouraki S, Strauss LG, Georgoulias V, Eisenhut M, Haberkorn U, Dimitrakopoulou-Strauss A (2006a) Comparison of the pharmacokinetics of ^{68}Ga-DOTATOC and [^{18}F]FDG in patients with metastatic neuroendocrine tumours scheduled for ^{90}Y-DOTA-TOC therapy. Eur J Nucl Med Mol Imaging 33:1115–1122

Koukouraki S, Strauss LG, Georgoulias V, Schuhmacher J, Haberkorn U, Karkavitsas N, Dimitrakopoulou-Strauss A (2006b) Evaluation of the pharmacokinetics of ^{68}Ga-DOTA-TOC in patients with metastatic neuroendocrine tumours scheduled for ^{90}Y-DOTATOC therapy. Eur J Nucl Med Mol Imaging 33:460–466

Kubas H, Schafer M, Bauder-Wust U, Eder M, Oltmanns D, Haberkorn U, Mier W, Eisenhut M (2010) Multivalent cyclic RGD ligands: influence of linker lengths on receptor binding. Nucl Med Biol 37:885–891

Kumar CC (2003) Integrin $\alpha_v\beta_3$ as a therapeutic target for blocking tumor-induced angiogenesis. Curr Drug Targets 4:123–131

Lang L, Li W, Guo N, Ma Y, Zhu L, Kiesewetter DO, Shen B, Niu G, Chen X (2011) Comparison study of [^{18}F]FAl-NOTA-PRGD2, [^{18}F]FPPRGD2, and [^{68}Ga]Ga-NOTA-PRGD2 for PET imaging of U87MG tumors in mice. Bioconjug Chem 22:2415–2422

Laverman P, McBride WJ, Sharkey RM, Eek A, Joosten L, Oyen WJ, Goldenberg DM, Boerman OC (2010) A novel facile method of labeling octreotide with (18)F-fluorine. J Nucl Med 51:454–461

Laverman P, D'Souza CA, Eek A, McBride WJ, Sharkey RM, Oyen WJ, Goldenberg DM, Boerman OC (2012a) Optimized labeling of NOTA-conjugated octreotide with F-18. Tumour Biol 33:427–434

Laverman P, Sosabowski JK, Boerman OC, Oyen WJ (2012b) Radiolabelled peptides for oncological diagnosis. Eur J Nucl Med Mol Imaging 39(Suppl 1):S78–S92

Li ZB, Wu Z, Chen K, Chin FT, Chen X (2007) Click chemistry for ^{18}F-labeling of RGD peptides and microPET imaging of tumor integrin $\alpha_v\beta_3$ expression. Bioconjug Chem 18:1987–1994

Li X, Link JM, Stekhova S, Yagle KJ, Smith C, Krohn KA, Tait JF (2008a) Site-specific labeling of annexin V with F-18 for apoptosis imaging. Bioconjug Chem 19:1684–1688

Li ZB, Chen K, Chen X (2008b) ^{68}Ga-labeled multimeric RGD peptides for microPET imaging of integrin $\alpha_v\beta_3$ expression. Eur J Nucl Med Mol Imaging 35:1100–1108

Li ZB, Wu Z, Chen K, Ryu EK, Chen X (2008c) ^{18}F-labeled BBN-RGD heterodimer for prostate cancer imaging. J Nucl Med 49:453–461

Li Y, Guo J, Tang S, Lang L, Chen X, Perrin DM (2013) One-step and one-pot-two-step radiosynthesis of cyclo-RGD-(18)F-aryltri-fluoroborate conjugates for functional imaging. Am J Nucl Med Mol Imaging 3:44–56

Liu S (2004) The role of coordination chemistry in the development of target-specific radiopharmaceuticals. Chem Soc Rev 33:445–461

Liu S (2005) 6-Hydrazinonicotinamide derivatives as bifunctional coupling agents for ^{99m}Tc-labeling of small biomolecules. In: Krause W (ed) Contrast agents III. Springer, Berlin, pp 117–153

Liu S (2006) Radiolabeled multimeric cyclic RGD peptides as integrin $\alpha_v\beta_3$ targeted radiotracers for tumor imaging. Mol Pharm 3:472–487

Liu S (2008) Bifunctional coupling agents for radiolabeling of biomolecules and target-specific delivery of metallic radionuclides. Adv Drug Deliv Rev 60:1347–1370

Liu S (2009) Radiolabeled cyclic RGD peptides as integrin $\alpha_v\beta_3$-targeted radiotracers: maximizing binding affinity via bivalency. Bioconjug Chem 20:2199–2213

Liu S, Chakraborty S (2011) ^{99m}Tc-centered one-pot synthesis for preparation of ^{99m}Tc radiotracers. Dalton Trans 40:6077–6086

Liu S, Edwards DS (2001) Bifunctional chelators for therapeutic lanthanide radiopharmaceuticals. Bioconjug Chem 12:7–34

Liu S, Cheung E, Ziegler MC, Rajopadhye M, Edwards DS (2001a) (90)Y and (177)Lu labeling of a DOTA-conjugated vitronectin receptor antagonist useful for tumor therapy. Bioconjug Chem 12:559–568

Liu S, Edwards DS, Ziegler MC, Harris AR, Hemingway SJ, Barrett JA (2001b) ^{99m}Tc-labeling of a hydrazinonicotinamide-conjugated vitronectin receptor antagonist useful for imaging tumors. Bioconjug Chem 12:624–629

Liu S, Hsieh WY, Kim YS, Mohammed SI (2005) Effect of coligands on biodistribution characteristics of ternary ligand ^{99m}Tc complexes of a HYNIC-conjugated cyclic RGDfK dimer. Bioconjug Chem 16:1580–1588

Liu S, He Z, Hsieh WY, Kim YS, Jiang Y (2006) Impact of PKM linkers on biodistribution characteristics of the ^{99m}Tc-labeled cyclic RGDfK dimer. Bioconjug Chem 17:1499–1507

Liu S, Hsieh WY, Jiang Y, Kim YS, Sreerama SG, Chen X, Jia B, Wang F (2007) Evaluation of a (99m)Tc-labeled cyclic RGD tetramer for noninvasive imaging integrin $\alpha_v\beta_3$-positive breast cancer. Bioconjug Chem 18:438–446

Liu S, Kim YS, Hsieh WY, Gupta Sreerama S (2008a) Coligand effects on the solution stability, biodistribution and metabolism of the ^{99m}Tc-labeled cyclic RGDfK tetramer. Nucl Med Biol 35:111–121

Liu Z, Wang F, Chen X (2008b) Integrin $\alpha_v\beta_3$-targeted cancer therapy. Drug Dev Res 69:329–339

Liu Z, Liu S, Wang F, Liu S, Chen X (2009a) Noninvasive imaging of tumor integrin expression using ^{18}F-labeled RGD dimer peptide with PEG (4) linkers. Eur J Nucl Med Mol Imaging 36:1296–1307

Liu Z, Niu G, Shi J, Liu S, Wang F, Liu S, Chen X (2009b) ^{68}Ga-labeled cyclic RGD dimers with Gly3 and PEG4 linkers: promising agents for tumor integrin $\alpha_v\beta_3$ PET imaging. Eur J Nucl Med Mol Imaging 36:947–957

Liu Z, Yan Y, Chin FT, Wang F, Chen X (2009c) Dual integrin and gastrin-releasing peptide receptor targeted tumor imaging using ^{18}F-labeled PEGylated RGD-bombesin heterodimer ^{18}F-FB-PEG3-Glu-RGD-BBN. J Med Chem 52:425–432

Liu Z, Yan Y, Liu S, Wang F, Chen X (2009d) ^{18}F, ^{64}Cu, and ^{68}Ga labeled RGD-bombesin heterodimeric peptides for PET imaging of breast cancer. Bioconjug Chem 20:1016–1025

Liu S, Liu Z, Chen K, Yan Y, Watzlowik P, Wester HJ, Chin FT, Chen X (2010) ^{18}F-labeled galacto and PEGylated RGD dimers for PET imaging of $\alpha_v\beta_3$ integrin expression. Mol Imaging Biol 12:530–538

Liu S, Liu H, Jiang H, Xu Y, Zhang H, Cheng Z (2011) One-step radiosynthesis of ^{18}F-AlF-NOTA-RGD(2) for tumor angiogenesis PET imaging. Eur J Nucl Med Mol Imaging 38:1732–1741

Liu SH, Lin TH, Cheng DC, Wang JJ (2015) Assessment of stroke volume from brachial blood pressure using arterial characteristics. IEEE Trans Bio-med Eng 62:2151–2157

Lorger M, Krueger JS, O'Neal M, Staflin K, Felding-Habermann B (2009) Activation of tumor cell integrin $\alpha_v\beta_3$ controls angiogenesis and metastatic growth in the brain. Proc Natl Acad Sci USA 106:10666–10671

Luo Y, Jiang F, Cole TB, Hradil VP, Reuter D, Chakravartty A, Albert DH, Davidsen SK, Cox BF, McKeegan EM, Fox GB (2012) A novel multi-targeted tyrosine kinase inhibitor, linifanib (ABT-869), produces functional and structural changes in tumor vasculature in an orthotopic rat glioma model. Cancer Chemother Pharmacol 69:911–921

Ma Q, Ji B, Jia B, Gao S, Ji T, Wang X, Han Z, Zhao G (2011) Differential diagnosis of solitary pulmonary nodules using ^{99m}Tc-3P(4)-RGD(2) scintigraphy. Eur J Nucl Med Mol Imaging 38:2145–2152

Ma Q, Chen B, Gao S, Ji T, Wen Q, Song Y, Zhu L, Xu Z, Liu L (2014) ^{99m}Tc-3P4-RGD2 scintimammography in the assessment of breast lesions: comparative study with ^{99m}Tc-MIBI. PLoS One 9:e108349

Maecke HR, Hofmann M, Haberkorn U (2005) ^{68}Ga-labeled peptides in tumor imaging. J Nucl Med 46(Suppl 1):172S–178S

Mankoff DA, Link JM, Linden HM, Sundararajan L, Krohn KA (2008) Tumor receptor imaging. J Nucl Med 49:149s–163s

Maschauer S, Haubner R, Kuwert T, Prante O (2014) ^{18}F-glyco-RGD peptides for PET imaging of integrin expression: efficient radiosynthesis by click chemistry and modulation of biodistribution by glycosylation. Mol Pharm 11:505–515

McBride WJ, Sharkey RM, Karacay H, D'Souza CA, Rossi EA, Laverman P, Chang CH, Boerman OC, Goldenberg DM (2009) A novel method of ^{18}F radiolabeling for PET. J Nucl Med 50:991–998

McBride WJ, D'Souza CA, Sharkey RM, Karacay H, Rossi EA, Chang CH, Goldenberg DM (2010) Improved ^{18}F labeling of peptides with a fluoride-aluminum-chelate complex. Bioconjug Chem 21:1331–1340

McBride WJ, D'Souza CA, Karacay H, Sharkey RM, Goldenberg DM (2012) New lyophilized kit for rapid radiofluorination of peptides. Bioconjug Chem 23:538–547

Meyer A, Auernheimer J, Modlinger A, Kessler H (2006) Targeting RGD recognizing integrins: drug development, biomaterial research, tumor imaging and targeting. Curr Pharm Des 12:2723–2747

Michalski MH, Chen X (2011) Molecular imaging in cancer treatment. Eur J Nucl Med Mol Imaging 38:358–377

Minn AJ, Kang Y, Serganova I, Gupta GP, Giri DD, Doubrovin M, Ponomarev V, Gerald WL, Blasberg R, Massague J (2005) Distinct organ-specific metastatic potential of individual breast cancer cells and primary tumors. J Clin Investig 115:44–55

Monferran S, Skuli N, Delmas C, Favre G, Bonnet J, Cohen-Jonathan-Moyal E, Toulas C (2008) $\alpha_v\beta_3$ and $\alpha_v\beta_5$ integrins control glioma cell response to ionising radiation through ILK and RhoB. Int J Cancer 123:357–364

Mu L, Hohne A, Schubiger PA, Ametamey SM, Graham K, Cyr JE, Dinkelborg L, Stellfeld T, Srinivasan A, Voigtmann U, Klar U (2008) Silicon-based building blocks for one-step ^{18}F-radiolabeling of peptides for PET imaging. Angew Chem 47:4922–4925

Müller G, Gurrath M, Kessler H, Timpl R (1992) Dynamic forcing, a method for evaluating activity and selectivity profiles of RGD (Arg-Gly-Asp) peptides. Angew Chem Int Ed Engl 31:326–328

Namavari M, Cheng Z, Zhang R, De A, Levi J, Hoerner JK, Yaghoubi SS, Syud FA, Gambhir SS (2009) A novel method for direct site-specific radiolabeling of peptides using [^{18}F]FDG. Bioconjug Chem 20:432–436

Navarro-Gonzalez N, Porrero MC, Mentaberre G, Serrano E, Mateos A, Cabal A, Dominguez L, Lavin S (2015) *Escherichia coli* O157:H7 in wild boars (*Sus scrofa*) and Iberian ibex (*Capra pyrenaica*) sharing pastures with free-ranging livestock in a natural environment in Spain. Vet Quart 35:102–106

Nwe K, Kim YS, Milenic DE, Baidoo KE, Brechbiel MW (2012) ^{111}In- and ^{203}Pb-labeled cyclic RGD peptide conjugate as an

$\alpha_v\beta_3$ integrin-binding radiotracer. J Labelled Compd Radiopharm 55:423–426

Omar O, Lenneras M, Svensson S, Suska F, Emanuelsson L, Hall J, Nannmark U, Thomsen P (2010) Integrin and chemokine receptor gene expression in implant-adherent cells during early osseointegration. J Mater Sci Mater Med 21:969–980

Pfaff M, Tangemann K, Muller B, Gurrath M, Muller G, Kessler H, Timpl R, Engel J (1994) Selective recognition of cyclic RGD peptides of NMR defined conformation by $\alpha_{IIb}\beta_3$, $\alpha_v\beta_3$, and $\alpha_5\beta_1$ integrins. J Biol Chem 269:20233–20238

Pilch J, Habermann R, Felding-Habermann B (2002) Unique ability of integrin $\alpha_v\beta_3$ to support tumor cell arrest under dynamic flow conditions. J Biol Chem 277:21930–21938

Pohle K, Notni J, Bussemer J, Kessler H, Schwaiger M, Beer AJ (2012) ^{68}Ga-NODAGA-RGD is a suitable substitute for ^{18}F-Galacto-RGD and can be produced with high specific activity in a cGMP/GRP compliant automated process. Nucl Med Biol 39:777–784

Roth P, Silginer M, Goodman SL, Hasenbach K, Thies S, Maurer G, Schraml P, Tabatabai G, Moch H, Tritschler I, Weller M (2013) Integrin control of the transforming growth factor-β pathway in glioblastoma. Brain 136:564–576

Schirrmacher R, Bernard-Gauthier V, Reader A, Soucy JP, Schirrmacher E, Wangler B, Wangler C (2013) Design of brain imaging agents for positron emission tomography: do large bioconjugates provide an opportunity for in vivo brain imaging? Fut Med Chem 5:1621–1634

Sengupta S, Chattopadhyay N, Mitra A, Ray S, Dasgupta S, Chatterjee A (2001) Role of $\alpha_v\beta_3$ integrin receptors in breast tumor. J Exp Clin Cancer Res 20:585–590

Shankar DB, Li J, Tapang P, Owen McCall J, Pease LJ, Dai Y, Wei RQ, Albert DH, Bouska JJ, Osterling DJ, Guo J, Marcotte PA, Johnson EF, Soni N, Hartandi K, Michaelides MR, Davidsen SK, Priceman SJ, Chang JC, Rhodes K et al (2007) ABT-869, a multitargeted receptor tyrosine kinase inhibitor: inhibition of FLT3 phosphorylation and signaling in acute myeloid leukemia. Blood 109:3400–3408

Sheldrake HM, Patterson LH (2014) Strategies to inhibit tumor associated integrin receptors: rationale for dual and multi-antagonists. J Med Chem 57:6301–6315

Shi J, Wang L, Kim YS, Zhai S, Liu Z, Chen X, Liu S (2008) Improving tumor uptake and excretion kinetics of ^{99m}Tc-labeled cyclic arginine-glycine-aspartic (RGD) dimers with triglycine linkers. J Med Chem 51:7980–7990

Shi J, Kim YS, Chakraborty S, Jia B, Wang F, Liu S (2009a) 2-Mercaptoacetylglycylglycyl (MAG2) as a bifunctional chelator for ^{99m}Tc-labeling of cyclic RGD dimers: effect of technetium chelate on tumor uptake and pharmacokinetics. Bioconjug Chem 20:1559–1568

Shi J, Kim YS, Zhai S, Liu Z, Chen X, Liu S (2009b) Improving tumor uptake and pharmacokinetics of ^{64}Cu-labeled cyclic RGD peptide dimers with Gly(3) and PEG(4) linkers. Bioconjug Chem 20:750–759

Shi J, Wang L, Kim YS, Zhai S, Jia B, Wang F, Liu S (2009c) ^{99m}TcO(MAG2-3G3-dimer): a new integrin $\alpha_v\beta_3$-targeted SPECT radiotracer with high tumor uptake and favorable pharmacokinetics. Eur J Nucl Med Mol Imaging 36:1874–1884

Shi J, Kim YS, Chakraborty S, Zhou Y, Wang F, Liu S (2011a) Impact of bifunctional chelators on biological properties of ^{111}In-labeled cyclic peptide RGD dimers. Amino Acids 41:1059–1070

Shi J, Zhou Y, Chakraborty S, Kim YS, Jia B, Wang F, Liu S (2011b) Evaluation of in-labeled cyclic RGD peptides: effects of peptide and linker multiplicity on their tumor uptake, excretion kinetics and metabolic stability. Theranostics 1:322–340

Shokeen M, Anderson CJ (2009) Molecular imaging of cancer with copper-64 radiopharmaceuticals and positron emission tomography (PET). Acc Chem Res 42:832–841

Siegel RL, Miller KD, Jemal A (2015) Cancer statistics, 2015. CA Cancer J Clin 65:5–29

Simecek J, Hermann P, Havlickova J, Herdtweck E, Kapp TG, Engelbogen N, Kessler H, Wester HJ, Notni J (2013) A cyclen-based tetraphosphinate chelator for the preparation of radiolabeled tetrameric bioconjugates. Chemistry 19:7748–7757

Sloan EK, Anderson RL (2002) Genes involved in breast cancer metastasis to bone. Cell Mol Life Sci 59:1491–1502

Sloan EK, Pouliot N, Stanley KL, Chia J, Moseley JM, Hards DK, Anderson RL (2006) Tumor-specific expression of $\alpha_v\beta_3$ integrin promotes spontaneous metastasis of breast cancer to bone. Breast Cancer Res 8:R20

Stollman TH, Ruers TJ, Oyen WJ, Boerman OC (2009) New targeted probes for radioimaging of angiogenesis. Methods 48:188–192

Sung V, Stubbs JT III, Fisher L, Aaron AD, Thompson EW (1998) Bone sialoprotein supports breast cancer cell adhesion proliferation and migration through differential usage of the $\alpha_v\beta_3$ and $\alpha_v\beta_5$ integrins. J Cell Physiol 176:482–494

Taherian A, Li X, Liu Y, Haas TA (2011) Differences in integrin expression and signaling within human breast cancer cells. BMC Cancer 11:293

Tannir NM, Wong YN, Kollmannsberger CK, Ernstoff MS, Perry DJ, Appleman LJ, Posadas EM, Cho D, Choueiri TK, Coates A, Gupta N, Pradhan R, Qian J, Chen J, Scappaticci FA, Ricker JL, Carlson DM, Michaelson MD (2011) Phase 2 trial of linifanib (ABT-869) in patients with advanced renal cell cancer after sunitinib failure. Eur J Cancer 47:2706–2714

Tateishi U, Oka T, Inoue T (2012) Radiolabeled RGD peptides as integrin $\alpha_v\beta_3$-targeted PET tracers. Curr Med Chem 19:3301–3309

Tsiapa I, Loudos G, Varvarigou A, Fragogeorgi E, Psimadas D, Tsotakos T, Xanthopoulos S, Mihailidis D, Bouziotis P, Nikiforidis GC, Kagadis GC (2013) Biological evaluation of an ornithine-modified ^{99m}Tc-labeled RGD peptide as an angiogenesis imaging agent. Nucl Med Biol 40:262–272

Tweedle MF (2009) Peptide-targeted diagnostics and radiotherapeutics. Acc Chem Res 42:958–968

Vaidyanathan G, White BJ, Zalutsky MR (2009) Propargyl 4-[F]fluorobenzoate: a putatively more stable prosthetic group for the fluorine-18 labeling of biomolecules via click chemistry. Curr Radiopharm 2:63–74

Vogetseder A, Thies S, Ingold B, Roth P, Weller M, Schraml P, Goodman SL, Moch H (2013) αv-Integrin isoform expression in primary human tumors and brain metastases. Int J Cancer 133:2362–2371

Wang J, Kim YS, Liu S (2008a) ^{99m}Tc-labeling of HYNIC-conjugated cyclic RGDfK dimer and tetramer using EDDA as coligand. Bioconjug Chem 19:634–642

Wang L, Shi J, Kim Y-S, Zhai S, Jia B, Zhao H, Liu Z, Wang F, Chen X, Liu S (2008b) Improving tumor-targeting capability and pharmacokinetics of ^{99m}Tc-labeled cyclic RGD dimers with PEG4 linkers. Mol Pharm 6:231–245

Wangler C, Schirrmacher R, Bartenstein P, Wangler B (2010) Click-chemistry reactions in radiopharmaceutical chemistry: fast and easy introduction of radiolabels into biomolecules for in vivo imaging. Curr Med Chem 17:1092–1116

Weigelt B, Peterse JL, van 't Veer LJ (2005) Breast cancer metastasis: markers and models. Nat Rev Cancer 5:591–602

Wong CI, Koh TS, Soo R, Hartono S, Thng CH, McKeegan E, Yong WP, Chen CS, Lee SC, Wong J, Lim R, Sukri N, Lim SE, Ong AB, Steinberg J, Gupta N, Pradhan R, Humerickhouse R, Goh BC (2009) Phase I and biomarker study of ABT-869, a multiple receptor tyrosine kinase inhibitor, in patients with refractory solid malignancies. J Clin Oncol 27:4718–4726

Wu Y, Zhang X, Xiong Z, Cheng Z, Fisher DR, Liu S, Gambhir SS, Chen X (2005) microPET imaging of glioma integrin $\alpha_v\beta_3$ expression using ^{64}Cu-labeled tetrameric RGD peptide. J Nucl Med 46:1707–1718

Wu Z, Li ZB, Chen K, Cai W, He L, Chin FT, Li F, Chen X (2007) MicroPET of tumor integrin $\alpha_v\beta_3$ expression using ^{18}F-labeled PEGylated tetrameric RGD peptide (^{18}F-FPRGD4). J Nucl Med 48:1536–1544

Yang Y, Ji S, Liu S (2014) Impact of multiple negative charges on blood clearance and biodistribution characteristics of ^{99m}Tc-labeled dimeric cyclic RGD peptides. Bioconjug Chem 25:1720–1729

Zhang X, Xiong Z, Wu Y, Cai W, Tseng JR, Gambhir SS, Chen X (2006) Quantitative PET imaging of tumor integrin $\alpha_v\beta_3$ expression with ^{18}F-FRGD2. J Nucl Med 47:113–121

Zhao Y, Bachelier R, Treilleux I, Pujuguet P, Peyruchaud O, Baron R, Clement-Lacroix P, Clezardin P (2007) Tumor $\alpha_v\beta_3$ integrin is a therapeutic target for breast cancer bone metastases. Cancer Res 67:5821–5830

Zhao D, Jin X, Li F, Liang J, Lin Y (2012) Integrin $\alpha_v\beta_3$ imaging of radioactive iodine-refractory thyroid cancer using ^{99m}Tc-3PRGD2. J Nucl Med 53:1872–1877

Zheng Y, Ji S, Czerwinski A, Valenzuela F, Pennington M, Liu S (2014) FITC-conjugated cyclic RGD peptides as fluorescent probes for staining integrin $\alpha_v\beta_3/\alpha_v\beta_5$ in tumor tissues. Bioconjug Chem 25:1925–1941

Zheng Y, Ji S, Tomaselli E, Yang Y, Liu S (2015) Comparison of biological properties of ^{111}In-labeled dimeric cyclic RGD peptides. Nucl Med Biol 42:137–145

Zhou J, Goh BC, Albert DH, Chen CS (2009) ABT-869, a promising multi-targeted tyrosine kinase inhibitor: from bench to bedside. J Hematol Oncol 2:33

Zhou Y, Chakraborty S, Liu S (2011a) Radiolabeled cyclic RGD peptides as radiotracers for imaging tumors and thrombosis by SPECT. Theranostics 1:58–82

Zhou Y, Kim YS, Chakraborty S, Shi J, Gao H, Liu S (2011b) ^{99m}Tc-labeled cyclic RGD peptides for noninvasive monitoring of tumor integrin $\alpha_v\beta_3$ expression. Mol Imaging 10:386–397

Zhou Y, Kim YS, Lu X, Liu S (2012) Evaluation of ^{99m}Tc-labeled cyclic RGD dimers: impact of cyclic RGD peptides and ^{99m}Tc chelates on biological properties. Bioconjug Chem 23:586–595

Zhu Z, Miao W, Li Q, Dai H, Ma Q, Wang F, Yang A, Jia B, Jing X, Liu S, Shi J, Liu Z, Zhao Z, Wang F, Li F (2012) ^{99m}Tc-3PRGD2 for integrin receptor imaging of lung cancer: a multicenter study. J Nucl Med 53:716–722

Zitzmann S, Ehemann V, Schwab M (2002) Arginine-glycine-aspartic acid (RGD)-peptide binds to both tumor and tumor-endothelial cells in vivo. Cancer Res 62:5139–5143

11

Identification of natural products as novel ligands for the human 5-HT$_{2C}$ receptor

Yao Peng[1,2,3,4], Simeng Zhao[3], Yiran Wu[3,5], Haijie Cao[6,7], Yueming Xu[3], Xiaoyan Liu[3], Wenqing Shui[3,5], Jianjun Cheng[3], Suwen Zhao[3,5], Ling Shen[3,5], Jun Ma[8], Ronald J. Quinn[8], Raymond C. Stevens[3,5], Guisheng Zhong[3,5✉], Zhi-Jie Liu[1,2,3,5✉]

[1] National Laboratory of Biomacromolecules, Institute of Biophysics, Chinese Academy of Sciences, Beijing 100101, China
[2] Institute of Molecular and Clinical Medicine, Kunming Medical University, Kunming 650500, China
[3] iHuman Institute, ShanghaiTech University, Shanghai 201210, China
[4] University of Chinese Academy of Sciences, Beijing 100049, China
[5] School of Life Science and Technology, ShanghaiTech University, Shanghai 201210, China
[6] College of Pharmacy, Nankai University, Tianjin 300071, China
[7] High-throughput Molecular Drug Discovery Center, Tianjin Joint Academy of Biotechnology and Medicine, Tianjin 300457, China
[8] Eskitis Institute for Drug Discovery, Griffith University, Brisbane, QLD 4111, Australia

Abstract G protein-coupled receptors (GPCRs) constitute the largest human protein family with over 800 members, which are implicated in many important medical conditions. Serotonin receptors belong to the aminergic GPCR subfamily and play important roles in physiological and psychological activities. Structural biology studies have revealed the structures of many GPCRs in atomic details and provide the basis for the identification and investigation of the potential ligands, which interact with and modulate the receptors. Here, an integrative approach combining a focused target-specific natural compound library, a thermal-shift-based screening method, affinity mass spectrometry, molecular docking, and *in vitro* as well as *in vivo* functional assay, was applied to identify (–)-crebanine and several other aporphine alkaloids as initial hits for a human serotonin receptor subtype, the 5-HT$_{2C}$ receptor. Further studies illuminated key features of their binding affinity, downstream signaling and tissue reaction, providing a molecular explanation for the interaction between (–)-crebanine and human 5-HT$_{2C}$ receptor.

Keywords GPCR, 5-HT$_{2C}$ receptor, Natural product, Alkaloids

Yao Peng and Simeng Zhao have contributed equally to this work.

✉ Correspondence: zhongsh@shanghaitech.edu.cn (G. Zhong), liuzhj@shanghaitech.edu.cn (Z.-J. Liu)

INTRODUCTION

G protein-coupled receptors (GPCRs) are cell surface receptors which are responsible for more than 80% of cell signal transduction across cell membranes. They are involved in a wide range of physiological as well as psychological activities and constitute the largest human protein family with over 800 members. GPCRs are implicated in many medical conditions such as heart disease, metabolic diseases, cancer, immune diseases and neurological disorders (Rask-Andersen et al. 2014). Drugs targeting GPCRs comprise as much as 33% of all marketed drugs approved by FDA (U.S. Food and Drug Administration) (Santos et al. 2017).

Serotonin, or 5-hydroxytryptamine (5-HT), is a neurotransmitter widely found in both the central nervous system (CNS) and the peripheral nervous system. It plays important roles in the functions of brain, gastrointestinal tract, cardiovascular system, and immune cells. In the CNS, serotonergic system regulates mood, perception, memory, food intake, sexual behaviors, and other functions. These physiological roles of serotonin are mediated by serotonin receptors, which are composed of fourteen subtypes in the mammalian system that are further grouped into seven subfamilies (5-HT$_{1-7}$). Except for 5-HT$_3$, which functions as a ligand-gated ion channel, all other serotonin receptors belong to the GPCR superfamily (McCorvy and Roth 2015).

Among serotonin receptors, the 5-HT$_{2C}$ receptor belongs to the 5-HT$_2$ subfamily. It shares 58% and 55% sequence similarity with the 5-HT$_{2A}$ and 5-HT$_{2B}$ receptors, respectively (Isberg et al. 2016). Activation of 5-HT$_{2C}$ receptor is believed to reduce appetite (Halford and Harrold 2012) and cure schizophrenia (Cheng et al. 2016). In 2012, lorcaserin, a selective 5-HT$_{2C}$ agonist, was approved by the FDA for the treatment of obesity (Narayanaswami and Dwoskin 2016). Its efficacy in treatment of nicotine addiction is currently being evaluated in clinical trials (Zeeb et al. 2015). Moreover, 5-HT$_{2C}$ has been demonstrated as a potential therapeutic target for treatment of mental disorders also (Englisch et al. 2016). Agomelatine, antagonist of 5-HT$_{2C}$ receptor, is used for treating depression and schizophrenia (Jacobson 2015).

The ligand-binding pockets of all serotonin receptors are very similar and, therefore, it is difficult to design an inhibitor exhibiting high degree of specificity for a particular serotonin receptor. Most of the marketed drugs targeting serotonin receptors have side effects which arise due to low specificity. For drug candidates targeting the 5-HT$_{2C}$ receptor, achieving high selectivity is very important because non-specific interaction of the drug with the other two 5-HT$_2$ receptors causes side effects or toxicities. Unwanted activation of 5-HT$_{2A}$ by drugs targeting the 5-HT$_{2C}$ receptor causes hallucinations (Nichols et al. 2002). Similarly, non-specific activation of 5-HT$_{2B}$ leads to valvular heart disease (Connolly et al. 1997).

Ligand identification for GPCRs is a tedious, time-consuming and resource intensive process. In the past few years, new assay methods have been developed to explore multidimensional chemical space in a much more efficient manner. These methods range from cell-based (Besnard et al. 2012), label-free (e.g., Surface Plasmon Resonance; SPR, and Isothermal Titration Calorimetry; ITC), new biosensors (Kroeze et al. 2015) to in silico screening (Huang et al. 2015), which has significantly increased the success rate of hit identification when compared to traditional methods.

Natural products contain chemical compounds or substances with pharmacological or biological activities, which can be harnessed for therapeutic benefit or treating diseases as exemplified by traditional Chinese medicine (All natural 2007). In the development of modern medicines, natural products are often used as starting points for drug discovery and have been considered as the most important resource for identification of lead compounds due to their diverse molecular architectures and a wide range of bioactivities (Ahn 2016). Consequently, during the past 30 years, natural products have been instrumental in the discovery of more than half of the approved drugs (Newman and Cragg 2016).

Given the challenges involved in GPCR ligand discovery; in particular, the need to screen a large chemical space, one efficient approach for identifying hits would be to design or select focused screening libraries to reduce the workload. In this study, as the 5-HT$_{2C}$ receptor belongs to monoamine type of receptors in the class A of GPCR family, we decided to screen novel ligands from alkaloids containing basic nitrogen atoms. The positively charged nitrogen atom of the ligand was expected to anchor to the highly conserved D$^{3.32}$ of the aminergic receptors. The initial screening was performed using thermal stability assay (Alexandrov et al. 2008) against a focused alkaloid library consisting of over 300 chemical components isolated from plants (Shang et al. 2010). (−)-Crebanine and several other aporphine alkaloids were identified as potential hits for the 5-HT$_{2C}$ receptor. The affinity mass spectrometry (MS) method was used to validate the hits and measure the binding affinities (Chen et al. 2015; Qin et al. 2015). The cell-based calcium influx assay was employed to characterize the function of the validated hits. Molecular docking studies coupled with site directed mutagenesis were used to predict the binding sites for the compound. A patch clamp experiment was also utilized to investigate the compound's physiological effects in neurons.

RESULTS

Aporphine alkaloids as potential ligands for 5-HT$_{2C}$ receptor

In order to obtain conformationally homogeneous, thermo-stable and highly pure protein samples for screening and characterization of potential ligand hits, the expression construct of 5-HT$_{2C}$ receptor was optimized as described previously (Lv *et al.* 2016). The final expression construct contains a BRIL (PDB ID 1M6T, MW 11.9 kDa) as a stabilizing fusion partner inserted in receptor's third intracellular loop (ICL3) between L246 and Q301. Additionally, the N- and C-terminals were truncated by 39 and 65 residues, respectively. The optimized construct was inserted into a pFastbac vector for expression in *Spodoptera frugiperda* (*Sf9*) cells.

A small pipetting workstation (Qiagility, Qigen) and a real-time fluorescence quantitative PCR were used to perform the high throughput CPM (the thiol-specific fluorochrome N-[4-(7-diethylamino-4-methyl-3-coumarinyl) phenyl] maleimide) screening (Alexandrov

et al. 2008). Aporphine alkaloids (1–5) were identified as potential hits for 5-HT$_{2C}$ receptors (Fig. 1A). Among them, (−)-crebanine (**1**) and (+)-isocorydine (**5**) showed more significant binding property to 5-HT$_{2C}$ receptor, with the thermal shift value (the difference between target A and B for melting temperature in CPM assay, ΔTm) of 9.25 and 4.82 °C, respectively, comparing to apo-protein (Fig. 1B). The other three alkaloids, (−)-dicentrine (**2**), (+)-magnoflorine (**3**), and didehydroglaucine (**4**) showed slight temperature shift (ΔTm < 1.00 °C) in the thermal shift experiments.

Ligand validation by ultrafiltration-based affinity mass spectrometry analysis

The ultrafiltration-based affinity MS technique has been established to search for ligands, verify binding and estimate affinity of specific ligands for given soluble protein targets (Chen *et al.* 2015; Qin *et al.* 2015). In this study, this technique was adapted to ligand-binding validation for the membrane 5-HT$_{2C}$ receptor. To confirm binding specificity of (−)-crebanine (**1**) and

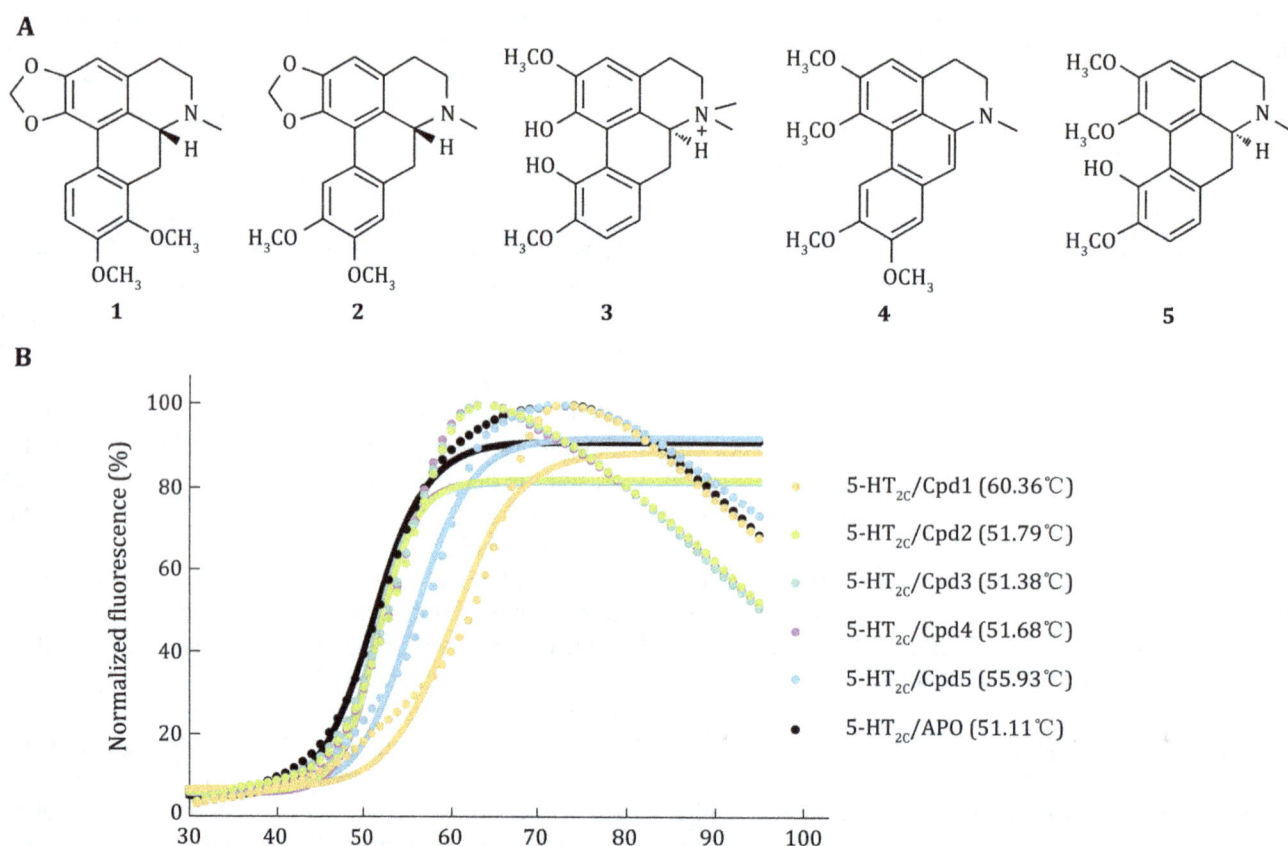

Fig. 1 A Structures of aporphine alkaloids. The five aporphine alkaloids: (−)-crebanine (**1**), (−)-dicentrine (**2**), (+)-magnoflorine (**3**), didehydroglaucine (**4**), (+)-isocorydine (**5**). **B** Thermo-stability values of aporphine alkalodis. Thermal stability ramping assay of 5-HT$_{2C}$ receptor combine with aporphine alkalodis. The Tm value of the 5-HT$_{2C}$/compound **1** (*yellow trace*) is higher than other combinations, indicating that 5-HT$_{2C}$/compound **1** combination improves the thermostability of 5-HT$_{2C}$ receptor

(+)-isocorydine (**5**) to 5-HT$_{2C}$ receptor, a negative control was prepared using another GPCR protein (hydroxycarboxylic acid receptor 2). An S/C ratio referring to the ratio of MS response of a given ligand detected in the 5-HT$_{2C}$ receptor incubation sample versus the control was used to assess specific enrichment of the ligand associated with 5-HT$_{2C}$ receptor. Previous study has shown that ligands with an S/C ratio > 2 are positive binders of the target protein (Chen *et al.* 2015). For (−)-crebanine (**1**) and (+)-isocorydine (**5**), their S/C ratios are significantly above the threshold and very close to a known high affinity 5-HT$_{2C}$ receptor antagonist ritanserin, indicating that they both showed obvious interactions with 5-HT$_{2C}$ receptor (Table 1). The high-resolution mass spectra for both compounds in the protein complex fraction confirmed their structural identification (Fig. 2). Then, a single-point K_d calculation method (Qin *et al.* 2015) was employed to estimate binding affinity of each ligand to the receptor (Table 1). It turned out that (−)-crebanine (**1**) displayed stronger affinity ($K_d \sim 0.34$ µmol/L) than its analog (+)-isocorydine (**5**) ($K_d \sim 11$ µmol/L) whereas the affinity of ritanserin (positive control) was in the high nmol/L range.

Calcium influx assay characterization of the potential hits

The 5-HT$_{2C}$ receptor mainly couples to G$_{\alpha q}$ proteins. The activated 5-HT$_{2C}$ receptor transmits the signals from the extracellular to the intercellular side using DAG/IP$_3$ (diacyl glycerol/inositol 1,4,5-trisphosphate) as its second massagers (Hannon and Hoyer 2008). IP$_3$ stimulates the endoplasmic reticulum to release calcium ions into the cytoplasm, which causes the increase of calcium concentration in cytosol. Therefore, calcium mobilization assay is commonly applied to understand how unknown small molecule ligands of 5-HT$_{2C}$ receptor modulate cellular signal transductions. In this case, Fluo-4 Direct dye was used as a fluorescence indicator to detect calcium flux in G protein-coupled receptor expressed cells. The fold of maximum response over average of basal reading is plotted against compound

Table 1 S/C ratios and estimated binding affinity of two new ligands and ritanserin

	Compound **1**	Compound **5**	Ritanserin
S/C	9.15 ± 1.43	7.89 ± 0.74	8.39 ± 1.80
K_d (µmol/L)	0.34 ± 0.07	11.0 ± 0.75	0.37 ± 0.05

For each ligand, S/C and K_d measurements were represented by the average values and standard deviations from experimental replicates ($n = 4$)

concentration to determine the potency of the compounds (Fig. 3). (−)-Crebanine (**1**) displayed antagonism for 5-HT$_2$ receptors (Fig. 3A). When 5-HT$_{2A/2B/2C}$ receptors were activated by 3 nmol/L of 5-HT, (−)-crebanine (**1**) inhibited the activation at IC_{50} at 564, 1693 and 149 nmol/L, respectively. (−)-Crebanine (**1**) showed higher efficacy with the IC_{50} value in 5-HT$_{2C}$ receptor compared to 5-HT$_{2A}$ and 5-HT$_{2B}$ receptors. (−)-Crebanine (**1**)'s analog (+)-isocorydine (**5**) showed very weak partial agonism towards 5-HT$_{2C}$ receptor ($E_{max} = 16.6\%$ of the effect of 1 µmol/L 5-HT) (Fig. 3B). While 5-HT$_{2C}$ receptor was activated ($EC_{50} = 2075$ nmol/L), 5-HT$_{2A/2B}$ receptors remained inactive with the addition of up to 30,000 nmol/L (+)-isocorydine (**5**). Hence, (−)-crebanine (**1**)'s analog, (+)-isocorydine (**5**), showed different pharmacology characters towards 5-HT$_{2C}$ receptor.

Identification of key interactions of (−)-crebanine with 5-HT$_{2C}$ receptor

Binding mode of (−)-crebanine (**1**) in 5-HT$_{2C}$ receptor model (built based on the crystal structure of 5HT$_{2B}$ receptor, PDB ID: 4IB4) is predicted by molecular docking (Fig. 4A). In the previous studies, the crystal structures of 5-HT$_{1B/2B}$ in complex with ergotamine and 5-HT2B with lysergic acid diethylamide (LSD) revealed similar orthosteric ligand-binding cavities defined by residues from helixes III, V, VI, VII, and ECL2 (Wacker *et al.* 2013, 2017; Wang *et al.* 2013). The binding pocket is embedded deep in the 7TM core of the receptor and (−)-crebanine (**1**) partially overlaps with the ergoline rings of ergotamine bound to 5-HT$_{1B/2B}$ receptor structures. Some key interactions are in common for (−)-crebanine and ergotamine: A salt bridge is formed between the positively charged nitrogen of (−)-crebanine (**1**) and the carboxylate of Asp134[3.32] (Venkatakrishnan *et al.* 2013) (Fig. 4B), which is fully conserved in 5-HT and other monoamine receptors. (−)-Crebanine (**1**) also forms π–π interaction with a benzene ring to Phe327[6.51], which resembles similar feature of ergotamine bound to 5-HT$_{1B}$ and 5-HT$_{2B}$ receptors. Due to the different shapes of aporphine and ergoline rings, (−)-crebanine also extends to space close to the entrance and forms hydrophobic interactions to Leu209 and Phe214 on ECL2. Hydrophobic interactions between (−)-crebanine (**1**) and Val135[3.33], Thr139[3.37], Gly218[5.42], Ala222[5.46], Trp324[6.48], Val354[7.39], and Tyr358[7.43] are also predicted. The different binding affinities and functions of (−)-crebanine (**1**) (antagonist) and (+)-isocorydine (**5**) (weak partial agonist) may due to the difference in chirality of the carbon atom or substitution groups in the aporphine scaffold. In this

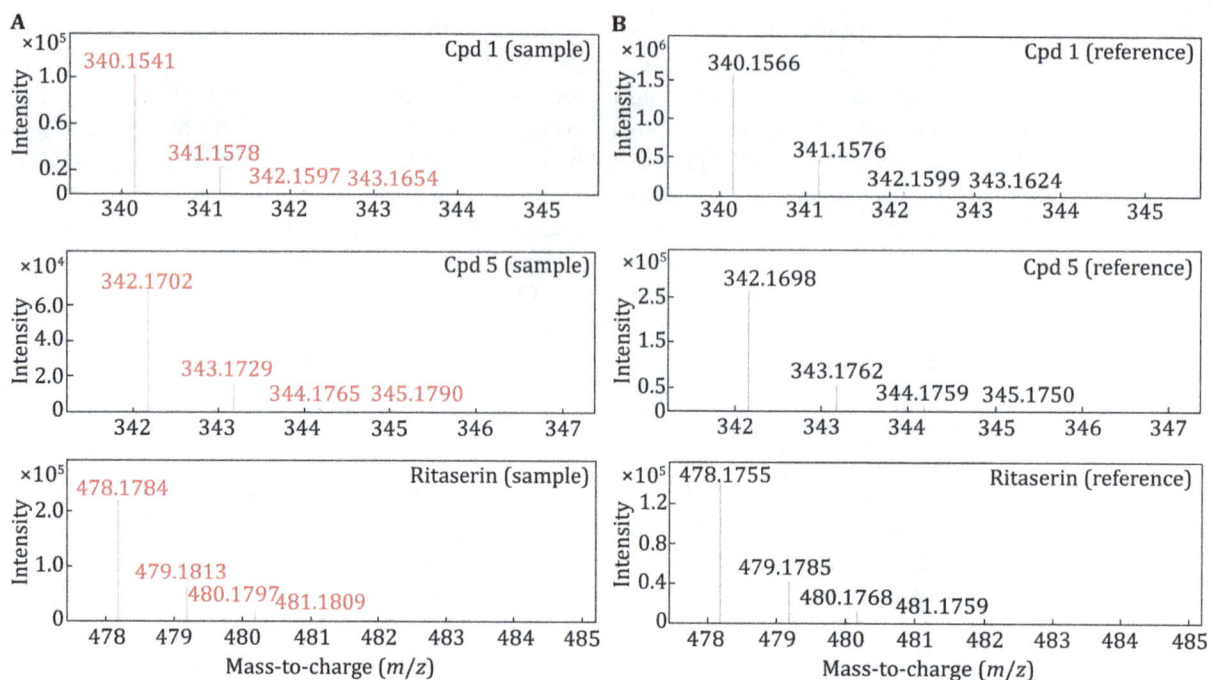

Fig. 2 Mass spectra of three compounds detected in the 5-HT$_{2C}$ receptor incubation sample (**A**) and the corresponding reference (**B**). Matching accurate mass and retention time of each ligand with the reference data are required for confident structural assignment

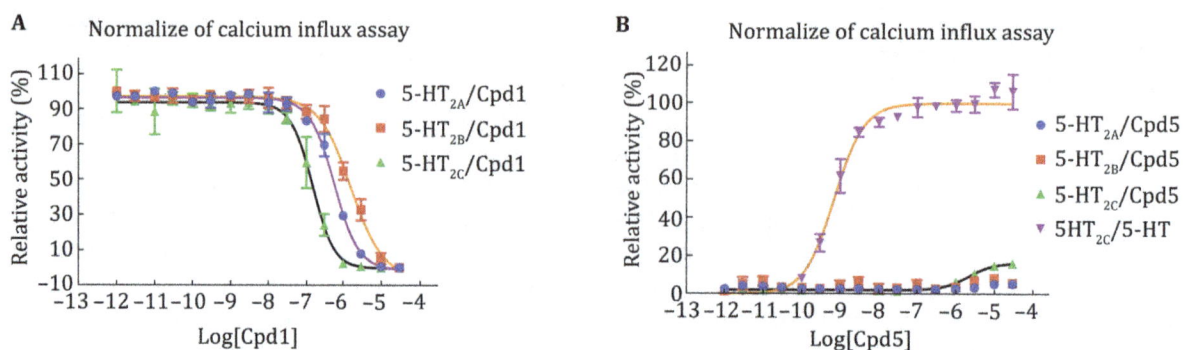

Fig. 3 Fluorescence-based calcium mobilization measurement of 5-HT$_{2A/2B/2C}$ mediated G$_q$ inhibition/activation by (−)-crebanine (**1**) and (+)-isocorydine (**5**), all in HEK 293 derived cells. **A** Normalized inhibition of compound **1** on human cloned 5-HT$_{2A/2B/2C}$ receptor-mediated Gq signaling. 3 nmol/L of 5-HT was used to induce initial activation. The 5-HT$_{2C}$ receptor shows the lowest IC_{50} among the tested receptors. **B** Normalized activation of compound **5** on human cloned 5-HT$_{2A/2B/2C}$ receptor-mediated G$_q$ signaling. Compound **5** shows partial agonism against 5-HT$_{2C}$ receptor

predicted binding mode, the orientation of aporphine scaffold is the same to previously published docking results of dihydrofuroaporphine in 5HT$_{1A}$ (Yuan *et al.* 2016).

Based on the binding poses predicted by molecular docking, single-point mutations were designed to test their impact on ligand binding. Mutations of several residues include the highly conserved Asp134$^{3.32}$, which is known to cause 5-HT$_{1B}$ or 5-HT$_{2B}$ receptors to abolish the monoamine ligand binding as reported in the literature (Wang *et al.* 2013). In this study, residues forming

hydrophobic interactions with the substitution groups on the aporphine scaffold of (−)-crebanine (**1**) were selected for mutation as well and they are Thr139$^{3.37}$, Gly218$^{5.42}$, Ala222$^{5.46}$, and Leu209^{ECL2}. Effects of those mutations were tested with the CPM assay and the affinity MS assay (Table 2). In the affinity MS experiment, relative ligand-binding capacity (represented by binding %) to different mutants and the wild-type receptor was compared. Among the six mutations tested, five of them reduced the binding of (−)-creba-nine (**1**) by more than 40% relative to binding to the

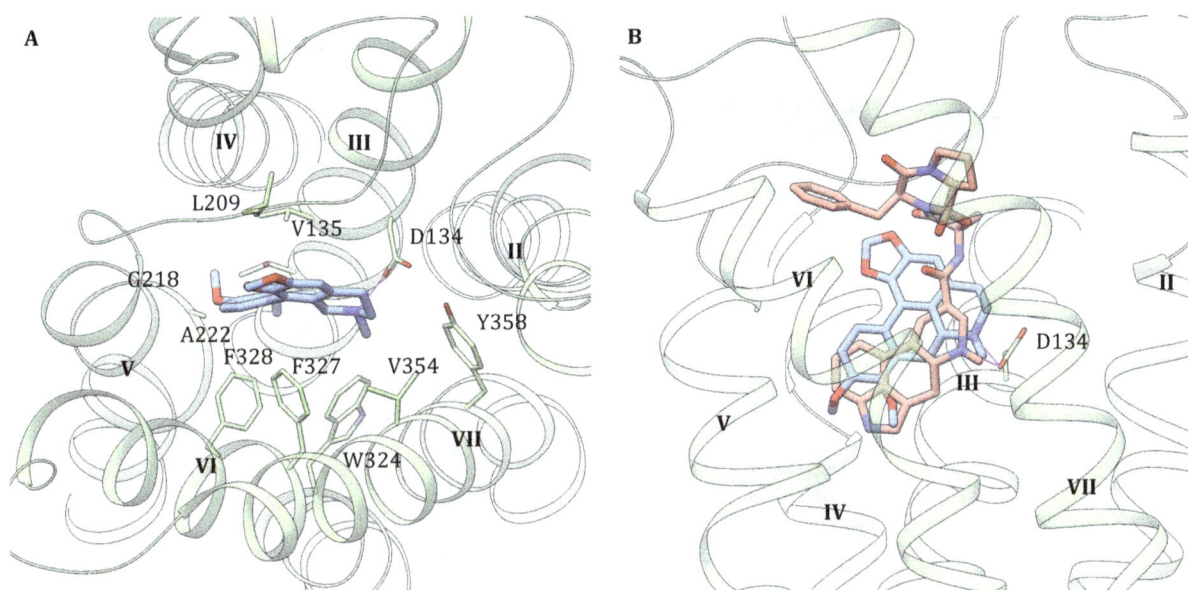

Fig. 4 Docked pose of (–)-crebanine in 5-HT$_{2C}$ model. **A** Ligand-receptor interactions. The key residues were labeled and showed in sticks. **B** Superposition with ergotamine in 5-HT$_{2B}$ crystal structure

Table 2 Mutations validation on the ligand binding pocket using thermostability and affinity MS assay

Mutants	(–)-Crebanine/CPM ($n = 2$) (°C)	(–)-Crebanine/MS ($n = 2$)[a] (%)
WT	59.93	27.57
G218$^{5.42}$A	53.92	9.82
G218$^{5.42}$S	53.24	6.04
A222$^{5.46}$V	59.55	25.84
A222$^{5.46}$F	55.70	15.87
T139$^{3.37}$A	54.63	14.64
L209^{ECL2}F	55.46	13.22

[a]Relative ligand binding (%) was calculated by the MS response of the ligand released from a specific mutant divided by the MS response of the total ligand present in the incubation sample

wild-type receptor, displaying the importance of hydrophobic interactions. Mutations Gly218$^{5.42}$Ala/Ser impacted the binding of (–)-crebanine the most, causing the Tm values decreased over 6.0 °C and ligand-binding capacity reduced by over 60%. These results fit the predicted compact interactions of (–)-crebanine (**1**) to Gly218$^{5.42}$. Thr139$^{3.37}$ is highly conserved residue in serotonin receptors, and Thr139$^{3.37}$Ala led to the Tm value decrease and ligand potency lost. Leu209^{ECL2}Phe also has significant influence on the binding of (–)-crebanine (**1**), supporting our prediction that (–)-crebanine (**1**) forms hydrophobic interactions to residues on ECL2. Ala222$^{5.46}$Phe has a similar effect to Thr139$^{3.37}$ and Leu209^{ECL2}Phe, while Ala222$^{5.46}$Val is the only mutation having little effect on the thermostability of (–)-crebanine (**1**) bound 5-HT$_{2C}$ receptor (Tm value

decreased by 0.38 °C) and ligand-binding capacity to the mutant inferred from the affinity MS assay. Different effects of mutations Ala222$^{5.46}$Phe/Val show that there is some space in the pocket around this site, but too bulky side chain collides with (–)-crebanine (**1**).

Crebanine attenuated the spontaneous synaptic current amplification induced by 5-HT

5-HT receptors play important roles in central and peripheral nervous systems. Malfunctions of these receptors have been linked to many neural disorders such as depression and psychosis. The effect of (–)-crebanine on changes in 5-HT induced synaptic transmission was evaluated by electrophysiological recording of mouse brain slices (Jang *et al.* 2012). The spontaneous synaptic current (SPSC) of cortical pyramidal neurons was recorded in whole cell mode. The SPSC was amplified after 5-HT treatment at the concentration of 10 µmol/L, while the effect of 5-HT induced SPSC was significantly attenuated in the presence of 30 µmol/L (–)-crebanine (**1**) incubation (Fig. 5).

DISCUSSION

Natural products contain a variety of components like alkaloids, peptides, lipids, glucosides, *etc.*. These diverse components are perfect candidates for serving as ligands to modulate the activity of G protein-coupled receptors. Furthermore, a co-evolutional relationship

A

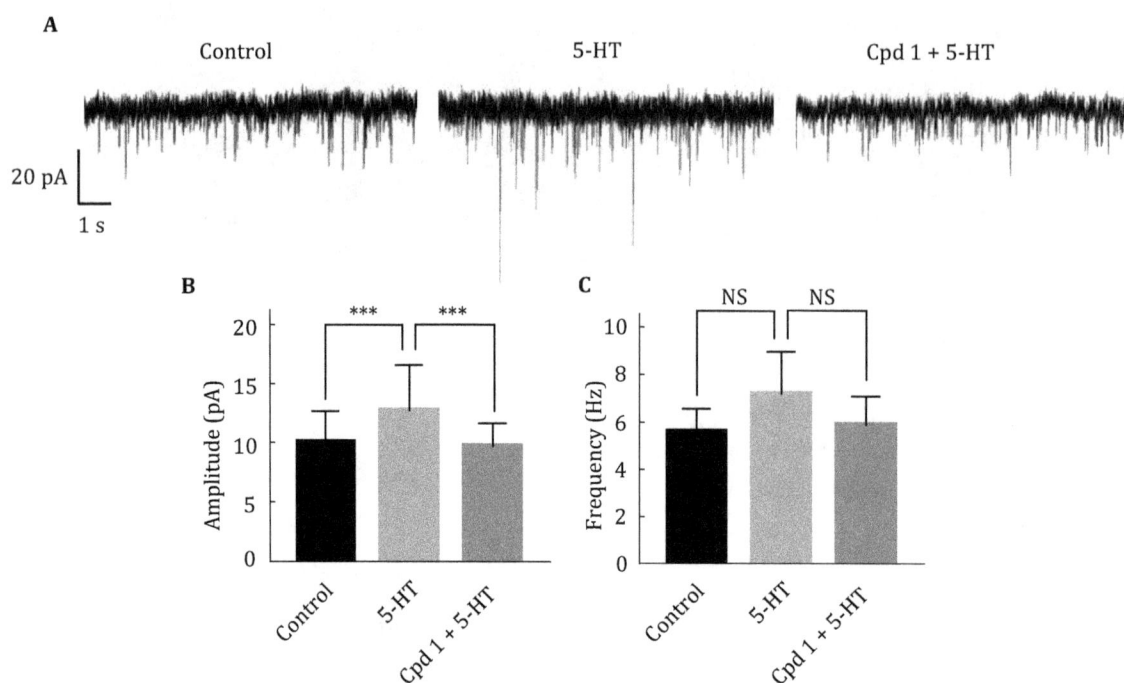

Fig. 5 Crebanine attenuated the SPSC amplification induced by 5-HT. **A** Representative traces of SPSC in control condition (*left*), and in response to 5-HT, with (*middle*) or without (*right*) (–)-crebanine. Amplitude histograms (**B**) and frequency histograms (**C**) of SPSC (***$p < 0.001$)

has been noted between receptors and plants. This is supported by the fact that many endogenous ligands are related to phyto-compounds and there is a certain degree of overlap between the chemical spaces of endogenous ligands and components of natural products. Therefore, screening of focused libraries containing a sub-set of natural products that show chemical similarity to the known ligands of the target receptors has the potential to significantly increase the success rate of discovery of new ligands. Serotonin (5-HT) shares its scaffold with many alkaloids from plants. To identify novel ligands for 5-HT$_{2C}$ receptor, we selected and screened a focused natural products-based library enriched in alkaloids. A potential hit, (–)-crebanine (**1**), increased the Tm of 5-HT$_{2C}$ receptor by 9.25 °C during thermal denaturation assays. Ultrafiltration-based affinity mass spectrometry validated the interaction between (–)-crebanine (**1**) and 5-HT$_{2C}$ receptor, and estimated the affinity of the ligand for the receptor ($K_d \sim 0.34$ μmol/L). Using cell-based calcium influx assay, we identified (–)-crebanine (**1**) to be an antagonist of 5-HT$_{2C}$ receptor. Furthermore, we predict its binding pose in 5-HT$_{2C}$ receptor by molecular docking and several key interactions of the ligand with the protein are proposed. The subsequent mutagenesis and binding experiments confirmed the predicted binding pose. In order to find out if (–)-crebanine (**1**) plays a

role in modulating the function of 5-HT receptors at tissue level, we recorded the electrical potential across the membrane of neurons from mice brain slices. (–)-Crebanine (**1**) inhibited the excitability caused by 5-HT in the brain slices, implying that (–)-crebanine (**1**) is effective in inhibiting the activity of 5-HT receptors and counteracting the excitatory role of 5-HT in the central nervous system of mice.

In summary, we have identified (–)-crebanine (**1**) as an antagonist for 5-HT$_{2A/2B/2C}$ receptors. The IC_{50} value of the compound was the lowest (149 nmol/L) for 5-HT$_{2C}$ receptor, suggesting some specificity in targeting the type of receptor. Structure-guided optimization of the crebanine scaffold is likely to further increase the receptor specificity. This is important because inhibition of 5-HT$_{2C}$ receptor activity has been shown to cure depression, schizophrenia and drug addiction. Results of our studies provide a framework for developing 5-HT$_{2C}$ receptor-specific antipsychotic drug. We show that natural products can be a viable source of novel ligands for 5-HT$_{2C}$ receptor. A comprehensive approach to screen and characterize novel ligands of GPCRs from the natural product library is also described in this study. We believe that the experimental protocols and analytical methods are general and could be used to facilitate biochemical and pharmacological studies for discovery of novel ligands targeting other GPCRs.

EXPERIMENTAL SECTION

Compounds library preparation

(−)-Crebanine (compound **1**) and (−)-dicentrine (compound **2**) were isolated from *A. scholaris* as previously described (Shang *et al.* 2010). Compounds (**3–5**) were purchased from Herbpurify Co., Ltd., Chengdu, China. Briefly, organic compounds from dried and powdered leaves of *A. scholaris* were extracted with ethanol and the solvent was evaporated under vacuum to obtain an extract. The extract was dissolved in 1% HCL and the constituents of the solution were considered as the alkaloid fraction. The solution was basified using ammonia water and extracted with ethyl acetate. The alkaloid extract was subjected to silica gel chromatography and eluted with chloroform–methanol (30:1–1:1) mixture in six fractions (I–VI). (−)-Crebanine and (−)-dicentrine were purified from fraction IV after column chromatography over silica gel (chloroform-acetone) and reverse phase RP$_{18}$ (methanol–water). The structures were validated by ^{1}H NMR spectrum analysis.

(−)-Crebanine (**1**) was obtained as colorless powder. ^{1}H NMR (CDCl$_3$, 500 MHz): δ 7.80 (1H, d, $J = 8.5$ Hz, H-11), 6.88 (1H, d, $J = 8.5$ Hz, H-10), 6.53 (1H, s, H-3), 6.06 (1H, d, $J = 1.4$ Hz, -OCH$_2$O-), 5.91 (1H, d, $J = 1.4$ Hz, -OCH$_2$O-), 3.90 (3H, s, -OCH$_3$), 3.81 (3H, s, -OCH$_3$), 3.67 (1H, dd, $J = 14.0, 4.0$ Hz, H-7), 3.12 (1H, m, H-6a), 3.05 (2H, m, H-4), 2.63 (1H, dd, $J = 16.0, 3.0$ Hz, H-5), 2.59 (3H, s, -NCH$_3$), 2.52 (1H, dd, $J = 16.0, 3.0$ Hz, H-5), 2.30 (1H, dd, $J = 14.5, 14.0$ Hz, H-7) (Bartley *et al.* 1994).

(−)-Dicentrine (**2**) was obtained as colorless powder. ^{1}H NMR (CDCl$_3$, 500 MHz): δ 7.67 (1H, s, H-11), 6.78 (1H, s, H-10), 6.52 (1H, s, H-3), 6.08 (1H, d, $J = 1.4$ Hz, -OCH$_2$O-), 5.93 (1H, d, $J = 1.4$ Hz, -OCH$_2$O-), 3.92 (3H, s, -OCH$_3$), 3.92 (3H, s, -OCH$_3$), 3.09 (1H, dd, $J = 14.0, 4.5$ Hz, H-7), 3.12 (1H, m, H-6a), 3.05 (2H, m, H-4), 2.67 (1H, dd, $J = 16.0, 3.0$ Hz, H-5), 2.56 (3H, s, -NCH$_3$), 2.62 (1H, dd, $J = 16.0, 3.0$ Hz, H-5), 2.53 (1H, dd, $J = 14.5, 14.0$ Hz, H-7) (Shang *et al.* 2010).

(+)-Magnoflorine (**3**) was obtained as brown powder. ^{1}H NMR (DMSO-d_6, 400 MHz): δ 6.60 (1H, d, $J = 8.0$ Hz, H-9), 6.36 (1H, d, $J = 8.0$ Hz, H-8), 6.51 (1H, s, H-3), 4.37 (1H, d, $J = 11.5$ Hz, H-6a), 3.69 (3H, s, -OCH$_3$), 3.66 (3H, s, -OCH$_3$), 3.67 (1H, m, overlap, H-5), 3.60 (1H, m, H-5), 3.33 (3H, H, -NCH$_3$), 3.12 (1H, dd, $J = 14.0, 13.0$ Hz, H-4), 3.11 (1H, dd, $J = 13.0, 3.5$ Hz, H-7), 2.90 (3H, s, -NCH$_3$), 2.82 (1H, dd, $J = 14.0, 3.0$ Hz, H-5), 2.61 (1H, dd, $J = 14.5, 13.0$ Hz, H-7) (Yin *et al.* 2016).

Didehydroglaucine (**4**) was obtained as light greenish powder. ^{1}H NMR (DMSO-d_6, 400 MHz): δ 8.91 (1H, s, H-11), 7.18 (1H, s, H-8), 7.18 (1H, s, H-7), 6.61 (1H, s,

H-3), 3.94 (3H, s, -OCH$_3$), 3.87 (3H, s, -OCH$_3$), 3.86 (3H, s, -OCH$_3$), 3.81 (3H, s, -OCH$_3$), 3.29 (1H, t, $J = 6.5$ Hz, H-4), 3.20 (1H, t, $J = 6.5$ Hz, H-7), 2.99 (3H, s, -NCH$_3$) (Xu *et al.* 2002).

(+)-Isocorydine (**5**) was obtained as colorless crystal. ^{1}H NMR (DMSO-d_6, 400 MHz): δ 8.64 (1H, s, -OH), 6.98 (1H, s, H-3), 6.97 (1H, d, $J = 8.0$ Hz, H-8), 6.88 (1H, d, $J = 8.0$ Hz, H-9), 4.06 (1H, m, H-6a), 3.86 (3H, s, -OCH$_3$), 3.80 (3H, s, -OCH$_3$), 3.65 (3H, s, -OCH$_3$), 3.36 (1H, m, H-7), 3.35 (1H, m, H-5), 3.05 (2H, m, H-4), 2.99 (1H, m, H-5), 2.98 (1H, m, H-7), 2.59 (3H, s, overlap, -NCH$_3$) (Zhong *et al.* 2015).

Cloning

The ΔN-5-HT$_{2C}$-BRIL-ΔC DNA was codon optimized, synthesized by DNA2.0 and subcloned into a modified pFastBac1 vector (Invitrogen). The construct had the following features: (1) Residues of the third intracellular loop of the wild-type human 5-HT$_{2C}$ receptor were replaced with Ala1-Leu106 of BRIL; (2) N-terminal residues before and including the glycosylation site and C-terminal residues after the helix 8 of 5-HT$_{2C}$ receptor were truncated. An apocytochrome b$_{562}$ RIL (BRIL) gene from *E. coli*, with M7 W, H102I, and R106L mutations is referred to as BRIL.

The vector designated as pFastBac1-830220 contained an expression cassette with a haemagglutinin (HA) signal sequence followed by a FLAG tag at the N-terminus, and a PreScission protease site followed by a 10× His tag at the C-terminus (Lv *et al.* 2016).

Virus generation and expression

High-titer recombinant baculovirus (>10^9 viral particles per ml) was obtained using the Bac-to-Bac Baculovirus Expression System (Invitrogen). Recombinant baculovirus was generated by transfecting 5–10 μg of recombinant bacmid into 2.5 ml *Spodoptera frugiperda* (*Sf9*) cells at a density of 10^6 cells per ml using 5 μl of FuGENE HD Transfection Reagent (Promega) and Transfection Medium (Expression Systems). After 4 days of shaking at 27 °C, P0 viral stock with ~10^9 virus particles per ml was harvested and used to generate high-titer baculovirus stock. Viral titers were determined by flow-cytometric analysis of cells stained with gp64-PE antibody (Expression Systems) (Hanson *et al.* 2007). Expression of the 5-HT$_{2C}$ receptor was carried out by infection of *Sf9* cells at a cell density of 2–3 × 10^6 cells/ml with P1 virus stock at multiplicity of infection (MOI) of five. Cells were harvested by centrifugation at 48 h post-infection and stored at −80 °C until further use.

Membrane purification

Insect cell membranes were disrupted by thawing frozen cell pellets in a hypotonic buffer containing 10 mmol/L HEPES, pH 7.5, 10 mmol/L MgCl$_2$, 20 mmol/L KCl and EDTA-free complete protease inhibitor cocktail tablets (Roche). Extensive washing of the isolated raw membranes was performed by repeated centrifugation in the same hypotonic buffer (two times), and then in a high osmotic buffer containing 1.0 mol/L NaCl, 10 mmol/L HEPES, pH 7.5, 10 mmol/L MgCl$_2$, 20 mmol/L KCl and EDTA-free complete protease inhibitor cocktail tablets (three times), to remove soluble and membrane-associated proteins. Purified membranes were directly flash-frozen in liquid nitrogen and stored at −80 °C until further use.

Protein purification

Purified membranes were resuspended in buffer containing 10 mmol/L HEPES, pH 7.5, 10 mmol/L MgCl$_2$, 20 mmol/L KCl, 150 mmol/L NaCl, and EDTA-free complete protease inhibitor cocktail tablets (Roche). Prior to solubilization, membranes were equilibrated at 4 °C and incubated for 30 min in the presence of 2 mg/mL iodoacetamide (Sigma). Membranes were then solubilized in 50 mmol/L HEPES, pH 7.5, 150 mmol/L NaCl, 1% (w/v) n-dodecyl-β-D-maltopyranoside (DDM, Anatrace), 0.2% (w/v) cholesteryl hemisuccinate (CHS, Sigma) and EDTA-free complete protease inhibitor cocktail tablets (Roche) for 2 h at 4 °C. Unsolubilized material was removed by centrifugation at 35,000 r/min for 30 min, and buffered imidazole and NaCl were added to the supernatant to adjust concentrations to 20 and 800 mmol/L, respectively. Proteins were bound to TALON IMAC resin (Clontech) overnight at 4 °C. The resin was then washed with 10 column volumes (cv) of Wash Buffer I (50 mmol/L HEPES, pH 7.5, 800 mmol/L NaCl, 0.1% (w/v) DDM, 0.02% (w/v) CHS, 20 mmol/L imidazole, 10% (v/v) glycerol), followed by 5 cv of Wash Buffer II (50 mmol/L HEPES, pH 7.5, 150 mmol/L NaCl, 0.05% (w/v) DDM, 0.01% (w/v) CHS, 10% (v/v) glycerol). Proteins were eluted in 5 cv of Wash Buffer II + 250 mmol/L imidazole. Protein purity and monodispersity were tested by SDS-PAGE and analytical size-exclusion chromatography (aSEC).

Thermal stability assay

Purified 5-HT$_{2C}$ receptor protein was thoroughly mixed with CPM fluorescent dyes. Protein solution containing the fluorescent dyes and compounds was dispensed into 0.2-ml tubes using a pipetting workstation. The tubes were incubated at 4 °C for 1 h prior to thermal denaturation. As the temperature was ramped from 25 to 95 °C, the protein began to unfold, exposing the cysteine residues that interact with the dye. Fluorescence signals, wavelength 387 nm (excitation) and 463 nm (emission), were monitored (Alexandrov et al. 2008). Analysis of 72 different samples was performed in parallel. The thermal denaturation of the samples was evaluated using the scatter plot curve, where the horizontal coordinates represented temperature and the vertical coordinates indicated normalized fluorescence intensity recorded for a particular temperature. Tm was obtained by fitting the data with a Boltzmann sigmoidal function using Prism (GraphPad Software).

Ligand-binding validation by affinity mass spectrometry analysis

The purified apo 5-HT$_{2C}$ receptor protein was incubated with each pure ligand (compounds **1**, **5** or ritanserin) at a final concentration of 500 nmol/L (protein) and 250 nmol/L (ligand) at 4 °C for 60 min. Individual 5-HT$_{2C}$ mutants were incubated with compound **1** under the same condition. Then, the incubated sample (∼1 μg) was filtered through 50 kDa MW cutoff ultrafiltration membrane (Sartorius, Germany) by centrifugation at 13,000 g for 10 min at 4 °C in the buffer containing 150 mmol/L ammonium acetate, 0.02% (w/v) DDM and 0.004% (w/v) CHS. Buffer exchange was repeated once. The protein complexes retained on the ultrafiltration membrane were transferred to a new centrifugal tube. The ligand was dissociated from the complexes with 90% methanol and separated from the denatured protein by centrifugation at 13,000 g for 20 min at 25 °C. Another purified GPCR protein (hydroxycarboxylic acid receptor 2) underwent the same process to serve as a negative control. The supernatant was dried out in speed vacuum, reconstituted in 50% methanol, diluted by tenfold prior to LC–MS analysis using Agilent 6530 TOF equipped with an Agilent 1260 HPLC system. The compound was eluted with 85% acetonitrile/0.1% formic acid from Eclipse Plus C18 column (2.1 mm × 100 mm, 3.5 μm, Agilent, USA) at a flow rate of 0.4 ml/min. Full-scan mass spectra were acquired in the range of 100–1000 m/z on Agilent 6530 TOF with ESI source settings: voltage 3000 V, gas temperature 350 °C, fragmentor 150 V.

Four experimental replicates were prepared for each pair of the test ligand and the negative control. LC–MS chromatograms for specific ligands were extracted using MassHunter software (Agilent, USA) based on the accurate mass measurement with a tolerance of 15 ppm and also matching RT of the reference compound. MS responses are represented by the peak heights of the corresponding extracted ion chromatograms. S/C ratios refer to the ratio of MS response of a specific ligand detected in the 5-HT_{2C} incubation sample versus the control. A single-point K_d calculation method established earlier for binding evaluation of pure ligands or simple ligand mixtures was then employed to estimate the affinity of each ligand to the receptor (Qin *et al.* 2015). Relative ligand-binding capacity (represented by binding %) to different mutants was calculated by the MS response of the ligand released from a specific mutant divided by the MS response of the total ligand present in the incubation sample.

Ca^{2+} mobilization assay

HEK 293T cells stably transfected with $5\text{-HT}_{2A/2B/2C}$ receptor were seeded in 384-well plates at a density of 15,000 cells/well in DMEM containing 1% dialyzed FBS 8 h before the calcium flux assay. After removing medium, cells were then incubated (20 µl/well) for 1 h at 37 °C with Fluo-4 Direct dye (Invitrogen) reconstituted in FLIPR buffer ($1\times$ HBSS, 2.5 mmol/L probenecid, and 20 mmol/L HEPES, pH 7.4). After the dye loaded, cells were placed in a FLIPR^{TETRA} fluorescence imaging plate reader (Molecular Devices); drug dilutions, prepared at $3\times$ final concentration in FLIPR buffer and aliquotted into 384-well plates, were also added to the FLIPR^{TETRA}. The fluidics module and plate reader of the FLIPR^{TETRA} were programmed to read baseline fluorescence for 10 s (1 read/s), then to add 10 µl of drug/well and to read for 6 min (1 read/s). Fluorescence in each well was normalized to the average of the first 10 reads (*i.e.*, baseline fluorescence). Then, the maximum-fold increase, which occurred within 60 s after drug addition, over baseline fluorescence elicited by vehicle or drug was determined. For positive allosteric modulator and antagonist candidates test, 5-HT EC20 (0.1 nmol/L) and EC80 (3 nmol/L) were used, respectively, to activate receptor.

Modeling of 5-HT_{2C} in complex with crebanine

Modeling of receptor–ligand complexes was carried out with Schrodinger Suite 2015-4. Homology model of 5-HT_{2C} receptor was built based on the crystal structure of 5-HT_{2B} (PDB entry: 4IB4) using the Advanced Homology Modeling tool. Processing of the protein structure was performed with the Protein Preparation Wizard. 3D structures of the compounds were first generated using the LigPrep tool, then optimized by quantum mechanics in B3LYP/6-31G** level using

Jaguar 9.0. The complex structures were generated in three steps: (1) Molecular docking of the compounds into 5-HT_{2C} homology model using Glide 6.9; (2) Structural refinements allowing movement of the compounds and protein atoms within 5 Å using Prime 4.2; (3) Re-scoring the binding modes of compounds in the receptors by the extra-precision score using Glide 6.9.

Brain slicing preparation

Cortical slices from ICR mouse of both genders (15–21 days) were prepared for electrophysiological recording. In brief, the animals were decapitated and brains were placed in cold sucrose artificial cerebrospinal fluid (sucrose-ASCF, in mmol/L: sucrose 213, KCl 2.5, NaH_2PO_4 1.25, $NaHCO_3$ 26, *D*-glucose 10, $MgSO_4$ 2, $CaCl_2$ 2, pH 7.4, 0–4 °C, saturated with 95% O_2 and 5% CO_2). The brains were cut into 280 or 300 µm slices with a vibratome (Leica VT1200S, Germany); slices were then incubated with ACSF (in mmol/L: NaCl 126, KCl 2.5, NaH_2PO_4 1.25, $NaHCO_3$ 26, *D*-glucose 25, $MgSO_4$ 2, $CaCl_2$ 2, pH 7.4, saturated with 95% O_2 and 5% CO_2) at room temperature for 1 h.

Electrophysiological recordings

A whole-cell patch clamp technique was used. One brain slice was transferred into the recording chamber, continually perfused with oxygenated ACSF and viewed under a DIC microscope ($60\times$ water immersion lens, Olympus, Japan). Activity of cortical pyramidal neurons was recorded using an amplifier (HEKA EPC 10 USB, Germany) in the voltage clamp mode. The electrode puller (Sutter P-1000, USA) was used to make electrodes with the resistance at 10–13 MΩ when filled with the pipette solution (in mmol/L: potassium gluconate 140, KCl 3, $MgCl_2$ 2, HEPES 10, EGTA 0.2, Na_2ATP 2, pH 7.25). Drug treatments were performed by switching the perfusion solution with different drugs (5-HT, compound **1**, compound **1** and 5-HT). To evaluate the effect of **1** against 5-HT, the brain slices were pretreated by **1** for 15–20 min.

Abbreviations

5-HT	5-hydroxytryptamine
CNS	Central nervous system
CPM	The thiol-specific fluorochrome N-[4-(7-diethylamino- 4-methyl-3-coumarinyl) phenyl] maleimide
GPCR	G protein-coupled receptor
ICL3	Intracellular loop 3
LSD	Lysergic acid diethylamide

MS   Mass spectrometry
Sf9   *Spodoptera frugiperda* (an insect cell line)
SPSC Spontaneous synaptic current

Acknowledgements This work was funded by the National Nature Science Foundation of China (31330019), the Shanghai Municipal Government, ShanghaiTech University and the Institute of Molecular and Clinical Medicine, Kunming Medical University. We thank the cloning, cell expression, protein purification, functional assay and bioImaging core facilities of iHuman Institute for their support. We thank Xiping Huang of Bryan Roth group in the University of North Carolina at Chapel Hill for the help on cell-based assay experiment. We thank Tian Hua, Houchao Tao, Dongsheng Liu, for the helpful discussions.

Compliance with Ethical Standards

Conflict of interest Yao Peng, Simeng Zhao, Yiran Wu, Haijie Cao, Yueming Xu, Xiaoyan Liu, Wenqing Shui, Jianjun Cheng, Suwen Zhao, Ling Shen, Raymond C. Stevens, Jun Ma, Ronald J. Quinn, Guisheng Zhong, and Zhi-Jie Liu declare that they have no conflict of interest.

Human and animal rights and informed consent All institutional and national guidelines for the care and use of laboratory animals were followed.

References

Ahn K (2016) Worldwide trend of botanical drug and strategies for developing global drugs. BMB Rep 50(3):111–116

Alexandrov AI, Mileni M, Chien EY, Hanson MA, Stevens RC (2008) Microscale fluorescent thermal stability assay for membrane proteins. Structure 16:351–359

All natural (2007) Nat Chem Biol 3: 351

Bartley JP, Baker LT, Carvalho CF (1994) Alkaloids of *Stephania bancroftii*. Phytochemistry 36:1327–1331

Besnard J, Ruda GF, Setola V, Abecassis K, Rodriguiz RM, Huang XP, Norval S, Sassano MF, Shin AI, Webster LA, Simeons FR, Stojanovski L, Prat A, Seidah NG, Constam DB, Bickerton GR, Read KD, Wetsel WC, Gilbert IH, Roth BL, Hopkins AL (2012) Automated design of ligands to polypharmacological profiles. Nature 492:215–220

Chen X, Qin S, Chen S, Li J, Li L, Wang Z, Wang Q, Lin J, Yang C, Shui W (2015) A ligand-observed mass spectrometry approach integrated into the fragment based lead discovery pipeline. Sci Rep 5:8361

Cheng J, McCorvy JD, Giguere PM, Zhu H, Kenakin T, Roth BL, Kozikowski AP (2016) Design and discovery of functionally selective serotonin 2C (5-HT2C) receptor agonists. J Med Chem 59:9866–9880

Connolly HM, Crary JL, McGoon MD, Hensrud DD, Edwards BS, Edwards WD, Schaff HV (1997) Valvular heart disease associated with fenfluramine-phentermine. N Engl J Med 337:581–588

Englisch S, Jung HS, Lewien A, Becker A, Nowak U, Braun H, Thiem J, Eisenacher S, Meyer-Lindenberg A, Zink M (2016) Agomelatine for the treatment of major depressive episodes in schizophrenia-spectrum disorders: an open-prospective proof-of-concept study. J Clin Psychopharmacol 36(6):597–607

Halford JC, Harrold JA (2012) 5-HT(2C) receptor agonists and the control of appetite. Handb Exp Pharmacol. https://doi.org/10.1007/978-3-642-24716-3_16

Hannon J, Hoyer D (2008) Molecular biology of 5-HT receptors. Behav Brain Res 195:198–213

Hanson MA, Brooun A, Baker KA, Jaakola VP, Roth C, Chien EY, Alexandrov A, Velasquez J, Davis L, Griffith M, Moy K, Ganser-Pornillos BK, Hua Y, Kuhn P, Ellis S, Yeager M, Stevens RC (2007) Profiling of membrane protein variants in a baculovirus system by coupling cell-surface detection with small-scale parallel expression. Protein Expr Purif 56:85–92

Huang XP, Karpiak J, Kroeze WK, Zhu H, Chen X, Moy SS, Saddoris KA, Nikolova VD, Farrell MS, Wang S, Mangano TJ, Deshpande DA, Jiang A, Penn RB, Jin J, Koller BH, Kenakin T, Shoichet BK, Roth BL (2015) Allosteric ligands for the pharmacologically dark receptors GPR68 and GPR65. Nature 527:477–483

Isberg V, Mordalski S, Munk C, Rataj K, Harpsøe K, Hauser AS, Vroling B, Bojarski AJ, Vriend G, Gloriam DE (2016) GPCRdb: an information system for G protein-coupled receptors. Nucleic Acids Res 44:D356–D364

Jacobson KA (2015) New paradigms in GPCR drug discovery. Biochem Pharmacol 98:541–555

Jang HJ, Cho KH, Park SW, Kim MJ, Yoon SH, Rhie DJ (2012) Layer-specific serotonergic facilitation of IPSC in layer 2/3 pyramidal neurons of the visual cortex. J Neurophysiol 107:407–416

Kroeze WK, Sassano MF, Huang XP, Lansu K, McCorvy JD, Giguère PM, Sciaky N, Roth BL (2015) PRESTO-Tango as an open-source resource for interrogation of the druggable human GPCRome. Nat Struct Mol Biol 22:362–369

Lv X, Liu J, Shi Q, Tan Q, Wu D, Skinner JJ, Walker AL, Zhao L, Gu X, Chen N, Xue L, Si P, Zhang L, Wang Z, Katritch V, Liu ZJ, Stevens RC (2016) *In vitro* expression and analysis of the 826 human G protein-coupled receptors. Protein Cell 7:325–337

McCorvy JD, Roth BL (2015) Structure and function of serotonin G protein-coupled receptors. Pharmacol Ther 150:129–142

Narayanaswami V, Dwoskin LP (2016) Obesity: current and potential pharmacotherapeutics and targets. Pharmacol Ther 170:116–147

Newman DJ, Cragg GM (2016) Natural Products as sources of new drugs from 1981 to 2014. J Nat Prod 79:629–661

Nichols DE, Frescas S, Marona-Lewicka D, Kurrasch-Orbaugh DM (2002) Lysergamides of isomeric 2,4-dimethylazetidines map the binding orientation of the diethylamide moiety in the potent hallucinogenic agent N, N-diethyllysergamide (LSD). J Med Chem 45:4344–4349

Qin S, Ren Y, Fu X, Shen J, Chen X, Wang Q, Bi X, Liu W, Li L, Liang G, Yang C, Shui W (2015) Multiple ligand detection and affinity measurement by ultrafiltration and mass spectrometry analysis applied to fragment mixture screening. Anal Chim Acta 886:98–106

Rask-Andersen M, Masuram S, Schioth HB (2014) The druggable genome: evaluation of drug targets in clinical trials suggests major shifts in molecular class and indication. Annu Rev Pharmacol Toxicol 54:9–26

Santos R, Ursu O, Gaulton A, Bento AP, Donadi RS, Bologa CG, Karlsson A, Al-Lazikani B, Hersey A, Oprea TI, Overington JP

(2017) A comprehensive map of molecular drug targets. Nat Rev Drug Discov 16:19–34

Shang JH, Cai XH, Feng T, Zhao YL, Wang JK, Zhang LY, Yan M, Luo XD (2010) Pharmacological evaluation of *Alstonia scholaris*: anti-inflammatory and analgesic effects. J Ethnopharmacol 129:174–181

Venkatakrishnan AJ, Deupi X, Lebon G, Tate CG, Schertler GF, Babu MM (2013) Molecular signatures of G-protein-coupled receptors. Nature 494:185–194

Wacker D, Wang C, Katritch V, Han GW, Huang XP, Vardy E, McCorvy JD, Jiang Y, Chu M, Siu FY, Liu W, Xu HE, Cherezov V, Roth BL, Stevens RC (2013) Structural features for functional selectivity at serotonin receptors. Science 340:615–619

Wacker D, Wang S, McCorvy JD, Betz RM, Venkatakrishnan AJ, Levit A, Lansu K, Schools ZL, Che T, Nichols DE, Shoichet BK, Dror RO, Roth BL (2017) Crystal structure of an LSD-bound human serotonin receptor. Cell 168(377–389):e312

Wang C, Jiang Y, Ma J, Wu H, Wacker D, Katritch V, Han GW, Liu W, Huang XP, Vardy E, McCorvy JD, Gao X, Zhou XE, Melcher K, Zhang C, Bai F, Yang H, Yang L, Jiang H, Roth BL, Cherezov V, Stevens RC, Xu HE (2013) Structural basis for molecular recognition at serotonin receptors. Science 340:610–614

Xu X, Wang ZT, Yu GD, Ruan BF, Li J (2002) Alkaloids from *Rhizoma corydalis*. Journal of China Pharm Univ 33:483–486

Yin X, Bai R, Guo Q, Su G, Wang J, Yang X, Li L, Tu P, Chai X (2016) Hendersine A, a novel isoquinoline alkaloid from *Corydalis hendersonii*. Tetrahedron Lett 57:4858–4862

Yuan S, Peng Q, Palczewski K, Vogel H, Filipek S (2016) Mechanistic studies on the stereoselectivity of the serotonin 5-HT1A receptor. Angew Chem Int Ed Engl 55:8661–8665

Zeeb FD, Higgins GA, Fletcher PJ (2015) The serotonin 2C receptor agonist lorcaserin attenuates intracranial self-stimulation and blocks the reward-enhancing effects of nicotine. ACS Chem Neurosci 6:1231–1240

Zhong M, Jiang YB, Chen YL, Yan Q, Liu JX, Di DL (2015) Asymmetric total synthesis of (*S*)-isocorydine. Tetrahedron 26:1145–1149

12

Using integrated correlative cryo-light and electron microscopy to directly observe syntaphilin-immobilized neuronal mitochondria *in situ*

Shengliu Wang[1], Shuoguo Li[2], Gang Ji[2], Xiaojun Huang[2], Fei Sun[1,2,3]✉

[1] National Key Laboratory of Biomacromolecules, CAS Center for Excellence in Biomacromolecules, Institute of Biophysics, Chinese Academy of Sciences, Beijing 100101, China

[2] Center for Biological Imaging, Institute of Biophysics, Chinese Academy of Sciences, Beijing 100101, China

[3] University of Chinese Academy of Sciences, Beijing, China

Abstract   Correlative cryo-fluorescence and cryo-electron microscopy (cryo-CLEM) system has been fast becoming a powerful technique with the advantage to allow the fluorescent labeling and direct visualization of the close-to-physiologic ultrastructure in cells at the same time, offering unique insights into the ultrastructure with specific cellular function. There have been various engineered ways to achieve cryo-CLEM including the commercial FEI iCorr system that integrates fluorescence microscope into the column of transmission electron microscope. In this study, we applied the approach of the cryo-CLEM-based iCorr to image the syntaphilin-immobilized neuronal mitochondria *in situ* to test the performance of the FEI iCorr system and determine its correlation accuracy. Our study revealed the various morphologies of syntaphilin-immobilized neuronal mitochondria that interact with microtubules and suggested that the cryo-CLEM procedure by the FEI iCorr system is suitable with a half micron-meter correlation accuracy to study the cellular organelles that have a discrete distribution and large size, *e.g.* mitochondrion, Golgi complex, lysosome, *etc.*

Keywords   Correlative cryo-light and electron microscopy, iCorr, Mitochondria, Primary hippocampal neuron cell, Syntaphilin

INTRODUCTION

Fluorescence light microscopy (FM) offers large-scale, time-resolved, and dynamic visualization of positions of interest (POIs) in cells and provides a variety of information in cellular processes. The resolution of FM is restricted to ~200 nm due to diffraction limit (Abbe 1873), which was overcome by the recently developed super-resolution fluorescence microscopy techniques including photon-activated localization microscopy (PALM)/stochastic optical reconstruction microscopy (STORM) (Betzig *et al.* 2006; Hess *et al.* 2006; Rust *et al.*

2006), stimulated emission depletion fluorescence microscopy (STED) (Hell and Wichmann 1994; Klar *et al.* 2001), structured illumination microscopy (SIM) (Heintzmann and Cremer 1999; Gustafsson 2000; Li *et al.* 2015), *etc.* However, FM could only provide localization information of labeled molecules, but with a lack of structural context in the cells.

Electron microscopy (EM) has been applied into the study of cellular ultrastructures for many years and can provide detailed structural information of cells in nanometer-scale resolution. In addition, besides chemical fixation, cryo-vitrification technique has provided a good immobilization of cellular ultrastructures in their native state, which can be imaged by cryo-electron

---

✉ Correspondence: feisun@ibp.ac.cn (F. Sun)

microscopy (cryo-EM) (Dubochet *et al.* 1988; Al-Amoudi *et al.* 2004). With the recent development of direct-detection devices, cryo-EM has been becoming one of the important tools in structural biology to resolve macromolecular structures in near-atomic resolution (Nogales and Scheres 2015). However, for cell biology study, EM technique is still restricted by a few limitations (Kukulski *et al.* 2011; Zhang 2013) due to (1) having a small field of view (FOV) and restricted specimen thickness; (2) being difficult to locate POI inside a cell; (3) being unable to capture dynamic state of events due to fixed specimen; and (4) being hard to distinguish specific molecules in a crowded cellular environment.

To combine the benefits of FM that provides localization information and EM that provides structural information, an emerging technique, termed correlative light and electron microscopy (CLEM), has been developed in the recent years (Mironov and Beznoussenko 2009; Hanein and Volkmann 2011). CLEM first localizes POI using FM and then transfers the localization information into EM and eventually generates two correlated images (fluorescence image and electron micrograph), within which the localizations of target molecules can be mapped onto their relevant structural contexts. To observe the biological structures in their native state, a particular technique, cryo-CLEM, obtained by correlating fluorescence cryo-microscopy (cryo-FM) with cryo-EM and imaging cryo-vitrified biological specimen, has received more and more attention in the recent years (Wolff *et al.* 2016). Several kinds of dedicated cryo-stages were developed to adapt standard fluorescence microscopes to perform cryo-FM (Sartori *et al.* 2007; Jun *et al.* 2011; Schorb *et al.* 2016). Fluorescent beads with enough electron density were used to correlate cryo-FM image into cryo-EM micrograph (Jun *et al.* 2011; Schorb *et al.* 2016). The cryo-vitrified specimen needs to be imaged by cryo-FM first and then transferred to the column of electron microscope for cryo-EM imaging.

Besides, to avoid specimen transfer that would cause severe ice contamination and specimen devitrification, an integrated cryo-CLEM workflow was developed, the so-called "iCorr" by the FEI Company, which integrates a fluorescence light microscope into the column of a transmission electron microscope (TEM). To use the iCorr system, the specimen loaded with a cryo-holder is rotated 90° to be imaged by fluorescence microscope (FM mode) and then rotated back to 0° for cryo-EM imaging (TEM mode). The iCorr system uses a LED illumination with the excitation wavelength ranging from 460 to 500 nm and with the peak at 470 nm. An optical filter in the iCorr system is used to pass through the emission light with the wavelength ranging from 510 to 560 nm. With a fixed objective lens having a numeric aperture (NA) of 0.5, the iCorr system can capture the fluorescence images at the green channel with the magnification of ×15 and the resolution of ~460 nm. Considering the early development stage of the iCorr system, there are a few cryo-CLEM applications of iCorr system in the literature. In the present work, we intend to test our prototype iCorr system integrated onto our Tecnai Spirit electron microscope by imaging cryo-vitrified rat hippocampal neuron cells.

Neuron cells are highly polarized and consist of three distinct structural and functional domains with unique morphologies: one with large and compact cell body that is called soma; one with thin and long axon; and one with numerous thick dendrites with branches. Axon and dendrites are also called neuronal processes. In neuron cells, mitochondria, as essential organelles for energy production, intracellular calcium homeostasis maintenance, and steroid and lipid synthesis (Nicholls and Budd 2000; Boldogh and Pon 2007), are transported between processes and soma according to energy and metabolic requirements at different regions (Hollenbeck and Saxton 2005). The transport of neuronal mitochondria has been extensively studied by using time-lapse fluorescence microscopy (Misgeld *et al.* 2007; Kang *et al.* 2008). Mitochondria can move a long distance without stopping or frequently changing direction, pausing or persistent dwelling (Hollenbeck and Saxton 2005; Misgeld *et al.* 2007; Kang *et al.* 2008). The movements of mitochondria are majorly along microtubules (Nangaku *et al.* 1994; Goldstein and Yang 2000). In mature neuron's axons, only approximately one-third of mitochondria are mobile while the remaining being stationary (Kang *et al.* 2008).

The molecular mechanism of mitochondrial station in neuron cell has been studied for many years. Syntaphilin (SNPH) has been found as a neuron-specific protein that locates on the mitochondrial outer membrane and binds to microtubule and would be a stationary factor for mitochondrial immobilization. Knock out SNPH gene resulted in a dramatic increase of mobile axonal mitochondria; however, overexpressing SNPH protein could immobilize almost all of the axonal mitochondria (Kang *et al.* 2008). Although SNPH has been revealed as a receptor for immobilizing mitochondria in axons, little is known about its molecular mechanism.

In the present study, we labeled SNPH with the fluorescent tag Dendra2 and mitochondria with the fluorescent marker TagRFP-mito in the rat hippocampal neuron cells and utilized the cryo-CLEM approach with iCorr to image the syntaphilin-immobilized neuronal mitochondria *in situ*.

RESULTS

Culturing hippocampal neuron cells on grids and cryo-vitrification

Rat hippocampal neuron cells can grow on the carbon film-coated gold EM grid with healthy morphology (Fig. 1A). After transfection with the plasmids of Dendra2-SNPH and TagRFP-mito, the cotransfected cells showed that all axonal mitochondria became immobilized as previously reported (Kang *et al.* 2008), and the fluorescence signals from SNPH and mitochondria were significantly colocalized (Fig. 1B). With the successful neuron cell culturing and transfection, we were ready for the further cryo-CLEM experiments.

Before applying cryo-CLEM to observe the SNPH-immobilized mitochondria, we first utilized those non-transfected neuron cells to explore an appropriate freezing method to vitrify the grids where the cells grow on. Plunge-freezing method has been successfully applied to cryo-preserve the neuron cells cultured on EM grids (Lucic *et al.* 2007). In the present study, we aimed to utilize FEI Vitrobot device to perform plunge freezing. However, we found that the double-side blotting method by FEI Vitrobot did not work well for freezing the fragile neuron cells that were easily broken down due to the blot force from the double sides (Fig. 2A). To overcome this problem, we removed one blot pad from one side, and let the filter paper from another side to blot the backside of the grid where no neuron cells were growing on. With this single-side blotting approach, we were able to vitrify the neurons in their native states (Fig. 2B). The neuronal processes, particularly the axons, could be embedded in thin (200–500 nm) layer of ice and readily be imaged by cryo-EM, and the double layer of mitochondria

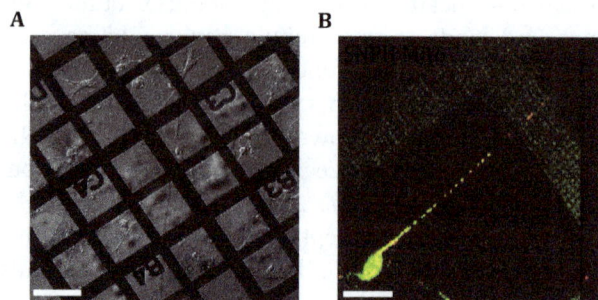

Fig. 1 Rat hippocampal neuron cells grown on EM grids. **A** Differential interference contrast light microscope image of rat primary hippocampus neurons cultured on EM grids for 9 d. Scale bar, 100 μm. **B** Fluorescent visualization of neurons cultured on EM grids that were cotransfected at DIV of 6 with Dendra2-SNPH (*green*) and TagRFP-mito (*red*). The colocalized regions are shown in *yellow*. Scale bar, 20 μm

membrane and the architecture of microtubule as well as trafficking vesicles were all clearly visible (Fig. 2B).

We noticed that mitochondria in the neuronal processes have variable morphologies including tubular, branched, and round. Moreover, we also observed the interaction between some mitochondria and microtubule and some not (Fig. 2B), which are actually relevant to different physiological states of mitochondrion. Whether SNPH overexpression could enhance such interaction and increase the distribution of the microtubule interacted mitochondrion needs to be further investigated by the subsequent cryo-CLEM experiments.

Cryo-CLEM of the SNPH transfected hippocampal neuron cells

Clonable fluorescent protein has been reported to have an unanticipated advantage in reducing the rate of fluorescence photo bleaching in cryogenic temperature and very suitable for cryo-CLEM experiments (Schwartz *et al.* 2007). In the present work, the nonactivated Dendra2, a GFP variant, which possesses excitation–emission maxima at 490 and 507 nm similar to EGFP and other green fluorescent proteins (Gurskaya *et al.* 2006), was selected to label the SNPH protein. Considering the excitation wavelength range (460–500 nm) and emission wavelength range (510–560 nm) of the iCorr system, Dendra2 is suitable (but not optimized) for cryo-CLEM work using the iCorr.

The entire process of cryo-CLEM is shown in Fig. 3. First, the reflective and fluorescent images with a large field of view were acquired and merged (Fig. 3A, left). Then, the region of interest was selected and magnified (Fig. 3A, right) for the subsequent cryo-EM imaging. A cryo-EM image in a medium magnification with a field of view ($\sim 10 \times 10$ μm^2) was acquired and automatically correlated with the cryo-FM image (Fig. 3B, left). With the benefit of holey carbon film, we manually slightly optimized the translation and rotation alignments between cryo-FM and low-magnification cryo-EM images. Thereafter, the subsequent cryo-EM images at a high magnification was acquired and aligned to the magnified cryo-FM images (Fig. 3B, right). The correlated images showed that the Dendra2-SNPH fluorescent signals had located along the neuron cell process. For each discrete fluorescence signal, one single mitochondria organelle was found nearby with elongated or round morphologies (Fig. 4A–D). We also found that the elongated mitochondria closely interact with the microtubules along its long axis (Fig. 4B–D), while the mitochondria in the round shape loosely interact with the microtubule via a small region (Fig. 4A).

Since SNPH proteins are colocalized with the discretely distributed mitochondria in the axonal process

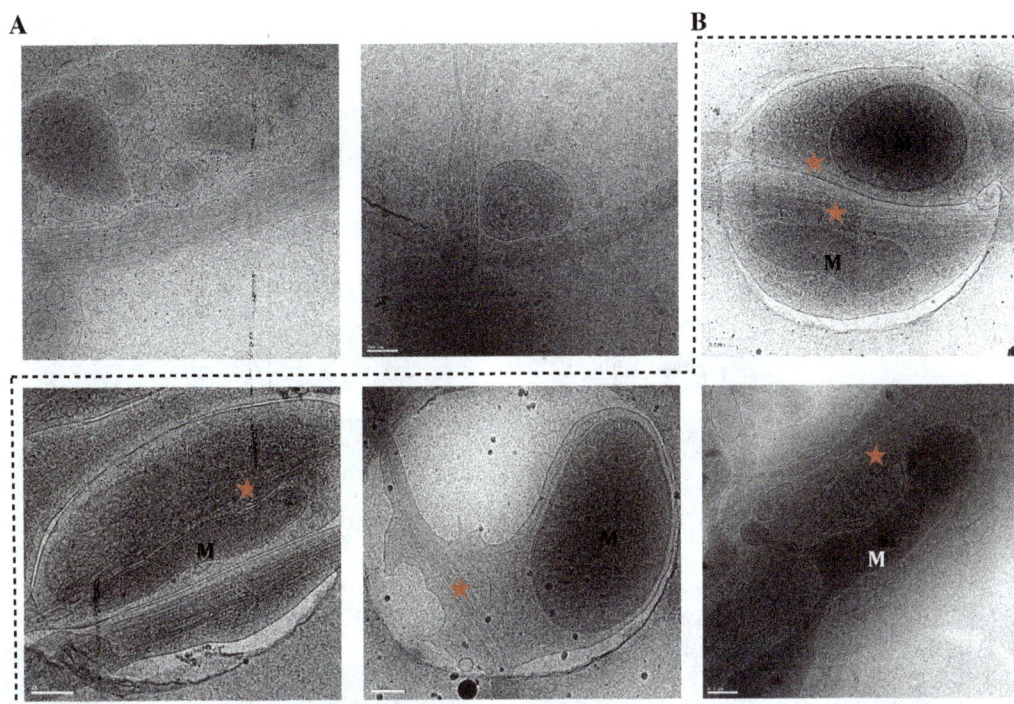

Fig. 2 Cryo-electron micrographs of vitrified rat hippocampal neuron cells (nontransfected) at process regions. **A** Broken cells caused by double-sided blotting method in plunge freezing. Scale bar, 200 nm. **B** Intact cells vitrified by single-side blotting method in plunge freezing. The mitochondria with round, elongated, and branched shapes are labeled in *yellow*, *orange*, and *white*, respectively. Microtubules are indicated with a *red star*. Scale bar, 200 nm

(Fig. 1B), we were confident to identify the fluorescent signals of SNPH-Dendra2 in the cryo-FM images, and the nearby-located mitochondria in the cryo-EM images were actually correlated (Fig. 4A–D). Upon this assumption, the correlation accuracy of this experiment could be estimated (see next section).

From the cryo-CLEM images of SNPH-transfected hippocampal neuron cells, it would be surmised that the SNPH-immobilized mitochondria in the neuronal axon vary in the morphologies from elongated to round, which is similar to the previous observation of the glutamine-induced mitochondrial immobilization (Rintoul *et al.* 2003).

Correlation accuracy estimation

The correlation accuracy between cryo-FM and cryo-EM depends on many factors including the resolution of cryo-FM itself, the distortion variations from different imaging systems, and the alignment accuracies of different sources of images. Here, we used the final correlated images to determine the overall correlation accuracy of our iCorr system.

The fluorescence signal of each correlated cryo-FM image (Fig. 4A–D) represents many SNPH molecules located on the outer membrane of mitochondria and thus the center of each fluorescence spot represents the central position of a cluster of SNPH molecules from one mitochondrion (Fig. 4E). Since SNPH mediates the direct interaction between mitochondria and microtubule and locates at their direct contacts, the center of the contact interface between mitochondria and microtubule also represents the central position of SNPH molecules from one mitochondrion (Fig. 4F). Comparing the shift between these two centers computed from two different sources of images would give an estimation of the correlation accuracy in this cryo-CLEM experiment. As shown in Table 1, from the four correlated images (Fig. 4A–D), the mean shift between these two centers was 488.5 ± 121.8 nm, which is close to the resolution (~460 nm) of the FM in the iCorr system, suggesting that the limitation factor of the correlation accuracy in the iCorr system is the resolution of its FM module.

CONCLUSION

In the present work, we tested the cryo-CLEM procedure based on the FEI iCorr system and applied this technique to the cryo-vitrified rat hippocampal neuron cells that were transfected with Dendra2-SNPH. We developed a successful protocol using FEI Vitrobot to

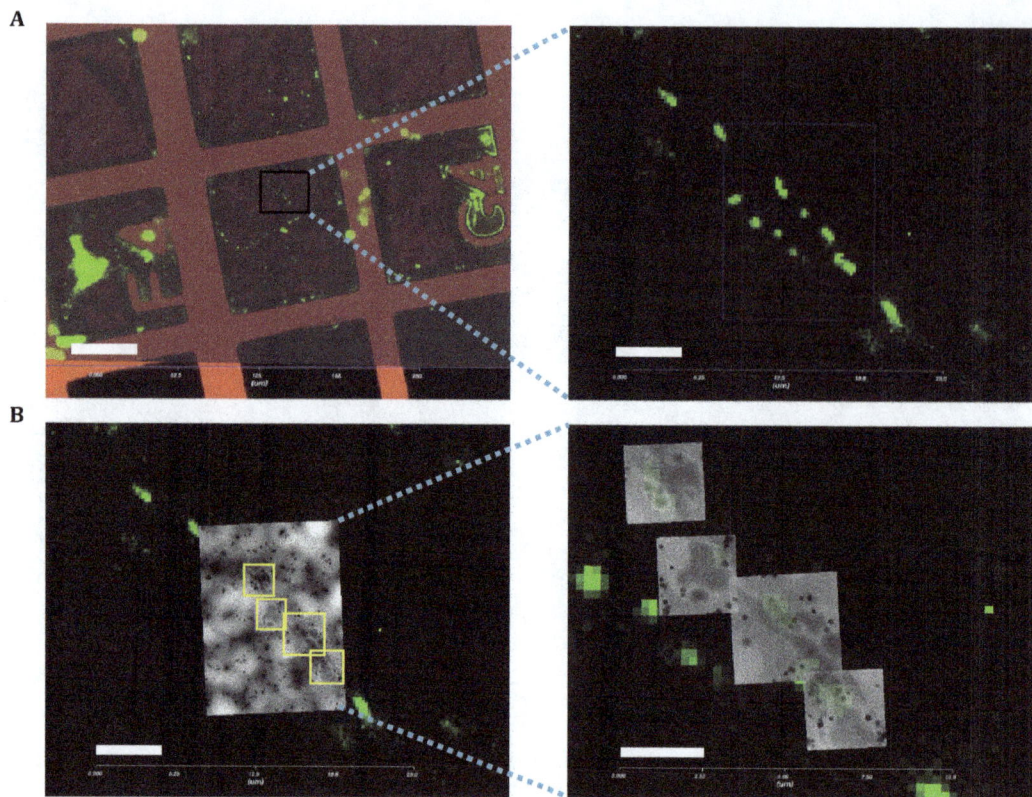

Fig. 3 Applying the cryo-CLEM procedure to the vitrified SNPH-transfected rat hippocampal neuron cells. **A** The reflection image (*red*) is merged with the corresponding fluorescence image (*green*). Scale bar, 50 μm. Region of interest (ROI) is selected by *blue box* and magnified at right. Scale bar, 5 μm. **B** Medium magnification transmission electron microscopy (TEM) image is first taken in the center of ROI, which is automatically correlated to the FM image with slight manual modification. Then, this correlated image is further used to select areas (indicated by *yellow box*) for high-magnification TEM imaging. Scale bar, 5 μm. The final merged and correlated TEM and FM images with high magnification are shown at right. Scale bar, 2.5 μm

freeze the fragile neuron cells grown on grids and preserve them in their native state. We directly visualized the SNPH-immobilized neuronal mitochondria *in situ* and successfully captured the varied morphologies of the SNPH-immobilized mitochondria as well as their interactions with microtubules. The estimated accuracy of the correlation between cryo-FM and cryo-EM was 488.5 ± 121.8 nm, suggesting that the current cryo-CLEM procedure by the FEI iCorr system would be suitable for cryo-CLEM study of cellular organelles like mitochondrion, Golgi complex, lysosome, and so on, which have a discrete distribution and large size.

MATERIALS AND METHODS

EM grids preparation

Gold EM-grids with Alpha-Numeric Finder (G200F1-G3, Gilder Grids Ltd.) were coated with holey carbon (2-μm round hole spaced by 2 μm), which was manufactured in house. All the grids were sterilized by ultraviolet light overnight before transferred into culture dishes that were coated with Matrigel matrix (Product #354248, Coring life science). The grids were put on the surface of Matrigel matrix with the carbon side on the top.

Neuron cell cultures and transfection

Primary hippocampal neuron cells were dissected from postnatal day 0–1 Spraque–Dawley rats in accordance with the procedures in Kang's lab, the Institute for Nutritional Sciences, SIBS, CAS. Briefly, hippocampal neurons were dissected in cold HBSS buffer (H2387, Sigma), digested in a DNase/trypsin solution (0.5 mg/mL DNase, 5 mg/mL trypsin, 25 mmol/L HEPES, 137 mmol/L NaCl, 5 mmol/L KCl, 7 mmol/L Na_2HPO_4, pH 7.2) for 5 min at 37 °C, and then dissociated into separated single cells in a DNase solution (0.5 mg/mL DNase, 25 mmol/L HEPES, 137 mmol/L NaCl, 5 mmol/L KCl, 7 mmol/L Na_2HPO_4, pH 7.2). After washing with HBSS and 10% FBS, neurons were plated on a Matrigel matrix coated culture dish with EM grids on the top. For cryo-CLEM application, low-density neuron cells were

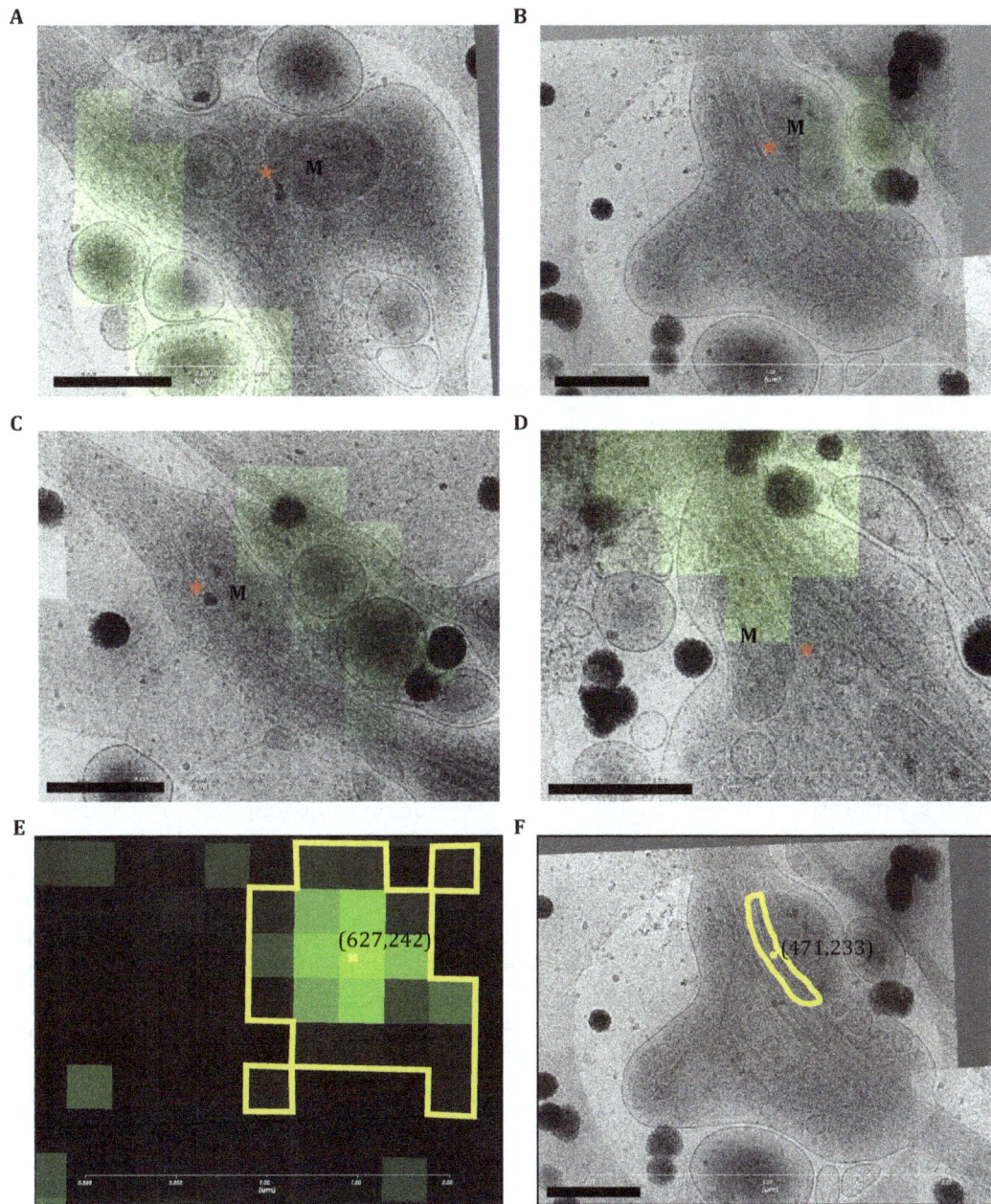

Fig. 4 Correlation between cryo-FM and cryo-EM images. **A–D** Aligned and overlaid cryo-FM and cryo-EM images located at the regions of interest in Fig. 3B from top to bottom. The mitochondria are labeled in *yellow,* and microtubules are indicated with *red stars.* Scale bar, 0.5 μm. (**E, F**) Correlation accuracy between cryo-FM and cryoEM images is measured on the basis of the center of the signals. In cryo-FM image (**E**), the contours in *yellow* represent the profile of the fluorescent signal. In addition, in cryo-EM image (**F**), the contours in *yellow* represent the interacting region between mitochondria and microtubule, where SNPH are localized. The *yellow* crosses represent the centers of the contours with the corresponding coordinates in pixel. Scale bar, 500 nm

kept at 37 °C in 5% CO_2. At days *in vitro* (DIV) of 6–9, the cells were transfected with the plasmids of Dendra2–SNPH and TagRFP-mito using the calcium phosphate method (Jiang and Chen 2006). After transfection, the cells were cultured further for additional 2–3 days before inspection using a laser scanning confocal fluorescence microscope (FV1000, Olympus), and the transfected cells were identified according to the fluorescence signals of Dendra2 and TagRFP. The regions of processes where SNPH are overexpressed were selected and marked using the nearest alpha-numeric finder in the grid.

Table 1 Signal shifts between the correlated cryo-FM and cryo-EM images that were captured in the iCorr system[a]

	Center of MM[b] interaction site in cryo-EM image		Center of SNPH[c] fluorescence in cryo-FM image		Pixel size (nm/pixel)	S^d (nm)
	X' (pixel)	Y' (pixel)	X (pixel)	Y (pixel)		
1	453	259	253	463	2.2	629.3
2	471	234	627	242	2.8	436.8
3	352	300	594	337	2.2	539.0
4	461	355	381	178	1.8	348.9
					S^d (nm)	488.5
					Sd^d (nm)	121.8

[a] FEI's iCorr technology that consists of a fluorescence light microscope module, which is integrated with FEI's Tecnai transmission electron microscope, and a software for automatically correlating FM and EM images. Location information 1–4 are from the four correlated images (Fig. 4A–D) respectively

[b] Mitochondria and microtubule

[c] Syntaphilin, which has a role for maintaining a large number of axonal mitochondria in a stationary state on microtubule, is labeled with Dendra2 fluorescent protein

[d] The shift between two centers of signals (see Materials and Methods)

Cryo-vitrification of hippocampal cells

The EM grids with neuron cells grown on were cryo-vitrified using FEI Vitrobot (Mark IV). To achieve single-side blotting that is important to keep the integrity of the fragile neuron cells, one blot pad of Vitrobot was removed before the vitrification process. The EM grids with the grown neuron cells were carefully picked up from the dishes using a Vitrobot tweezer (FEI), and washed once in a dish with warm HBSS buffer (H2387, Sigma). After applying additional 3 μL HBSS buffer onto the side where neurons have grown, the tweezer with the grid was mounted onto Vitrobot by allowing the side of neurons to be facing the side of the removed blot pad. As a result, the excess liquid was blotted by the filter paper from the backside of the grid, and there is no direct contact between the filter paper and the cells. The following parameters were set up during blotting: blot force 8, blot time 8 s, temperature 25 °C, and humidity 100%. After blotting, the grid was rapidly frozen in liquid ethane that was precooled by liquid nitrogen and transferred to liquid nitrogen for storage.

Cryo-CLEM of the vitrified cells

The vitrified grid was mounted into a cryo-holder (Model 626, Gatan) that was precooled in liquid nitrogen. Then the cryo-holder was loaded into the column of the transmission electron microscopy Tecnai Spirit that is supplied with the FEI iCorr module. The regions with the marked finders, which were selected prior to vitrification, were searched and centered in the TEM low-magnification mode. Then, the subsequent cryo-CLEM operations were performed in the software of the iCorr system.

The FM mode was first selected, and the stage was tilted to the angle of 90°. Since most of the liquid nitrogen stored in the cryo-holder Dewar spilled out when tilting at 90°, the image acquisition in FM model should be finished within 30 min to prevent warming up the specimen. Both reflection and fluorescence images were recorded by adjusting the Z-focus value and optimizing the light illumination intensity. Then, the stage was tilted back to 0° for EM model operation. Positions of interest (POIs) with fluorescent signals were selected, and the cryo-EM images with medium magnification and FOV of $\sim 10 \times 10$ μm^2 were acquired. The cryo-EM images were automatically correlated to the cryo-FM images by the software using the precalibrated parameters for translation, rotation, and scaling. This initial correlation might not be sufficiently correct due to the mechanical error of the stage and the drift of the specimen. To optimize the correlation between the cryo-EM and cryo-FM images, repositioning manually according to the features of carbon holes in both reflection and cryo-EM images was performed. After correlation optimization, the final cryo-EM micrographs targeted at the higher magnification were acquired by clicking the POIs in the correlated FM–EM image using the software of the iCorr system. All the procedure for cryo-EM imaging were controlled in a low-dose condition.

Quantification of the correlation accuracy

The correlated cryo-FM and cryo-EM images were separately saved in a PNG format. Then, these images were uploaded into Image J software (Schneider et al. 2012). For the cryo-FM image, a polygon selection tool was used to contour the profile of the entire fluorescence spot, and then the center of the fluorescence signal was calculated using the tool of "measuring the center of mass". The coordinates of the center were denoted by (X_i, Y_i) in pixels. For the cryo-EM image, a freehand selection tool was used to contour the interacting site between mitochondria and microtubule, which was the position where SNPH proteins are localized. The tool of "measuring the center of mass" was also used to calculate the center of the SNPH-localized region with the coordinates denoted by (X_i', Y_i') in pixels. Thus, after pixel size correction, the shift S_i of the correlated signals between cryo-FM and cryo-EM images were calculated as follows:

$$S_i = \sqrt{(X_i - X_i')^2 + (Y_i - Y_i')^2}$$

The mean shift S and its standard deviation Sd were calculated as follows:

$$\bar{S} = \frac{1}{n} \sum_{i=1}^{n} S_i$$

$$Sd = \sqrt{\frac{\sum (S - \bar{S})^2}{n-1}}$$

ACKNOWLEDGEMENTS The authors would like to thank Prof. Jiansheng Kang and Dr. Chunfeng Liu in Kang's lab from the Institute for Nutritional Sciences, SIBS, CAS, for extending their valuable help during culturing of the rat hippocampus neuron cells and providing the plasmid of Dendra2–SNPH. This work was supported by grants from the Strategic Priority Research Program of Chinese Academy of Sciences (XDB08030202) and the National Basic Research Program ("973" Program) of the Ministry of Science and Technology of China (2014CB910700). All the EM works were performed at the Center for Biological Imaging (CBI, http://cbi.ibp.ac.cn), the Institute of Biophysics, Chinese Academy of Sciences.

Compliance with Ethical Standards

Conflict of interest Shengliu Wang, Shuoguo Li, Gang Ji, Xiaojun Huang, and Fei Sun declare that they have no conflict of interest.

Human and Animal Rights and Informed Consent All the institutional and national guidelines for the care and use of laboratory animals were followed.

REFERENCES

Abbe E (1873) Beitrage zur theorie des mikroskops und der mikroskopischen wahrnehmung. Arch Mikr Anat 9:413–468

Al-Amoudi A, Chang JJ, Leforestier A, McDowall A, Salamin LM, Norlen LP, Richter K, Blanc NS, Studer D, Dubochet J (2004) Cryo-electron microscopy of vitreous sections. The EMBO journal 23:3583–3588

Betzig E, Patterson GH, Sougrat R, Lindwasser OW, Olenych S, Bonifacino JS, Davidson MW, Lippincott-Schwartz J, Hess HF (2006) Imaging intracellular fluorescent proteins at nanometer resolution. Science 313:1642–1645

Boldogh IR, Pon LA (2007) Mitochondria on the move. Trends Cell Biol 17:502–510

Dubochet J, Adrian M, Chang JJ, Homo JC, Lepault J, McDowall AW, Schultz P (1988) Cryo-electron microscopy of vitrified specimens. Q Rev Biophys 21:129–228

Goldstein LS, Yang Z (2000) Microtubule-based transport systems in neurons: the roles of kinesins and dyneins. Annu Rev Neurosci 23:39–71

Gurskaya NG, Verkhusha VV, Shcheglov AS, Staroverov DB, Chepurnykh TV, Fradkov AF, Lukyanov S, Lukyanov KA (2006) Engineering of a monomeric green-to-red photoactivatable fluorescent protein induced by blue light. Nat Biotechnol 24:461–465

Gustafsson MG (2000) Surpassing the lateral resolution limit by a factor of two using structured illumination microscopy. J Microsc 198:82–87

Hanein D, Volkmann N (2011) Correlative light-electron microscopy. Adv Protein Chem Struct Biol 82:91–99

Heintzmann R, Cremer CG (1999) In: Laterally modulated excitation microscopy: improvement of resolution by using a diffraction grating, pp 185-196

Hell SW, Wichmann J (1994) Breaking the diffraction resolution limit by stimulated emission: stimulated-emission-depletion fluorescence microscopy. Opt Lett 19:780–782

Hess ST, Girirajan TP, Mason MD (2006) Ultra-high resolution imaging by fluorescence photoactivation localization microscopy. Biophys J 91:4258–4272

Hollenbeck PJ, Saxton WM (2005) The axonal transport of mitochondria. J Cell Sci 118:5411–5419

Jiang M, Chen G (2006) High Ca^{2+}-phosphate transfection efficiency in low-density neuronal cultures. Nat Protoc 1:695–700

Jun S, Ke D, Debiec K, Zhao G, Meng X, Ambrose Z, Gibson GA, Watkins SC, Zhang P (2011) Direct visualization of HIV-1 with correlative live-cell microscopy and cryo-electron tomography. Structure 19:1573–1581

Kang JS, Tian JH, Pan PY, Zald P, Li C, Deng C, Sheng ZH (2008) Docking of axonal mitochondria by syntaphilin controls their mobility and affects short-term facilitation. Cell 132:137–148

Klar TA, Engel E, Hell SW (2001) Breaking Abbe's diffraction resolution limit in fluorescence microscopy with stimulated emission depletion beams of various shapes. Phys Rev E 64:066613

Kukulski W, Schorb M, Welsch S, Picco A, Kaksonen M, Briggs JA (2011) Correlated fluorescence and 3D electron microscopy with high sensitivity and spatial precision. J Cell Biol 192:111–119

Li D, Shao L, Chen BC, Zhang X, Zhang M, Moses B, DE Milkie, Beach JR, Hammer JA 3rd, Pasham M, Kirchhausen T, Baird MA, Davidson MW, Xu P, Betzig E (2015) Extended-resolution structured illumination imaging of endocytic and cytoskeletal dynamics. Science 349:aab3500

Lucic V, Kossel AH, Yang T, Bonhoeffer T, Baumeister W, Sartori A (2007) Multiscale imaging of neurons grown in culture: from light microscopy to cryo-electron tomography. J Struct Biol 160:146–156

Mironov AA, Beznoussenko GV (2009) Correlative microscopy: a potent tool for the study of rare or unique cellular and tissue events. J Microsc 235:308–321

Misgeld T, Kerschensteiner M, Bareyre FM, Burgess RW, Lichtman JW (2007) Imaging axonal transport of mitochondria *in vivo*. Nat Methods 4:559–561

Nangaku M, Sato-Yoshitake R, Okada Y, Noda Y, Takemura R, Yamazaki H, Hirokawa N (1994) KIF1B, a novel microtubule plus end-directed monomeric motor protein for transport of mitochondria. Cell 79:1209–1220

Nicholls DG, Budd SL (2000) Mitochondria and neuronal survival. Physiol Rev 80:315–360

Nogales E, Scheres SH (2015) Cryo-EM: a unique tool for the visualization of macromolecular complexity. Mol Cell 58:677–689

Rintoul GL, Filiano AJ, Brocard JB, Kress GJ, Reynolds IJ (2003) Glutamate decreases mitochondrial size and movement in primary forebrain neurons. J Neurosci 23:7881–7888

Rust MJ, Bates M, Zhuang X (2006) Sub-diffraction-limit imaging by stochastic optical reconstruction microscopy (STORM). Nat Methods 3:793–795

Sartori A, Gatz R, Beck F, Rigort A, Baumeister W, Plitzko JM (2007) Correlative microscopy: bridging the gap between fluorescence light microscopy and cryo-electron tomography. J Struct Biol 160:135–145

Schneider CA, Rasband WS, Eliceiri KW (2012) NIH image to imageJ: 25 years of image analysis. Nat Methods 9:671–675

Schorb M, Gaechter L, Avinoam O, Sieckmann F, Clarke M, Bebeacua C, Bykov YS, Sonnen AF, Lihl R, Briggs JA (2016) New hardware and workflows for semi-automated correlative cryo-fluorescence and cryo-electron microscopy/tomography. J Struct Biol. doi:10.1016/j.jsb.2016.06.020

Schwartz CL, Sarbash VI, Ataullakhanov FI, Mcintosh JR, Nicastro D (2007) Cryo-fluorescence microscopy facilitates correlations between light and cryo-electron microscopy and reduces the rate of photobleaching. J Microsc-Oxford 227:98–109

Wolff G, Hagen C, Grunewald K, Kaufmann R (2016) Towards correlative super-resolution fluorescence and electron cryo-microscopy. Biol cell. doi:10.1111/boc.201600008

Zhang P (2013) Correlative cryo-electron tomography and optical microscopy of cells. Curr Opin Struct Biol 23:763–770

13

Targeting tumor cells with antibodies enhances anti-tumor immunity

Zhichen Sun[1,2], Yang-Xin Fu[1,3], Hua Peng[1✉]

[1] Key Laboratory of Infection and Immunity, Institute of Biophysics, Chinese Academy of Sciences, Beijing 100101, China
[2] University of Chinese Academy of Sciences, Beijing 100049, China
[3] Department of Pathology, University of Texas Southwestern Medical Center, Dallas, TX 75390, USA

Abstract Tumor-targeting antibodies were initially defined as a group of therapeutic monoclonal antibodies (mAb) that recognize tumor-specific membrane proteins, block cell signaling, and induce tumor-killing through Fc-driven innate immune responses. However, in the past decade, ample evidence has shown that tumor-targeting mAb (TTmAb) eradicates tumor cells via activation of cytotoxic T cells (CTLs). In this review, we specifically focus on how TTmAbs induce adaptive anti-tumor immunity and its potential in combination therapy with immune cytokines, checkpoint blockade, radiation, and enzyme-targeted small molecule drugs. Exploring the mechanisms of these preclinical studies and retrospective clinical data will significantly benefit the development of highly efficient and specific TTmAb-oriented anti-tumor remedies.

Keywords Tumor antigen, Targeting antibody, Innate immunity, Adaptive immunity, Cytokine, Tumor microenvironment

INTRODUCTION OF CLASSICAL TUMOR-TARGETING ANTIBODIES

The first generation of tumor-targeting antibodies approved by US Food and Drug Administration (FDA), including trastuzumab, cetuximab, and rituximab, were initially known as signal blockers to target oncogenic receptors of tumor cells and have great potential for effective cancer immunotherapy (Hynes and Lane 2005; Li *et al.* 2005). Later, Fc receptor (FcR) in immune cells was found to play an essential role in Ab-dependent cell cytotoxicity to tumor cells *in vivo* (Clynes *et al.* 2000; Musolino *et al.* 2008) and Complement-dependent cytotoxicity (CDC) (Teeling *et al.* 2004; van Meerten *et al.* 2006). Recently, we and others have reported the function of adaptive immunity in the Ab-mediated tumor regression (Abes *et al.* 2010; Mortenson *et al.*

2013; Park *et al.* 2010; Ren *et al.* 2017; Stagg *et al.* 2011; Yang *et al.* 2013b, 2014). With further understanding of tumor microenvironment (TEM), tumor-targeting Ab has been used to construct bispecific Ab and Ab–cytokine. In combination with checkpoint blockade Abs, radiation therapy, and traditional small molecule chemo- and targeted drugs, TTmAbs can be widely applied to break the immune tolerance and acquired resistance to long-term anti-tumor treatments for the final tumor elimination.

ROLES FOR TUMOR-TARGETING AB TO BRIDGE INNATE IMMUNITY TO CTL

Anti-HER2 antibody

The human epidermal growth factor receptor 2 (HER2) is the homologue of the rat neu oncogene (HER2/neu).

✉ Correspondence: hpeng@moon.ibp.ac.cn (H. Peng)

Along with HER1 (EGFR), HER3 and HER4, these proteins belong to the type I growth receptor family (Rajkumar and Gullick 1994). These transmembrane receptors, with molecular weights of approximately 170–185 kDa, are receptor tyrosine kinases that undergo homo- or hetero-dimerization when activated by related ligands (Olayioye et al. 2000; Rajkumar and Gullick 1994; Shepard et al. 2008), but the specific ligand for HER2/neu (ErbB2) is still not identified (Yarden and Pines 2012). This receptor signaling plays essential roles in multiple normal cellular processes of cell proliferation, differentiation, adhesion, motility, and apoptosis (Menard et al. 2003; Quaglino et al. 2008). However, HER2 has been found to be overexpressed in a quarter of breast cancers and is connected to higher malignancy, relapse rates, and mortality (Hudis 2007; Kiessling et al. 2002; Meric-Bernstam and Hung 2006; Slamon et al. 1987). In 1998, the FDA approved trastuzumab (Herceptin) for clinical trial in treating human breast cancer patients. Trastuzumab is a humanized monoclonal antibody containing the complementary regions of the murine antibody (clone 4D5) and Fc region of human IgG1 (Carter et al. 1992; Nahta and Esteva 2006). It binds to HER2 on the cell surface and has been proven to be an effective treatment for HER2/neu-positive breast cancers in multiple animal studies and clinical trials (Abramson and Arteaga 2011; Hudis 2007; Moasser 2007).

In vitro mechanistic studies have shown that the anti-HER2/neu antibody inhibits HER2/neu+ tumor growth, mainly via inducing G1 cell cycle arrest (Le et al. 2005; Mittendorf et al. 2010). Antibody-dependent cellular cytotoxicity (ADCC) is also required for the antitumor effects of anti-HER2 therapy. NK cells induce ADCC after the anti-HER2/neu antibody engagement, resulting in HMGB-1 release and MyD88-dependent TLR stimulation (Clynes et al. 2000; Musolino et al. 2008). Park et al. firstly revealed that the adaptive immune system plays critical roles in anti-HER2/neu-mediated anti-tumor therapeutic effect (Park et al. 2010). They demonstrate that (1) the therapeutic effect of HER2/neu antibody is tumor-specific T-cell dependent; (2) effective anti-HER2/neu treatment achieves immune memory that resists the subsequent high dose tumor rechallenge; (3) anti-HER2/neu treatment results in enhanced CD8+ cell infiltration in the tumor tissue of mouse models and clinical patients.

Anti-EGFR antibody

Epidermal growth factor receptor (EGFR) is another member of the type I growth receptor family. Although HER2 is characterized by lacking an identified ligand,

EGFR has been found to have several other ligands besides EGF, such as transforming growth factor alpha (TGF-α) (Citri and Yarden 2006; Yarden and Sliwkowski 2001). The first FDA-approved anti-EGFR monoclonal antibodies, including cetuximab (Erbitux, a human–murine chimeric. Ennis et al. 1991) and panitumumab (a human monoclonal antibody), have been successfully used for the treatment of EGFR-expressing cancers (Yang et al. 2001). These monoclonal antibodies bind to the extracellular domain of EGFR, inhibiting receptor dimerization and downstream signaling (Burgess et al. 2003), and inducing receptor internalization, ubiquitinization, and degradation (Sunada et al. 1986). ADCC and CDC are also the tumor-killing mechanisms resulting from the binding of the monoclonal antibody (Kimura et al. 2007). Primary and acquired resistance becomes increasingly challenging for targeted therapy (Bardelli and Siena 2010; Cobleigh et al. 1999). Ab resistance mediated by mutations within targeted oncogenes or in genes related to oncogenic pathways has been broadly investigated (Misale et al. 2012; Yonesaka et al. 2011). These studies will provide important information for developing drugs that target the increasing intrinsic resistance in tumor cells after antioncogenic Ab treatment (Bostrom et al. 2009; Fayad et al. 2013; Hurvitz et al. 2013; Krop et al. 2012; Yoon et al. 2011). Nevertheless, Yang et al. focus on tumor-extrinsic resistance and propose a tumor-extrinsic strategy to bypass intrinsic Ab resistance by reactivating both innate and adaptive immune cells inside the tumor. Using Ab-sensitive TUBO (HER2/neu+) and Ab-resistant EGFR-transduced B16 mouse tumor models (Rovero et al. 2000), they found increased production of type I IFNs in the Ab-sensitive tumor model, in contrast to the Ab-resistant tumor model. These data suggest that enhanced type I IFN production was caused by Ab-induced oncogenic receptor and stress. Yang et al. are the first to point out that Type I IFNs are the cytokines essential for Ab-mediated tumor regression and tumor-targeting delivery of type I IFNs may induce stronger anti-tumor immune responses to overcome antibody resistance or tumor immune tolerance.

Anti-CD20 antibody

Rituximab is a murine–human chimeric antibody that recognizes the human B-cell CD20 antigen featuring primary response rates up to 70% (Maloney et al. 1992, 1994, 1997a, b). Similar to cetuximab and panitumumab, rituximab can induce tumor-cell death via ADCC, CDC, and apoptosis of tumor cells *in vitro* and in animal models (Clynes et al. 2000; Maloney 2012). Rituximab was approved by the FDA in 1997 for treating non-Hodgkin B-cell lymphoma. Effective control of B-cell lymphoma by anti-CD20 in xenograft models indicates the importance of direct tumor killing or innate-mediated killing induced by this antibody. However, the function of the adaptive immune response was later discovered to also play essential roles in lymphoma clearance. Using the huCD20-EL4 tumor model, a murine T-cell lymphoma transfected with the human CD20 molecule, Abes and Xuan reported that by the CD4+, but not CD8+, T-cell immune responses may contribute to long-lasting protection by anti-human CD20 treatment (Abes *et al.* 2010; Xuan *et al.* 2010). Conversely, Ren *et al.* reported in 2016 that CD8+ T cells alone, but not CD4+ T cells, contributed to the effective anti-mouse CD20 Ab therapy in a syngeneic A20 B-cell lymphoma mouse model. In this study, they characterized how anti-CD20 treatment initiated a potent tumor-specific T-cell response for tumor control. Ab could kill some tumor cells through ADCC by macrophages that produce type I IFNs for cross-priming; IFN binds to interferon α/β receptor (IFNAR) and activates Dendritic Cells (DCs) to better process tumor antigens for cross-priming T cells in the DLN; tumor-specific CTLs travel back to the tumor site for tumor control. They further demonstrated the role of CTLA-4 in Tregs within advanced B-cell lymphoma in limiting anti-CD20-mediated tumor regression. Thus, anti-CTLA-4 and anti-CD20 combined treatment is a possible new clinical strategy in overcoming adaptive resistance and preventing relapse of B-cell lymphoma.

Overall, these studies reveal the essential contribution of adaptive immune responses in the early elimination and late resistance of TTmAb therapy. Most importantly, these studies have demonstrated that DCs are the major tolerized immune cells in tumors and DCs determine the immune-active or immunosuppressive status in TME. Targeting DCs will be another important strategy for improving the efficacy of cancer immunotherapy. This can be achieved by providing type I IFNs, the key players linking innate and adaptive antitumor immunity, to induce Ab-mediated tumor regression. Moreover, these studies raise the potential of using checkpoint blockades to overcome adaptive resistance in the future.

ANTIBODY ARMED WITH CYTOKINES TO FURTHER PROMOTE ADAPTIVE ANTI-TUMOR IMMUNITY

Type I IFNs, including IFN-α, IFN-β, IFN-ε, IFN-κ, and IFN-ν, are a family of monomeric cytokines with multiple functions (Pestka *et al.* 2004). Type I IFNs are involved in regulating many aspects of innate and adaptive immune responses by affecting the activation, migration, differentiation, and survival of macrophages, NK cells, DCs, monocytes, and B/T cells. Type I IFNs have been demonstrated to improve the antigen cross-presentation ability of CD8a+ DCs (Diamond *et al.* 2011; Lorenzi *et al.* 2011).

Due to their direct anti-proliferation and pro-apoptosis effects, Type I IFNs have been used for melanoma and lymphoma treatments. Recently, our laboratory and other research groups have found that endogenous type I IFNs perform significant functions in various antitumor therapies (Deng *et al.* 2014; Sistigu *et al.* 2014; Woo *et al.* 2014).

However, administration of type I IFNs has not been efficient in clinical cancer therapy, because of its limited potency and severe side effects (Trinchieri 2010). It has been suggested that the delivery of exogenous type I IFN may dramatically impact both immune responses and tumor cell proliferation/survival (Tang *et al.* 2016; Trinchieri 2010).

Yang *et al.* proposed that type I IFNs play an essential and sufficient role to bridge innate and adaptive antitumor immune responses during Ab-based antitumor therapy (Yang *et al.* 2014). They developed Ab–IFNβ as an advanced therapeutic strategy for those Ab-resistant tumors. Their study demonstrates that Ab-resistant tumors can be efficiently controlled by employing Ab–IFNβ fusion proteins. IFNAR on DCs, but not on tumor cells or T cells was required for the targeting effect of Ab–IFNβ which increases DC cross-presentation and antitumor CTL function.

Another study demonstrated that Anti-CD20–IFNα fusion protein was more effective than anti-CD20 Ab alone for direct killing of type IFNα receptor (IFNAR)-positive lymphoma (Xuan *et al.* 2010). However, the function of IFNAR on host immune cells was not addressed. Using a syngeneic immunocompetent mouse model, Liao *et al.* observed that targeting lymphoma with IFNα abolished resistance of B-cell lymphoma to anti-CD20 Ab while also limiting interferon (IFN)-associated systemic toxicity in the host (Liao *et al.* 2017). Anti-CD20–IFNα fusion protein-mediated tumor control is dependent on existing tumor-infiltrating CD8+ T cells. Although resistant to direct killing induced by IFNα, IFN-exposed A20 lymphoma cells become the dominant APCs for the reactivation of CTLs in the tumor. Anti-CD20–IFNα also abolishes checkpoint blockade resistance in advanced B-cell lymphoma. Thus, this study indicates that anti-CD20–IFNα eradicates B-cell lymphoma by employing tumor cells as APCs to reactivate tumor-infiltrating CD8+ T cells and synergizing with anti-PD-L1 treatment.

LIGHT ("the homologous to lymphotoxin that exhibits inducible expression and competes with HSV glycoprotein D for binding to herpesvirus entry mediator, a receptor expressed on T lymphocytes"), also known as tumor necrosis factor superfamily member 14 (TNFSF14), is one of the co-stimulatory molecules for T-cell activation (Lee *et al.* 2006; Wang *et al.* 2009). LIGHT is predominantly expressed on the surface of immature dendritic cells (DCs) and activated T cells. Several studies have indicated that LIGHT signaling might increase lymphocyte infiltration in the tumor (Yu *et al.* 2004, 2007; Zou *et al.* 2012). Based on LIGHT activities, Tang *et al.* constructed an anti-EGFR–LIGHT (Ab–LIGHT) targeting into EGFR+ but anti-PD-L1-resistant tumor tissues (Tang *et al.* 2016). In the study, they demonstrated that LIGHT indeed induces lymphocyte infiltration and antitumor immunity in both mouse and human tumor model. They also proved that additional LIGHT treatment can promote the efficacy of checkpoint blockade therapies by enhancing lymphocyte infiltration. These data suggest that LIGHT could be used to increase the responsiveness to checkpoint blockades and other immunotherapies in non-T-cell-inflamed tumors.

Interleukin-2 (IL-2) is a pleiotropic cytokine that promotes proliferation of NK and T cells induced by antigen stimulation (Liao *et al.* 2013; Morgan *et al.* 1976). IL-2 was one of the first FDA-approved immunotherapy drugs for metastatic melanoma and renal cell cancer (Rosenberg 2014; Rosenberg *et al.* 1998). However, IL-2 immunotherapy has not been widely employed because of its short half-life *in vivo* and severe toxicity at a therapeutic dose (Chavez *et al.* 2009; Panelli *et al.* 2004; Skrombolas and Frelinger 2014). In addition, IL-2 induces proliferation of regulatory T cells (Tregs) through binding to IL-2R alpha that is preferentially expressed on Tregs (Ahmadzadeh and Rosenberg 2006; Jensen *et al.* 2009; Sim *et al.* 2014), which might be a major barrier for IL-2-mediated expansion of CTL. To limit systemic toxicity, antibody-based delivery of IL-2 (Ab-IL2) has been investigated (Becker *et al.* 1996; Du *et al.* 2013; Gillies *et al.* 2011; Gutbrodt *et al.* 2013, 2014; Yang *et al.* 2013a). Systemic delivery of IL-2 may activate T cells in lymphoid and non-lymphoid tissues. The anti-tumor function of IL-2 therapy is known to directly activate CTLs (Gutbrodt *et al.* 2014; Jackaman *et al.* 2003). However, IL-2 induced Treg inhibitory effects are also significant. Thus, there are tremendous research efforts in constructing the mutant IL-2 with tumor targeting that increases binding to CD8+ T cells and reduces affinity to Tregs, which can alter antitumor response in tumor microenvironment and activate tumor-specific CTLs, thus leading to

significantly improved anti-tumor responses (De Luca *et al.* 2017; Hartimath *et al.* 2018; Zhu *et al.* 2015). Similarly, IL-21, IL-12, and IL-15 cytokines are under investigation for possibilities in achieving potent tumor-targeting effects with minor peripheral side effects.

In summary, these studies established new concepts for the Ab-based treatment, such as the Ab-IL2, Ab–IFNβ, and Ab–LIGHT fusion proteins, which stimulate or augment the tumor-specific CTL responses to deal with Ab resistance and relapse more effectively. Killing more tumor cells by enhanced CTLs can then create a positive feedback loop for anti-tumor immune responses. In addition, all these studies conclude that blocking inhibitory PD-L1 upregulated by Ab–cytokine treatment may further improve the antitumor effect via recruiting more Ab–cytokine molecules and open new avenues for future clinical cancer treatment.

ANTIBODIES TARGETING TO IMMUNE INHIBITORY RECEPTORS AND COMBINATION THERAPIES

Anti-CD47

CD47 is a major player of the 'donot eat me' signal (McCracken *et al.* 2015). The inhibitory phagocytic signaling is transduced when cell-surface CD47 binds signal-regulatory protein alpha (SIRPα) on phagocytic cells (Willingham *et al.* 2012). CD47 is overexpressed on tumor cells and considered as a marker for cancer prognosis (Chan *et al.* 2009; Jaiswal *et al.* 2009; Majeti *et al.* 2009). The characteristic overexpression of CD47 on tumor cells also makes it a potential target for antibody-driven immunotherapy. There are several ongoing clinical trials for anti-CD47 monoclonal antibodies that have yet to yield conclusions (Russ *et al.* 2018). Risk for toxicity and anti-tumor effectiveness from a CD47 blockade is still under observation. Tumor-associated macrophages were commonly reported to be the major anti-tumor phagocytes in xenograft transplantation models, where the role of adaptive immunity is excluded (Chao *et al* 2010, 2011; Willingham *et al.* 2012). However, using syngeneic immune complete cancer mouse models instead of transplanted xenografts, Liu *et al.* first demonstrated that the anti-tumor effects mediated by the CD47 blockade was mostly dependent on the tumor-specific CD8+ T cells. CD11c+ DCs, but not macrophages, are the major APCs for the cross-priming of CD8+ T cells in a STING signaling-dependent manner to further drive type I IFN production and CTL activation. This DC activation was not induced by the MyD88 Toll-like receptor signaling as previously reported. The discovery of adaptive antitumor immunity mediated by CD47-targeting Ab blockade sheds light on designing new strategies with anti-CD47 in conjunction with traditional chemotherapeutics and other targeted therapies.

Ab in combination with checkpoint blockade

Cytotoxic T-lymphocyte-associated antigen-4 (CTLA4) and programmed death-1 (PD1) are two of the major co-inhibitory immune checkpoint molecules expressed on T cells. PD1 ligand (PDL1) is another immune inhibitory molecule expressed on dendritic cells, activated T cells, and tumor cells. Anti-CTLA4 and anti-PD1/PD-L1 monoclonal antibodies have been developed as checkpoint blockades to inhibit the suppressive function of CTLA4 or PD-1/PD-L1 (Pardoll 2012). Physiologically, these inhibitory molecules play important roles in protecting the host from autoimmune diseases (Keir *et al.* 2008; Nishimura *et al.* 2001). However, tumors take advantage of these pathways to escape antitumor immune responses (Dong *et al.* 2002; Iwai *et al.* 2002; Shin and Ribas, 2015). Nivolumab and pembrolizumab, two monoclonal antibodies targeting PD1, have shown impressive anti-tumor efficacy in melanoma and non-small cell lung cancer (NSCLC) (Brahmer *et al.* 2012; Robert *et al.* 2015a, b; Topalian *et al.* 2012). Ipilimumab, a first class anti-CTLA4 monoclonal antibody (Hodi *et al.* 2010; O'Day *et al.* 2007; Wolchok et al. 2010), is effective in about 10%–20% of patients (Maio *et al.* 2015; Schadendorf *et al.* 2015).

Studies have shown that checkpoint blockades can reverse T-cell suppression and improve the therapeutic effects. However, only a small number of patients are sensitive to such therapy. There is a pressing need for mechanism studies on this subject and potential mechanisms for synergistic effects. In the past 10 years, we and other laboratories have shown that anti-EGFR, anti-HER2, anti-CD20, and anti-CD47 in combination with immune checkpoint blockade could have synergistic effects in host adaptive anti-immune responses and tumor eradication. The mechanisms of such synergistic effect may be complicated. Antibodies targeting to tumor cells can not only reduce tumor burden but also change tumor microenvironment. Some antibodies, such as anti-Her2/neu, can increase IFN through MyD88, while others, such as anti-CD47, induce IFN through STING pathway. Our study also demonstrates that CTLA4 is a major immune suppressor in A20 tumors and leads to anti-CD20 resistance. Anti-CTLA-4 and anti-CD20 combined treatment overcomes adaptive resistance and prevents relapse in the mouse B-cell lymphoma model (Ren *et al.* 2017). But the mechanism is unclear. In another combination therapy, PD-L1

blockade could enhance the antitumor efficacy of anti-CD20–IFNα and reduce relapse rates for advanced large tumors that are resistant to either anti-CD20 Ab or the anti-CD20–IFNα fusion protein (Liao *et al.* 2017). IFN can increase PD-L1 expression on tumor cells and antigen presentation, each of which represents distinct mechanism of tumor control. PD-L1 on tumor cells may prevent T-cell-mediated killing, while PD-L1 on antigen-presenting cells may suppress T-cell reactivation. The role of PD-L1 on draining LN is unclear.

In our most recent study, Tang *et al.* demonstrated that anti-PD-L1 is significantly accumulated in tumors after systemic treatment and could be utilized to deliver immunomodulatory molecules (Tang *et al.* 2018), such as IFN–anti-PD-L1, specifically into tumor tissues (unpublished data from Yang-Xin Fu's group). IFN–anti-PD-L1 may elicit a positive feedback loop to enhance targeting effects by upregulating PD-L1 expression, which is beneficial for treating tumors with lower levels of PD-L1. This strategy of PD-L1 antibody armed with IFN can overcome resistance to checkpoint blockade therapy in advanced tumors.

Local radiotherapy can overcome PD-L1 resistance

Radiotherapy (RT) is a cancer treatment that employs high dose of radiation to kill tumor cells and reduce tumor burden. Not only tumor cells but also stromal cells at the tumor site can be affected by radiation. Radiation-damaged cancer cells release tumor antigens, DNA and RNA that are captured by antigen-presenting cells to promote the priming and activation of cytotoxic T cells, and to facilitate further infiltration of immune cells. Radiation-induced inflammatory response may result in multiple IR-resistant signaling that facilitate tumor relapse (Barcellos-Hoff *et al.* 2005), including enhancement of PD-L1/PD-1. Thus, application of radiation therapy with TTmAb and checkpoint blockade may result in synergistic effects for cancer regression. Animal studies and clinical trials have been taken to explore the most effective combinations, which may depend on the tumor type, specific immunotherapy, and optimal timing. In a study of radiation therapy combined with checkpoint blockades, anti-CTLA4 functions most effectively when administrated ahead of radiation, but anti-OX40 must be given after radiation to improve treatment efficacy in a mouse tumor model (Young *et al.* 2016). Another study showed that PDL1 blockade could overcome radiotherapy resistance only when anti-PDL1 was given concurrently with radiation, not a week before or after radiation (Dovedi *et al.* 2014).

Our previous study shows that upregulation of the PD-L1/PD-1 in tumor after IR inhibits radiation-induced anti-tumor immune responses and facilitates tumor relapse. Combination of IR with PD-L1 blockade results in the elimination of MDSCs by T-cell-derived TNF (Deng *et al.* 2014). This rational design of RT in combination with anti-PD-L1 should be referred to and applied in clinical cancer treatment. The optimization of radiation dose and timing is critical for enhancing the effectiveness of individual-based combination treatment.

TTmAb combined with small molecule chemo- and targeted therapy

Antibody treatment in combination with multiple chemo-therapeutic agents has been investigated in a number of tumor models (Pegram *et al.* 1999, 2004). There is no conclusive result for whether chemotherapy synergizes with anti-HER2/neu antibody (Hudis 2007; Piccart-Gebhart *et al.* 2005; Romond *et al.* 2005). Our previous studies have shown that the timing of chemotherapy administration is critical for synergistic effects when combined with anti-Her2 and anti-CD47 therapy (Liu *et al.* 2015; Park *et al.* 2010). Pretreatment with chemotherapy can enhance the anti-tumor effect of anti-Her2 and anti-CD47 therapy, whereas chemotherapy applied following anti-Her2 or anti-CD47 treatment abolishes the tumor regression induced by either single treatment and even destroyed the tumor-specific T-cell memory responses (Liu *et al.* 2015; Park *et al.* 2010).

Molecular targeted therapy functions through blocking the growth of cancer cells by interfering with tumor-specific targeted molecules needed for oncogenesis and tumor growth. Tyrosine kinase inhibitors that target EGFR family members are among the most successful targeted cancer therapies (Arteaga and Engelman 2014; Scaltriti and Baselga 2006; Wieduwilt and Moasser 2008). Although the standard HyperTKI (low dose with a high frequency) regimen has shown a promising initial anti-tumor effect, tumor relapse happens to almost all patients eventually within about 1 year. Besides improving response rates, significant reducing tumor relapse rates have been a big challenge for EGFR TKIs treatment.

Immunotherapy, such as nivolumab or pembrolizumab, has been approved as the standard second-line or third-line treatment for TKI-resistant patients with high-level PD-L1 expression in tumors. Unfortunately, recent clinical trials were prematurely terminated due to significant side effects from checkpoint blockade combined with prolonged treatment of EGFR TKIs, which has achieved higher response rates (Ahn *et al.* 2017). Therefore, safety and efficacy must be carefully evaluated for development of the combination of EGFR TKI and immunotherapy.

SUMMARY AND PERSPECTIVES

Recent work exploring the role of the tumor-targeting Ab in cancer therapy has yielded significant progress. Our laboratory's research and the work of other research groups show that tumor-targeting Ab can be effectively combined with traditional small molecule chemo- and targeted therapy as well as newly developed checkpoint blockades, thus providing new potential for creating highly specific therapies. By targeting tumors, these new therapies would address the fundamental mechanism behind tumor growth and metastasis, namely breaking immune tolerance. It is the TTmAb that bridges the innate and adaptive immunity to modulate tumor microenvironment, overcomes drug resistance, and induces effective anti-tumor response to achieve tumor regression. This is the underlying principle behind the ongoing interest in combining targeting Ab with radiation and small molecule chemo- and targeted therapies.

Although this review mainly focuses on the functional studies and the clinical application of tumor-targeting antibodies, recent evidence has shown further development of such antibody fusion proteins armed with inflammation cytokines, including IFN, LIGHT, and IL-2. Moreover, immune checkpoints have been deemed critical for the cancer immune tolerance. As such, tumor-targeting checkpoint blockade would be a worthy direction for extensive future study. Exploring the synergistic treatment effect between targeting antibodies, immune cytokines, checkpoint blockade, radiotherapy, and small molecule chemo- and targeted therapies holds great promise for elucidating the underlying immunology mechanisms as well as developing effective therapies for a broad range of malignant diseases.

Compliance with ethical standards

Conflict of interest Hua Peng, Zhichen Sun, and Yang-Xin Fu declare that they have no conflict of interest.

Human and animal rights and informed consent This article does not contain any studies with human or animal subjects performed by the any of the authors.

References

Abes R, Gelize E, Fridman WH, Teillaud JL (2010) Long-lasting antitumor protection by anti-CD20 antibody through cellular immune response. Blood 116:926–934

Abramson V, Arteaga CL (2011) New strategies in HER2-overexpressing breast cancer: many combinations of targeted drugs available. Clin Cancer Res 17:952–958

Ahmadzadeh M, Rosenberg SA (2006) IL-2 administration increases CD4+ CD25(hi) Foxp3+ regulatory T cells in cancer patients. Blood 107:2409–2414

Ahn MJ, Sun JM, Lee SH, Ahn JS, Park K (2017) EGFR TKI combination with immunotherapy in non-small cell lung cancer. Expert Opin Drug Saf 16:465–469

Arteaga CL, Engelman JA (2014) ERBB receptors: from oncogene discovery to basic science to mechanism-based cancer therapeutics. Cancer Cell 25:282–303

Barcellos-Hoff MH, Park C, Wright EG (2005) Radiation and the microenvironment—tumorigenesis and therapy. Nat Rev Cancer 5:867–875

Bardelli A, Siena S (2010) Molecular mechanisms of resistance to cetuximab and panitumumab in colorectal cancer. J Clin Oncol 28:1254–1261

Becker JC, Varki N, Gillies SD, Furukawa K, Reisfeld RA (1996) An antibody-interleukin 2 fusion protein overcomes tumor heterogeneity by induction of a cellular immune response. Proc Natl Acad Sci USA 93:7826–7831

Bostrom J, Yu SF, Kan D, Appleton BA, Lee CV, Billeci K, Man W, Peale F, Ross S, Wiesmann C, Fuh G (2009) Variants of the antibody herceptin that interact with HER2 and VEGF at the antigen binding site. Science 323:1610–1614

Brahmer JR, Tykodi SS, Chow LQ, Hwu WJ, Topalian SL, Hwu P, Drake CG, Camacho LH, Kauh J, Odunsi K, Pitot HC, Hamid O, Bhatia S, Martins R, Eaton K, Chen S, Salay TM, Alaparthy S, Grosso JF, Korman AJ, Parker SM, Agrawal S, Goldberg SM, Pardoll DM, Gupta A, Wigginton JM (2012) Safety and activity of anti-PD-L1 antibody in patients with advanced cancer. N Engl J Med 366:2455–2465

Burgess AW, Cho HS, Eigenbrot C, Ferguson KM, Garrett TP, Leahy DJ, Lemmon MA, Sliwkowski MX, Ward CW, Yokoyama S (2003) An open-and-shut case? Recent insights into the activation of EGF/ErbB receptors. Mol Cell 12:541–552

Carter P, Presta L, Gorman CM, Ridgway JB, Henner D, Wong WL, Rowland AM, Kotts C, Carver ME, Shepard HM (1992) Humanization of an anti-p185HER2 antibody for human cancer therapy. Proc Natl Acad Sci USA 89:4285–4289

Chan KS, Espinosa I, Chao M, Wong D, Ailles L, Diehn M, Gill H, Presti J Jr, Chang HY, van de Rijn M, Shortliffe L, Weissman IL (2009) Identification, molecular characterization, clinical prognosis, and therapeutic targeting of human bladder tumor-initiating cells. Proc Natl Acad Sci USA 106:14016–14021

Chao MP, Alizadeh AA, Tang C, Myklebust JH, Varghese B, Gill S, Jan M, Cha AC, Chan CK, Tan BT, Park CY, Zhao F, Kohrt HE, Malumbres R, Briones J, Gascoyne RD, Lossos IS, Levy R, Weissman IL, Majeti R (2010) Anti-CD47 antibody synergizes with rituximab to promote phagocytosis and eradicate non-Hodgkin lymphoma. Cell 142:699–713

Chao MP, Alizadeh AA, Tang C, Jan M, Weissman-Tsukamoto R, Zhao F, Park CY, Weissman IL, Majeti R (2011) Therapeutic antibody targeting of CD47 eliminates human acute lymphoblastic leukemia. Cancer Res 71:1374–1384

Chavez AR, Buchser W, Basse PH, Liang X, Appleman LJ, Maranchie JK, Zeh H, de Vera ME, Lotze MT (2009) Pharmacologic administration of interleukin-2. Ann N Y Acad Sci 1182:14–27

Citri A, Yarden Y (2006) EGF-ERBB signalling: towards the systems level. Nat Rev Mol Cell Biol 7:505–516

Clynes RA, Towers TL, Presta LG, Ravetch JV (2000) Inhibitory Fc receptors modulate in vivo cytotoxicity against tumor targets. Nat Med 6:443–446

Cobleigh MA, Vogel CL, Tripathy D, Robert NJ, Scholl S, Fehrenbacher L, Wolter JM, Paton V, Shak S, Lieberman G, Slamon DJ (1999) Multinational study of the efficacy and safety of humanized anti-HER2 monoclonal antibody in women who have HER2-overexpressing metastatic breast cancer that has progressed after chemotherapy for metastatic disease. J Clin Oncol 17:2639–2648

De Luca R, Soltermann A, Pretto F, Pemberton-Ross C, Pellegrini G, Wulhfard S, Neri D (2017) Potency-matched dual cytokine-antibody fusion proteins for cancer therapy. Mol Cancer Ther 16:2442–2451

Deng L, Liang H, Xu M, Yang X, Burnette B, Arina A, Li XD, Mauceri H, Beckett M, Darga T, Huang X, Gajewski TF, Chen ZJ, Fu YX, Weichselbaum RR (2014) STING-dependent cytosolic dna sensing promotes radiation-induced type I interferon-dependent antitumor immunity in immunogenic tumors. Immunity 41:843–852

Diamond MS, Kinder M, Matsushita H, Mashayekhi M, Dunn GP, Archambault JM, Lee H, Arthur CD, White JM, Kalinke U, Murphy KM, Schreiber RD (2011) Type I interferon is selectively required by dendritic cells for immune rejection of tumors. J Exp Med 208:1989–2003

Dong H, Strome SE, Salomao DR, Tamura H, Hirano F, Flies DB, Roche PC, Lu J, Zhu G, Tamada K, Lennon VA, Celis E, Chen L (2002) Tumor-associated B7-H1 promotes T-cell apoptosis: a potential mechanism of immune evasion. Nat Med 8:793–800

Dovedi SJ, Adlard AL, Lipowska-Bhalla G, McKenna C, Jones S, Cheadle EJ, Stratford IJ, Poon E, Morrow M, Stewart R, Jones H, Wilkinson RW, Honeychurch J, Illidge TM (2014) Acquired resistance to fractionated radiotherapy can be overcome by concurrent PD-L1 blockade. Cancer Res 74:5458–5468

Du YJ, Lin ZM, Zhao YH, Feng XP, Wang CQ, Wang G, Wang CD, Shi W, Zuo JP, Li F, Wang CZ (2013) Stability of the recombinant anti-erbB2 scFv-Fc-interleukin-2 fusion protein and its inhibition of HER2-overexpressing tumor cells. Int J Oncol 42:507–516

Ennis BW, Lippman ME, Dickson RB (1991) The EGF receptor system as a target for antitumor therapy. Cancer Invest 9:553–562

Fayad L, Offner F, Smith MR, Verhoef G, Johnson P, Kaufman JL, Rohatiner A, Advani A, Foran J, Hess G, Coiffier B, Czuczman M, Gine E, Durrant S, Kneissl M, Luu KT, Hua SY, Boni J, Vandendries E, Dang NH (2013) Safety and clinical activity of a combination therapy comprising two antibody-based targeting agents for the treatment of non-Hodgkin lymphoma: results of a phase I/II study evaluating the immunoconjugate inotuzumab ozogamicin with rituximab. J Clin Oncol 31:573–583

Gillies SD, Lan Y, Hettmann T, Brunkhorst B, Sun Y, Mueller SO, Lo KM (2011) A low-toxicity IL-2-based immunocytokine retains antitumor activity despite its high degree of IL-2 receptor selectivity. Clin Cancer Res 17:3673–3685

Gutbrodt KL, Schliemann C, Giovannoni L, Frey K, Pabst T, Klapper W, Berdel WE, Neri D (2013) Antibody-based delivery of interleukin-2 to neovasculature has potent activity against acute myeloid leukemia. Sci Transl Med 5:201ra118

Gutbrodt KL, Casi G, Neri D (2014) Antibody-based delivery of IL2 and cytotoxics eradicates tumors in immunocompetent mice. Mol Cancer Ther 13:1772–1776

Hartimath SV, Manuelli V, Zijlma R, Signore A, Nayak TK, Freimoser-Grundschober A, Klein C, Dierckx R, de Vries EFJ

(2018) Pharmacokinetic properties of radiolabeled mutant Interleukin-2v: a PET imaging study. Oncotarget 9:7162–7174

Hodi FS, O'Day SJ, McDermott DF, Weber RW, Sosman JA, Haanen JB, Gonzalez R, Robert C, Schadendorf D, Hassel JC, Akerley W, van den Eertwegh AJ, Lutzky J, Lorigan P, Vaubel JM, Linette GP, Hogg D, Ottensmeier CH, Lebbe C, Peschel C, Quirt I, Clark JI, Wolchok JD, Weber JS, Tian J, Yellin MJ, Nichol GM, Hoos A, Urba WJ (2010) Improved survival with ipilimumab in patients with metastatic melanoma. N Engl J Med 363:711–723

Hudis CA (2007) Trastuzumab–mechanism of action and use in clinical practice. N Engl J Med 357:39–51

Hurvitz SA, Dirix L, Kocsis J, Bianchi GV, Lu J, Vinholes J, Guardino E, Song C, Tong B, Ng V, Chu YW, Perez EA (2013) Phase II randomized study of trastuzumab emtansine versus trastuzumab plus docetaxel in patients with human epidermal growth factor receptor 2-positive metastatic breast cancer. J Clin Oncol 31:1157–1163

Hynes NE, Lane HA (2005) ERBB receptors and cancer: the complexity of targeted inhibitors. Nat Rev Cancer 5:341–354

Iwai Y, Ishida M, Tanaka Y, Okazaki T, Honjo T, Minato N (2002) Involvement of PD-L1 on tumor cells in the escape from host immune system and tumor immunotherapy by PD-L1 blockade. Proc Natl Acad Sci USA 99:12293–12297

Jackaman C, Bundell CS, Kinnear BF, Smith AM, Filion P, van Hagen D, Robinson BWS, Nelson DJ (2003) IL-2 intratumoral immunotherapy enhances CD8+ T cells that mediate destruction of tumor cells and tumor-associated vasculature: a novel mechanism for IL-2. J Immunol 171:5051–5063

Jaiswal S, Jamieson CH, Pang WW, Park CY, Chao MP, Majeti R, Traver D, van Rooijen N, Weissman IL (2009) CD47 is upregulated on circulating hematopoietic stem cells and leukemia cells to avoid phagocytosis. Cell 138:271–285

Jensen HK, Donskov F, Nordsmark M, Marcussen N, von der Maase H (2009) Increased intratumoral FOXP3-positive regulatory immune cells during interleukin-2 treatment in metastatic renal cell carcinoma. Clin Cancer Res 15:1052–1058

Keir ME, Butte MJ, Freeman GJ, Sharpe AH (2008) PD-1 and its ligands in tolerance and immunity. Annu Rev Immunol 26:677–704

Kiessling R, Wei WZ, Herrmann F, Lindencrona JA, Choudhury A, Kono K, Seliger B (2002) Cellular immunity to the Her-2/neu protooncogene. Adv Cancer Res 85:101–144

Kimura H, Sakai K, Arao T, Shimoyama T, Tamura T, Nishio K (2007) Antibody-dependent cellular cytotoxicity of cetuximab against tumor cells with wild-type or mutant epidermal growth factor receptor. Cancer Sci 98:1275–1280

Krop IE, LoRusso P, Miller KD, Modi S, Yardley D, Rodriguez G, Guardino E, Lu M, Zheng M, Girish S, Amler L, Winer EP, Rugo HS (2012) A phase II study of trastuzumab emtansine in patients with human epidermal growth factor receptor 2-positive metastatic breast cancer who were previously treated with trastuzumab, lapatinib, an anthracycline, a taxane, and capecitabine. J Clin Oncol 30:3234–3241

Le XF, Lammayot A, Gold D, Lu Y, Mao W, Chang T, Patel A, Mills GB, Bast RC Jr (2005) Genes affecting the cell cycle, growth, maintenance, and drug sensitivity are preferentially regulated by anti-HER2 antibody through phosphatidylinositol 3-kinase-AKT signaling. J Biol Chem 280:2092–2104

Lee Y, Chin RK, Christiansen P, Sun Y, Tumanov AV, Wang J, Chervonsky AV, Fu YX (2006) Recruitment and activation of naive T cells in the islets by lymphotoxin beta receptor-dependent tertiary lymphoid structure. Immunity 25:499–509

Li S, Schmitz KR, Jeffrey PD, Wiltzius JJ, Kussie P, Ferguson KM (2005) Structural basis for inhibition of the epidermal growth factor receptor by cetuximab. Cancer Cell 7:301–311

Liao W, Lin JX, Leonard WJ (2013) Interleukin-2 at the crossroads of effector responses, tolerance, and immunotherapy. Immunity 38:13–25

Liao J, Luan Y, Ren Z, Liu X, Xue D, Xu H, Sun Z, Yang K, Peng H, Fu YX (2017) Converting lymphoma cells into potent antigen-presenting cells for interferon-induced tumor regression. Cancer Immunol Res 5:560–570

Liu X, Pu Y, Cron K, Deng L, Kline J, Frazier WA, Xu H, Peng H, Fu YX, Xu MM (2015) CD47 blockade triggers T cell-mediated destruction of immunogenic tumors. Nat Med 21:1209–1215

Lorenzi S, Mattei F, Sistigu A, Bracci L, Spadaro F, Sanchez M, Spada M, Belardelli F, Gabriele L, Schiavoni G (2011) Type I IFNs control antigen retention and survival of CD8alpha(+) dendritic cells after uptake of tumor apoptotic cells leading to cross-priming. J Immunol 186:5142–5150

Maio M, Grob JJ, Aamdal S, Bondarenko I, Robert C, Thomas L, Garbe C, Chiarion-Sileni V, Testori A, Chen TT, Tschaika M, Wolchok JD (2015) Five-year survival rates for treatment-naive patients with advanced melanoma who received ipilimumab plus dacarbazine in a phase III trial. J Clin Oncol 33:1191–1196

Majeti R, Chao MP, Alizadeh AA, Pang WW, Jaiswal S, Gibbs KD Jr, van Rooijen N, Weissman IL (2009) CD47 is an adverse prognostic factor and therapeutic antibody target on human acute myeloid leukemia stem cells. Cell 138:286–299

Maloney DG (2012) Anti-CD20 antibody therapy for B-cell lymphomas. N Engl J Med 366:2008–2016

Maloney DG, Brown S, Czerwinski DK, Liles TM, Hart SM, Miller RA, Levy R (1992) Monoclonal anti-idiotype antibody therapy of B-cell lymphoma: the addition of a short course of chemotherapy does not interfere with the antitumor effect nor prevent the emergence of idiotype-negative variant cells. Blood 80:1502–1510

Maloney DG, Liles TM, Czerwinski DK, Waldichuk C, Rosenberg J, Grillo-Lopez A, Levy R (1994) Phase I clinical trial using escalating single-dose infusion of chimeric anti-CD20 monoclonal antibody (IDEC-C2B8) in patients with recurrent B-cell lymphoma. Blood 84:2457–2466

Maloney DG, Grillo-Lopez AJ, Bodkin DJ, White CA, Liles TM, Royston I, Varns C, Rosenberg J, Levy R (1997a) IDEC-C2B8: results of a phase I multiple-dose trial in patients with relapsed non-Hodgkin's lymphoma. J Clin Oncol 15:3266–3274

Maloney DG, Grillo-Lopez AJ, White CA, Bodkin D, Schilder RJ, Neidhart JA, Janakiraman N, Foon KA, Liles TM, Dallaire BK, Wey K, Royston I, Davis T, Levy R (1997b) IDEC-C2B8 (Rituximab) anti-CD20 monoclonal antibody therapy in patients with relapsed low-grade non-Hodgkin's lymphoma. Blood 90:2188–2195

McCracken MN, Cha AC, Weissman IL (2015) Molecular pathways: activating T cells after cancer cell phagocytosis from blockade of CD47 'Don't Eat Me' signals. Clin Cancer Res 21:3597–3601

Menard S, Pupa SM, Campiglio M, Tagliabue E (2003) Biologic and therapeutic role of HER2 in cancer. Oncogene 22:6570–6578

Meric-Bernstam F, Hung MC (2006) Advances in targeting human epidermal growth factor receptor-2 signaling for cancer therapy. Clin Cancer Res 12:6326–6330

Misale S, Yaeger R, Hobor S, Scala E, Janakiraman M, Liska D, Valtorta E, Schiavo R, Buscarino M, Siravegna G, Bencardino K, Cercek A, Chen CT, Veronese S, Zanon C, Sartore-Bianchi A, Gambacorta M, Gallicchio M, Vakiani E, Boscaro V, Medico E, Weiser M, Siena S, Di Nicolantonio F, Solit D, Bardelli A (2012)

Emergence of KRAS mutations and acquired resistance to anti-EGFR therapy in colorectal cancer. Nature 486:532–536

Mittendorf EA, Liu Y, Tucker SL, McKenzie T, Qiao N, Akli S, Biernacka A, Liu Y, Meijer L, Keyomarsi K, Hunt KK (2010) A novel interaction between HER2/neu and cyclin E in breast cancer. Oncogene 29:3896–3907

Moasser MM (2007) Targeting the function of the HER2 oncogene in human cancer therapeutics. Oncogene 26:6577–6592

Morgan DA, Ruscetti FW, Gallo R (1976) Selective *in vitro* growth of T lymphocytes from normal human bone marrows. Science 193:1007–1008

Mortenson ED, Park S, Jiang Z, Wang S, Fu YX (2013) Effective anti-neu-initiated antitumor responses require the complex role of CD4+ T cells. Clin Cancer Res 19:1476–1486

Musolino A, Naldi N, Bortesi B, Pezzuolo D, Capelletti M, Missale G, Laccabue D, Zerbini A, Camisa R, Bisagni G, Neri TM, Ardizzoni A (2008) Immunoglobulin G fragment C receptor polymorphisms and clinical efficacy of trastuzumab-based therapy in patients with HER-2/neu-positive metastatic breast cancer. J Clin Oncol 26:1789–1796

Nahta R, Esteva FJ (2006) Herceptin: mechanisms of action and resistance. Cancer Lett 232:123–138

Nishimura H, Okazaki T, Tanaka Y, Nakatani K, Hara M, Matsumori A, Sasayama S, Mizoguchi A, Hiai H, Minato N, Honjo T (2001) Autoimmune dilated cardiomyopathy in PD-1 receptor-deficient mice. Science 291:319–322

O'Day SJ, Hamid O, Urba WJ (2007) Targeting cytotoxic T-lymphocyte antigen-4 (CTLA-4): a novel strategy for the treatment of melanoma and other malignancies. Cancer 110:2614–2627

Olayioye MA, Neve RM, Lane HA, Hynes NE (2000) The ErbB signaling network: receptor heterodimerization in development and cancer. EMBO J 19:3159–3167

Panelli MC, White R, Foster M, Martin B, Wang E, Smith K, Marincola FM (2004) Forecasting the cytokine storm following systemic interleukin (IL)-2 administration. J Transl Med 2:17

Pardoll DM (2012) The blockade of immune checkpoints in cancer immunotherapy. Nat Rev Cancer 12:252–264

Park S, Jiang Z, Mortenson ED, Deng L, Radkevich-Brown O, Yang X, Sattar H, Wang Y, Brown NK, Greene M, Liu Y, Tang J, Wang S, Fu YX (2010) The therapeutic effect of anti-HER2/neu antibody depends on both innate and adaptive immunity. Cancer Cell 18:160–170

Pegram M, Hsu S, Lewis G, Pietras R, Beryt M, Sliwkowski M, Coombs D, Baly D, Kabbinavar F, Slamon D (1999) Inhibitory effects of combinations of HER-2/neu antibody and chemotherapeutic agents used for treatment of human breast cancers. Oncogene 18:2241–2251

Pegram MD, Konecny GE, O'Callaghan C, Beryt M, Pietras R, Slamon DJ (2004) Rational combinations of trastuzumab with chemotherapeutic drugs used in the treatment of breast cancer. J Natl Cancer Inst 96:739–749

Pestka S, Krause CD, Walter MR (2004) Interferons, interferon-like cytokines, and their receptors. Immunol Rev 202:8–32

Piccart-Gebhart MJ, Procter M, Leyland-Jones B, Goldhirsch A, Untch M, Smith I, Gianni L, Baselga J, Bell R, Jackisch C, Cameron D, Dowsett M, Barrios CH, Steger G, Huang CS, Andersson M, Inbar M, Lichinitser M, Lang I, Nitz U, Iwata H, Thomssen C, Lohrisch C, Suter TM, Ruschoff J, Suto T, Greatorex V, Ward C, Straehle C, McFadden E, Dolci MS, Gelber RD (2005) Trastuzumab after adjuvant chemotherapy in HER2-positive breast cancer. N Engl J Med 353:1659–1672

Quaglino E, Mastini C, Forni G, Cavallo F (2008) ErbB2 transgenic mice: a tool for investigation of the immune prevention and treatment of mammary carcinomas. Curr Protoc Immunol 82(1):20–29

Rajkumar T, Gullick WJ (1994) The type I growth factor receptors in human breast cancer. Breast Cancer Res Treat 29:3–9

Ren Z, Guo J, Liao J, Luan Y, Liu Z, Sun Z, Liu X, Liang Y, Peng H, Fu YX (2017) CTLA-4 limits anti-CD20-mediated tumor regression. Clin Cancer Res 23:193–203

Robert C, Long GV, Brady B, Dutriaux C, Maio M, Mortier L, Hassel JC, Rutkowski P, McNeil C, Kalinka-Warzocha E, Savage KJ, Hernberg MM, Lebbe C, Charles J, Mihalcioiu C, Chiarion-Sileni V, Mauch C, Cognetti F, Arance A, Schmidt H, Schadendorf D, Gogas H, Lundgren-Eriksson L, Horak C, Sharkey B, Waxman IM, Atkinson V, Ascierto PA (2015a) Nivolumab in previously untreated melanoma without BRAF mutation. N Engl J Med 372:320–330

Robert C, Schachter J, Long GV, Arance A, Grob JJ, Mortier L, Daud A, Carlino MS, McNeil C, Lotem M, Larkin J, Lorigan P, Neyns B, Blank CU, Hamid O, Mateus C, Shapira-Frommer R, Kosh M, Zhou H, Ibrahim N, Ebbinghaus S, Ribas A (2015b) Pembrolizumab versus ipilimumab in advanced melanoma. N Engl J Med 372:2521–2532

Romond EH, Perez EA, Bryant J, Suman VJ, Geyer CE Jr, Davidson NE, Tan-Chiu E, Martino S, Paik S, Kaufman PA, Swain SM, Pisansky TM, Fehrenbacher L, Kutteh LA, Vogel VG, Visscher DW, Yothers G, Jenkins RB, Brown AM, Dakhil SR, Mamounas EP, Lingle WL, Klein PM, Ingle JN, Wolmark N (2005) Trastuzumab plus adjuvant chemotherapy for operable HER2-positive breast cancer. N Engl J Med 353:1673–1684

Rosenberg SA (2014) IL-2: the first effective immunotherapy for human cancer. J Immunol 192:5451–5458

Rosenberg SA, Yang JC, White DE, Steinberg SM (1998) Durability of complete responses in patients with metastatic cancer treated with high-dose interleukin-2: identification of the antigens mediating response. Ann Surg 228:307–319

Rovero S, Amici A, Di Carlo E, Bei R, Nanni P, Quaglino E, Porcedda P, Boggio K, Smorlesi A, Lollini PL, Landuzzi L, Colombo MP, Giovarelli M, Musiani P, Forni G (2000) DNA vaccination against rat her-2/Neu p185 more effectively inhibits carcinogenesis than transplantable carcinomas in transgenic BALB/c mice. J Immunol 165:5133–5142

Russ A, Hua AB, Montfort WR, Rahman B, Riaz IB, Khalid MU, Carew JS, Nawrocki ST, Persky D, Anwer F (2018) Blocking "don't eat me" signal of CD47-SIRPalpha in hematological malignancies, an in-depth review. Blood Rev. https://doi.org/10.1016/j.blre.2018.04.005

Scaltriti M, Baselga J (2006) The epidermal growth factor receptor pathway: a model for targeted therapy. Clin Cancer Res 12:5268–5272

Schadendorf D, Hodi FS, Robert C, Weber JS, Margolin K, Hamid O, Patt D, Chen TT, Berman DM, Wolchok JD (2015) Pooled analysis of long-term survival data from phase II and phase III trials of ipilimumab in unresectable or metastatic melanoma. J Clin Oncol 33:1889–1894

Shepard HM, Brdlik CM, Schreiber H (2008) Signal integration: a framework for understanding the efficacy of therapeutics targeting the human EGFR family. J Clin Investig 118:3574–3581

Shin DS, Ribas A (2015) The evolution of checkpoint blockade as a cancer therapy: what's here, what's next? Curr Opin Immunol 33:23–35

Sim GC, Martin-Orozco N, Jin L, Yang Y, Wu S, Washington E, Sanders D, Lacey C, Wang Y, Vence L, Hwu P, Radvanyi L (2014) IL-2 therapy promotes suppressive ICOS+ Treg expansion in melanoma patients. J Clin Investig 124:99–110

Sistigu A, Yamazaki T, Vacchelli E, Chaba K, Enot DP, Adam J, Vitale I, Goubar A, Baracco EE, Remedios C, Fend L, Hannani D,

Aymeric L, Ma Y, Niso-Santano M, Kepp O, Schultze JL, Tuting T, Belardelli F, Bracci L, La Sorsa V, Ziccheddu G, Sestili P, Urbani F, Delorenzi M, Lacroix-Triki M, Quidville V, Conforti R, Spano JP, Pusztai L, Poirier-Colame V, Delaloge S, Penault-Llorca F, Ladoire S, Arnould L, Cyrta J, Dessoliers MC, Eggermont A, Bianchi ME, Pittet M, Engblom C, Pfirschke C, Preville X, Uze G, Schreiber RD, Chow MT, Smyth MJ, Proietti E, Andre F, Kroemer G, Zitvogel L (2014) Cancer cell-autonomous contribution of type I interferon signaling to the efficacy of chemotherapy. Nat Med 20:1301–1309

Skrombolas D, Frelinger JG (2014) Challenges and developing solutions for increasing the benefits of IL-2 treatment in tumor therapy. Expert Rev Clin Immunol 10:207–217

Slamon DJ, Clark GM, Wong SG, Levin WJ, Ullrich A, McGuire WL (1987) Human breast cancer: correlation of relapse and survival with amplification of the HER-2/neu oncogene. Science 235:177–182

Stagg J, Loi S, Divisekera U, Ngiow SF, Duret H, Yagita H, Teng MW, Smyth MJ (2011) Anti-ErbB-2 mAb therapy requires type I and II interferons and synergizes with anti-PD-1 or anti-CD137 mAb therapy. Proc Natl Acad Sci USA 108:7142–7147

Sunada H, Magun BE, Mendelsohn J, MacLeod CL (1986) Monoclonal antibody against epidermal growth factor receptor is internalized without stimulating receptor phosphorylation. Proc Natl Acad Sci USA 83:3825–3829

Tang H, Wang Y, Chlewicki LK, Zhang Y, Guo J, Liang W, Wang J, Wang X, Fu YX (2016) Facilitating T cell infiltration in tumor microenvironment overcomes resistance to PD-L1 blockade. Cancer Cell 30:500

Tang H, Liang Y, Anders RA, Taube JM, Qiu X, Mulgaonkar A, Liu X, Harrington SM, Guo J, Xin Y, Xiong Y, Nham K, Silvers W, Hao G, Sun X, Chen M, Hannan R, Qiao J, Dong H, Peng H, Fu YX (2018) PD-L1 on host cells is essential for PD-L1 blockade-mediated tumor regression. J Clin Investig 128:580–588

Teeling JL, French RR, Cragg MS, van den Brakel J, Pluyter M, Huang H, Chan C, Parren PW, Hack CE, Dechant M, Valerius T, van de Winkel JG, Glennie MJ (2004) Characterization of new human CD20 monoclonal antibodies with potent cytolytic activity against non-Hodgkin lymphomas. Blood 104:1793–1800

Topalian SL, Hodi FS, Brahmer JR, Gettinger SN, Smith DC, McDermott DF, Powderly JD, Carvajal RD, Sosman JA, Atkins MB, Leming PD, Spigel DR, Antonia SJ, Horn L, Drake CG, Pardoll DM, Chen L, Sharfman WH, Anders RA, Taube JM, McMiller TL, Xu H, Korman AJ, Jure-Kunkel M, Agrawal S, McDonald D, Kollia GD, Gupta A, Wigginton JM, Sznol M (2012) Safety, activity, and immune correlates of anti-PD-1 antibody in cancer. N Engl J Med 366:2443–2454

Trinchieri G (2010) Type I interferon: friend or foe? J Exp Med 207:2053–2063

van Meerten T, van Rijn RS, Hol S, Hagenbeek A, Ebeling SB (2006) Complement-induced cell death by rituximab depends on CD20 expression level and acts complementary to antibody-dependent cellular cytotoxicity. Clin Cancer Res 12:4027–4035

Wang Y, Zhu M, Miller M, Fu YX (2009) Immunoregulation by tumor necrosis factor superfamily member LIGHT. Immunol Rev 229:232–243

Wieduwilt MJ, Moasser MM (2008) The epidermal growth factor receptor family: biology driving targeted therapeutics. Cell Mol Life Sci 65:1566–1584

Willingham SB, Volkmer JP, Gentles AJ, Sahoo D, Dalerba P, Mitra SS, Wang J, Contreras-Trujillo H, Martin R, Cohen JD, Lovelace P, Scheeren FA, Chao MP, Weiskopf K, Tang C, Volkmer AK, Naik TJ, Storm TA, Mosley AR, Edris B, Schmid SM, Sun CK, Chua MS, Murillo O, Rajendran P, Cha AC, Chin RK, Kim D,

Adorno M, Raveh T, Tseng D, Jaiswal S, Enger PO, Steinberg GK, Li G, So SK, Majeti R, Harsh GR, van de Rijn M, Teng NN, Sunwoo JB, Alizadeh AA, Clarke MF, Weissman IL (2012) The CD47-signal regulatory protein alpha (SIRPa) interaction is a therapeutic target for human solid tumors. Proc Natl Acad Sci USA 109:6662–6667

Wolchok JD, Neyns B, Linette G, Negrier S, Lutzky J, Thomas L, Waterfield W, Schadendorf D, Smylie M, Guthrie T Jr, Grob JJ, Chesney J, Chin K, Chen K, Hoos A, O'Day SJ, Lebbe C (2010) Ipilimumab monotherapy in patients with pretreated advanced melanoma: a randomised, double-blind, multicentre, phase 2, dose-ranging study. Lancet Oncol 11:155–164

Woo SR, Fuertes MB, Corrales L, Spranger S, Furdyna MJ, Leung MY, Duggan R, Wang Y, Barber GN, Fitzgerald KA, Alegre ML, Gajewski TF (2014) STING-dependent cytosolic DNA sensing mediates innate immune recognition of immunogenic tumors. Immunity 41:830–842

Xuan C, Steward KK, Timmerman JM, Morrison SL (2010) Targeted delivery of interferon-alpha via fusion to anti-CD20 results in potent antitumor activity against B-cell lymphoma. Blood 115:2864–2871

Yang XD, Jia XC, Corvalan JR, Wang P, Davis CG (2001) Development of ABX-EGF, a fully human anti-EGF receptor monoclonal antibody, for cancer therapy. Crit Rev Oncol Hematol 38:17–23

Yang RK, Kalogriopoulos NA, Rakhmilevich AL, Ranheim EA, Seo S, Kim K, Alderson KL, Gan J, Reisfeld RA, Gillies SD, Hank JA, Sondel PM (2013a) Intratumoral treatment of smaller mouse neuroblastoma tumors with a recombinant protein consisting of IL-2 linked to the hu14.18 antibody increases intratumoral CD8+ T and NK cells and improves survival. Cancer Immunol Immunother 62:1303–1313

Yang X, Zhang X, Mortenson ED, Radkevich-Brown O, Wang Y, Fu YX (2013b) Cetuximab-mediated tumor regression depends on innate and adaptive immune responses. Mol Ther 21:91–100

Yang X, Zhang X, Fu ML, Weichselbaum RR, Gajewski TF, Guo Y, Fu YX (2014) Targeting the tumor microenvironment with interferon-beta bridges innate and adaptive immune responses. Cancer Cell 25:37–48

Yarden Y, Pines G (2012) The ERBB network: at last, cancer therapy meets systems biology. Nat Rev Cancer 12:553–563

Yarden Y, Sliwkowski MX (2001) Untangling the ErbB signalling network. Nat Rev Mol Cell Biol 2:127–137

Yonesaka K, Zejnullahu K, Okamoto I, Satoh T, Cappuzzo F, Souglakos J, Ercan D, Rogers A, Roncalli M, Takeda M, Fujisaka Y, Philips J, Shimizu T, Maenishi O, Cho Y, Sun J, Destro A, Taira K, Takeda K, Okabe T, Swanson J, Itoh H, Takada M, Lifshits E, Okuno K, Engelman JA, Shivdasani RA, Nishio K, Fukuoka M, Varella-Garcia M, Nakagawa K, Janne PA (2011) Activation of ERBB2 signaling causes resistance to the EGFR-directed therapeutic antibody cetuximab. Sci Transl Med 3:99ra86

Yoon J, Koo KH, Choi KY (2011) MEK1/2 inhibitors AS703026 and AZD6244 may be potential therapies for KRAS mutated colorectal cancer that is resistant to EGFR monoclonal antibody therapy. Cancer Res 71:445–453

Young KH, Baird JR, Savage T, Cottam B, Friedman D, Bambina S, Messenheimer DJ, Fox B, Newell P, Bahjat KS, Gough MJ, Crittenden MR (2016) Optimizing timing of immunotherapy improves control of tumors by hypofractionated radiation therapy. PLoS ONE 11:e0157164

Yu P, Lee Y, Liu W, Chin RK, Wang J, Wang Y, Schietinger A, Philip M, Schreiber H, Fu YX (2004) Priming of naive T cells inside tumors leads to eradication of established tumors. Nat Immunol 5:141–149

Yu P, Lee Y, Wang Y, Liu X, Auh S, Gajewski TF, Schreiber H, You Z, Kaynor C, Wang X, Fu YX (2007) Targeting the primary tumor to generate CTL for the effective eradication of spontaneous metastases. J Immunol 179:1960–1968

Zhu EF, Gai SA, Opel CF, Kwan BH, Surana R, Mihm MC, Kauke MJ, Moynihan KD, Angelini A, Williams RT, Stephan MT, Kim JS, Yaffe MB, Irvine DJ, Weiner LM, Dranoff G, Wittrup KD (2015) Synergistic innate and adaptive immune response to combination immunotherapy with anti-tumor antigen antibodies and extended serum half-life IL-2. Cancer Cell 27:489–501

Zou W, Zheng H, He TC, Chang J, Fu YX, Fan W (2012) LIGHT delivery to tumors by mesenchymal stem cells mobilizes an effective antitumor immune response. Cancer Res 72:2980–2989

14

Dye-based mito-thermometry and its application in thermogenesis of brown adipocytes

Tao-Rong Xie[1], Chun-Feng Liu[1], Jian-Sheng Kang[1]✉

[1] CAS Key Laboratory of Nutrition and Metabolism, Institute for Nutritional Sciences, Shanghai Institutes for Biological Sciences, Graduate School of the Chinese Academy of Sciences, Chinese Academy of Sciences, Shanghai 200031, China

Abstract   Mitochondrion is the main intracellular site for thermogenesis and attractive energy expenditure targeting for obesity therapy. Here, we develop a method of mitochondrial thermometry based on Rhodamine B methyl ester, which equilibrates as a thermosensitive mixture of nonfluorescent and fluorescent resonance forms. Using this approach, we are able to demonstrate that the efficacy of norepinephrine-induced thermogenesis is low, and measure the maximum transient rate of temperature increase in brown adipocytes.

Keywords   Mitochondrial thermometry, Nanothermometry, Thermogenesis, Brown adipocytes

INTRODUCTION

Temperature probing for live cells is challenging, and a lot of efforts have been made to develop nanothermometry to monitor temperatures of living cells (Ye *et al.* 2011; Jaque and Vetrone 2012; Li and Liu 2012; Kiyonaka *et al.* 2013; Kucsko *et al.* 2013; Arai *et al.* 2014, 2015; Homma *et al.* 2015). Recently, all such works have been challenged and criticized by Baffou *et al.*, who have claimed no detectable temperature heterogeneities in living cells (Baffou *et al.* 2014). Apparently, Baffou *et al.* have neglected well-known facts in biology, such as the thermogenesis of brown adipocytes (BA) and mitochondrial role in thermogenesis (Cannon and Nedergaard 2004).

Sympathetic neurotransmitter norepinephrine (NE) can mobilize free fatty acids stored in lipid droplets of BA, and dissipate electrochemical potential energy stored in mitochondrial proton gradient to product heat (Cannon and Nedergaard 2004; Fedorenko *et al.* 2012). Not only as the energy factory of the cells, mitochondrion is the main intracellular site for thermogenesis, which has been targeted for therapy to reduce obesity (Lowell and Spiegelman 2000; Tseng *et al.* 2010). Here, we demonstrate a method of mitochondrial thermometry (mito-thermometry) based on the thermosensitive characteristics of Rhodamine B methyl ester (RhB-ME). With this mito-thermometry, we revealed the low efficacy of NE-induced thermogenesis and the maximum transient rate of temperature increase in BA, and indicated the improper critique of Baffou both practically and theoretically.

RESULTS

Evaluation of RhB-ME-based mito-thermometry in HeLa cells

Our technique for mito-thermometry employs the thermosensitive and mitochondrial targeting properties

---

Tao-Rong Xie and Chun-Feng Liu have contributed equally to this work.

✉ Correspondence: jskang@gmail.com (J.-S. Kang)

of RhB-ME (Fig. 1, synthesis method available online). RhB-ME (Fig. 1A and Supplementary Fig. S1A), a cationic dye like Rhodamine 800 (Rh800, a mitochondrial marker, Fig. 1B and Supplementary Fig. S1B) and many others (Sakanoue *et al.* 1997; Johnson *et al.* 1981), can redistribute within subcellular compartments in response to the negative electric potential, especially in mitochondria (Fig. 1A–D). Mitochondria with large negative membrane potentials lead to spontaneous accumulation of thermosensitive RhB-ME and thermoneutral Rh800 (Fig. 1E) in their matrixes, until reaching equilibrium in accordance with the Nernst distribution law (Sakanoue *et al.* 1997).

The thermochromic transformation of RhB-ME in aqueous solution results in a simple temperature profile, which can be fitted with Arrhenius equation, a single exponential model for the temperature dependence of reaction rates (Fig. 1F). The Arrhenius plot indicates that activation energy of RhB-ME thermochromic transformation is about -4.4 kcal/mol. In living cells, to cancel out the influence of mitochondrial membrane potentials on RhB-ME concentration, the fluorescent intensity ratio of Rh800 to RhB-ME is used to represent mitochondrial thermal profile (Fig. 1G). Both RhB-ME and Rh800 are insensitive to pH, Ca^{2+} or Mg^{2+} (Supplementary Fig. S2). This RhB-ME-based mito-thermometry enables us to acquire the mitochondrial thermal map of HeLa cells at room temperature (RT). The image in Fig. 1G shows mitochondrial

temperature gradients in HeLa cells with higher temperature at the center, which can be explained by the geometry of the cells (Fig. 1C).

The mechanism of RhB-ME thermochromic transformation

RhB-ME is a methyl ester derivative of RhB. RhB is cell membrane impermeable (Johnson *et al.* 1981), also thermosensitive (Supplementary Fig. S3), and exists as a mixture of nonfluorescent lactone and fluorescent zwitterion (Hinckley and Seybold 1988). Unlike RhB, our data suggest that RhB-ME exists as an equilibrium mixture of nonfluorescent mesomerism and fluorescent resonance form (Fig. 2A–C), and thermal energy ($\sim$7.5 k_BT at 25 °C) can convert single RhB-ME fluorescent molecule to its nonfluorescent form (Fig. 2A). The two diethylamino groups are symmetric in the nonfluorescent form, but asymmetric in the fluorescent form (Fig. 2A). Proton nuclear magnetic resonance (^{1}H-NMR) can distinguish between the symmetric form and the asymmetric form by the ^{1}H-NMR spectral difference of the 12 protons from four methyl moieties in two diethylamino groups (Fig. 2B–C). The symmetric nonfluorescent form is dominant at higher temperature, which is evidenced by the undistinguishable chemical shifts of 12 protons showing single triplet in the ^{1}H-NMR spectra at RT (Fig. 2B and Supplementary Fig. S4).

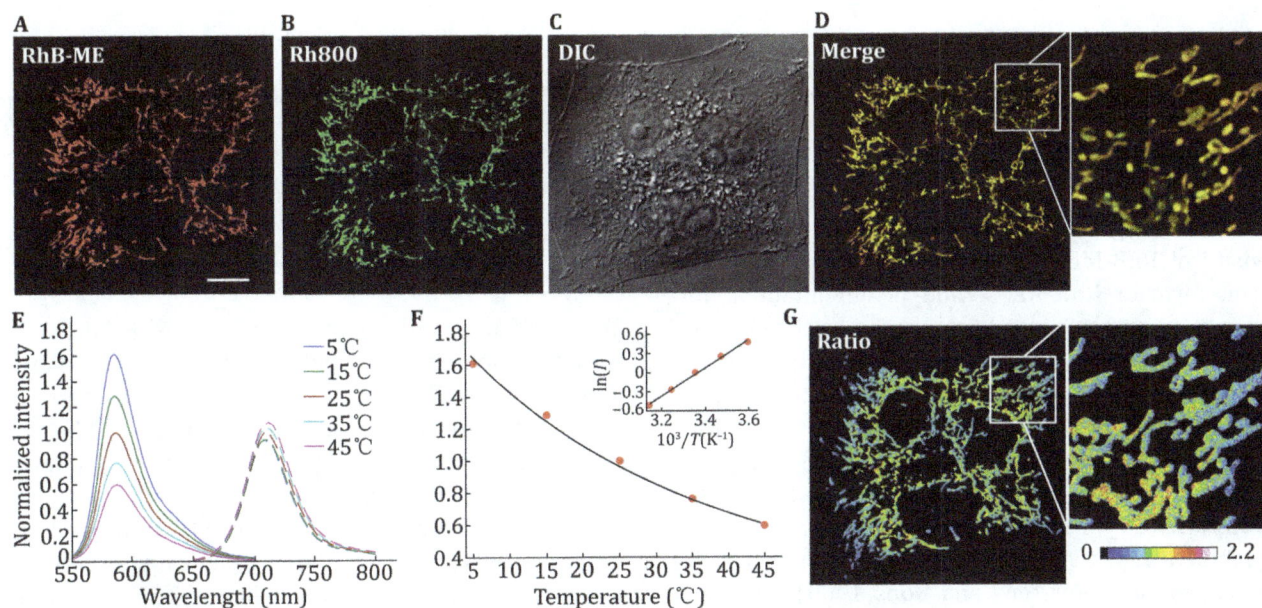

Fig. 1 RhB-ME based mito-thermometry in living cells. **A–D** Confocal images of RhB-ME channel (**A**), Rh800 channel (**B**), differential interference contrast (DIC) image (**C**) and merged image (**D**) of stained HeLa cells. *Scale bar*, 10 μm. **E** Emission spectra of 10 μmol/L RhB-ME (*solid lines*) and 10 μmol/L Rh800 (*dashed lines*) from 5 to 45 °C, respectively. **F** Arrhenius fitting for peak values (*red dots*) of RhB-ME spectra. The *black line* indicates a fitting with Arrhenius equation. The *inset* shows the Arrhenius plot. **G** Mitochondrial thermal map of HeLa cells represented by the pseudocolor image of intensity ratio of Rh800 to RhB-ME

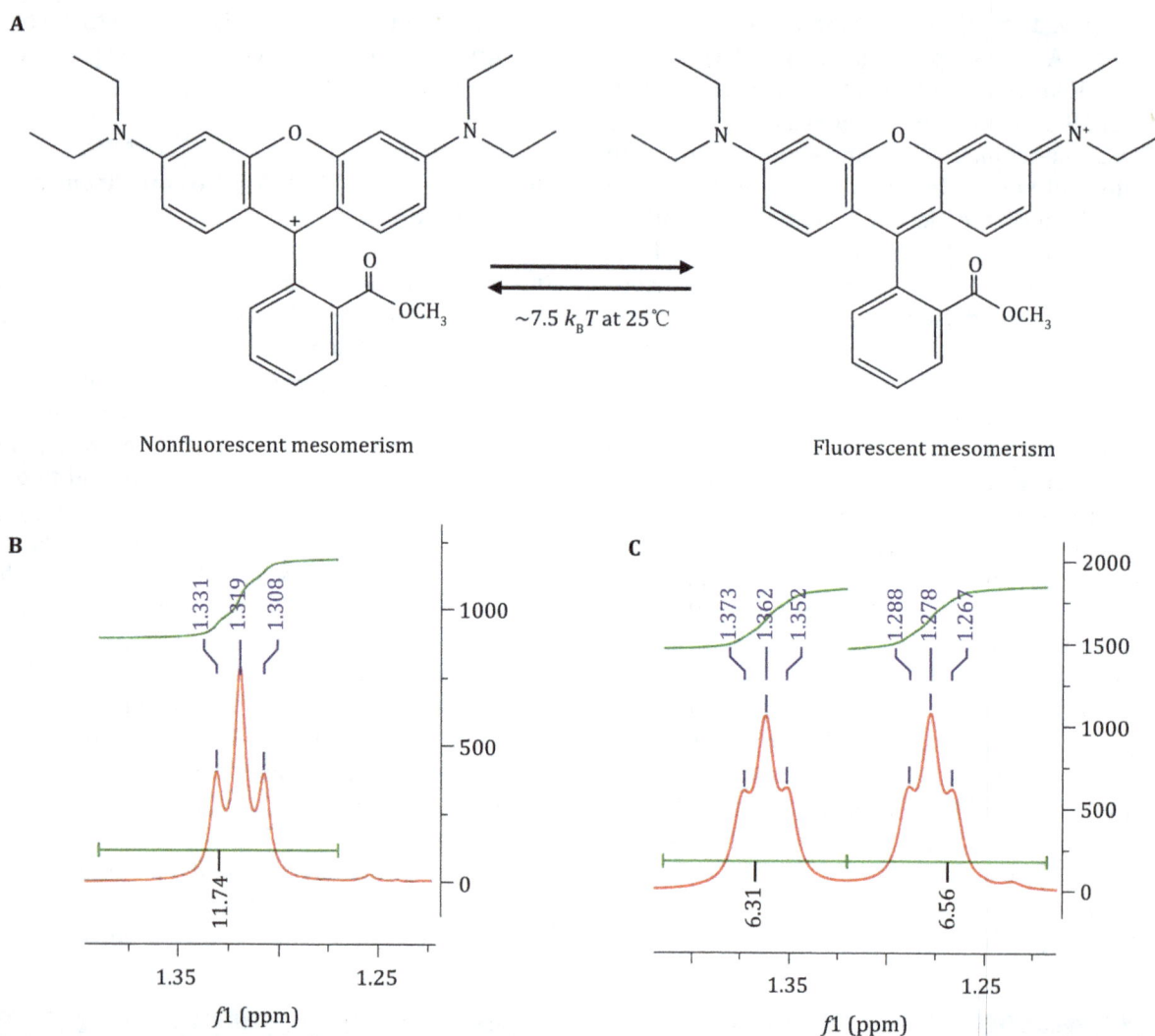

Fig. 2 The mechanism of RhB-ME thermochromic transformation. **A** RhB-ME exits as an equilibrium mixture of nonfluorescent and fluorescent resonance forms. **B** and **C** The ^{1}H-NMR spectra of four methyl moieties in two diethylamino groups at RT (**B**) and −40 °C (**C**). **B** and **C** show the chemical shifts of 12 protons in the symmetric nonfluorescent form and the asymmetrical fluorescent form, respectively

Since decreasing temperature increases the fluorescent intensity of RhB-ME (Fig. 1E, F), the asymmetric fluorescent form of RhB-ME should be dominant at low temperature, and this idea is confirmed by the chemical shifts of 12 protons splitting into double triplets in the spectral data of ^{1}H-NMR at −40 °C (Fig. 2C and Supplementary Fig. S5). Compared to thermoneutral Rhodamine 110 (Rh110, Supplementary Fig. S6) and Rh800 (Supplementary Fig. S1B), RhB (Supplementary Fig. S3) or RhB-ME (Supplementary Fig. S1A) has extra four ethyl moieties for two amino groups, which may free for torsional motion (Karstens and Kobs 1980). Thus, our findings demonstrate that thermochromic activation energy of RhB and RhB-ME is contributed to the torsional motion of diethylamino groups, which stabilizes the nonfluorescent form.

Study the thermogenesis of BA with RhB-ME-based mito-thermometry

To evaluate and make use of RhB-ME based mito-thermometry, we applied it to study the thermogenesis of BA. For thermogenic studies of BAT or BA, calorimeter and oxygen consumption rate (OCR) have been frequently used to evaluate heat production (Clark *et al.* 1986; Wikstrom *et al.* 2014), but both are indirect and might be cumbersome (Cannon and Nedergaard 2004). Although the genetically coded thermometry is versatile for organelle targeting, due to low transfection efficiency of BA, it is difficult to be used for collecting large-scale datasets, which are usually necessary for experiments with large variations. Since there is no need for transfection, injection, or elaborate equipment, dyes (Arai *et al.* 2015; Homma *et al.* 2015) and RhB-ME-based mito-thermometry demonstrated in this study make it easier for large-scale data acquisition, and are capable of detecting thermogenic responses at mitochondrial level.

As illustrated in Fig. 3A–D, RhB-ME, like mitochondrial marker Rh800, is accumulated in numerous mitochondria of primary cultured BA. The thermogenic responses of BA, represented by the fluorescent intensity ratio of Rh800 to RhB-ME, are evoked by 0.1 µmol/L NE in minutes (Fig. 3E–G, Supplementary Video S1

and S2). Compared to 10 µmol/L carbonyl cyanide m-chlorophenylhydrazone (CCCP, a proton uncoupling agent, Fig. 3H, I and Supplementary Video S3), 0.1 µmol/L NE shows markedly lower thermogenic efficacy (Fig. 3J). In addition, 0.1 µmol/L NE-induced thermogenesis in BA shows large cell-to-cell variation (Fig. 3E, F, Supplementary Video S1 and S2). Only $59.4 \pm 15.9\%$ (69 of 118, mean $\pm$ S.D.) of NE treated BA show thermogenic responses, which accounts for the low efficacy of 0.1 µmol/L NE-induced thermogenesis in BA (Fig. 3E, F).

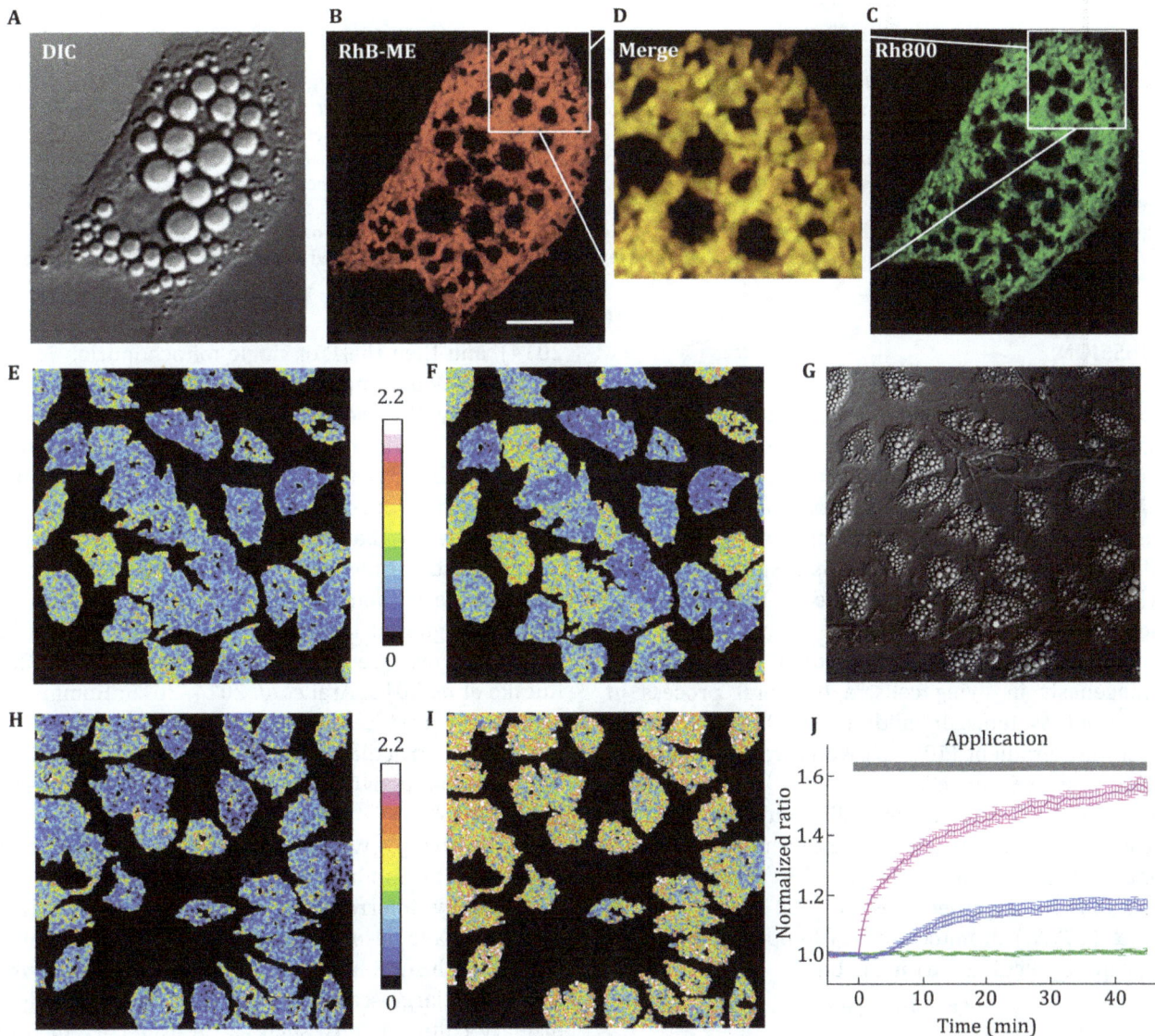

Fig. 3 Mito-thermometry reveals low efficacy of NE-induced thermogenesis in BA. **A–C** Confocal images of DIC image (**A**), RhB-ME (**B**), Rh800 (**C**) in stained BA. *Scale bar*, 10 µm. **D** Zoomed and merged image shows crowded mitochondria in BA. **E** and **F** Representative thermal ratio images are shown for the moments before and after NE treatment respectively. *Scale bar*, 20 µm. **G** DIC image of BA. **H** and **I** Representative thermal ratio images are shown for the moments before and after CCCP treatment, respectively. **J** NE (0.1 µmol/L, *blue line*, $n = 118$ cells in four experiments) and CCCP (10 µmol/L, *magenta line*, $n = 88$ cells in three experiments) induced thermogenesis in BA. The *green line* shows the control results of the solvent treatment in BA ($n = 93$ cells in three experiments). All data points in figures represent mean $\pm$ S.E.M

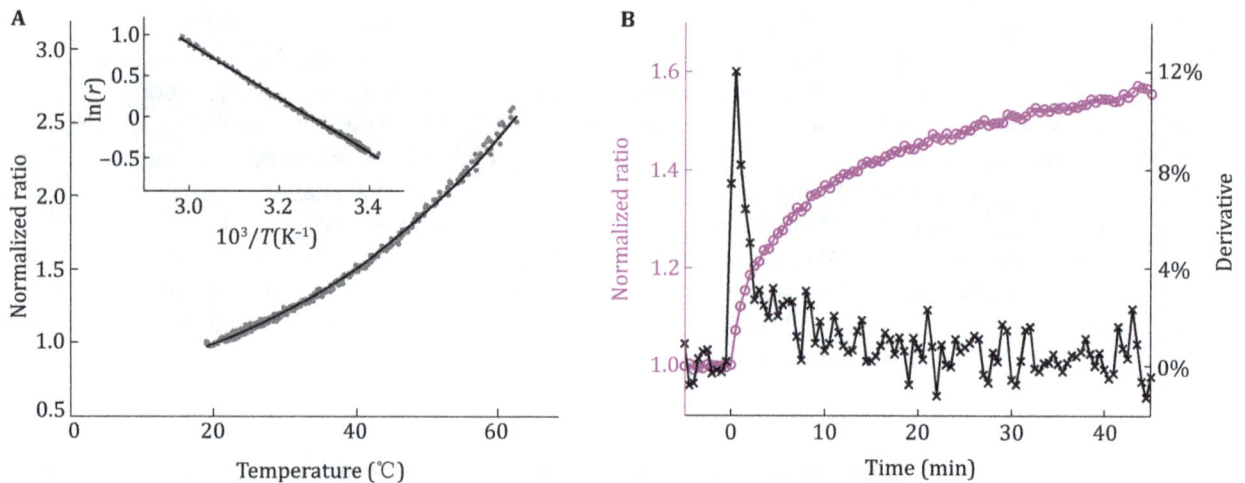

Fig. 4 Maximum transient rate of mitochondrial temperature increase in BA induced by 10 μmol/L CCCP. **A** Ratiometric confocal imaging data of 10 μmol/L RhB-ME and 10 μmol/L Rh800 mixed in aqueous solution (from 18.9 to 62.9 °C). The *scatter plot* of fluorescent ratios (*gray dots*) of Rh800 to RhB-ME versus temperature monitored with thermocouple (Pt100A). The *black solid line* is a fit with Arrhenius equation. The *inset* shows the Arrhenius plot. **B** Maximum transient rate of BA-mitochondrial temperature increase induced by 10 μmol/L CCCP. The ratio (*magenta line*) is the same raw dataset in Fig. 3G. The derivative (*black line*) of the ratio data is calculated by numerical differentiation, and the peak value is about 12%/min. Since all ratio data are normalized with the data acquired at 33 °C (306.15 K), the change rate of normalized ratio at this temperature has been theoretically (see details in **Method**) and experimentally determined as 3.5%/K (**A**); thus, the maximum transient rate of mitochondrial temperature increase in BA treated with 10 μmol/L CCCP is ~0.057 K/s at 306.15 K

DISCUSSION

Baffou *et al.* have criticized all methods for temperature imaging in living cells (Baffou *et al.* 2014), which have based on a conclusion "temperature increase should be on the order of $\Delta T \sim 10^{-5}$ K" governed by the heat diffusion equation at one-dimensional steady-state conditions for cell. However, "ΔT" is a spatial temperature gradient independent of time in Baffou's model rather than "temperature increase" (over time), so it is inappropriate to apply a spatial gradient "ΔT" to discuss thermogenesis in living cells, a temporal process in time-variant systems. In addition, such small spatial temperature gradient (10^{-5} K) would result in negligible heat flows within cell.

Actually, heat sources in cell would be able to increase their temperature at a transient rate, or at a lasting rate (under a condition with negligible heat dissipation) in the order of $\sim 10^{-2}$–10^{-1} K/s, such as the rate ($\Delta T/\Delta t$) of mitochondrial temperature change (ΔT), can be calculated with the Eq. 1:

$$\Delta T/\Delta t = P/C_{P,v}v, \qquad (1)$$

which is deriving from the definition of heat capacity, and where $C_{P,v}$ is the volumetric heat capacity of water (4.1796 J/(cm³K)); v is a mitochondrial volume of 1 μm³; the average heat power (P) for a cell is 100 pW, which can reach 1 nW for brown fat cells (Baffou *et al.*

2014), and then the P of single mitochondrion is taken as 0.1–1 pW assuming ~1000 mitochondria per cell.

To verify the theoretical calculation, we have estimated the maximum transient rate of mitochondrial temperature increase in BA induced by 10 μmol/L CCCP, which is indeed on the order of $\sim 10^{-2}$–10^{-1} K/s (Fig. 4). Clearly, Baffou's model is too simple to give accurate estimations since they have wrongly used the steady state without temporal factor. In addition, they have also ignored that there are various internal heat sources in cell (Ye *et al.* 2011; Kiyonaka *et al.* 2013; Kucsko *et al.* 2013; Arai *et al.* 2014, 2015; Homma *et al.* 2015), and that cell is nonhomogeneous and rich in membrane structures where lipid bilayer has a low thermal conductivity (0.25 W/(mK)) (Nakano *et al.* 2010).

In summary, we have practically demonstrated uneven mitochondrial thermal maps in living cells, theoretically inferred detectable heat sources (mitochondria), and also pointed out the error of Baffou's critique. RhB-ME-based mito-thermometry makes it easier for large-scale data acquisition, especially for primary cultured cells, such as BA. Our current observations raise open questions about diversely thermogenic responses of individual BA evoked by 0.1 μmol/L NE, for instance, what is the *in vivo* regulatory mechanism to increase the efficacy of NE-induced thermogenesis in BA?

METHODS

RhB-ME synthesis

A mixture of RhB (500 mg, 1.04 mmol) and thionyl chloride (2 ml) in chloroform (20 ml) was heated to 60 °C and stirred for 10 min. After cooling to room temperature, the mixture was quenched with methanol. The solvent was removed under reduced pressure and purified by prep-HPLC to give compound 375 mg, yield 73%. ^{1}H-NMR (600 MHz, CDCl$_3$) δ 8.30 (d, $J = 7.2$ Hz, 1H), 8.23 (brs, 2H), 7.79–7.82 (m, 1H), 7.73–7.76 (m, 1H), 7.31 (d, $J = 7.2$ Hz, 1H), 7.06 (d, $J = 9.6$ Hz, 2H), 6.82–6.83 (m, 4H), 3.68 (s, 3H), 3.60 (q, $J = 7.2$ Hz, 8H), 1.32 (t, $J = 7.2$ Hz, 12H); ESI-HRMS exact mass calcd for $[M]^+$ requires m/z 457.2486; found m/z 457.2484.

The change rate of normalized intensity ratio (Rh800 to RhB-ME) at a temperature

According to Arrhenius Eq. 2, the local/pixel (j) fluorescent intensity ratio $r(j)$ of Rh800 to RhB-ME at temperature $T(j)$ can be fitted with:

$$r(j) = Ae^{-\frac{E_a}{k_B T(j)}}, \tag{2}$$

where k_B is the Boltzmann constant; E_a is the observed activation energy, which is experimentally estimated ($E_a = 6.55$ kcal/mol, Fig. 4); A is a parameter related to imaging setup and experimental settings.

The parameter A in Eq. 2 can be canceled out by dividing the ratio value r_{ref} of a reference with a measurable temperature T_{ref}. Thus, the normalized ratio nr is a function of T and determined by Eq. 3:

$$nr = \frac{r}{r_{ref}} = e^{-\frac{E_a\left(\frac{1}{T}-\frac{1}{T_{ref}}\right)}{k_B}}. \tag{3}$$

The change rate of nr at a reference temperature T_{ref} can be deduced from the derivative of Eq. 3 with respect to T, and determined by Eq. 4:

$$\frac{\Delta nr}{\Delta T} = \frac{E_a}{k_B}\frac{1}{T_{ref}^2}. \tag{4}$$

Full Methods and any associated references are available in the online version of the paper.

Acknowledgements We thank the following people for their help: S.-L. You and X.-W. Liu for ^{1}H-NMR and discussions; Y. Chen, H.-B. Cai, Z.-H. Sheng, D.-S. Li, Q.-W. Zhai, H. Ying and H.-X. Qi for critical reading of the manuscript; S.-L. You, X.-W. Liu, K. Hou, C. Chen and J.-J. Hao for chemical synthesis. Y.-Y. Le for the gift of HeLa cell line. This work was partially supported by National Basic Research Program of China (2011CB910903 and 2010CB912001), National Natural Science Foundation of China (31171369), Chinese Academy of Sciences (Hundred Talents Program and 2009OHTP10).

Compliance with Ethical Standards

Conflict of interest Tao-Rong Xie, Chun-Feng Liu, Jian-Sheng Kang declare that they have no conflict of interest.

Human and Animal Rights and Informed Consent All institutional and national guidelines for the care and use of laboratory animals were followed.

References

Arai S, Lee S-C, Zhai D, Suzuki M, Chang YT (2014) A molecular fluorescent probe for targeted visualization of temperature at the endoplasmic reticulum. Sci Rep 4:6701

Arai S, Suzuki M, Park S-J, Yoo JS, Wang L, Kang N-Y, Ha H-H, Chang Y-T (2015) Mitochondria-targeted fluorescent thermometer monitors intracellular temperature gradient. Chem Commun 51:8044–8047

Baffou G, Rigneault H, Marguet D, Jullien L (2014) A critique of methods for temperature imaging in single cells. Nat Methods 11:899–901

Cannon B, Nedergaard J (2004) Brown adipose tissue: function and physiological significance. Physiol Rev 84:277–359

Clark DG, Brinkman M, Neville SD (1986) Microcalorimetric measurements of heat production in brown adipocytes from control and cafeteria-fed rats. Biochem J 235:337–342

Fedorenko A, Lishko PV, Kirichok Y (2012) Mechanism of fatty-acid-dependent UCP1 uncoupling in brown fat mitochondria. Cell 151:400–413

Hinckley DA, Seybold PG (1988) A spectroscopic/thermodynamic study of the rhodamine B lactone $\rightleftharpoons$ zwitterion equilibrium. Spectrochim Acta Part A 44:1053–1059

Homma M, Takei Y, Murata A, Inoue T, Takeoka S (2015) A ratiometric fluorescent molecular probe for visualization of mitochondrial temperature in living cells. Chem Commun 51:6194–6197

Jaque D, Vetrone F (2012) Luminescence nanothermometry. Nanoscale 4:4301–4356

Johnson LV, Walsh ML, Bockus BJ, Chen LB (1981) Monitoring of relative mitochondrial membrane potential in living cells by fluorescence microscopy. J Cell Biol 88:526–535

Karstens T, Kobs K (1980) Rhodamine B and rhodamine 101 as reference substances for fluorescence quantum yield measurements. J Phys Chem 84:1871–1872

Kiyonaka S, Kajimoto T, Sakaguchi R, Shinmi D, Omatsu-Kanbe M, Matsuura H, Imamura H, Yoshizaki T, Hamachi I, Morii T, Mori Y (2013) Genetically encoded fluorescent thermosensors visualize subcellular thermoregulation in living cells. Nat Methods 10:1232–1238

Kucsko G, Maurer PC, Yao NY, Kubo M, Noh HJ, Lo PK, Park H, Lukin MD (2013) Nanometre-scale thermometry in a living cell. Nature 500:54–58

Li K, Liu B (2012) Polymer encapsulated conjugated polymer nanoparticles for fluorescence bioimaging. J Mater Chem 22:1257–1264

Lowell BB, Spiegelman BM (2000) Towards a molecular understanding of adaptive thermogenesis. Nature 404:652–660

Nakano T, Kikugawa G, Ohara T (2010) A molecular dynamics study on heat conduction characteristics in DPPC lipid bilayer. J Chem Phys 133:154705

Sakanoue J, Ichikawa K, Nomura Y, Tamura M (1997) Rhodamine 800 as a probe of energization of cells and tissues in the near-infrared region: a study with isolated rat liver mitochondria and hepatocytes. J Biochem (Tokyo) 121:29–37

Tseng Y-H, Cypess AM, Kahn CR (2010) Cellular bioenergetics as a target for obesity therapy. Nat Rev Drug Discov 9:465–482

Wikstrom JD, Mahdaviani K, Liesa M, Sereda SB, Si Y, Las G, Twig G, Petrovic N, Zingaretti C, Graham A, Cinti S, Corkey BE, Cannon B, Nedergaard J, Shirihai OS (2014) Hormone-induced mitochondrial fission is utilized by brown adipocytes as an amplification pathway for energy expenditure. EMBO J 33:418–436

Ye F, Wu C, Jin Y, Chan Y-H, Zhang X, Chiu DT (2011) Ratiometric Temperature Sensing with Semiconducting Polymer Dots. J Am Chem Soc 133:8146–8149

15

Implications for directionality of nanoscale forces in bacterial attachment

Jan J. T. M. Swartjes[1✉]**, Deepak H. Veeregowda**[2]

[1] University of Groningen and University Medical Center Groningen, Department of Biomedical Engineering, Antonius Deusinglaan 1, 9713 AV Groningen, the Netherlands
[2] Ducom Instruments Europe B.V, Center for Innovation, 9713 GX Groningen, the Netherlands

Abstract   Adhesion and friction are closely related and play a predominant role in many natural processes. From the wall-clinging feet of the gecko to bacteria forming a biofilm, in many cases adhesion is a necessity to survive. The direction in which forces are applied has shown to influence the bond strength of certain systems tremendously and can mean the difference between adhesion and detachment. The spatula present on the extension of the feet of the gecko can either attach or detach, based on the angle at which they are loaded. Certain proteins are known to unfold at different loads, depending on the direction at which the load is applied and some bacteria have specific receptors which increase their bond strength in the presence of shear. Bacteria adhere to any man-made surface despite the presence of shear forces due to running fluids, air flow, and other causes. In bacterial adhesion research, however, adhesion forces are predominantly measured perpendicularly to surfaces, whereas other directions are often neglected. The angle of shear forces acting on bacteria or biofilms will not be at a 90° angle, as shear induced by flow is often along the surface. Measuring at different angles or even lateral to the surface will give a more complete overview of the adhesion forces and mechanism, perhaps even resulting in alternative means to discourage bacterial adhesion or promote removal.

Keywords   Bacteria, Bacterial adhesion, Friction, Anisotropy, Shear

Both friction and adhesion play a key role in many natural phenomena. Along with the important role in all kinds of processes, the notion that both friction and adhesion can depend on the applied direction and angle, has intrigued scientists. One well-known example is the occurrence of high and low friction and adhesion cycles in the attachment and detachment of the gecko toe (Tian et al. 2006). Containing millions of small extensions, called spatula, all exerting nanoscale forces to the surface, the gecko can climb even upside down. By rolling its toe, the gecko changes the angle between its spatula and the surface, allowing it to shift between increasing the normal adhesion force and the frictional component (Autumn et al. 2006). At a molecular level, these changes in the angle of the spatula influence the Van der Waals forces in such a way that the attractive force between the spatula and the surface is altered to switch between high and low values (Tian et al. 2006). Simplified, by changing the direction of loading, either the normal adhesion force is high and the friction is low, or the frictional component is high and the normal adhesion force is low.

Whereas geckos can actively choose the loading angle, allowing them to either stay attached or detached, less autonomous systems like molecules and proteins do not have this option. Nevertheless, these systems display forces that highly depend on direction as well. The E2lip3 protein, for example, which is high in beta-sheet content, displays a resistance to pulling that strongly depends on the angle of the applied force

✉ Correspondence: janjtmswartjes@gmail.com (J. J. T. M. Swartjes),

(Brockwell et al. 2003). Similar behavior is found in the unfolding of Ubiquitin by mechanical stretching (Carrion-Vazquez et al. 2003). The direction of the applied force determines to a large extent the protein's stability. In these cases, the different angles in which the force is applied are believed to cause a change in the way the hydrogen bonds of inner beta-sheets rupture. As in the case of the gecko, the angle of the force determines whether bonds are broken by shearing or peeling (Brockwell et al. 2003). The regulation of bond organization by mechanical force has been simplified by describing it either as parallel distribution of forces, where each bond aids in resisting a mechanical force, or a zipper-like distribution (Fig. 1) in which one bond after another is required to oppose detachment (Albrecht et al. 2003; Hess 2006; Isabey et al. 2013). Based on the organization of the bonds, changing the loading direction will shift the distribution of forces, switching from parallel to zipper-like, or the other way around. In the parallel scenario, the collective bond behavior is able to resist much larger forces as the contribution of each bond is added up. A popular example of this is Velcro, which is easily loosened when pulled up from

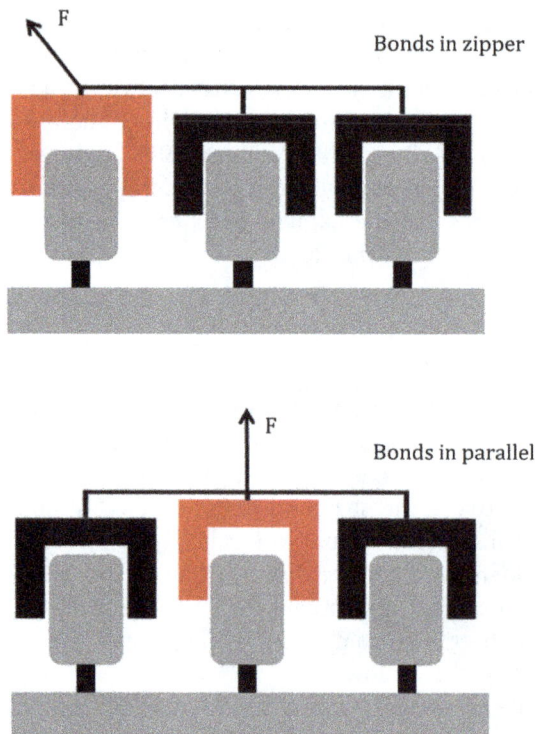

Fig. 1 Distribution of forces over multiple bonds. In the zipper-like distribution (*top*) each bond is loaded consecutively, passing the load on to the next bond after breaking. While in the parallel distribution (*bottom*) the load is distributed over all included bonds and after breaking of one bond, the load is redistributed over the remaining ones. Adapted with permissions from Isabey et al. (2013)

one side, but displays a highly increased resistance to detachment when pulled sideways (Matouschek and Bustamante 2003).

Unzipping of proteins is, for example, also observed in the amyloid-like interactions within clusters of adhesins called Als, which contribute in the adhesion of *Candida albicans* (Alsteens et al. 2010, 2012). Force analysis of the adhesion shows mechanical unzipping of β-sheet interactions between Als proteins upon being pulled from an amyloid coated surface at 90°. One step beyond the scale of inter-protein bindings, shear, or lateral, force dependency of adhesion is observed in ligand-receptor complexes. The FimH adhesin expressed by *Escherichia coli* binds to mannose and is found to enhance adhesion under high shear conditions (Thomas et al. 2004; Aprikian et al. 2007). Whereas the behavior of the previously mentioned proteins is generally ascribed to bond organization, it has been suggested that in the case of *E. coli* the typical behavior stems from allosteric regulation, causing bond enhancement by mechanical force (Yakovenko et al. 2008).

From animals to proteins it is clear that the direction of an applied force can make the difference between adhesion or detachment, structural integrity or unfolding. For certain microorganisms, like bacteria, surface attachment is the preferred mode of survival, as stable surface bound communities offer protection to antibiotics and mechanical removal (O'Toole et al. 2000; Dunne 2002; Vlamakis et al. 2013). At the same time, whether the surface comprises the inner-lining of the human body, or an implant, the formation of these biofilm communities is often highly unwanted (Cegelski et al. 2008; Löfling et al. 2011; Foster et al. 2014). For decades, researchers are trying to deal with bacterial adhesion by an almost endless effort to create non-fouling surfaces which can withstand adhesion of bacteria. On the other hand, there is also a tremendous amount of work put into new strategies of effective removal of adhered bacteria. In both cases, fundamental knowledge of bacterial adhesion and the mechanisms behind it are of crucial importance, and since this knowledge is limited, new information can hold the key to breakthroughs in either field.

Especially in the medical field where bacterial adhesion and the forthcoming biofilms cause life-threatening infections that are continuing to be more difficult to treat with antibiotics, detachment as well as adhesion prevention strategies are well sought after (Busscher et al. 2012; Campoccia et al. 2013; Swartjes et al. 2013; Swartjes et al. 2014a). To find out more about the mechanisms of bacterial adhesion, atomic force microscopy (AFM) has proven to be the tool of preference in order to determine the forces by which bacteria attach to surfaces and keep themselves adhered (Dufrene 2002; Dorobantu and Gray 2010; Müller and Dufrêne 2011; Dorobantu et al. 2012). The vertical motion of the AFM cantilever is often used to determine the force necessary to pull a bacterium from a cell or surface. This force, which has a magnitude of several nano Newton, is considered as the adhesion force (Dorobantu and Gray 2010). Whenever AFM is used to measure bacterial adhesion, the angle of the direction in which the bacterial adhesion force is measured and the substrate is approximately 90°. However, the amount of work (W) to overcome adhesion is a function of the pull-off angle θ, i.e., $W = F \cdot d \cdot \cos(\theta)$, which indicates that the force required for detachment might change for different pull-off angles. Additionally, bacteria in most situations adhere from a flowing condition, in which the angle of approach leads to friction between a bacterium and the surface (Swartjes et al. 2014b). In fact, there is a relationship between adhesion and friction at nanoscale often used to describe the contact between two solid bodies, $F_f = \mu (F_n + F_{adh})$, stating that the friction force F_f, equals the coefficient of friction (μ) multiplied by the sum of the normal force F_n and the adhesion force F_{adh} (Gao et al. 2004). In relation to these directional influences on bacterial adhesion, several methods have been applied to determine the lateral forces between bacteria and surfaces.

A distinction can be made between two types of lateral forces; first, the shear adhesion force depending on the strength of the bond between an adhered bacterium and a surface, which breaks by moving the bacterium along the surface after it has adhered, and second, the lateral force arising between a bacterium and a surface when initial contact is made by a bacterium approaching the surface at an angle, representing the friction (Swartjes et al. 2014b). By challenging the shear strength of the adhesion bond using different flow rates of the liquid carrying the bacteria (Gazzola et al. 2015), or by detachment induced by passage of a liquid-air interface (Perera-Costa et al. 2014), estimations of the first type of lateral force have been made. However, since perpendicular to the surface the adhesion force of a single bacterium can be measured directly using the AFM, it is desirable to achieve a similar mode of action to measure the adhesion force at a different angle. Several attempts have recently been made to determine the lateral forces occurring between bacteria and surfaces using AFM (Verran et al. 2010; Zhang et al. 2011; Swartjes et al. 2014b). Quantification of the shear strength of bacterial adhesion has been achieved by imaging of bacteria; as the AFM cantilever moves along the surface in contact mode, the lateral

movement of the cantilever can displace bacteria by pushing them away (Verran et al. 2010; Zhang et al. 2011). To measure lateral forces more directly, single-cell force spectroscopy (SCFS), in which a single bacterial cell is attached to the AFM cantilever, can be applied to probe the forces between this single bacterium and a surface. Kweon et al. modified an AFM cantilever with a bacterial spore and rubbed the spore against a silica surface to retrieve the values of occurring friction forces (Kweon et al. 2011). Even though this only involved a bacterial product, rather than an actual bacterium, the principle has also been performed using bacteria instead of a spore. The friction between polymer brush-modified surfaces and bacteria attached to a cantilever showed that friction was correlated to the amount of bacteria adhering to the surface, suggesting that friction forces play a role in attachment (Swartjes et al. 2014b). Interestingly, the friction and adhesion forces did not relate to each other as per the previously stated equation describing friction forces, indicating that bacterial friction and adhesion is more complex and challenging.

Most of these studies on lateral forces involved whole bacteria, however, based on the behavior of single proteins when subjected to forces at different angles, direction-dependent adhesion can also be studied by looking at components of bacterial adhesion complexes. SCFS has taken a flight over the last years and has expanded the insights about bacterial adhesion mechanisms considerably (Helenius et al. 2008; Müller and Dufrêne 2011; Isabey et al. 2013; Beaussart et al. 2014). Additionally, the technique has been extended to the use of single molecules, offering the possibility of isolating specific adhesion structures of bacteria and identifying their sole contribution in adhesion (Benoit et al. 2000; Sullan et al. 2015). Interestingly, single-molecule force spectroscopy (SMFS) using specific bacterial adhesion complexes reveals peaks in the force-distance curves due to breakage of multiple bonds, suggested to be caused by unfolding of the protein (Fig. 2C, D) and closely resembling the unzipping of previously mentioned proteins displaying anisotropic behavior (Fig. 2A, B) (El-Kirat-Chatel et al. 2014; Sullan et al. 2015). Additionally, measurements of a whole *Streptococcus mutans* cell suggest the presence of up to 10 ligand-receptor complexes being responsible for the binding of a single bacterial cell (Sullan et al. 2015).

Even though there are many examples showing that bond strength and adhesion phenomena in certain cases display properties highly depending on the direction of the applied force, direct measurements in the case of bacteria are scarce. In the quest for solutions to the bacterial adhesion problem, attention for shear and friction forces is present, as based on the previously

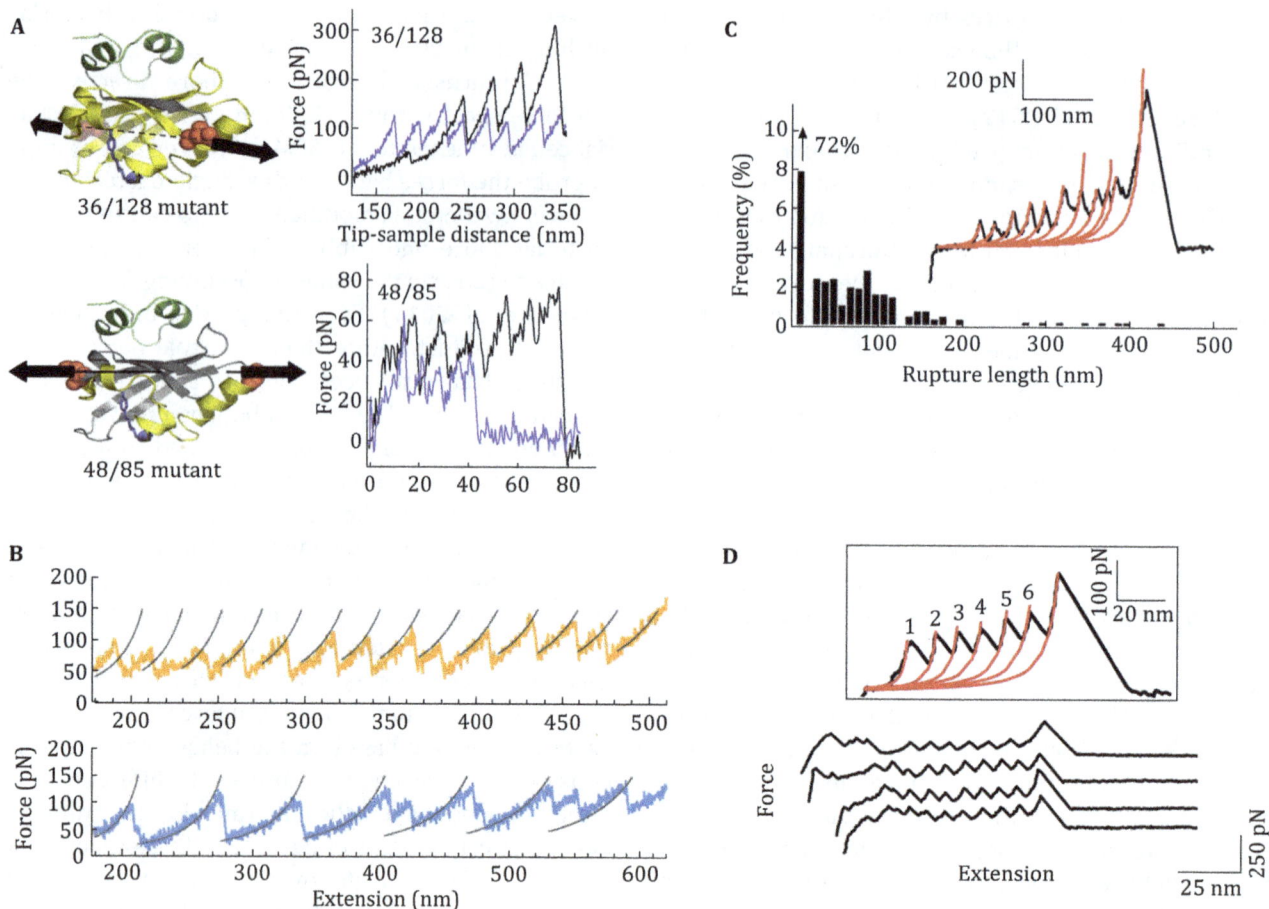

Fig. 2 Unfolding behavior of proteins shown to have anisotropic responses to loading (**A**, **B**) and bacterial adhesion proteins displaying similar force patterns (**C**, **D**). **A** The distinct differences in force curves upon stretching of PYP by pulling at different axis. **B** Force-extension curves of unfolding of GFP displaying a distinct unzipping pattern for different directions of loading. **C** Force curves for the interaction between *S. mutans* adhesin P1 and fibronectin-coated solid substrates, exhibiting similar peaks observed for anisotropic proteins. **D** Unfolding force patterns of Als5p adhesion proteins closely resemble those of proteins known to respond differently to different loading directions. Adapted with permissions from Dietz et al. (2006), Nome et al. (2007), Alsteens et al. (2009) and Sullan et al. (2015)

mentioned examples and studies developing antifouling super-slippery surfaces (Wong et al. 2011; Epstein et al. 2012; Li et al. 2013; MacCallum et al. 2015), however, the fundamental role it plays in adhesion is not known. As the use of bacterial probes in AFM is increasing, unraveling the role of directionality could reveal completely new information on the adhesion mechanism of bacteria. Besides the added value it could have in non-specific adhesion, the example of a shear-dependent specific adhesion mechanism in *E. coli* suggests that specific interactions between ligands and receptors, which are present in the majority of bacteria, are able to exhibit different strengths based on how their unity is challenged. Given that specific patterns of unzipping and unfolding of individual proteins involved in bacterial adhesion are observed (Table 1), it is very well possible that the same directional dependent

behavior seen in multiple types of proteins would also apply in the case of proteins associated with adhesion of bacteria. Additionally, for several pathogens it is suggested that zipper-like sequences are involved in host cell invasion, implicating that structures known to show directional dependent strength are partly responsible not only for general adhesion, but also for bacterial pathogenesis (Schwarz-Linek et al. 2003). Altogether, anisotropic adhesion behavior could not only stem from an array of adhesion complexes acting as individual bonds that can be loaded in a parallel or a zipper-like fashion, but also from within adhesion proteins or complexes where unfolding of single proteins might depend on the loading direction.

As shear is present almost everywhere inside the human body, e.g., in the oral cavity, blood vessels, intestine, and lungs, it suggests that firm attachment by

Table 1 Overview of different proteins whose behavior depends on the loading axis and of bacterial associated proteins whose unfolding characteristics imply the possibility of similar anisotropic behavior

Protein/structure	Interaction model	References
Ubiquitin	Direction-dependent unfolding	Carrion-Vazquez et al. (2003)
GCN4 protein	Anisotropic response to pulling	Gao et al. (2011)
Src SH3 protein	Anisotropic response to applied force	Jagannathan et al. (2012)
Green fluorescent protein (GFP)	Anisotropic deformation response	Dietz et al. (2006)
E2Lip3	Anisotropic unfolding	Brockwell et al. (2003)
Photoactive yellow protein (PYP)	Anisotropic unfolding	Nome et al. (2007)
Escherichia coli FimH	Shear-enhanced adhesion	Thomas et al. (2004), Aprikian et al. (2007), Yakovenko et al. (2008)
Candida albicans Als5p cell adhesion protein	Sequential unfolding	Alsteens et al. (2009)
Bordetella pertussis adhesin FHA	Sequential unfolding	Alsteens et al. (2013)
Pili from *Lactobacillus rhamnosus* GG	Zipper-like	Tripathi et al. (2013)
Streptococcus mutans P1 adhesin	Zipper-like	Sullan et al. (2015)
Als amyloids of *Candida albicans*	Zipper-like	Alsteens et al. (2012)

resisting shear forces is mandatory for bacteria in order to persist in an adhered state. Proteins highly involved in the adhesion of different bacterial strains have shown to exhibit similar unfolding patterns compared to other proteins, which are highly anisotropic in their unfolding. Additionally, the shear strengthening of FimH in *E. coli* supports the implications for the possibility of direction-dependent adhesion mechanisms in bacteria, similar to those suggested for mammalian cells (Isabey et al. 2013). By probing friction forces and the adhesion forces lateral to the surface, specific information can be obtained that possibly provide new clues for anti-adhesive, or easy to clean surfaces. It is impossible to say which type of lateral force has more impact on bacterial adhesion, and the frictional forces probably contribute most to the transitions from unbound to surface attached, while it is likely that shear adhesion forces are most important in remaining an adhered state. As such, the frictional forces seem most interesting for design of non-fouling surface, while the shear adhesion force could help in designing strategies for bacterial removal. Nano-topographic surfaces could perhaps alter the direction in which bacteria experience shear forces, making them less likely to adhere, or easier to be removed.

Bacteria have outsmarted mankind by adapting resistance to a major part of our antibiotic spectrum, resulting in an increase in infections which are extremely hard to resolve (Spellberg et al. 2013; Wellington et al. 2013). In order to prevent infection there are many aspects of bacterial adhesion and biofilm formation requiring our utmost attention. The many sugges-

tions for anisotropy of bacterial adhesion forces therefore imply that studying forces between bacteria and surfaces in multiple directions are desirable, as it might reveal precious information that can help in making crucial steps toward the development of new and more efficient anti-bacterial strategies.

Compliance with Ethical Standards

Conflict of Interest Jan J. T. M. Swartjes and Deepak H. Veeregowda declare that they have no conflict of interest. Deepak H. Veeregowda is a manager of Ducom Instruments in the Netherlands and also a research scientist at the University Medical Center Groningen.

Human and Animal Rights and Informed Consent This article does not contain any studies with human or animal subjects performed by any of the authors.

References

Albrecht C, Blank K, Lalic-Mülthaler M, Hirler S, Mai T, Gilbert I, Schiffmann S, Bayer T, Clausen-Schaumann H, Gaub HE (2003) DNA: a programmable force sensor. Science 301:367–370

Alsteens D, Dupres V, Klotz SA, Gaur NK, Lipke PN, Dufrêne YF (2009) Unfolding individual Als5p adhesion proteins on live cells. ACS Nano 3:1677–1682

Alsteens D, Garcia MC, Lipke PN, Dufrêne YF (2010) Force-induced formation and propagation of adhesion nanodomains in living fungal cells. Proc Natl Acad Sci USA 107:20744–20749

Alsteens D, Ramsook CB, Lipke PN, Dufrêne YF (2012) Unzipping a functional microbial amyloid. ACS Nano 6:7703–7711

Alsteens D, Martinez N, Jamin M, Jacob-Dubuisson F (2013) Sequential unfolding of beta helical protein by single-molecule atomic force microscopy. PLoS One 8:e73572

Aprikian P, Tchesnokova V, Kidd B, Yakovenko O, Yarov-Yarovoy V, Trinchina E, Vogel V, Thomas W, Sokurenko E (2007) Interdomain interaction in the FimH adhesin of *Escherichia coli* regulates the affinity to mannose. J Biol Chem 282:23437-23446

Autumn K, Dittmore A, Santos D, Spenko M, Cutkosky M (2006) Frictional adhesion: a new angle on gecko attachment. J Exp Biol 209:3569-3579

Beaussart A, El-Kirat-Chatel S, Sullan RMA, Alsteens D, Herman P, Derclaye S, Dufrêne YF (2014) Quantifying the forces guiding microbial cell adhesion using single-cell force spectroscopy. Nat Protoc 9:1049-1055

Benoit M, Gabriel D, Gerisch G, Gaub HE (2000) Discrete interactions in cell adhesion measured by single-molecule force spectroscopy. Nat Cell Biol 2:313-317

Brockwell DJ, Paci E, Zinober RC, Beddard GS, Olmsted PD, Smith DA, Perham RN, Radford SE (2003) Pulling geometry defines the mechanical resistance of a beta-sheet protein. Nat Struct Biol 10:731-737

Busscher HJ, van der Mei HC, Subbiahdoss G, Jutte PC, van den Dungen JJAM, Zaat SAJ, Schultz MJ, Grainger DW (2012) Biomaterial-associated infection: locating the finish line in the race for the surface. Sci Transl Med 4:153rv10

Campoccia D, Montanaro L, Arciola CR (2013) A review of the biomaterials technologies for infection-resistant surfaces. Biomaterials 34:8533-8554

Carrion-Vazquez M, Li H, Lu H, Marszalek PE, Oberhauser AF, Fernandez JM (2003) The mechanical stability of ubiquitin is linkage dependent. Nat Struct Biol 10:738-743

Cegelski L, Marshall GR, Eldridge GR, Hultgren SJ (2008) The biology and future prospects of antivirulence therapies. Nat Rev Microbiol 6:17-27

Dietz H, Berkemeier F, Bertz M, Rief M (2006) Anisotropic deformation response of single protein molecules. Proc Natl Acad Sci USA 103:12724-12728

Dorobantu LS, Gray MR (2010) Application of atomic force microscopy in bacterial research. Scanning 32:74-96

Dorobantu LS, Goss GG, Burrell RE (2012) Atomic force microscopy: a nanoscopic view of microbial cell surfaces. Micron 43:1312-1322

Dufrene YF (2002) Atomic force microscopy, a powerful tool in microbiology. J Bacteriol 184:5205-5213

Dunne WM (2002) Bacterial adhesion: seen any good biofilms lately? Society 15:155-166

El-Kirat-Chatel S, Beaussart A, Boyd CD, O'Toole GA, Dufreîne YF (2014) Single-cell and single-molecule analysis deciphers the localization, adhesion, and mechanics of the biofilm adhesin LapA. ACS Chem Biol 9:485-494

Epstein AK, Wong TS, Belisle RA, Boggs EM, Aizenberg J (2012) Liquid-infused structured surfaces with exceptional anti-biofouling performance. Proc Natl Acad Sci USA 109:13182-13187

Foster TJ, Geoghegan JA, Ganesh VK, Höök M (2014) Adhesion, invasion and evasion: the many functions of the surface proteins of *Staphylococcus aureus*. Nat Rev Microbiol 12:49-62

Gao J, Luedtke WD, Gourdon D, Ruths M, Israelachvili JN, Landman U (2004) frictional forces and amontons' law: from the molecular to the macroscopic scale. J Phys Chem B 108:3410-3425

Gao Y, Sirinakis G, Zhang Y (2011) Highly anisotropic stability and folding kinetics of a single coiled coil protein under mechanical tension. J Am Chem Soc 133:12749-12757

Gazzola G, Habimana O, Murphy CD, Casey E (2015) Comparison of biomass detachment from biofilms of two different *Pseudomonas* spp. under constant shear conditions. Biofouling 31:13-18

Helenius J, Heisenberg C-P, Gaub HE, Muller DJ (2008) Single-cell force spectroscopy. J Cell Sci 121:1785-1791

Hess H (2006) Self-assembly driven by molecular motors. Soft Mater 2:669

Isabey D, Féréol S, Caluch A, Fodil R, Louis B, Pelle G (2013) Force distribution on multiple bonds controls the kinetics of adhesion in stretched cells. J Biomech 46:307-313

Jagannathan B, Elms PJ, Bustamante C, Marqusee S (2012) Direct observation of a force-induced switch in the anisotropic mechanical unfolding pathway of a protein. Proc Natl Acad Sci USA 109:17820-17825

Kweon H, Yiacoumi S, Tsouris C (2011) Friction and adhesion forces of *Bacillus thuringiensis* spores on planar surfaces in atmospheric systems. Langmuir 27:14975-14981

Li J, Kleintschek T, Rieder A, Cheng Y, Baumbach T, Obst U, Schwartz T, Levkin PA (2013) Hydrophobic liquid-infused porous polymer surfaces for antibacterial applications. ACS Appl Mater Interfaces 5:6704-6711

Löfling J, Vimberg V, Battig P, Henriques-Normark B (2011) Cellular interactions by LPxTG-anchored pneumococcal adhesins and their streptococcal homologues. Cell Microbiol 13:186-197

MacCallum N, Howell C, Kim P, Sun D, Friedlander R, Ranisau J, Ahanotu O, Lin JJ, Vena A, Hatton B, Wong TS, Aizenberg J (2015) Liquid-infused silicone as a biofouling-free medical material. ACS Biomater Sci Eng 1:43-51

Matouschek A, Bustamante C (2003) Finding a protein' s Achilles heel. Nat Struct Biol 10:674-676

Müller DJ, Dufrêne YF (2011) Atomic force microscopy: a nanoscopic window on the cell surface. Trends Cell Biol 21:461-469

Nome RA, Zhao JM, Hoff WD, Scherer NF (2007) Axis-dependent anisotropy in protein unfolding from integrated nonequilibrium single-molecule experiments, analysis, and simulation. Proc Natl Acad Sci USA 104:20799-20804

O'Toole GA, Kaplan HB, Kolter R (2000) Biofilm formation as microbial development. Annu Rev Microbiol 54:49-79

Perera-Costa D, Bruque JM, González-Martín ML, Gómez-García AC, Vadillo-Rodríguez V (2014) Studying the influence of surface topography on bacterial adhesion using spatially organized microtopographic surface patterns. Langmuir 30:4633-4641

Schwarz-Linek U, Werner JM, Pickford AR, Gurusiddappa S, Kim JH, Pilka ES, Briggs JAG, Gough TS, Höök M, Campbell ID, Potts JR (2003) Pathogenic bacteria attach to human fibronectin through a tandem beta-zipper. Nature 423:177-181

Spellberg B, Bartlett JG, Gilbert DN (2013) The future of antibiotics and resistance. N Engl J Med 368:299-302

Sullan RMA, Li JK, Crowley PJ, Brady LJ, Dufrêne YF (2015) Binding forces of *Streptococcus mutans* P1 Adhesin. ACS Nano 9:1448-1460

Swartjes JJTM, Das T, Sharifi S, Subbiahdoss G, Sharma PK, Krom BP, Busscher HJ, Van der Mei HC (2013) A functional DNase I coating to prevent adhesion of bacteria and the formation of biofilm. Adv Funct Mater 23:2843-2849

Swartjes JJTM, Sharma PK, van Kooten T, Van der Mei HC, Mahmoudi M, Busscher HJ, Rochford ETJ (2014a) Current developments in antimicrobial surface coatings for biomedical applications. Curr Med Chem 21:2116-2129

Swartjes JJTM, Veeregowda DH, Van der Mei HC, Busscher HJ, Sharma PK (2014b) Normally oriented adhesion versus friction forces in bacterial adhesion to polymer-brush functionalized surfaces under fluid flow. Adv Funct Mater 24:4435–4441

Thomas WE, Nilsson LM, Forero M, Sokurenko EV, Vogel V (2004) Shear-dependent 'stick-and-roll' adhesion of type 1 fimbriated *Escherichia coli*. Mol Microbiol 53:1545–1557

Tian Y, Pesika N, Zeng H, Rosenberg K, Zhao B, McGuiggan P, Autumn K, Israelachvili J (2006) Adhesion and friction in gecko toe attachment and detachment. Proc Natl Acad Sci USA 103:19320–19325

Tripathi P, Beaussart A, Alsteens D, Dupres V, Claes I, Von Ossowski I, De Vos WM, Palva A, Lebeer S, Vanderleyden J, Dufreîne YF (2013) Adhesion and nanomechanics of pili from the probiotic *Lactobacillus rhamnosus* GG. ACS Nano 7:3685–3697

Verran J, Packer A, Kelly PJ, Whitehead KA (2010) Use of the atomic force microscope to determine the strength of bacterial attachment to grooved surface features. J Adhes Sci Technol 24:2271–2285

Vlamakis H, Chai Y, Beauregard P, Losick R, Kolter R (2013) Sticking together: building a biofilm the *Bacillus subtilis* way. Nat Rev Microbiol 11:157–168

Wellington EMH, Boxall ABA, Cross P, Feil EJ, Gaze WH, Hawkey PM, Johnson-Rollings AS, Jones DL, Lee NM, Otten W, Thomas CM, Williams AP (2013) The role of the natural environment in the emergence of antibiotic resistance in Gram-negative bacteria. Lancet Infect Dis 13:155–165

Wong TS, Kang SH, Tang SKY, Smythe EJ, Hatton BD, Grinthal A, Aizenberg J (2011) Bioinspired self-repairing slippery surfaces with pressure-stable omniphobicity. Nature 477:443–447

Yakovenko O, Sharma S, Forero M, Tchesnokova V, Aprikian P, Kidd B, Mach A, Vogel V, Sokurenko E, Thomas WE (2008) FimH forms catch bonds that are enhanced by mechanical force due to allosteric regulation. J Biol Chem 283:11596–11605

Zhang T, Chao Y, Shih K, Li XY, Fang HHP (2011) Quantification of the lateral detachment force for bacterial cells using atomic force microscope and centrifugation. Ultramicroscopy 111:131–139

16

Energy coupling mechanisms of AcrB-like RND transporters

Xuejun C. Zhang[1,2✉], Min Liu[1,2], Lei Han[1,2]

[1] National Laboratory of Biomacromolecules, CAS Center for Excellence in Biomacromolecules, Institute of Biophysics, Chinese Academy of Sciences, Beijing 100101, China
[2] College of Life Science, University of Chinese Academy of Sciences, Beijing 100049, China

Abstract   Prokaryotic AcrB-like proteins belong to a family of transporters of the RND superfamily, and as main contributing factor to multidrug resistance pose a tremendous threat to future human health. A unique feature of AcrB transporters is the presence of two separate domains responsible for carrying substrate and generating energy. Significant progress has been made in elucidating the three-dimensional structures of the homo-trimer complexes of AcrB-like transporters, and a three-step functional rotation was identified for this class of transporters. However, the detailed mechanisms for the transduction of the substrate binding signal, as well as the energy coupling processes between the functionally distinct domains remain to be established. Here, we propose a model for the interdomain communication in AcrB that explains how the substrate binding signal from the substrate-carrier domain triggers protonation in the transmembrane domain. Our model further provides a plausible mechanism that explains how protonation induces conformational changes in the substrate-carrier domain. We summarize the thermodynamic principles that govern the functional cycle of the AcrB trimer complex.

Keywords   AcrB, RND transporter, Membrane potential

INTRODUCTION

In Gram-negative bacteria, the periplasm functions as a protective buffer zone against cytotoxic substances from the extracellular environment (Li *et al.* 2015). The cell actively expels toxic compounds that have penetrated the outer membrane and that have accumulated in the periplasmic space, thus preventing the toxins from either harming the cell from within the periplasm, or further penetrating into the cytosol. In addition, metabolic waste and other cytotoxic compounds transported from cytosol to periplasm should also be expelled promptly. Resistance-nodulation-cell division (RND) type of transporters plays essential roles in these toxin-extrusion processes vital for cell survival (Higgins 2007).

RND transporters form a superfamily of membrane proteins that play crucial roles in diverse biological processes from multidrug resistance in bacteria to trafficking of lipid molecules in eukaryotic cells (Higgins 2007). RND proteins unidirectionally transport substrates between two distinct micro-environments not necessarily separated by a biological membrane. In particular, RND-mediated transport often occurs against the substrate chemical potential, *e.g.*, from a low-concentration to a high-concentration environment, or from a high-affinity carrier to a low-affinity carrier. Such transport requires external energy, and RND transporters are known to utilize the electrochemical potential of protons, *i.e.*, the proton motive force (PMF), to drive their substrate-transport processes (Yamaguchi *et al.* 2015).

The AcrAB–TolC tripartite complex, a prototypical member of the RND family, is constitutively expressed in many Gram-negative bacteria, is responsible for efflux transport of lipophilic and/or amphipathic compounds (a common property of many cytotoxic substances such as antibiotics), and thus plays important roles in multidrug resistance (Nikaido and Takatsuka 2009; Yamaguchi *et al.* 2015). In this complex, AcrB captures a wide variety of structurally dissimilar drugs and toxic compounds from either periplasm or the periplasmic leaflet of the inner membrane and delivers them to a conduit formed by a TolC trimer which connects to the exterior of the cell. Analysis of amino acid sequences shows that AcrB belongs to the Hydrophobe/Amphiphile Efflux-1 (HAE1) family of the RND superfamily (Perrin *et al.* 2010). Hereafter, we refer to members of the HAE1 family and other phylogenetically related RND families as AcrB-like transporters (Fig. S1). From the view of energy coupling, AcrB-like transporters are distinct from ATP-binding cassette (ABC) transporters. In most cases, ABC transporters utilize cytosolic energy source (*i.e.*, ATP hydrolysis) to drive cross-membrane transport of substrates. In contrast, AcrB-like transporters utilize PMF across the bacterial inner membrane to transport substrates, either within an aqueous environment or at the aqueous–membrane interface.

Structures, substrate recognition mechanisms, and the functional cycle of AcrB-like transporters have been extensively reviewed previously (Higgins 2007; Nikaido and Takatsuka 2009; Li *et al.* 2015; Yamaguchi *et al.* 2015). Here, we focus on the structural basis of the energy coupling mechanisms of AcrB-like transporters. We will first summarize the structural features of the AcrB complex for the sake of discussion of the mechanistic principles, and then propose a model on the energy coupling mechanism of this family of transporters between the substrate-carrier domain and the transmembrane (TM) PMF-driven motor domain. All arguments discussed below are based on insights from *Escherichia coli* AcrB (Ec-AcrB), unless stated otherwise.

BASIC STRUCTURES OF ACRB

The AcrAB–TolC tripartite complex from *E. coli* is the most extensively studied RND transporter (Yamaguchi *et al.* 2015). Recent cryo-electron microscopy studies revealed that this complex is of a stoichiometry of $AcrB_3{:}AcrA_6{:}TolC_3$ (Du *et al.* 2014; Jeong *et al.* 2016). As shown schematically in Fig. 1, AcrB serves as the PMF-driven transporter and forms a homo-trimer; the TolC trimer forms a conduit connecting the exit chamber of the AcrB trimer to exterior of the cell; and the AcrA hexamer serves as an adaptor between the AcrB and TolC trimers.

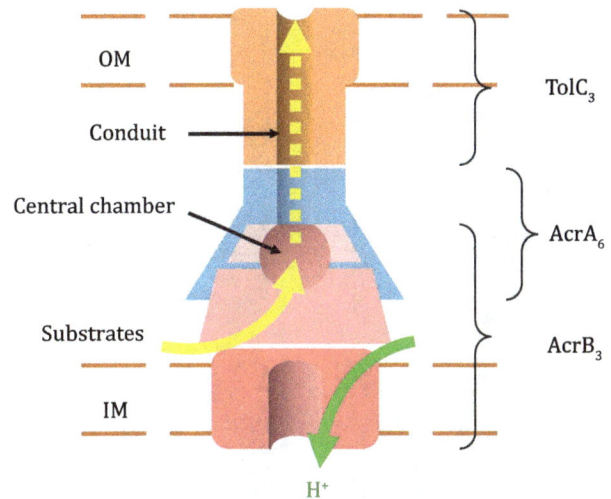

Fig. 1 Schematic diagram of the AcrAB–TolC complex. "OM" and "IM" stand for outer and inner membranes, respectively

Among the three component proteins in the *E. coli* AcrAB–TolC complex, AcrB (GenBank access ID: ANK05811.1), consisting of 1,049 amino acid residues, is the most extensively studied protein both structurally and functionally. Currently, the Protein Data Bank (PDB) database (Berman *et al.* 2000) contains over 40 full-length AcrB crystal structures, demonstrating that the stable homo-trimer is the functional form of AcrB. Each protomer is composed of three major domains, namely the TM domain; the substrate-carrier, porter domain (formerly called port domain); and the TolC-docking domain (Murakami *et al.* 2002) (Fig. 2). The TM domain contains 12 TM helices (TMs 1–12) (Fig. 2C; Fig. S1A) and resides in the inner membrane with both its N- and C-termini located in the cytosol. This domain assumes a pseudo twofold symmetry between the N-terminal subdomain (termed N_{TM}, composed of TMs 1–6) and C-terminal subdomain (C_{TM}, TMs 7–12), with the rotation axis perpendicular to the membrane. Around the symmetry axis, TM4 and TM10 form a parallel central helix-pair. Functionally important residues D407, D408, and K940 (numbered as in Ec-AcrB) are located in the TM4–TM10 interface (Fig. 3A) and form a rather extensive hydrogen (H)-bond network (Middlemiss and Poole 2004; Eicher *et al.* 2012). In particular, D407 is located in a highly conserved sequence motif, GX_3DX_6E, of AcrB-like transporters (Perrin *et al.* 2010) (Fig. S1). Integrity of this H-bond network is essential for the transport function of AcrB-like transporters, and the D407–D408 acidic pair is identified to be the protonation site required for PMF-energy conversion (Murakami *et al.* 2002). Furthermore, between N_{TM} and C_{TM} (particularly between TMs 6 and 7), a long amphipathic helix (~26 residues, termed α6-7) represents a

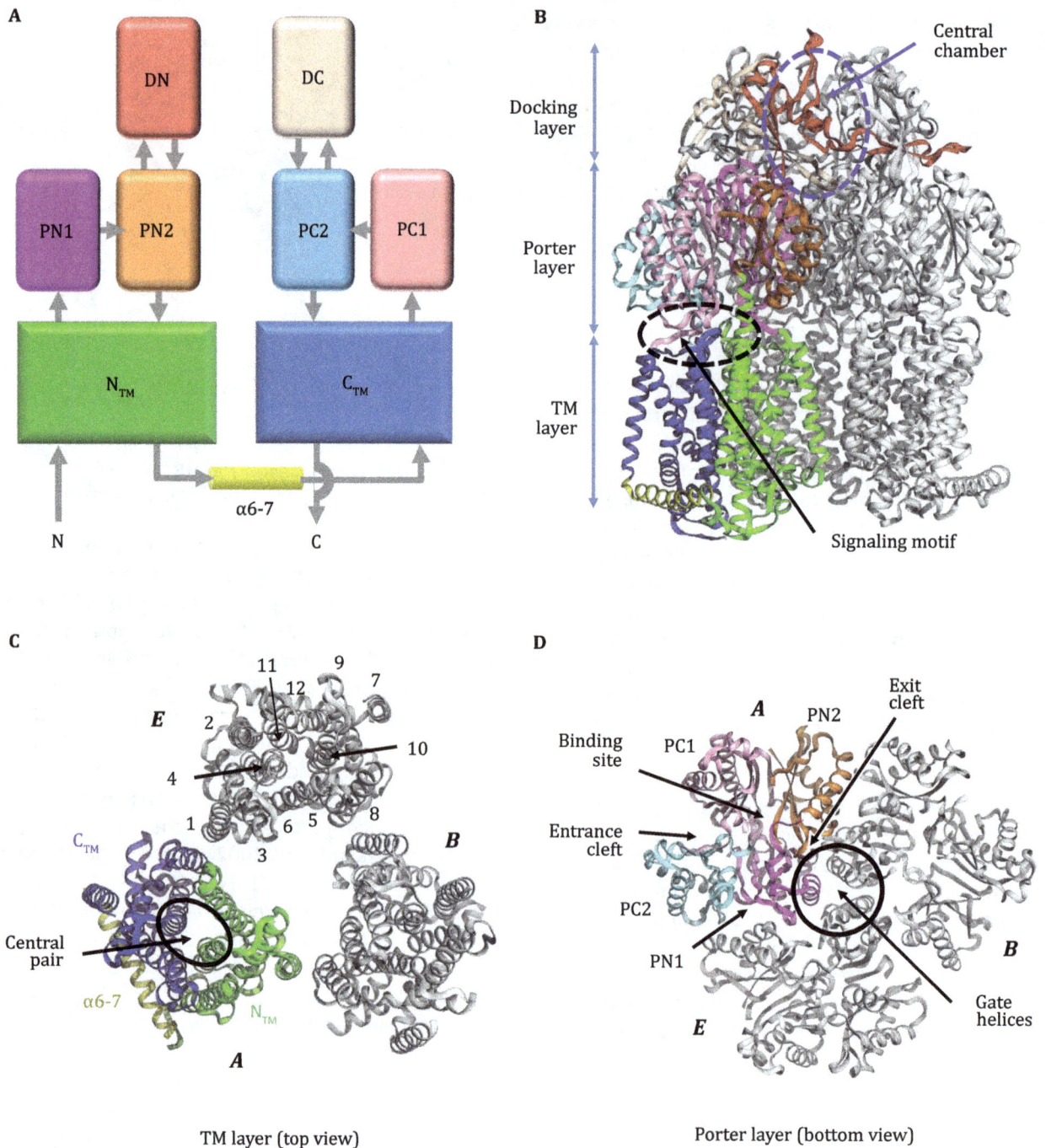

Fig. 2 Crystal structures of the AcrB asymmetric trimer. **A** The topology of the AcrB protomer. **B–D** The AcrB trimer (PDB ID: 4DX5) is shown in ribbon diagrams. In each panel, the access protomer is shown in color, while the other two protomers are displayed in gray. In **C** and **D**, the access, binding, and extrusion protomers are labeled as "*A*", "*B*", and "*E*," respectively. In the top view of the TM layer (**C**), TM helices are labeled as "1" through "12"

conserved structural feature of RND transporters (Fig. 2) (Murakami *et al.* 2002; Sennhauser *et al.* 2009; Gong *et al.* 2016). It presumably lies on the cytosolic surface of the inner membrane, forming part of the rim of the AcrB trimer complex. A highly conserved tyrosine residue, Y527 (Fig. S1), anchors α6-7 to another conserved region at the C-terminal end of TM12.

The porter domain contains four topologically homologous subdomains, namely PN1, PN2, PC1, and PC2, each of which consists of two β-α-β repeats (Fig. 2). The four β-strands in each subdomain form a β-sheet which participates in the formation of the substrate-transport path. In the primary sequence, PN1 and PN2 are tandemly inserted between TMs 1 and 2 of

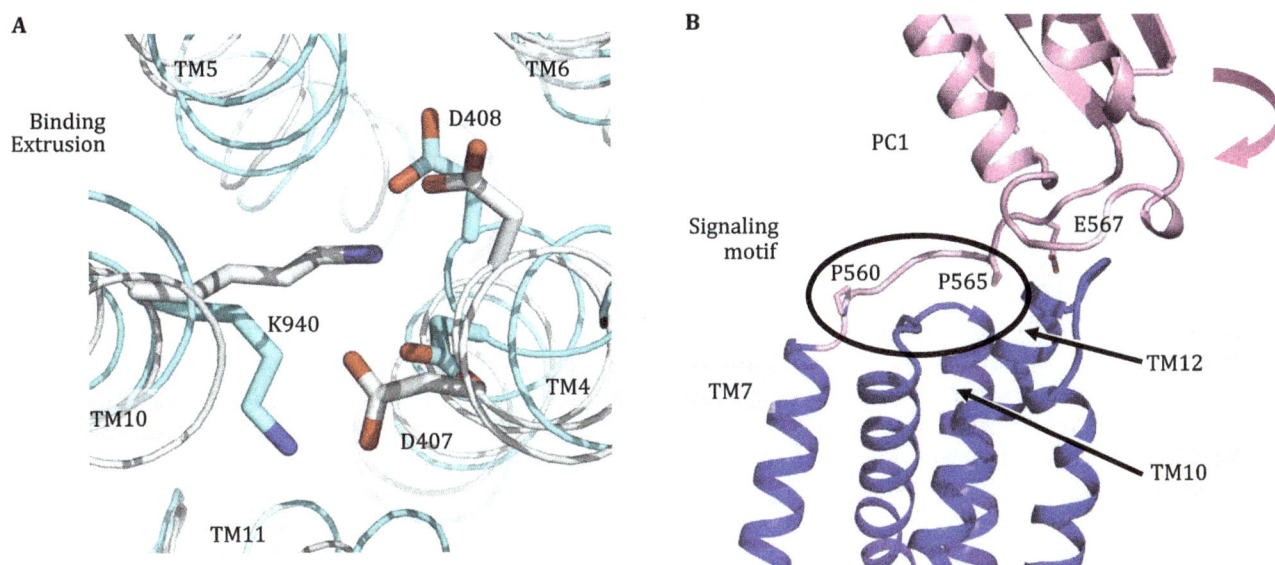

Fig. 3 Protonation switch in the TM domain. **A** The D407-D408-K940 cluster, viewed from the periplasm direction. The TM domains of the binding (gray) and extrusion (cyan) protomers are superimposed, with an overall RMSD of 1.3 Å for 314 Cα-atom pairs. **B** Side view of the signaling motif between C_{TM} (blue) and PC1 (pink). The substrate binding induced clockwise rotation of PC1 is postulated to further induce a clockwise rotation of TM10, which triggers protonation in TM4

N_{TM}, and PC1 and PC2 between TMs 7 and 8 of C_{TM} (Fig. 2A; Fig. S1A). Furthermore, in the 3D structure, PN1 and PC2 pack together via a shared β-sheet, forming a rigid body (Eicher *et al.* 2014). So do PN2 and PC1, albeit significantly more flexible (when comparing different protomers). Thus, the porter domain in each AcrB protomer may be considered to be composed of two structural units, termed PN1-PC2 and PN2-PC1. The PN2-PC1 unit is observed to possess one of the major substrate binding sites (referred to as distal/deep binding site) (Murakami *et al.* 2006). The substrate entrance (also called proximal binding site) is located between the subdomains PC1 and PC2, laterally facing the periplasmic space. The substrate exit is located between the subdomains PN1 and PN2, connecting to a central exit chamber of the AcrB trimer. Thus, the entire substrate-transport path (including the entrance cleft, "major" binding site, and exit cleft) is formed between the two units, PN1-PC2 and PN2-PC1 (Li *et al.* 2015) (Fig. 2D). Importantly, exchanging the porter domains of AcrB and AcrD (an AcrB homolog) changes substrate specificity of the corresponding chimeric transporters (Elkins and Nikaido 2002). This result and other related studies firmly establish that the porter domain is the substrate-carrier domain of AcrB and determines the substrate specificity. In addition, the docking domain of an AcrB protomer is composed of two subdomains of similar folding, termed as DN (protruded from PN2) and DC (protruded from PC2), each of which also consists of two β-α-β repeats. Taken together, at the level of folding topology, both the porter domain and the docking

domain of an AcrB protomer possess an internal pseudo twofold symmetry, which is an extension of the twofold symmetry of the TM domain (Fig. 2A).

In the homo-trimer, AcrB molecules form three layers parallel to the membrane plane (Fig. 2): (1) The three TM domains form a loosely packed ring structure located in the inner membrane. (2) The three porter domains form a tightly packed, membrane-proximal layer in the periplasm. A three-helix bundle is formed around the pseudo threefold-symmetry axis, with each protomer contributing one helix (termed gate helix, formerly called pore helix) from its PN1 subdomain. (3) The docking domains interlock with each other using a protruding β-hairpin (known as "peg") from each of the DN and DC subdomains, forming a rigid, threefold symmetrical, membrane distal layer. Folding homology between DN and DC subdomains further renders the layer an approximate sixfold symmetrical architecture, which may facilitate interactions of the docking layer with the homo-hexamer of the adapter AcrA in the assembled complex (Jeong *et al.* 2016). Furthermore, the docking-domain trimer forms the central exit chamber (formerly called funnel), which extends into the cellular exterior via the TolC conduit (Fig. 1).

PERISTALTIC MECHANISM OF ACRB

Based on asymmetric tripartite structures of AcrB reported from multiple laboratories (Murakami *et al.* 2006; Seeger *et al.* 2006; Sennhauser *et al.* 2007), a peristaltic mechanism containing a three-step functional rotation has been proposed. In this functional cycle, each AcrB protomer successively runs through the access (A), binding (B), and extrusion (E) phases (also known as loose, tight, and open states). Furthermore, in a given crystal structure of the asymmetric trimer of AcrB, each protomer assumes one of the three different phases, with a counter-clockwise phase rotation (viewed from the TolC direction). Thus, although containing no physically rotating parts, the periplasmic porter domain cycles through three sequential phases, (1) capturing a lipophilic substrate from the peripheral periplasm–membrane interface, (2) transferring the substrate to the deep binding pocket, and (3) squeezing the substrate through the exit cleft into the central chamber.

In the functional cycle, the subdomains PN1, PN2, PC1, and PC2 change their packing conformations (Eicher et al. 2014). First, the entrance cleft between PC1 and PC2 opens laterally to the periplasm in the access phase, but gets partially closed in the binding phase. It is completely closed in the extrusion phase, with a relative rotation of 25° between PC1 and PC2 during the binding-to-extrusion (B-E) transition. Secondly, the "major" binding site between PN2 and PC1 is of a more extended conformation in the access and binding phases, but contracts in the extrusion phase (with a relative rotation of 21°) essentially eliminating the substrate binding. Thirdly, the exit cleft between PN1 and PN2 is closed in the binding phase, but becomes open to the central chamber in the extrusion phase. This opening is partially caused by an 18° relative rotation between PN1 and PN2 during the B-E transition, and partially by a gating movement of the gate helix from a neighboring promoter that is under the extrusion-to-access (E-A) transition.

This functional cycle requires external energy to drive, which is provided by the PMF-driven motors located in the three TM domains, in a fashion analogous to a three-cylinder car engine. As substrate transport and power generation are two functions physically separated (in the porter and TM domains, respectively), reliable communication and energy coupling between the two corresponding layers are essential for ensuring proper energy-transduction efficiency and thus high transport rate. Questions remaining to be addressed concerning the peristaltic mechanism include "How does substrate binding in the porter domain activate protonation of the TM domain?" and "How does protonation in the TM domain in turn drive the conformational change in the porter domain?" Finding answers to these questions will allow for a better understanding of the mechanistic details of RND transporters.

PROTONATION IN ACRB TRANSPORTER

As pointed out earlier, the protonation site in each Ec-AcrB protomer is composed of residues D407, D408 in TM4, and K940 in TM10. The asymmetric trimer structure of AcrB provides a number of clues as to how this putative protonation site may change its local conformation between distinct phases. The D407–D408 pair is located in the middle of the lipid bilayer and is insulated from bulk solvent. Because of the low dielectric constant of their environment, either of these two acidic residues may potentially be protonated. (See Ref. (Hanz et al. 2016) for an NMR study on the protonation of acidic residues in a TM helix.) On the other hand, by forming salt-bridge bonds, the nearby basic residue K940 may reduce the protonation ability (i.e., pK_a) of the acidic pair. In both the access and binding protomers, the positive charge of the K940 side chain is located between the two negatively charged side chains of D407 and D408, preventing the acid residues from being protonated. The equilibrium conformation of such a deprotonated state (including its relative position to the membrane) will be considered as a reference for conformational changes upon protonation (see below). In contrast, in the extrusion protomer, the lysine side chain moves away from the acidic residues (Fig. 3A) and is stabilized by a H-bond formed with another functionally important residue T978 in TM11 (Takatsuka and Nikaido 2006). Therefore, in the extrusion phase pK_a values of both aspartate residues increase, so that one of them (presumably of higher pK_a) will become protonated (Li et al. 2015). Once one of the two residues is protonated, the pK_a of the other residue will decrease. Thus, it is unlikely that these two acidic residues are protonated simultaneously.

Next, we will discuss the sources and sinks of proton flux. Based on available structural evidence we put forward the following argument: Upon activation of the TM domain, a proton (carrying a positive charge) moves from the periplasmic side of the inner membrane, following the negative-inside electrostatic membrane potential ($\Delta\Psi$), to the proton binding site at the D407–D408 pair. As illustrated in the 1.9 Å-resolution crystal structure of AcrB (PDB ID: 4DX5), the proton-transfer wire, through which protons move, is likely to be formed by side chains of polar residues in TM helices as well as by membrane-embedded water molecules (Eicher et al. 2012). For examples, the highly conserved residues P906, G930, N941, R971, and G1010 (Fig. S1) are either directly involved in the formation of the proton-transfer wires, or facilitate binding of water molecules. Thus, the proton is likely to be transferred from a periplasmic pool to the D407–D408 pair via a mechanism similar to the Newton's cradle. Moreover, once the protomer passes the E-A transition state, the positive charge of the K940 side chain returns to the vicinity of the D470–D408 pair, and deprotonation occurs. The released proton will continue its cross-membrane journey, presumably releasing the rest of its PMF energy. As part of the proton release path, conserved R971 in TM11 is located on the cytosolic side of the D407–D408 pair. In particular, in the access and binding phases D407 is connected to R971 via a cluster of four water molecules, while in the extrusion phase this proton path is blocked by the two side chains of V411 and L944 which are conserved as hydrophobic residues in the AcrB-like family (Fig. S1). Taken together, depending on the protonation status, two distinct constellations of H-bond networks are found inside the TM domain, thus establishing the pathway for proton loading and release.

A similar mechanism of substrate binding-induced protonation has been proposed for some transporters of the major facilitator superfamily (MFS) (Zhao *et al.* 2014). In the bacterial proton-dependent oligopeptide transporters (POTs), which import di- or tri-peptides by utilizing PMF energy, a cluster of conserved polar residues is present in the middle of the transmembrane region nearby the substrate binding site. This cluster typically includes one (or two) acidic residue(s), one (or two) positively charge residue(s), and other uncharged polar residues. For example, in the *E. coli* peptide-H$^+$ symporter, YbgH, residues Q18, E21, Y22, and K118 form such a cluster. Upon binding a substrate peptide, the negatively charged carboxyl terminal group of the substrate attracts the positively charged side chain of K118 away from the clusters, thus promoting protonation of E21. Once E21 becomes protonated, $\Delta\Psi$ exerts an inward force to the transporter, triggering the outward-to-inward conformational change of the symporter. Regarding AcrB-like transporters, D408 is less conserved than D407 (Fig. S1A). Thus, many members from this transporter family have only one acidic residue as the protonation site. As suggested in the above YbgH case, a single acidic residue may function effectively as a protonation site, provided that some electronegative groups are present in its vicinity. In addition, as shown in Fig. S1, the presence of a basic residue in TM10 is not absolutely essential for alternating the protonation status of TM4. Nevertheless, removing a positive charge (*e.g.*, K940 in Ec-AcrB) is likely accompanied by eliminating a negative charge (*e.g.*, D408), and thus the protonation ability of the polar cluster is maintained. In such a case, the pK_a value of the remaining buried acidic residue might be reduced by being approached by an H-bond donor. A similar

deprotonation mechanism has been proposed recently for activation of class-A GPCRs (Zhang *et al.* 2014).

Results from several previous studies support our hypothesis. For instance, the D407-D408-K940 cluster of MexB (an AcrB homolog) has been shown to be very sensitive to point mutations. Any mutation that causes a charge change (*e.g.*, D407N and D408N) results in loss of transport activity (Guan and Nakae 2001; Middlemiss and Poole 2004). In addition, a chemical modification assay using DCCD (dicyclohexylcarbodiimide) as a probe, which identifies protonatible acidic residue(s) in a hydrophobic environment, showed that D408 of Ec-AcrB can be protonated (Seeger *et al.* 2009). Recently, mutations in the proposed proton release path (residues V411, L944, and R971) have been shown to be lethal (Liu and Zhang 2017). These results support the notion that the D407–D408 pair forms part of the proton-transfer wire for the PMF-driven mechanism.

ENERGY COUPLING IN ACRB

For Ec-AcrB, the relative movement K940 in TM10 away from the D407–D408 acidic pair in TM4 seems to represent the trigger for the protonation event occurring within the TM domain. In order to induce such a relative movement between TMs 4 and 10, the substrate binding signal must be transduced from the porter layer to the TM layer. The communication between the two layers is speculated to be long-distance in nature (Eicher *et al.* 2014; Yamaguchi *et al.* 2015), and possible mechanisms of signaling through direct physical transduction are currently ignored. By inspiring available crystal structures of AcrB, however, we proposed an alternative, interdomain contact-mediated mechanism for the signaling. We noticed that a short β-strand located inside an S-shaped loop connecting TM9 and TM10 forms a three H-bond, irregular, parallel β-sheet with another short β-strand in the loop connecting C_{TM} to PC1 (Fig. 3B). We proposed that this β-sheet provides an information path for the substrate binding signal transduced from the porter layer to the TM layer, and therefore we term this sheet PC1-C_{TM} signaling motif. Based on previously reported structures (Murakami *et al.* 2006; Seeger *et al.* 2006; Sennhauser *et al.* 2007), we proposed that this motif allows movement of PC1 upon substrate binding to be sensed by TM10, inducing a $\sim 12°$ rotation of TM10 relative to the remaining part of C_{TM} during the B-E transition. Furthermore, during the B-E transition, a similar though smaller movement is also observed for TM4. In particular, a short β-strand connecting TM3 and TM4 forms three H-bonds with a β-strand inside the N_{TM}-PN1 connecting loop, and we term this parallel β-sheet PN1-N_{TM} signaling motif. Because of the presence of this motif, a movement of PN1 upon substrate binding may be sensed by TM4. Comparing the binding and extrusion phases reveals a ~8° rotation of the N-terminal (periplasmic) half of TM4 relative to the C-terminal (cytosolic) half, resulting in a helix bulge near the conserved G403 (of the GX_3-DX_6E motif) in the extrusion phase. As a consequence of this deformation, the side chain of D407 interacts with the backbone carbonyl oxygen of G403, presumably increasing the pK_a value of D407. Interestingly, both the short β-strands in the N_{TM}-PN1 and C_{TM}-PC1 loops are flanked by two conserved proline residues (Fig. 3B; Fig. S1A). Together, the relative movement (with a net 14° rotation) between TM10 and the C-terminal half of TM4 likely triggers protonation of the D407–D408 pair, using the abovementioned mechanism. Recently, it has been shown that mutations in the signaling motifs indeed reduce or abolish AcrB activity (Liu and Zhang 2017). In addition, although not directly located in the signaling motifs, several random point mutations in the porter–TM interface region were found to reduce the transport activity of MexB (e.g., S462F in the loop connecting TMs 5 and 6, E864K, V928M, and G1002D in the N-terminal (periplasmic) regions of TMs 8, 10, and 12, respectively) (Middlemiss and Poole 2004). More importantly, PC1 is involved in substrate binding in both the access and binding phases. During the B-E transition, PC1 exhibits the largest movement, compared to other subdomains in the porter layer (Seeger et al. 2006). An AcrB inhibitor that binds between PN2 and PC1 in the B-state seems to block this movement of PC1 (Nakashima et al. 2013), thus breaking the communication between the porter and TM layers. Taken together, we proposed that the signaling motifs, especially that of PC1-C_{TM}, are essential for transducing substrate binding signal from the porter layer to TM layer.

Effective proton transfer across the membrane does not necessarily translate into efficient energy usage. For example, slippage in energy coupling might result in futile energy dissipation in theoretical models (Hill 1989). Thus, for an effective AcrB-like transporter, a mechanism of transducing the PMF energy into mechanical rearrangements in the porter layer likely exists. Similar to the case of PMF-driven MSF transporters (Zhang et al. 2015), the protonation at D407–D408 pair is likely to exert an extra inward force to the N_{TM}-subdomain relative to the abovementioned reference state, causing a movement of the N_{TM}-subdomain towards the cytosolic direction. The strength of the electrostatic force is of an order of 5 pN, if the electrostatic field of the membrane potential is uniformly distributed across the 30-Å thickness of the membrane

(Zhang et al. 2015). A force of such a magnitude is sufficiently large to conduct many molecular biology events, such as moving of a motor protein along microtubules, separating of a dsDNA helix, or packaging a DNA molecule into a phage shell (Schnitzer et al. 2000; Cocco et al. 2001; Liu et al. 2014). Furthermore, embedding a proton-transfer wire into the membrane is equivalent to applying a focused membrane potential across the membrane, and thus the effective membrane thickness is reduced. Therefore, the extra electrostatic force upon protonation may be significantly larger than 5 pN. This extra electrostatic force may further be transduced from the TM domain to the porter domain via routes additional to the signaling motif, e.g., the structural connection between TM2 and PN2. Consequently, the corresponding subdomains in the porter domain are likely to move also in an inward direction. Such a movement will result in further conformational rearrangement necessary for the B-E transition, including a tilting of the gate helix in PN1 to block the exit cleft of the neighboring porter domain (Seeger et al. 2006).

As the TM domain moves out of the lipid bilayer, the hydrophobic mismatch between them will dramatically increase. This hydrophobic mismatch generates forces that partially balance the electrostatic force, such that no further out-membrane movement would be possible. Upon deprotonation, the energy stored in the form of a hydrophobic mismatch is released, returning the TM domain to its "reference" status (i.e., the access phase). In addition, the amphipathic helix α6-7, which is strategically located in the peripheral of the C_{TM}-subdomain, keeps C_{TM} aligned with the membrane surface upon movement of the N_{TM}-subdomain towards the cytosol. Thus, α6-7 serves as a pivot point and may facilitate tilting of the TM domain as a whole. Disrupting this amphipathic helix (including the anchoring tyrosine residue) results in loss of the AcrB activity (Liu and Zhang 2017). Similar functional roles of amphipathic helices have been proposed earlier for GPCR activation (Zhang et al. 2014), conformational switching of MFS transporters (Jiang et al. 2013), and sensing of membrane tension in mechanosensitive channels (Zhang et al. 2016). Moreover, the loosely packed TM layer in the AcrB trimer (Fig. 2B) allows the three TM domains tilt independently in response of their individual protonation status. In the AcrB trimer, a differentially phased tilting process of the TM domains is likely to correlate with the packing rearrangement inside the porter layer.

In the high-resolution structures of asymmetric AcrB trimer (e.g., 4DX5), the TMs 1–4 in N_{TM} form a groove facing the lipid bilayer. A detergent molecule was found to bind in this groove, suggesting that this site is a ligand- or inhibitor-binding site that may potentially interfere with the coupling between the TM domain and PN1. In the same crystal structure, a similar ligand binding site was also observed in the symmetric position in C$_{TM}$. Interestingly, in a recently reported crystal structure of human cholesterol transporter, NPC1 (PDB ID: 5I31, a member of the RND superfamily), a sterol-sensing domain (SSD) is formed by helices TMs 3–5 (equivalent to TMs 2–4 in AcrB). Both genetic and structural analyses suggest that this region is important for NPC1 function. The head group of the cholesterol molecule bound to the SSD may interfere with the putative signaling β-sheet between the TM domain and the exo-membrane domain. Therefore, the signaling motifs that we proposed here for Ec-AcrB may indeed be a general feature for RND transporters.

We would like to further emphasize that the putative protonation-induced movement requires the presence of a cross-membrane electrochemical potential of protons, particularly the negative-inside $\Delta\Psi$. Such precondition was not taken into account in most *in vitro* experiments reported so far, neither for AcrB functional nor for structural studies. Even in molecular dynamic simulations, $\Delta\Psi$ was usually not included as part of the environmental condition (Schulz *et al.* 2015). The absence of membrane potential may explain why the observed transport rate of AcrB was slow (approx. one substrate molecule per minute) in a proteoliposome-based transport assay (Nikaido and Takatsuka 2009), in contrast to a much higher efficiency during *in vitro* transport (*e.g.*, ~100/s for substrate ampicillin) (Li *et al.* 2015).

FREE-ENERGY LANDSCAPE

Functional separation of substrate efflux and proton influx in distinct domains is a characteristic feature of AcrB-like RND transporters (Yamaguchi *et al.* 2015). Because of this separation, Jardetzky's classical alternating access model (Jardetzky 1966) appears to be not directly applicable to RND transporters. However, the AcrB-like transporters maintain an important feature of most transporters, *i.e.*, switching back and forth between conformations of high and low affinities towards a given substrate. In this sense, the two structural units of the AcrB porter domain, namely PN1-PC2 and PN2-PC1, may be analogous to the N- and C-transmembrane domain in an MFS transporter. Whereas the relative movement between the two domains in MFS results in an in-membrane conformational change

in agreement to the Jardetzky model, the relative movement between PN1-PC2 and PN2-PC1 results in a conceptually similar conformation alternation, though in the periplasm. A major difference between these two types of transporters is the energy coupling mechanism utilized. For instance, in the case of an MFS antiporter, a direct competition between the substrate and the driving substance, proton or Na$^+$, is postulated (Zhang *et al.* 2015). In contrast, the energy coupling mechanism appears to be more complicated in the case of AcrB-like transporters (Eicher *et al.* 2014). As a tool to dissect the energy coupling mechanism of AcrB, we introduce its free-energy landscape plot, as shown in Fig. 4.

By definition, substrates of AcrB are compounds that are able (1) to bind to the access state and (2) to induce subsequent transitions. These rather general criteria allow AcrB to accommodate a broad substrate specificity, *e.g.*, that observed in multidrug resistance (Yamaguchi *et al.* 2015). As discussed earlier, the second criterion is equivalent to the ability of the substrate to induce a rotation of the PC1 subdomain, thus triggering protonation in the TM domain. However, for a substrate to meet the first requirement, the loading-site dissociation constant $K_{d,L}$ should be smaller than the releasing-site dissociation constant $K_{d,R}$, so that the substrate can be captured from a low-concentration environment and be released into a high-concentration environment. Such a difference in substrate binding affinity between two conformations of a transporter is associated with the free-energy term, previously named differential binding energy, *i.e.*, ΔG_D ($\equiv RT \cdot \ln(K_{d,R}/K_{d,L}) > 0$) (Zhang *et al.* 2015) (see Box 1). Without an external energy input, the transporter would not proceed sustainably to transform from a high-affinity state into a lower-affinity state. In order for the functional cycle of substrate export to proceed, the positive ΔG_D must be compensated by the driving energy, *e.g.*, PMF. In the three-step functional cycle, ΔG_D for the substrate may be further divided into two terms associated with the A-B and B-E transitions (Fig. 4B). Furthermore, a good substrate usually reduces the height of the energy barrier of the transition state(s) (indicated as green horizontal lines in Fig. 4B), but poor substrates or inhibitors lack this ability. Such a reduction can be achieved through stronger binding of the substrate to the transporter at the transition state relative to the 'ground' state. This mechanism of reducing the transition-state barrier is likely to also contribute to the substrate specificity observed for transporters (Zhang and Han 2016).

The three transitions between successive phases are associated with conformational energy changes, termed

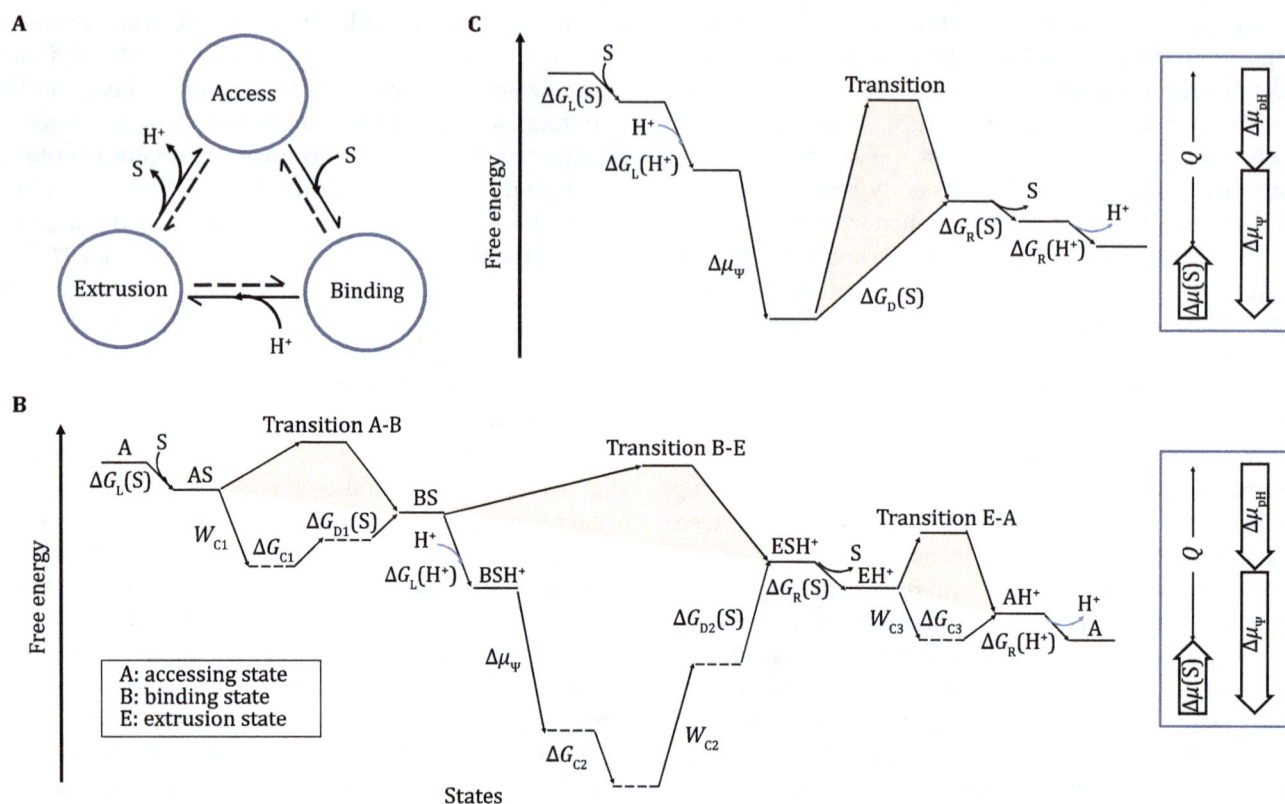

Fig. 4 Free-energy landscape of the AcrB protomer. **A** King–Altman plot of the three-state functional cycle of AcrB. The dominant directions of the reactions are shown in solid half arrows. **B** Energy landscape of the AcrB protomer. The plot must satisfy the First and Second Laws of thermodynamics (right-side Box). The vertical axis represents the Gibbs free energy. Horizontal lines represent states, and the horizontal axis is essentially an expansion of the King–Altman plot. Thus, tilted lines represent transitions between states. Transitions associated with the proton binding are indicated in blue, those with the substrate in red, and those with $\Delta\Psi$ in purple. For simplicity, the electrostatic energy is shown as one package, although it could be separated into two parts associated with proton loading and releasing. In addition, kinetic terms are shown in green. Subscripts "L", "R", and "D" stand for energy terms associated with loading, releasing, and differential binding, respectively. The intrinsic conformational energy terms, $\Delta G_{C1/2/3}$, and energy terms for cooperativity work, $W_{C1/2/3}$, are discussed in the main text. In principle, since the transport process cycles, the choice of the starting point is arbitrary. Therefore, the starting and ending states are considered identical, only differing by the amount of heat (Q) dissipated during one transport cycle. Thus, the end state must be located below the starting state. Neighboring states may be tightly coupled energetically. In such a case, their sequential order would be arbitrary. Locally, any transition of positive G is likely to be driven by a coupled transition of a negative G (also see Box 1). **C** Energy landscape plot of the AcrB trimer. Because of cooperativity between the three protomers, the intrinsic conformational energy terms, $\Delta G_{C1/2/3}$, and energy terms for cooperativity work, $W_{C1/2/3}$, shown in **B** cancel each other out. Thus, the plot is simplified. Only one-third of the functional cycle is shown, and the remaining two-thirds are its precise repeats

ΔG_{C1}, ΔG_{C2}, and ΔG_{C3}, which are intrinsic property of the transporter and are independent of the substrate binding. As a hypothetical example, as shown in Fig. 4B, the B (binding) state is of higher conformational energy than the A (access) state ($\Delta G_{C1} > 0$). The positive ΔG_{C1} suggests that in the absence of external energy, the A-B transition is thermodynamically unfavorable. There are many factors that may contribute to these differential conformational energy terms. For instance, the above-mentioned hydrophobic mismatch may be considered as a contributor of the conformational energy in the extrusion phase, thus affecting both ΔG_{C2} and ΔG_{C3}. Since the promoter must return to its initial state after a complete functional cycle, the sum of these energy

terms associated with the three conformational changes should be zero:

$$\Delta G_{C1} + \Delta G_{C2} + \Delta G_{C3} = 0.$$

Conceptually, if the protomers were isolated from the trimer, such conformational energy terms could be determined by measuring its population distribution in different states at equilibrium. Due to the functional coupling between the protomers, however, these energy terms (ΔG_{C1}, ΔG_{C2}, and ΔG_{C3}) may not be determined experimentally for an AcrB-like transporter.

One characteristic feature of the AcrB trimer is the absolute requirement of functional cooperation between the three protomers (Eicher *et al.* 2014). It has been

Box 1 Definition of the electrochemical potential terms

(i) Free-energy terms of the substrate

$$\Delta\mu(S) \equiv -RT \cdot \ln\left([S]_L/[S]_R\right)$$
$$= \Delta G_L(S) + \Delta G_{D1}(S) + \Delta G_{D2}(S) + \Delta G_R(S)$$
$$\Delta G_L(S) = -RT \cdot \ln\left([S]_L/K_{d1}\right)$$
$$\Delta G_{D1}(S) = -RT \cdot \ln\left(K_{d1}/K_{d2}\right)$$
$$\Delta G_{D2}(S) = -RT \cdot \ln\left(K_{d2}/K_{d3}\right)$$
$$\Delta G_R(S) = -RT \cdot \ln\left(K_{d3}/[S]_R\right)$$

(ii) Chemical potential of protons

$$\Delta\mu_{pH} \equiv 2.3 \cdot RT\Delta pH$$
$$= \Delta G_L(H^+) + \Delta G_R(H^+)$$
$$\Delta G_L(H^+) = 2.3 \cdot RT \cdot (pH_L - pK_a)$$
$$\Delta G_R(H^+) = 2.3 \cdot RT \cdot (pK_a - pH_R)$$

Box 2 Cooperativity

Assumption: The cooperative complex contains N copies of identical protomers, and each of them is in a distinct transition-state (or phase) of the functional cycle

(i) Energy conservation

$$\sum_i^N W_{Ci} = 0$$

where W_{Ci} is the cooperative work that is output by the protomer in the i-th step. A negative W_{Ci} would indicate that the i-th step absorbs energy from the cooperative complex

(ii) Arrhenius theorem

$$k_i e^{-W_{ci}/RT} = k_j e^{-W_{cj}/RT}$$

where k_i is the (pseudo) first-order rate constant of the i-th step in a hypothetical non-cooperative protomer

(iii) Cooperative work

$$W_{Ci} = \frac{-RT}{N} \ln \frac{\prod_i^N k_j}{k_i^N} = RT\ln \frac{k_i}{k_c}$$

$$k_c = \left(\prod_j^N k_j\right)^{\frac{1}{N}}$$

where k_c is the pseudo first-order rate constant of the cooperative complex. For instance, for the AcrB trimer, $k_C = (k_1.k_2.k_3)^{1/3}$

shown that, in a genetically fused AcrB trimer, deactivating one single protomer abolishes the activity of the entire trimer complex (Takatsuka and Nikaido 2009). Furthermore, not all external energy (*i.e.*, $\Delta\mu_{pH} + F\Delta\Psi$, where F is the Faraday constant and $\Delta\mu_{pH}$ is explained in Box 1) is released as one package to drive the B-E transition. Part of the input energy (W_{C2}) may be considered as output work for "cooperativity." In other words, input energy generated by proton influx in one protomer is partially used to allow the other two protomers to overcome their transition barriers as well as thermodynamically unfavorable ΔG_C terms. Like $\Delta G_{C1/2/3}$ terms, the energy terms associated with cooperativity is also in balance:

$$W_{C1} + W_{C2} + W_{C3} = 0.$$

For a hypothetical, isolated, monomeric transporter, relative free-energy levels (and the associated rate constants) of each state together would determine the population probability among different states. This distribution in turn would affect the overall transport rate of the transporter (Hill 1989). The rates of three transitions, A-B, B-E, and E-A, would be different. However, because of the cooperativity within one AcrB trimer, each protomer is present in one of the three major states. Therefore, the population probability of each major state is virtually 1/3. This imposes a strong restriction both on the free-energy landscape and on the kinetics of the transitions between successive states (see Box 2). By automatically adjusting energy terms of cooperative work (W_{C1}, W_{C2}, and W_{C3}), rates at the three transitions are maintained to be the same. For instance, if the substrate concentration at the loading site increases, the pseudo first-order rate constant ($[S]_L k_{on}^0$) will increase (Hill 1989). This increased rate constant would shift the population probability from the empty access state (A) to the substrate-bound access state (AS)

and would make the A-B transition run faster. Thus, the external energy required for this transition decreases (*i.e.*, W_{C1} becomes less negative) in order to maintain all three transitions at the same rate. In contrast, in the absence of any substrate at the loading site, the A-B transition as well as the functional cycle stops. In short, the input PMF energy will be distributed dynamically among the three cooperative promoters of the transporter trimer, depending on the types and concentrations of substrate molecules associated with each protomer. For a given type of substrate at a fixed concentration gradient, the cooperativity works W_{C1}, W_{C2}, and W_{C3} from all three protomers cancel each other out at any given moment. Furthermore, according to the Arrhenius theorem, the pseudo first-order rate constant of a cooperative transporter complex will change by a factor of $\exp(-W_{Ci}/RT)$ (where $i = 1, 2, 3$), relative to the hypothetical rate of an isolated monomeric transporter (Box 2). In addition, ΔG_{C1}, ΔG_{C2}, and ΔG_{C3} from the three protomers balance each other, albeit they are intrinsic properties of the transporter. Thus, the energy landscape plot of a cooperative AcrB trimer can be presented in a simplified manner, as shown in Fig. 4C. A key feature of this simplified plot is that PMF, including energy terms associated with both ΔpH and $\Delta\Psi$, remains to be utilized to compensate the positive $\Delta G_D(S)$ in thermodynamic terms and to overcome the transition state energy barrier in kinetic terms.

PERSPECTIVE

The United Nations General Assembly recently unanimously approved a declaration committing nations to a more active battle against threat of microbial resistance to existing drugs, which is one of the most severe and rapidly growing threats to contemporary public health. Massive intergovernmental efforts will be required to achieve such an ambitious undertaking. Since bacteria develop drug resistance naturally in an environment containing antibiotics, studies on the mechanism of multidrug resistance, including those of AcrB-like transporters, are an essential part of our fight against multidrug resistance (Li *et al.* 2015). Based on the progress in structural and functional studies on AcrB-like transporters in the last 15 years, a new energy coupling mechanism is proposed in the current review for this major type of multidrug-resistance transporters. If verified, this mechanism is expected to permit us to de-couple between their substrate binding and energy converting. Experimental approaches stemmed from this mechanistic hypothesis may thus open new avenues for developing novel inhibitors to attack the multidrug resistance mediated by AcrB-like transporters.

Abbreviations

PMF   Proton motive force
RND   Resistance-nodulation-cell division (transporter)
TM   Transmembrane (helix)

Acknowledgement We thank Dr. Torsten Juelich for linguistic assistance during the preparation of this manuscript. This work was supported by the Ministry of Science and Technology (China) (2015CB910104 to XCZ) and National Natural Science Foundation of China (31470745 to XCZ).

Compliance with Ethical Standards

Conflict of interest Xuejun C. Zhang, Min Liu, and Lei Han declare that they have no conflict of interest.

Human and animal rights and informed consent This article does not contain any studies with human or animal subjects performed by any of the authors.

References

Berman HM, Westbrook J, Feng Z, Gilliland G, Bhat TN, Weissig H, Shindyalov IN, Bourne PE (2000) The protein data bank. Nucleic Acids Res 28:235–242

Cocco S, Monasson R, Marko JF (2001) Force and kinetic barriers to unzipping of the DNA double helix. Proc Natl Acad Sci USA 98:8608–8613

Du D, Wang Z, James NR, Voss JE, Klimont E, Ohene-Agyei T, Venter H, Chiu W, Luisi BF (2014) Structure of the AcrAB-TolC multidrug efflux pump. Nature 509:512–515

Eicher T, Cha HJ, Seeger MA, Brandstatter L, El-Delik J, Bohnert JA, Kern WV, Verrey F, Grutter MG, Diederichs K, Pos KM (2012) Transport of drugs by the multidrug transporter AcrB involves an access and a deep binding pocket that are separated by a switch-loop. Proc Natl Acad Sci USA 109:5687–5692

Eicher T, Seeger MA, Anselmi C, Zhou W, Brandstatter L, Verrey F, Diederichs K, Faraldo-Gomez JD, Pos KM (2014) Coupling of remote alternating-access transport mechanisms for protons and substrates in the multidrug efflux pump AcrB. Elife. doi:10.7554/eLife.03145

Elkins CA, Nikaido H (2002) Substrate specificity of the RND-type multidrug efflux pumps AcrB and AcrD of Escherichia coli is determined predominantly by two large periplasmic loops. J Bacteriol 184:6490–6498

Gong X, Qian H, Zhou X, Wu J, Wan T, Cao P, Huang W, Zhao X, Wang X, Wang P, Shi Y, Gao GF, Zhou Q, Yan N (2016) Structural insights into the Niemann-Pick C1 (NPC1)-mediated cholesterol transfer and ebola infection. Cell 165:1467–1478

Guan L, Nakae T (2001) Identification of essential charged residues in transmembrane segments of the multidrug transporter MexB of *Pseudomonas aeruginosa*. J Bacteriol 183:1734–1739

Hanz SZ, Shu NS, Qian J, Christman N, Kranz P, An M, Grewer C, Qiang W (2016) Protonation-driven membrane insertion of a pH-low insertion peptide. Angew Chem Int Ed Engl 55:12376–12381

Higgins CF (2007) Multiple molecular mechanisms for multidrug resistance transporters. Nature 446:749–757

Hill TL (1989) Free energy transduction and biochemical cycle kinetics. Springer, New York

Jardetzky O (1966) Simple allosteric model for membrane pumps. Nature 211:969–970

Jeong H, Kim JS, Song S, Shigematsu H, Yokoyama T, Hyun J, Ha NC (2016) Pseudoatomic structure of the tripartite multidrug efflux pump AcrAB-TolC reveals the intermeshing cogwheel-like interaction between AcrA and TolC. Structure 24:272–276

Jiang D, Zhao Y, Wang X, Fan J, Heng J, Liu X, Feng W, Kang X, Huang B, Liu J, Zhang XC (2013) Structure of the YajR transporter suggests a transport mechanism based on the conserved motif A. Proc Natl Acad Sci USA 110:14664–14669

Li XZ, Plesiat P, Nikaido H (2015) The challenge of efflux-mediated antibiotic resistance in Gram-negative bacteria. Clin Microbiol Rev 28:337–418

Liu M, Zhang XC (2017) Energy-coupling mechanism of the multidrug resistance transporter AcrB: evidence for membrane potential-driving hypothesis through mutagenic analysis. Protein Cell 8(8):623–627

Liu S, Chistol G, Hetherington CL, Tafoya S, Aathavan K, Schnitzbauer J, Grimes S, Jardine PJ, Bustamante C (2014) A viral packaging motor varies its DNA rotation and step size to preserve subunit coordination as the capsid fills. Cell 157:702–713

Middlemiss JK, Poole K (2004) Differential impact of MexB mutations on substrate selectivity of the MexAB-OprM multidrug efflux pump of *Pseudomonas aeruginosa*. J Bacteriol 186:1258–1269

Murakami S, Nakashima R, Yamashita E, Yamaguchi A (2002) Crystal structure of bacterial multidrug efflux transporter AcrB. Nature 419:587–593

Murakami S, Nakashima R, Yamashita E, Matsumoto T, Yamaguchi A (2006) Crystal structures of a multidrug transporter reveal a functionally rotating mechanism. Nature 443:173–179

Nakashima R, Sakurai K, Yamasaki S, Hayashi K, Nagata C, Hoshino K, Onodera Y, Nishino K, Yamaguchi A (2013) Structural basis for the inhibition of bacterial multidrug exporters. Nature 500:102–106

Nikaido H, Takatsuka Y (2009) Mechanisms of RND multidrug efflux pumps. Biochim Biophys Acta 1794:769–781

Perrin E, Fondi M, Papaleo MC, Maida I, Buroni S, Pasca MR, Riccardi G, Fani R (2010) Exploring the HME and HAE1 efflux systems in the genus Burkholderia. BMC Evol Biol 10:164

Schnitzer MJ, Visscher K, Block SM (2000) Force production by single kinesin motors. Nat Cell Biol 2:718–723

Schulz R, Vargiu AV, Ruggerone P, Kleinekathofer U (2015) Computational study of correlated domain motions in the AcrB efflux transporter. Biomed Res Int 2015:487298

Seeger MA, Schiefner A, Eicher T, Verrey F, Diederichs K, Pos KM (2006) Structural asymmetry of AcrB trimer suggests a peristaltic pump mechanism. Science 313:1295–1298

Seeger MA, von Ballmoos C, Verrey F, Pos KM (2009) Crucial role of Asp408 in the proton translocation pathway of multidrug transporter AcrB: evidence from site-directed mutagenesis and carbodiimide labeling. Biochemistry 48:5801–5812

Sennhauser G, Amstutz P, Briand C, Storchenegger O, Grutter MG (2007) Drug export pathway of multidrug exporter AcrB revealed by DARPin inhibitors. PLoS Biol 5:e7

Sennhauser G, Bukowska MA, Briand C, Grutter MG (2009) Crystal structure of the multidrug exporter MexB from *Pseudomonas aeruginosa*. J Mol Biol 389:134–145

Takatsuka Y, Nikaido H (2006) Threonine-978 in the transmembrane segment of the multidrug efflux pump AcrB of Escherichia coli is crucial for drug transport as a probable component of the proton relay network. J Bacteriol 188:7284–7289

Takatsuka Y, Nikaido H (2009) Covalently linked trimer of the AcrB multidrug efflux pump provides support for the functional rotating mechanism. J Bacteriol 191:1729–1737

Yamaguchi A, Nakashima R, Sakurai K (2015) Structural basis of RND-type multidrug exporters. Front Microbiol 6:327

Zhang XC, Han L (2016) Uniporter substrate binding and transport: reformulating mechanistic questions. Biophys Rep 2:45–54

Zhang XC, Cao C, Zhou Y, Zhao Y (2014) Proton transfer-mediated GPCR activation. Protein Cell 6:12–17

Zhang XC, Zhao Y, Heng J, Jiang D (2015) Energy coupling mechanisms of MFS transporters. Protein Sci 24:1560–1579

Zhang XC, Liu Z, Li J (2016) From membrane tension to channel gating: a principal energy transfer mechanism for mechanosensitive channels. Protein Sci. doi:10.1002/pro.3017

Zhao Y, Mao G, Liu M, Zhang L, Wang X, Zhang XC (2014) Crystal structure of the *E. coli* peptide transporter YbgH. Structure 22:1152–1160

17

Fabrication and modification of implantable optrode arrays for *in vivo* optogenetic applications

Lulu Wang[1,2], Kang Huang[1,2], Cheng Zhong[1], Liping Wang[1✉], Yi Lu[1✉]

[1] Shenzhen Key Lab of Neuropsychiatric Modulation and Collaborative Innovation Center for Brain Science, CAS Center for Excellence in Brain Science and Intelligence Technology, The Brain Cognition and Brain Disease Institute, Shenzhen Institutes of Advanced Technology, Chinese Academy of Sciences, Shenzhen 518055, China
[2] Shenzhen College of Advanced Technology, University of Chinese Academy of Sciences, Shenzhen 518055, China

Graphical Abstract

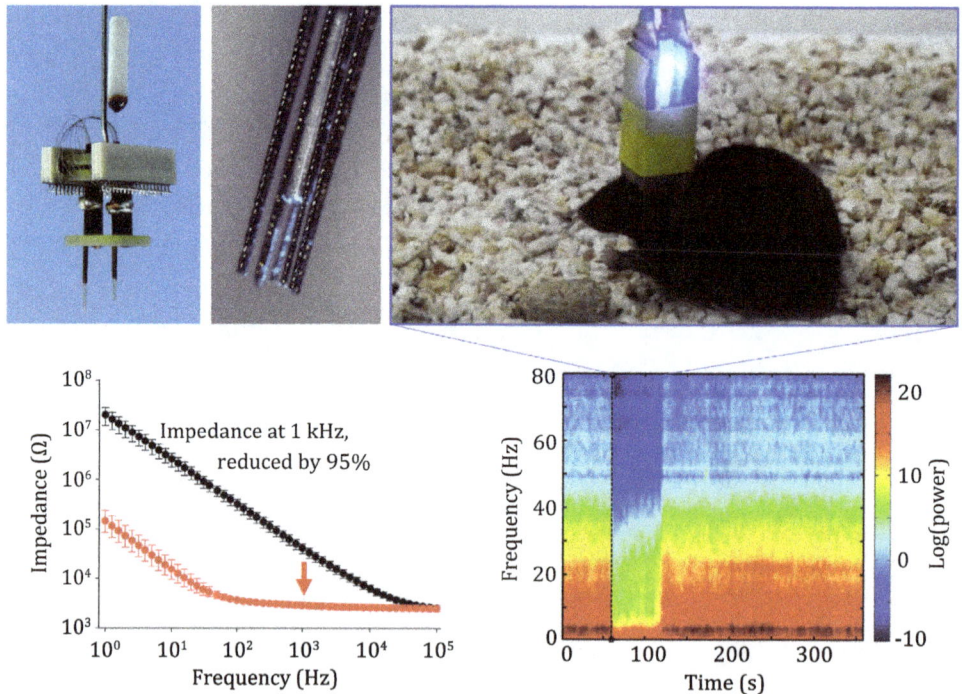

Abstract   Recent advances in optogenetics have established a precisely timed and cell-specific methodology for understanding the functions of brain circuits and the mechanisms underlying neuropsychiatric disorders. However, the fabrication of optrodes, a key functional element in optogenetics, remains a great challenge. Here, we report reliable and efficient fabrication strategies for chronically

Lulu Wang and Kang Huang have contributed equally to this work.

✉ Correspondence: lp.wang@siat.ac.cn (L. Wang),
luyi@siat.ac.cn (Y. Lu)

implantable optrode arrays. To improve the performance of the fabricated optrode arrays, surfaces of the recording sites were modified using optimized electrochemical processes. We have also demonstrated the feasibility of using the fabricated optrode arrays to detect seizures in multiple brain regions and inhibit ictal propagation *in vivo*. Furthermore, the results of the histology study imply that the electrodeposition of composite conducting polymers notably alleviated the inflammatory response and improved neuronal survival at the implant/neural-tissue interface. In summary, we provide reliable and efficient strategies for the fabrication and modification of customized optrode arrays that can fulfill the requirements of *in vivo* optogenetic applications.

Keywords Optogenetics, Optrode, Neural electrode, Electrodeposition, Optical stimulation, Electrophysiological recording

INTRODUCTION

Optogenetics is a technology that combines optical control and genetic targeting using cell-type-specific and optically sensitive proteins for the precise manipulation of neuronal functions with millisecond precision (Yizhar *et al.* 2011a; Zhang *et al.* 2007). A major advantage of optogenetic methodology is that it enables the excitation or inhibition of specific neuron subtypes and the electrophysiological recording of neuronal activity simultaneously without causing large stimulus artifacts that may overlap the recording results (Anikeeva *et al.* 2012). Benefiting from these advantages, optogenetics has been widely applied to investigate the mechanisms of neuropsychiatric diseases as well as the functions of brain networks through precisely timed control of specific neuron groups in a neural circuit (Brown 2012; Gradinaru 2009; Kravitz *et al.* 2010; Li 2015; Lu *et al.* 2016; Otis *et al.* 2017; Schmitt *et al.* 2017; Tovote *et al.* 2016; Tye *et al.* 2013; Yizhar *et al.* 2011b; Zimmerman *et al.* 2016).

In order to investigate complex brain processing mechanisms at a functional level, researchers need to use optogenetics to detect the activity from specific neural circuits during the manipulation of target neuron populations. In many cases, chronic studies in freely moving animals are necessary for dissecting the characteristics of a particular neural circuit control animal behavior and understanding the intrinsic neural basis of a specific behavioral phenomenon. Therefore, chronically implantable optrode arrays that integrate optical stimulation with large-scale electrophysiological readout methods are greatly desired for *in vivo* applications (Anikeeva *et al.* 2012; Lu *et al.* 2012). Recent advances in Micro-Electro-Mechanical Systems (MEMS) have shed new light on the development of optrodes (Chen *et al.* 2013; Dehkhoda *et al.* 2018; Iseri and Kuzum 2017; Kwon *et al.* 2015). However, a reliable and efficient fabrication technology for customized chronically

implantable optrodes still needs to be developed to meet the requirements of different experimental purposes.

Optrode arrays are typically composed of an optical waveguide for light delivery and multiple electrode channels for recording. The electrode is recognized to be a crucial readout element in optrodes, which determines the signal-recording quality required for neural circuit dissection (Anikeeva *et al.* 2012; Lu *et al.* 2012). However, achieving effective and stable long-term electrophysiological recording, with the electrode *in vivo*, continues to be a great challenge. Therefore, another desirable research goal related to optrodes is to develop novel materials for biocompatible and low-impedance electrode/neural-tissue interfaces.

With the development of microfabrication technologies, high-density microelectrode arrays have been widely used to increase spatial precision and minimize insertion trauma. However, the impedance of the electrode increases drastically with a decrease in its size, which subsequently results in high thermal-background noise and a low signal-to-noise ratio during recording. To solve this problem, electrochemical deposition of high double-layer capacitance or Faradaic-capacitance materials, such as platinum particles, titanium nitride, and iridium oxide, has been frequently used to reduce the electrode impedance (Cogan 2008; Kotov *et al.* 2009; Lu *et al.* 2009b). Another major bottleneck that unfortunately hinders the applications of neural probes is the inconsistent performance caused by elicited inflammatory response (Grill *et al.* 2009; Kotov *et al.* 2009; Schwartz *et al.* 2006). The inflammatory response usually results in a dense astroglial encapsulation of the implant, which isolates the neural probe from the surrounding neurons, and the response also leads to a loss of neurons adjacent to the electrode/neural-tissue interface, which further deteriorates the performance of the electrode in chronic studies. A commonly applied strategy for improving both the electrochemical performance and biocompatibility of neural electrodes is surface modification with conducting polymers (CPs). Previous studies have reported that CPs such as polypyrrole (PPy) and poly(3,4-ethylenedioxythiophene) (PEDOT) can notably increase the electrode capacitance, reduce surface impedance, and alleviate inflammatory responses (Abidian and Martin 2008; George *et al.* 2005; Lu *et al.* 2010; Zhong *et al.* 2017). However, the electrodeposition process and characteristics of these materials have not been adequately investigated with respect to the requirements of optogenetics.

In this work, we report on the fabrication of two types of optrode arrays for *in vivo* optogenetic applications, including a microwire optrode array for investigating single brain regions and a drivable optrode array for simultaneously targeting multiple brain structures. In order to improve the performance of the fabricated optrodes during electrophysiological recording, we also demonstrate three different strategies for decreasing the electrode impedance: a simplified cathodic electrochemical process for the deposition of platinum particles, a cyclic voltammetric process for depositing iridium oxide, and a multi-step modification process for composite CP deposition. The electrochemical characteristics and surface morphology of the deposited optrodes were determined by performing cyclic voltammetry (CV), electrochemical impedance spectroscopy (EIS), and scanning electron microscopy (SEM) analyses. Furthermore, we examined the feasibility of using customized optrode arrays to investigate the neuronal activity during the ictal period and inhibit seizure propagation in multiple brain regions *in vivo*. Finally, after chronic implantation in rat brains, the astrocyte intensity and neuronal survival around the modified implant sites were assessed by analyzing the immunoreactivity of glial fibrillary acidic protein (GFAP) and neuronal nuclei (NeuN), respectively. All of these results were evaluated and discussed with respect to the requirements of optogenetic applications.

EXPERIMENTAL SECTION

Animals

All experiments were conducted in accordance with the protocols approved by the Ethics Committee for Animal Research, Shenzhen Institute of Advanced Technology, Chinese Academy of Sciences. Male Sprague–Dawley (SD) rats weighting 240–280 g were used for optrode implantation and histological study. Adult male VGAT-ChR2(H134R)-EYFP bacterial artificial chromosome (BAC) transgenic mice were used for optogenetic stimulation and electrophysiological recordings. Animals were housed in controlled conditions (ambient

temperature 24 ± 1 °C, humidity 50%–60%, 12-h light/dark cycle) with food and water given *ad libitum*.

Fabrication of optogenetic probes

Fabrication of microwire optrode arrays

Microwire optrode arrays were fabricated from optical fibers (diameter 200 μm, NA 0.37, Thorlabs, USA) and formvar-coated nickel–chromium (diameter 35.6 μm, Stablohm 650, California Fine Wire, USA) or platinum–iridium (diameter 35.6 μm, Pt/Ir 90/10, California Fine Wire, USA) microwires using a custom-made optrode mold (Fig. 1A). Eight microwires in an optrode array were arranged in two parallel rows for electrophysiological recordings, with each row containing four wires. The spacing between neighboring microwires was 180 μm. Approximately 2 mm of the formvar coating was removed from one end of each microwire by brief exposure to a flame, and each microwire was soldered into separate slots of a standard electrode connector (Omnetics, USA). Two pairs of uninsulated silver microwires (diameter 100 μm) were soldered into the electrode connector as the reference and ground electrodes, respectively. An optical fiber was arranged to deliver light using the optrode mold. One end of the optical fiber was fixed to an optical ceramic ferrule (inner diameter = 220 μm) and stabilized onto the electrode connector using epoxy resin (Fig. 1B). The tip of the optrode was coated with polyethylene glycol (PEG, M_W = 2000 g/mol, Sigma–Aldrich, USA) to enhance its mechanical strength (Fig. 1C). Prior to implantation, phosphate-buffered saline (PBS) was applied to dissolve and remove the PEG coatings. In order to ensure illumination of the neurons to be recorded, the tips of the microwires were placed approximately 500 μm deeper than the optical fiber (Fig. 1D). This microwire optrode array is easy to fabricate and is suitable for targeting single regions in relatively shallow brain structures (Fig. 1E).

Fabrication of multisite drivable optrode arrays

The drivable optrode arrays consisted of screw-based microdrive scaffolds (Fig. 2A) and optrode bundles (Fig. 2B). Each optrode bundle, which contains one stimulation channel and eight twisted tetrodes (32 electrophysiological recording channels in total), was fabricated from optical fibers (diameter 200 μm, NA 0.37, Thorlabs, USA) and formvar-coated nickel–chromium (diameter 12.7 or 17.8 μm, Stablohm 650, California Fine Wire, USA) or platinum–iridium (diameter 12.7 or 17.8 μm, Pt/Ir 80/20, California Fine Wire, USA)

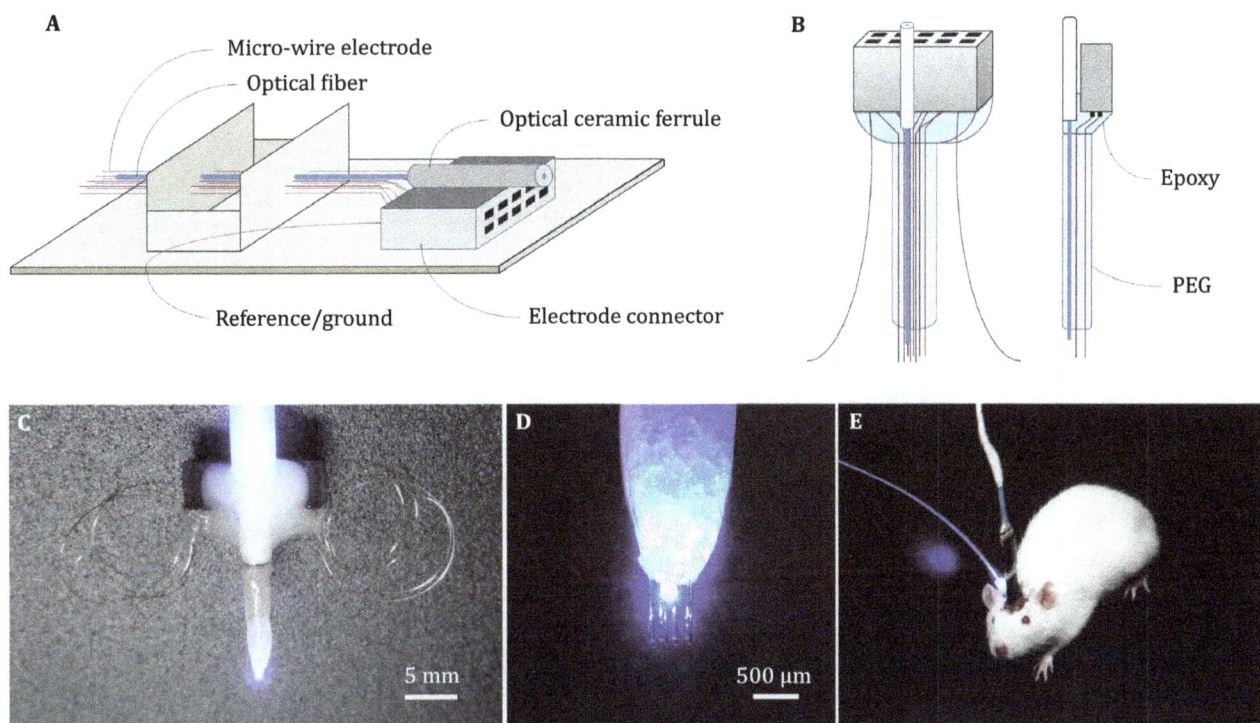

Fig. 1 Fabrication of microwire optrode arrays. **A** Fabrication process for an eight-channel microwire optrode array using a custom-designed mold. **B, C** A schematic diagram (**B**) and photo (**C**) of an eight-channel microwire optrode array. **D** Tip of the microwire optrode array (enwrapped by PEG). **E** Optogenetic stimulation and electrophysiological recording using a microwire optrode array implanted in a free-moving rat

microwires. Each tetrode was threaded through a silica tube (inner diameter 100 μm, outer diameter 164 μm, Polymicro, USA). In an optrode bundle, eight tetrodes were arranged around an optical fiber, and the center-to-center spacing between each tetrode and the optical fiber was approximately 200 μm (Fig. 2C). Then, the optrode bundles were glued onto the movable screw nut on the microdrive scaffolds. Additional optrode bundles can be integrated with the microdrive system to construct a multisite drivable optrode array (Fig. 2D–F).

Electrodeposition

Electrodeposition of Pt

All electrochemical experiments were performed using a potentiostat (Gamry Reference 600, USA). The electrodes were mounted in a three-electrode cell with a saturated calomel electrode (SCE) as the reference electrode and a large-area platinum electrode as the counter electrode. The electrodeposition solution was prepared by dissolving $PtCl_4$ in 100 mmol/L HCl solution to a concentration of 5 mmol/L. The prepared solution can be used for at least six months after preservation at 4 °C. Platinum

nanoparticles were deposited on the recording sites of the optrode in the deposition solution via a simplified cathodic reaction at a potential of −0.25 V (vs. SCE) applied by a Gamry potentiostat. The electrodeposition process can be described by the following equation:

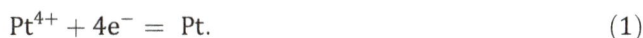

$$Pt^{4+} + 4e^- = Pt. \tag{1}$$

Electrodeposition of iridium oxide (IrO₂)

The electrodeposition solution was prepared as described in previous studies (Lu *et al.* 2009b; Zhong *et al.* 2014). Briefly, oxalic acid ($H_2C_2O_4$) was slowly dissolved to a concentration of 100 mmol/L in a 3 mmol/L hydrogen hexachloroiridate (H_2IrCl_6) aqueous solution, and H_2O_2 (2% *v/v*) was added to form a stable complex of iridium oxide. Then, the pH of the solution was adjusted to 10.5 by gradually adding K_2CO_3. The resulting solution was allowed to age for one week at room temperature until it turned dark blue. The deposition solution can be used for more than two months after preservation at 4 °C. Thin IrO_2 films were formed on the recording sites in the stabilized electrodepositing solution by cyclic voltammetry scanning between 0.05 and 0.55 V (vs. SCE) at 100 mV/s with a Gamry potentiostat

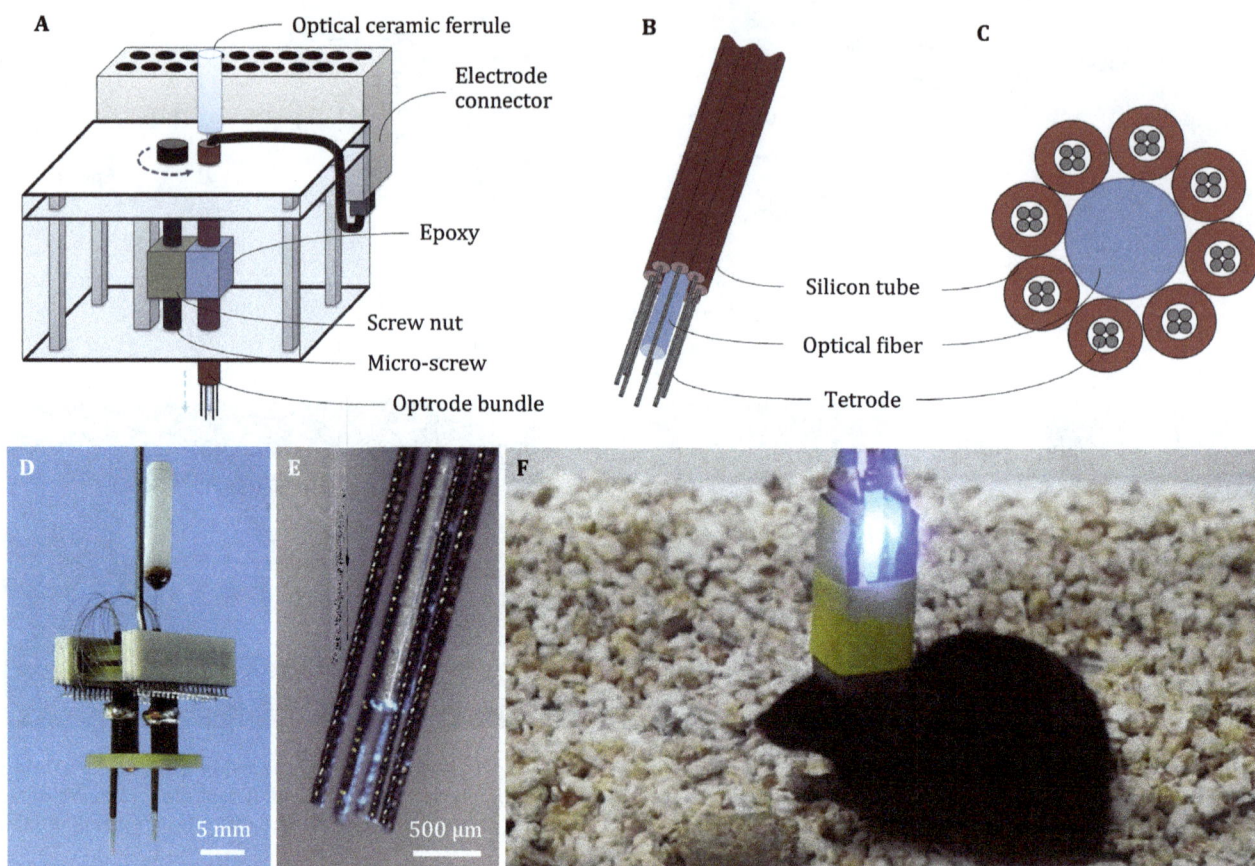

Fig. 2 Design of multisite drivable optrode arrays. **A** Schematic diagram of a 32-channel drivable optrode array. **B**, **C** Detailed schematic diagram (**B**) and top view (**C**) of the optrode tip. **D**, **E** Photos of a 64-channel multisite drivable optrode array (**D**) and the optrode tip (**E**). **F** Optogenetic stimulation and electrophysiological recording using a multisite drivable optrode array implanted in a freely moving mouse

(Fig. 3). The electrodeposition process can be described by the following equation (Lu *et al.* 2009b):

$$\left[\mathrm{Ir(OH)_4C_2O_4}\right]^{2-} - 2e^- = \mathrm{IrO_2} + 2\mathrm{CO_2} + 2\mathrm{H_2O}. \qquad (2)$$

Electrodeposition of composite conducting polymers (CPs)

Poly(vinyl alcohol)/poly(acrylic acid) interpenetrating polymer networks (PVA/PAA IPNs) were synthesized following the approach described in previous studies (Lu *et al.* 2012; Zhong *et al.* 2017). Briefly, an aqueous PVA solution (2 wt%, M_W 89,000–98,000 g/mol, Sigma–Aldrich, USA) was placed in a round-bottom flask, and acrylic acid (AA) monomer was added under magnetic stirring to a ratio of 60 mol% (moles of AA monomer per mole of PVA repeat unit). Ammonium persulfate (Degussa-AJ, China) was added to the flask to a concentration of 1000 ppm as an initiator. Then, the mixed solution was purged with Ar to remove O_2, and the flask was sealed and immersed in an oil bath at 80 °C. The reaction was allowed to take place for 72 h, and the

resulting polymer solution was filtered to remove undissolved solids and bubbles prior to use. The samples were dipped into the homogeneous polymer solution and dried at 60 °C for 2 h to form a thin hydrogel film on the surface. The aqueous electrodeposition solution was prepared by dissolving 10 mmol/L

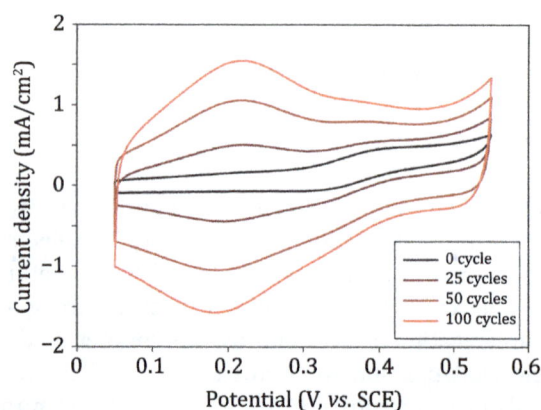

Fig. 3 The deposition of IrO_2 under repetitive potential cycling between 0.05 and 0.55 V (vs. SCE) in an iridium complex ($[\mathrm{Ir(OH)_4C_2O_4}]^{2-}$) solution

3,4-ethylenedioxythiophene monomer (EDOT, Sigma–Aldrich, USA), 250 mmol/L poly(sodium 4-styrenesulfonate) (PSSNa, M_W 70,000 g/mol, Sigma–Aldrich, USA), 10 µg/mL nerve growth factor (NGF, Sigma–Aldrich, USA), and 10 mmol/L dexamethasone sodium phosphate (DEX, NIFDC, China). The coated substrates were then immersed in the electrodeposition solution for at least 1 h prior to use. PEDOT/PSS/NGF/DEX films were electrodeposited onto the hydrogel-coated substrates using a Gamry potentiostat at 0.9 V (vs. SCE), forming the composite CP films.

Physicochemical characterizations

Cyclic voltammetry (CV) measurements were performed within the safe potential window in a testing solution at a scan rate of 50 mV/s using a Gamry potentiostat. Electrochemical impedance spectra (EIS) of the recording sites were measured at their open circuit potentials in artificial cerebrospinal fluid (ACSF) using a Gamry potentiostat with a 25-mV (rms) AC sinusoid signal in the frequency range 100 kHz to 1 Hz. The morphologies of the electrodeposited surfaces were examined using an extremely high-resolution scanning electron microscope (XHRSEM, FEI Magellan 400L, USA) operated at 5 kV. In order to facilitate the observation of the modified surfaces, the recording sites of the optrodes were coated with epoxy. The samples were sputtered with a layer of iridium prior to use.

Characterization of Pt-deposited electrodes

After Pt electrodeposition, the redox currents in the CV curves were significantly increased (Fig. 4A), suggesting an enlarged double-layer capacitance. The electrochemical impedance at 1 kHz is a critical characteristic parameter for neural electrodes, as this frequency is relevant to the electrical activity of neurons. The average impedance value at 1 kHz was decreased from 1.52 MΩ to 11.96 kΩ, ~99% lower than the unmodified electrode (Fig. 4B), possibly due to the increase in pseudo-capacitance after modification. The surface of the Pt-deposited layer contains numerous nanoparticles that aggregated into sub-micrometer structures. This rough surface significantly increased the effective surface area at the electrode/electrolyte interface and, as a consequence, decreased the electrochemical impedance of the electrode (Fig. 4C).

Characterization of IrO₂-deposited electrodes

No significant current peak was observed in the background CV curve of the undeposited electrode. After cyclic voltammetric scanning in the electrodeposition solution for 100 cycles, two pairs of peaks correlating to redox reactions of iridium oxides appeared at 100 and 500 mV, respectively (Fig. 5A). Besides, the electrode impedance at 1 kHz decreased from 2.71 MΩ to 148 kΩ, ~95% lower than that of the unmodified electrode (Fig. 5B). The surface of the electrode substrate presented a homogeneous and flat morphology (Fig. 5C, D). By comparison, the deposited IrO₂ film was composed of numerous 20–50 nm particles, and the film exhibited a more porous structure and increased roughness (Fig. 5E, F), which is beneficial for the improvement of the electrochemical performance.

Characterization of composite CP-deposited electrodes

The CV measurement of the composite CP-modified electrode shows the emergence of an anodic peak and a cathodic peak at −300 and −400 mV (vs. SCE), respectively (Fig. 5A). In the high-frequency range (10^2–10^5 Hz), the impedance of the composite CP-modified electrode was almost independent of frequency. The average impedance value at 1 kHz was ~95% lower than the impedance of the unmodified electrode (Fig. 5B). The drastic increase in charge-storage capacity and reduction in impedance after modification is possibly due to the increase of the pseudo-capacitance and effective surface area (Fig. 5C) at the electrode/electrolyte interface.

Optogenetic study

Seizure detection and inhibition

VGAT-ChR2(H134R)-EYFP BAC transgenic mice were anesthetized with urethane (1.5 g/kg) and fixed in a standard stereotaxic frame. Holes were drilled through the skull, and the multisite optrodes were implanted. Optrode bundles were directed toward the dentate gyrus/hilus (DGH, stereotaxis atlas coordinates: AP −2.06 mm, ML +1.35 mm, DV +2.10 mm) and primary motor cortex (M1, AP +1.80 mm, ML +1.80 mm, DV +1.80 mm), respectively.

Kainic acid (KA) was dissolved in PBS at 0.30 mg/mL, and then 550 nl of the KA solution was unilaterally injected into the dorsal hippocampus using a micropump to induce an acute status epilepticus. Electrophysiological recordings were performed using a 64-channel neural acquisition processor (Plexon, USA). After the onset of seizures, 473 nm blue light pulses (10 mW/mm², 5 ms pulses at 130 Hz, 1 min on and 5 min off) were directed into the implanted optrode array for optogenetic stimulation.

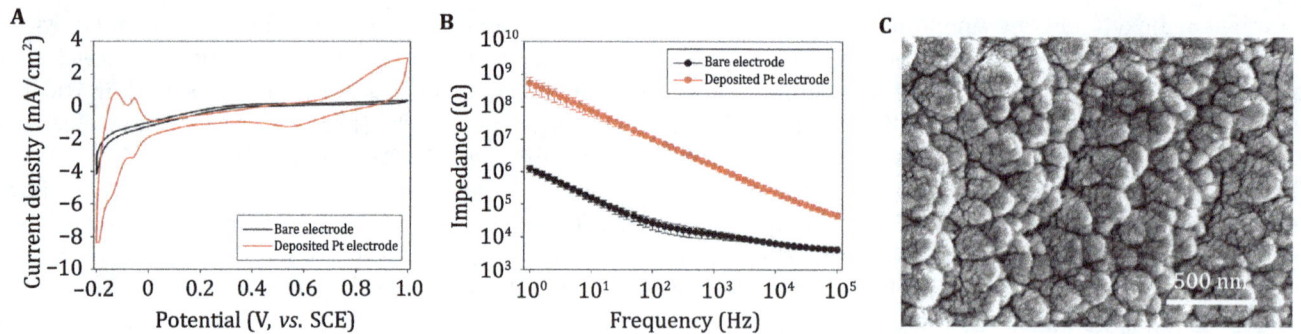

Fig. 4 Characterization of platinum particles. **A** Cyclic voltammogram of an optrode before (*black*) and after platinum deposition (*red*) at a sweep rate of 50 mV/s in 0.1 mol/L HCl. **B** Bode plot of electrochemical impedance spectra of an optrode before (*black*, $n = 4$) and after platinum deposition (*red*, $n = 4$) in ACSF, data are shown as mean $\pm$ SD. **C** SEM image of the optrode tip after platinum deposition

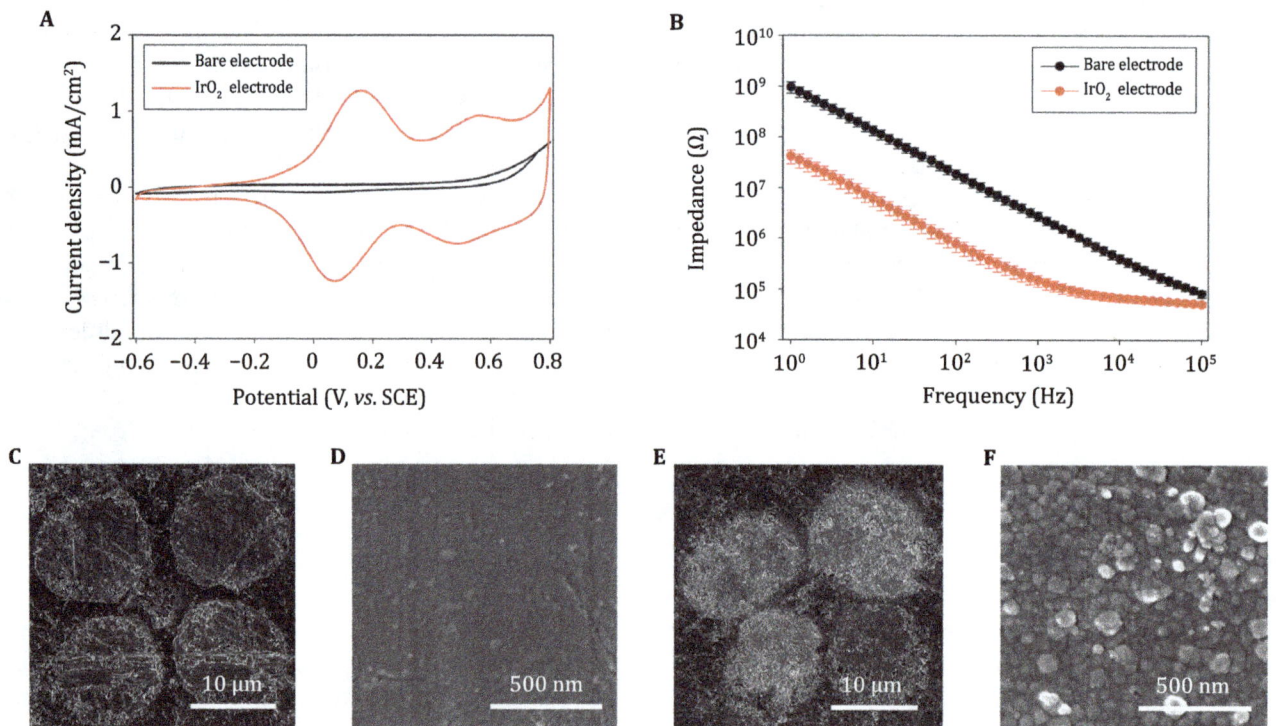

Fig. 5 Characterization of IrO$_2$. **A** Cyclic voltammogram of an optrode before (*black*) and after IrO$_2$ deposition (*red*) at a sweep rate of 50 mV/s in PBS. **B** Bode plot of electrochemical impedance spectra of an optrode before (*black*, $n = 4$) and after IrO$_2$ deposition (*red*, $n = 4$) in ACSF, data are shown as mean $\pm$ SD. **C–F** SEM images of the optrode tip before (**C**, **D**) and after IrO$_2$ deposition (**E**, **F**)

Data analysis

Neural electrophysiological data for all recording channels were bandpass filtered. Multi-unit recordings were high-pass filtered (300 Hz) with a Bessel filter for the detection of spikes. The threshold for spike detection was set to -4.5 SD (standard deviations) and the spike waveforms were measured in a time window 1400 µs long and beginning 150 µs before threshold crossing.

Local field potentials (LFPs) were sampled at 1 kHz and bandpass filtered at 1–300 Hz. Data analyses were performed with a custom software written in MATLAB (MathWorks, USA) and NeuroExplorer (Nex Technologies, USA), and the mean-squared values of the multi-unit activities, 60 s before and 300 s after optogenetic stimulation (10 s bins), were calculated. LFPs were filtered in beta (12–30 Hz) and gamma bands (30–80 Hz). The squared values of the filtered signals, 60 s before and 300 s after stimulation (10 s bins), were calculated for each band. The power measured before stimulation was averaged (6 bins, 60 s). The power in each bin measured after stimulation was compared with the average power before stimulation by a one-side paired *t* test to investigate whether optical stimulation reduced LFP power.

Histological study

Sample preparation and implantation procedures

After being anesthetized with 1% phenobarbital sodium solution (1 ml/100 g), the rats were immobilized in a stereotaxic frame for sample implantation. A midsagittal incision was made in the scalp, and two holes were carefully bored in each animal at locations −4.0 mm anterior and ±3.0 mm lateral to the bregma. The brains were slowly implanted with the samples: Pt/Ir dummy probes (diameter 100 μm, uninsulated) as the control group and the composite CP-deposited Pt/Ir dummy probes as the testing group. Finally, the implants were fixed to the skull by gluing the cap to the skull surface with a medical adhesive (Fuaile, China), and the skin was sutured shut with monofilament nylon.

Tissue preparation and histological analysis

Six weeks after implantation, the rats were sacrificed and immunohistological analyses were performed as described previously (Lu *et al.* 2009a, 2010; Zhong *et al.* 2017). Briefly, the rats were perfused transcardially with 0.1 mol/L PBS followed by 4% paraformaldehyde in 0.1 mol/L PBS, and the brains were removed and fixed at 4 °C for two days. Subsequently, the block tissue around the implant was paraffin-embedded, and horizontal sections (30 μm thick) were prepared from all the brain samples. Antibodies against GFAP and NeuN (both from Abcam, USA) were used to label astrocytes and mature neurons, respectively. Fluorescence images were obtained using an Olympus IX71 inverted fluorescence microscope. Quantitative analysis was performed using custom software developed in MATLAB. The staining intensities of GFAP and NeuN were calculated as a function of distance up to the implant surface. The results shown are the average intensity profiles of the analyzed area within a distance of 250 μm from the implant/neural-tissue interface. All data are presented as mean ± standard error of the mean. The differences in staining intensities of various implants at the same distance were analyzed by performing independent sample *t* tests using SPSS 16.0 (SPSS, USA).

RESULTS AND DISCUSSION

Fabrication and surface modification of optrode arrays

In order to integrate optogenetic stimulation with electrophysiological readout methods, chronically implantable optrode arrays were designed and fabricated. An optrode mold was custom made for arranging the microwires and optical fibers (Fig. 1). Aided by this mold, the recording and stimulation channels can be customized to different experimental designs, which greatly simplified the fabrication process. In order to record neural activity from a large number of individual cells in relatively deep brain areas, drivable optrode arrays were designed based on previous work (Lin *et al.* 2006). The position of optrode bundles on the microdrive scaffold can be conveniently formatted according to different experimental needs (Fig. 2). This lightweight (typically <2 g) optrode array can be used to explore the causal, temporally precise, and behaviorally relevant interactions of neurons in multiple brain regions of freely moving mice.

As the electrode (recording site) is a key readout element of an optrode, its performance is crucial for optogenetic investigations. For improved spatial resolution and signal quality during electrophysiological recording, neural electrodes must meet the requirements of small size and low interface impedance (Cogan 2008). However, because the electrochemical impedance at the interface is inversely proportional to the square of the electrode diameter, balancing the impedance and the size of a neural electrode is a difficult task.

One of the most commonly applied strategies involves using high-capacitance electroactive materials to decrease the charge-transfer resistance at the electrode/electrolyte (electrode/neural-tissue) interface. Therefore, we modified the recording sites of optrodes with Pt nanoparticles, IrO_2, and composite CPs, respectively (Figs. 4, 5, 6). After electrodeposition, the redox currents in the cyclic voltammograms of the recording sites were significantly increased, while the electrochemical impedances at 1 kHz were 95%–99% lower than the unmodified sites. This is possibly due to the increase in the pseudo-capacitance after modification, which may be beneficial for decreasing background thermal noise and improving signal quality during electrical recording.

It should be mentioned that the impedance at 100 kHz (Z_{100kHz}) for the Pt-deposited electrode was approximately 90% lower than that for the undeposited electrode. As the electrode impedance at particularly

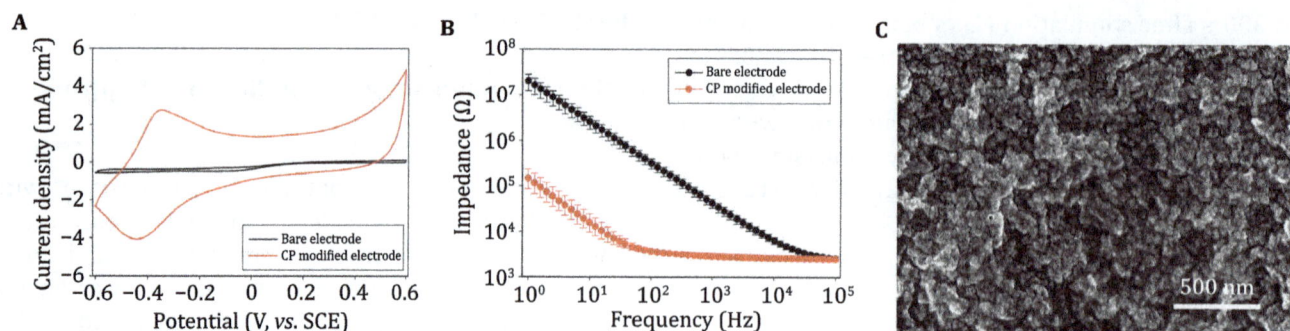

Fig. 6 Characterization of conducting polymer (CP) films. **A** Cyclic voltammogram of an optrode before (*black*) and after CP deposition (*red*) at a sweep rate of 50 mV/s in PBS. **B** Bode plot of electrochemical impedance spectra of an optrode before (*black, n* = 6) and after CP deposition (*red, n* = 6) in ACSF. **C** SEM image of the optrode tip after CP deposition

high frequencies mainly depends on the geometric area of the conductive site, the reduction of Z_{100kHz} is probably due to the expanded electrode surface covered with Pt particles after electrodeposition. The expansion of the deposited Pt layers may lead to a short circuit of the neighboring recording sites (especially on tetrodes or other high-density electrode arrays). However, the impedance of the IrO_2 electrode and the CP electrode at 100 kHz were comparable with that of the undeposited electrode, implying that the IrO_2 and CP films were not over-expanded during electrodeposition. The electrodepositions of the IrO_2 and CP films are much more controllable than Pt deposition, which is especially suitable for high-density microelectrode arrays.

In vivo seizure detection and inhibition

We next verified the feasibility of using the fabricated optrode arrays to investigate the circuit-level mechanisms controlling seizure activities under the modulation of specific neuron subtypes at high temporal resolution *in vivo*. A customized two-site optrode array was implanted simultaneously into the DGH and M1 of the VGAT-ChR2(H134R)-EYFP transgenic mice. Neural activity in the M1 was recorded as an indicator of behavioral seizures. After the onset of ictal seizures, spontaneous multi-unit bursts and large-amplitude spikes and LFPs were frequently detected in the DGH and M1 (Fig. 7), suggesting hypersynchronization of the affected neurons. We found that selectively activating DGH GABAergic interneurons not only significantly decreased the activity of local neurons (*n* = 27 from 4 mice, Fig. 7A), but also inhibited multi-unit firings in the M1 (*n* = 22 from 4 mice, Fig. 7E). During ictal seizures, the LFP power in the DGH and M1 all remarkably increased within a broad frequency range. Optical stimulation of the DGH GABAergic interneurons caused a significant decrease in local LFP activity, especially at

the beta–gamma band (*n* = 4 mice, Fig. 7B–D), as well as in the M1 (*n* = 4 mice, Fig. 7F–H). The neuronal activity levels reduced to control baselines (relative LFP powers are close to 1) during optogenetic modulation, which indicates that activating the DGH inhibitory neurons is sufficient to rescue ongoing behavioral seizures.

Histological analysis

After a neural electrode is implanted into the central nervous system, a host response is subsequently elicited, which results in the encapsulation of the implant by a dense astrocyte layer that separates the implant from the targeted neurons and hinders charge transfer at the electrode/neural-tissue interface. Consequently, the electrode impedance is increased substantially, which causes a decline in signal quality. Furthermore, the inflammatory response might also lead to a loss of neurons adjacent to the implant/neural-tissue interface, which would further deteriorate the performance of the neural electrodes. Therefore, we used Pt/Ir implants as electrodeposition substrates (the control group); these implants feature large electrode areas and can be used conveniently for investigating the responses of the electrode interface to the host tissue. Previous studies demonstrated that the iridium oxide modified implants exhibited a more significant reactive response compared to platinum substrates *in vivo* (Ereifej *et al.* 2013). Therefore, to improve biocompatibility at the implant/neural-tissue interface, we modified the implants using composite CP films, and then tested each of the deposited and control implants. We performed immunochemical analysis on tissue sections of rat brain at six weeks after implantation and evaluated the chronic performance of the composite CP films.

The inflammatory response at the implant/neural-tissue interface was characterized by the expression of

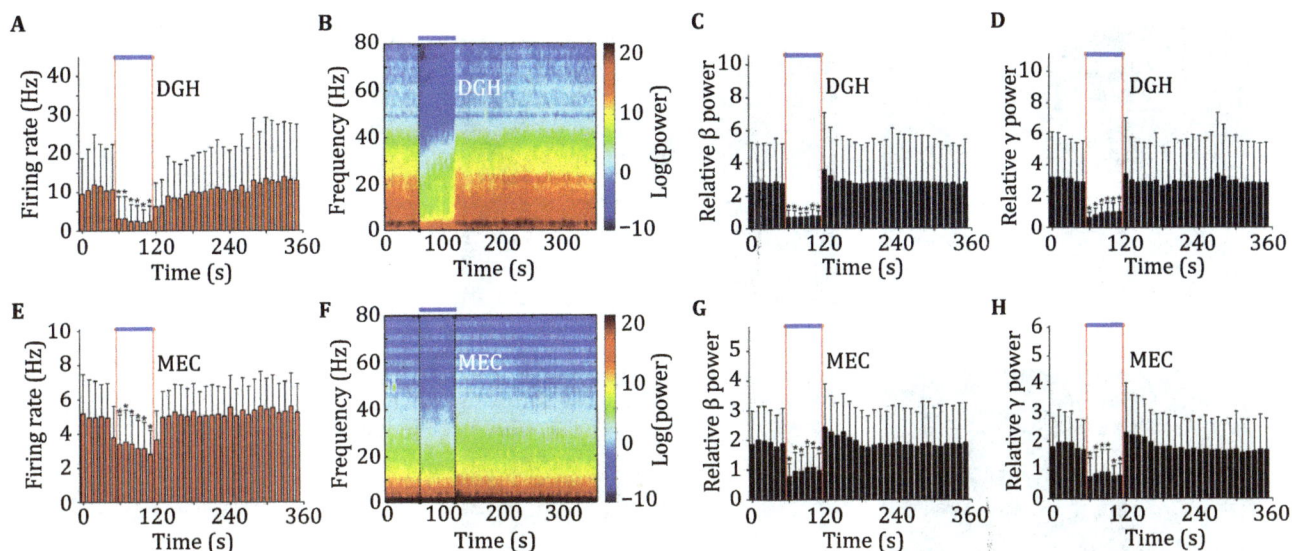

Fig. 7 Optogenetic inhibition of seizures in the intrahippocampal kainate-injected mice using a custom-made multisite optrode array. **A, E** Averaged multi-unit firing rates recorded in the DG (**A**, n = 27 from 4 mice) and M1 (**E**, n = 22 from 4 mice). **B, F** Representative examples of the spectrograms of LFPs in the DG (**B**) and M1 (**F**). **C, D, G, H** Power quantification of DG (**C, D**, n = 4 mice) and M1 (**G, H**, n = 4 mice) LFPs, the averaged powers were shown in the beta (**C, G**) and gamma bands (**D, H**), and the values were normalized to the total power in the pre-KA period. Light pulses (473 nm, 5-ms pulse duration at 130 Hz) were delivered into the DG/hilus at time 0. The *thick blue line* denotes the 60-s stimulation period. All data are shown as mean ± SEM (*p < 0.05, t test)

GFAP, an astrocyte-specific marker. We found that reactivated astrocytes accumulated in and occupied the zone around the implantation site in the control group (Fig. 8A). By comparison, GFAP expression was considerably weaker around the implantation site in the composite CP group (Fig. 8B). Figure 8C shows the quantified GFAP immunohistological intensity profiles of the control and composite CP-modified implants as a function of distance from the interface (n = 7 in each group). The average thickness of the astroglial encapsulation in the control group was ∼150 μm, whereas it was only ∼40 μm in the CP-modified group. Statistical results show that the average intensity of GFAP immunostaining in the CP-modified group was significantly lower (p < 0.05) than that in the control group up to a distance of ∼200 μm from the implant interface. Besides, neuronal loss was observed around the implantation site, and this loss was particularly severe in the control group (Fig. 8D). Interestingly, no notable loss of neurons was detected in the composite CP group (Fig. 8E). The quantified NeuN intensity profiles (Fig. 5F, n = 7 in each group) reveal a severe loss of neurons within an average distance of ∼60 μm from the implantation site in the control group. However, in the CP-modified groups, the average distance was decreased to ∼15 μm. Statistical analysis suggests that the average intensity of NeuN immunostaining in the modified group was markedly higher than that in the control group (p < 0.05 within ∼60 μm). Given the

inherent merits of the deposited CP film, our data suggest that modification of the electrode with the composite CP film can drastically alleviate the inflammatory response (reduced astroglial intensity) and promote neuronal viability (increased NeuN intensity) around the implant site. This implies that the composite CP film can improve the implant/neural-tissue interface and is suitable for long-term implantation.

CONCLUSION

In this study, we have demonstrated the feasibility and advantages of using customized optrode arrays for *in vivo* optogenetic applications. We designed and fabricated a microwire optrode array and a drivable optrode array, which are suitable for targeting single sites in relatively shallow brain structures and multiple sites in relatively deep structures, respectively. Besides, electrochemical deposition strategies for improving the performance of optrode arrays were also presented, and the physicochemical characteristics of the deposited surfaces were investigated. Platinum particles were deposited by a simplified cathodic electrochemical process, which rapidly decreased the electrode impedance (reduced by 99% at 1 kHz) by increasing the double-layer capacitance at the electrode interface. Low impedance and stable IrO_2 films were formed on the recording sites under a slow-sweep-rate cyclic

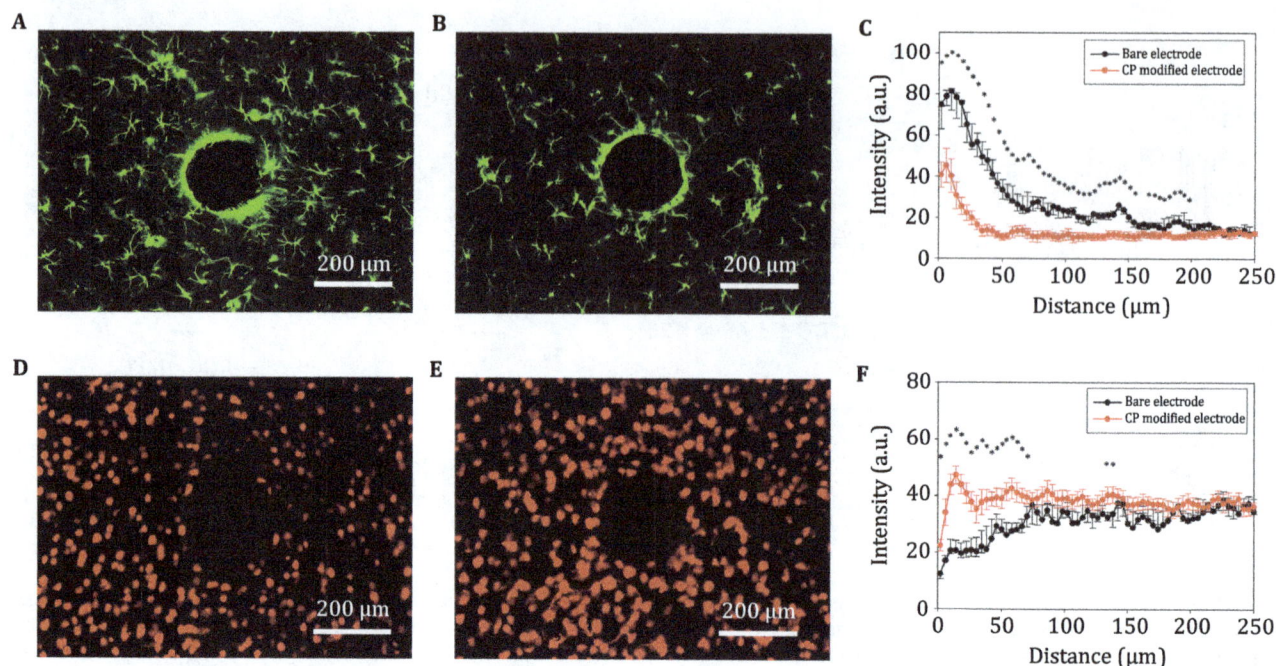

Fig. 8 Histological studies of the composite conducting polymer (CP) films' modified surfaces. Inflammatory response (**A–C**, GFAP) and neuronal survival (**D–F**, NeuN) around implants at six weeks post implantation in the rat brain before (**A, D**) and after composite CP deposition (**B, E**). Quantitative comparisons of the immunoreactivity between control ($n = 7$) and CP-modified implants ($n = 7$) were performed by using intensity profiles as a function of distance from the implant interface, shown as mean $\pm$ SEM (*$p < 0.05$, t test)

voltammetric scanning process. The electrodeposition of IrO$_2$ films is highly controllable, which is suitable for high-density optrode arrays. Furthermore, our data suggest that the electrodeposited composite CP films can significantly improve both the electrochemical performance and the biocompatibility of implants. With the aid of our customized optrode arrays, we successfully analyzed the neuronal activity during seizures and inhibited ictal propagation in multiple brain regions *in vivo*. All of these characteristics demonstrated and discussed here are crucial for the chronic optrode arrays used for precisely timed analyses of neuron subpopulations in freely moving animals. Most importantly, our results provide simple and reliable strategies for the fabrication and surface modification of optrode arrays. Combinations of these aforementioned strategies could help to construct high-performance optrode arrays optimized to the requirements of different experimental conditions. Other new fabrication strategies and novel interface materials, as well as future applications of the optrode arrays, are avenues for further investigations.

Acknowledgements This work was partially sponsored by the NSFC Program (81425010, 31630031, 31700921, 81471164, 31471109), the "Strategic Priority Research Program (XDB0205 0003)", the "Youth Innovation Promotion Association" and the Key Research Program of Frontier Sciences (QYZDB-SSW-SMC056) of the CAS, the Shenzhen Governmental Research Grants (JSGG20160429184327274, JSGG20160428140402911, JCYJ201 60428164440255), the "Guangdong Key Lab of Brain Connectome", the "Shenzhen Engineering Lab of Brain Activity Mapping Technologies", and the "Shenzhen Discipline Construction Project for Neurobiology".

Compliance with Ethical Standards

Conflict of interest Lulu Wang, Kang Huang, Cheng Zhong, Liping Wang, and Yi Lu declare that they have no conflict of interest.

Human and animal rights and informed consent All institutional and national guidelines for the care and use of laboratory animals were followed.

References

Abidian MR, Martin DC (2008) Experimental and theoretical characterization of implantable neural microelectrodes modified with conducting polymer nanotubes. Biomater 29:1273–1283

Anikeeva P, Andalman AS, Witten I, Warden M, Goshen I, Grosenick L, Gunaydin LA, Frank LM, Deisseroth K (2012) Optetrode: a multichannel readout for optogenetic control in freely moving mice. Nat Neurosci 15:163–170

Brown MTC, Tan KR, O'Connor EC, Nikonenko I, Muller D, Luescher C (2012) Ventral tegmental area GABA projections pause accumbal cholinergic interneurons to enhance associative learning. Nature 492:452–456

Chen S, Pei W, Gui Q, Chen Y, Zhao S, Wang H, Chen H (2013) A fiber-based implantable multi-optrode array with contiguous optical and electrical sites. J Neural Eng 10:046020

Cogan SR (2008) Neural stimulation and recording electrodes. Annu Rev Biomed Eng 10:275–309

Dehkhoda F, Soltan A, Ponon N, Jackson A, O'Neill A, Degenaar P (2018) Self-sensing of temperature rises on light emitting diode based optrodes. J Neural Eng 15:026012

Ereifej ES, Khan S, Newaz G, Zhang JS, Auner GW, VandeVord PJ (2013) Comparative assessment of iridium oxide and platinum alloy wires using an in vitro glial scar assay. Biomed Microdevices 15:917–924

George PM, Lyckman AW, LaVan DA, Hegde A, Leung Y, Avasare R, Testa C, Alexander PM, Langer R, Sur M (2005) Fabrication and biocompatibility of polypyrrole implants suitable for neural prosthetics. Biomater 26:3511–3519

Gradinaru V, Mogri M, Thompson KR, Henderson JM, Deisseroth K (2009) Optical Deconstruction of parkinsonian neural circuitry. Science 324:354–359

Grill WM, Norman SE, Bellamkonda RV (2009) Implanted neural interfaces: biochallenges and engineered solutions. Annu Rev Biomed Eng 11:1–24

Iseri E, Kuzum D (2017) Implantable optoelectronic probes for in vivo optogenetics. J Neural Eng 14:031001

Kotov NA, Winter JO, Clements IP, Jan E, Timko BP, Campidelli S, Pathak S, Mazzatenta A, Lieber CM, Prato M, Bellamkonda RV, Silva GA, Kam NWS, Patolsky F, Ballerini L (2009) Nanomaterials for neural interfaces. Adv Mater 21:3970–4004

Kravitz AV, Freeze BS, Parker PRL, Kay K, Thwin MT, Deisseroth K, Kreitzer AC (2010) Regulation of parkinsonian motor behaviours by optogenetic control of basal ganglia circuitry. Nature 466:U622–U627

Kwon KY, Lee H-M, Ghovanloo M, Weber A, Li W (2015) Design, fabrication, and packaging of an integrated, wirelessly-powered optrode array for optogenetics application. Front Syst Neurosci 9:69–69

Li N, Chen T-W, Guo ZV, Gerfen CR, Svoboda K (2015) A motor cortex circuit for motor planning and movement. Nature 519:U51–U88

Lin L, Chen G, Xie K, Zaia KA, Zhang S, Tsien JZ (2006) Large-scale neural ensemble recording in the brains of freely behaving mice. J Neurosci Method 155:28–38

Lu Y, Wang D, Li T, Zhao X, Cao Y, Yang H, Duan YY (2009a) Poly(vinyl alcohol)/poly(acrylic acid) hydrogel coatings for improving electrode-neural tissue interface. Biomater 30:4143–4151

Lu Y, Wang T, Cai Z, Cao Y, Yang H, Duan YY (2009b) Anodically electrodeposited iridium oxide films microelectrodes for neural microstimulation and recording. Sens Actuat B Chem 137:334–339

Lu Y, Li T, Zhao X, Li M, Cao Y, Yang H, Duan YY (2010) Electrodeposited polypyrrole/carbon nanotubes composite films electrodes for neural interfaces. Biomater 31:5169–5181

Lu Y, Li Y, Pan J, Wei P, Liu N, Wu B, Cheng J, Lu C, Wang L (2012) Poly(3,4-ethylenedioxythiophene)/poly(styrenesulfonate)-poly(vinyl alcohol)/poly(acrylic acid) interpenetrating polymer networks for improving optrode-neural tissue interface in optogenetics. Biomater 33:378–394

Lu Y, Zhong C, Wang LL, Wei PF, He W, Huang K, Zhang Y, Zhan Y, Feng GP, Wang LP (2016) Optogenetic dissection of ictal propagation in the hippocampal-entorhinal cortex structures. Nat Commun 7:10962

Otis JM, Namboodiri VMK, Matan AM, Voets ES, Mohorn EP, Kosyk O, McHenry JA, Robinson JE, Resendez SL, Rossi MA, Stuber GD (2017) Prefrontal cortex output circuits guide reward seeking through divergent cue encoding. Nature 543:103–107

Schmitt LI, Wimmer RD, Nakajima M, Happ M, Mofakham S, Halassa MM (2017) Thalamic amplification of cortical connectivity sustains attentional control. Nature 545:219–223

Schwartz AB, Cui XT, Weber DJ, Moran DW (2006) Brain-controlled interfaces: movement restoration with neural prosthetics. Neuron 52:205–220

Tovote P, Esposito MS, Botta P, Haudun FC, Fadok JP, Markovic M, Wolff SBE, Ramakrishnan C, Fenno L, Deisseroth K, Herry C, Arber S, Luthi A (2016) Midbrain circuits for defensive behaviour. Nature 534:206–212

Tye KM, Mirzabekov JJ, Warden MR, Ferenczi EA, Tsai H-C, Finkelstein J, Kim S-Y, Adhikari A, Thompson KR, Andalman AS, Gunaydin LA, Witten IB, Deisseroth K (2013) Dopamine neurons modulate neural encoding and expression of depression-related behaviour. Nature 493:537–541

Yizhar O, Fenno LE, Davidson TJ, Mogri M, Deisseroth K (2011a) Optogenetics in neural systems. Neuron 71:9–34

Yizhar O, Fenno LE, Prigge M, Schneider F, Davidson TJ, O'Shea DJ, Sohal VS, Goshen I, Finkelstein J, Paz JT, Stehfest K, Fudim R, Ramakrishnan C, Huguenard JR, Hegemann P, Deisseroth K (2011b) Neocortical excitation/inhibition balance in information processing and social dysfunction. Nature 477:171–178

Zhang F, Wang LP, Brauner M, Liewald JF, Kay K, Watzke N, Wood PG, Bamberg E, Nagel G, Gottschalk A, Deisseroth K (2007) Multimodal fast optical interrogation of neural circuitry. Nature 446:U633–U634

Zhong C, Zhang Y, He W, Wei P, Lu Y, Zhu Y, Liu L, Wang L (2014) Multi-unit recording with iridium oxide modified stereotrodes in Drosophila melanogaster. J Neurosci Method 222:218–229

Zhong C, Ke D, Wang L, Lu Y, Wang L (2017) Bioactive interpenetrating polymer networks for improving the electrode/neural-tissue interface. Electrochem Commun 79:59–62

Zimmerman CA, Lin Y-C, Leib DE, Guo L, Huey EL, Daly GE, Chen Y, Knight ZA (2016) Thirst neurons anticipate the homeostatic consequences of eating and drinking. Nature 537:680–684

18

Particle segmentation algorithm for flexible single particle reconstruction

Qiang Zhou[1,2]✉, **Niyun Zhou**[2], **Hong-Wei Wang**[2]✉

[1] State Key Laboratory of Biomembrane and Membrane Biotechnology, Center for Structural Biology, School of Life Sciences, Tsinghua University, Beijing 100084, China
[2] Ministry of Education Key Laboratory of Protein Science, Tsinghua-Peking Joint Center for Life Sciences, Center for Structural Biology, School of Life Sciences, Tsinghua University, Beijing 100084, China

Abstract   As single particle cryo-electron microscopy has evolved to a new era of atomic resolution, sample heterogeneity still imposes a major limit to the resolution of many macromolecular complexes, especially those with continuous conformational flexibility. Here, we describe a particle segmentation algorithm towards solving structures of molecules composed of several parts that are relatively flexible with each other. In this algorithm, the different parts of a target molecule are segmented from raw images according to their alignment information obtained from a preliminary 3D reconstruction and are subjected to single particle processing in an iterative manner. This algorithm was tested on both simulated and experimental data and showed improvement of 3D reconstruction resolution of each segmented part of the molecule than that of the entire molecule.

Keywords   Single particle reconstruction, Cryo-EM, Particle segmentation, Local reconstruction

INTRODUCTION

Single particle cryo-electron microscopy (cryo-EM) is a powerful structural biology tool being developed in the past several decades and becoming more matured in recent years (Bai *et al.* 2015a; Carazo *et al.* 2015; Cheng 2015; Cheng *et al.* 2015; Nogales and Scheres 2015). By quickly freezing biological macromolecules in a thin film of vitreous ice, cryo-EM preserves the molecules as they are in solution immediately before the freezing. This stipulates cryo-EM the unique advantage to reveal the molecular structure in their close-to-native states and the possibility to examine structures in action. The most recent development of new direct-electron detection device and image processing algorithms has dramatically boosted the capability of this technique so that three-dimensional (3D) structures of biological macromolecules can be solved to near atomic resolution from averaging many individual images without crystallization (Bai *et al.* 2013; Liao *et al.* 2013; Bartesaghi *et al.* 2015). This has led to a resolution revolution of the cryo-EM technology and is transforming the field of structural biology (Kuhlbrandt 2014).

Despite the major technical progresses, compositional and conformational heterogeneity still imposes a major obstacle on high-resolution single particle cryo-EM structural determination. Different from crystallography where the macromolecules are constrained within a crystalline lattice, single particle molecules in solution are more flexible in changing their ternary and quaternary structures which may cause conformational

Qiang Zhou and Niyun Zhou have contributed equally to this work.

✉ Correspondence: zhouqiang00@tsinghua.org.cn (Q. Zhou), hongweiwang@tsinghua.edu.cn (H.-W. Wang)

or compositional heterogeneity among the molecules. In cases where the heterogeneity is relatively subtle and localized, single particle 3D reconstruction of a macro-molecule complex is an averaged structure of the common region of all the molecules but with a low resolution at the flexible region. Algorithms based on multivariate statistical analysis were developed to classify molecules into different states (van Heel and Frank 1981). The maximum likelihood algorithm was developed to classify molecule images with low signal to noise ratio (Scheres et al. 2007). Methods such as random conical tilt and orthogonal tilt reconstruction were developed to obtain 3D models of different molecular states (Radermacher et al. 1987; Leschziner and Nogales 2006). Using statistical classification approach, these algorithms sort the heterogeneous particle images into different classes based on the level of similarity among them and treat each class of images as a homogeneous set of molecules. The classification thus generates multiple structures each reflecting a different state of the biological sample in vitreous ice. The above methods all assume common structure within the same class of molecules. While these methods have been proved to be very successful on the structural studies of many macromolecular complexes and revealed important mechanistic insight to the conformational switch of important molecular machines, there are still a lot of complexes with more complicated conformational heterogeneity that cannot be easily studied. In a severe conformational heterogeneity such as a global variation within the molecule or a continuous domain–domain movement at large scale, a correct 3D reconstruction cannot even be obtained using the conventional classification approach.

Several algorithms without classification strategy have been introduced to single particle analysis of macromolecular complexes with continuous conformational changes. These include the normal-mode analysis (Ma and Karplus 1997; Brink et al. 2004; Ma 2005; Jin et al. 2014), energy landscape analysis and manifold embedding (Dashti et al. 2014; Frank and Ourmazd 2016), 3D variance analysis (Penczek et al. 2006; Zhang et al. 2008), covariance analysis (Anden et al. 2015; Katsevich et al. 2015; Liao et al. 2015), and eigen analysis-based methods (Penczek et al. 2011; Tagare et al. 2015). These algorithms can provide quantitative description of the conformational variation mode in the complex to guide further processing of the dataset. More recently, local masking technique was used in reconstructing the rigid body within a complex or further classifying local subtle conformational heterogeneity in a focused region of the molecule. This has been quite successful in improving the local resolution significantly

of different rigid portions within a complex (Amunts et al. 2014; Brown et al. 2014; Chang et al. 2015; Yan et al. 2015).

Further implementation of algorithms that can separate the relative mobile parts within a flexible molecule and reconstruct the different parts separately will be more useful. Because the electron micrograph of a molecule reflects the 2D projection of the molecule along the electron beam illumination direction, different parts of the complex superimpose with each other in the 2D image. So simply masking the 2D image or 3D model does not eliminate the influence by the signal of the mobile portion on the 3D reconstruction. A clearer way should be to remove the signal of mobile portion from the 2D image entirely so a reconstruction of the interesting part can be done with greater fidelity. Such kind of separation has been realized in Fourier–Bessel space for the reconstruction of a double-layered helical assembly of tubulin (Wang and Nogales 2005). Recently, separation and reconstruction of icosahedral viral genomic structure from the capsid structure were achieved by subtracting the capsid signal from the raw images of virus particles (Liu and Cheng 2015; Zhang et al. 2015). In our most recent work, we have developed a segmentation algorithm to separate the SNAP–SNARE structure from 20S particle by subtracting the hexameric NSF complex in the raw image of 20S particle and thus overcome the symmetry mismatch and severe conformational heterogeneity in the 20S particles. This allowed us to reconstruct the SNAP–SNARE complex with higher resolution than using the whole particle images (Zhou et al. 2015). At nearly the same time, Bai et al. (2015b), Ilca et al. (2015), and Shan et al. (2016) developed similar algorithms independently. A recent development in RELION software (Scheres 2012a, b) makes it possible to subtract certain portions within a complex from the raw 2D images without introducing major artifact. This allowed much better classification of the interested portion to further sort the heterogeneous particle images to even higher resolution than the overall average (Bai et al. 2015b).

In this work, we further expand the particle segmentation algorithm that we have developed for the analysis of 20S particles to other samples. The successful application of this algorithm to different systems with conformational heterogeneity indicated its generality. We also incorporated the image subtraction algorithm at micrograph level so it not only overcomes the potential artifact from interpolation and contrast transfer function, but more importantly also provides new opportunities to analyze micrographs of crowding particle images.

THEORY AND ALGORITHM

Particles segmentation

In the current algorithm, we consider a scenario where the being-studied macromolecule is composed of two rigid bodies that are relatively mobile with each other. In a cubic volume with $N \times N \times N$ voxels, the 3D densities of the two rigid bodies are V_1 and V_2, respectively. For a certain conformation of the macromolecule, its 3D density V thus can be written as

$$V = V_1 \cdot E_1 + V_2 \cdot E_2, \qquad (1)$$

where E_1 and E_2 are the Euler matrix of V_1 and V_2, respectively. The Euler matrices are functions of Euler angles and translational vectors

$$E_k = f(\Phi_k, \vec{r_k}), k = 1, 2. \qquad (2)$$

The different combinations of E_1 and E_2 define a heterogeneous conformation among the molecules. Our goal is to determine the high-resolution structure of the two rigid bodies, V_1 and V_2. During the process, we should also be able to reveal all the E_1 and E_2 combinations therefore the conformational distribution within the specimen.

For a particle i in a transmission electron microscope, its 2D image as a $N \times N$ array is

$$X_i = F^{-1}\left[CTF_i \cdot \left(A^{E_{1,i}} \cdot F(V_1) + A^{E_{2,i}} \cdot F(V_2)\right) + N_i\right], \qquad (3)$$

where F and F^{-1} are Fourier transform and reverse Fourier transform operation, respectively; CTF_i is the contrast transfer function for particle i; $A^{E_{k,i}}$ is the slicing operation on the 3D Fourier transform according to $E_{k,i}$, $k = 1,2$; N_i is the noise of the particle i.

In this 2D image, the projection of V_1 or V_2 is

$$P_{k,i} = F^{-1}\left[CTF_i \cdot A^{E_{k,i}} \cdot F(V_k)\right], k = 1, 2. \qquad (4)$$

If we know V_1 and V_2 and their exact corresponding Euler matrices, we should be able to subtract the signal of either V_1 or V_2 from the raw particle or micrograph and then segment the other part according to its location for further analysis (Fig. 1A).

$$X_{h,i} = Win(r_{h,i}, b)(X_i - P_{k,i}), k, h = 1, 2, k \neq h, \qquad (5)$$

where $r_{h,i}$ is the location of V_h and $Win(r_{h,i}, b)$ is a function to re-window an image with box size b at $r_{h,i}$, $h = 1,2$. This operation thus calculates a new image with most of the signal of V_k removed.

In situations where the flexibility between the two rigid bodies is within certain range, i.e., the 20S particle, a global low-resolution reconstruction from all the images may serve as a starting model. The initial V_k can be obtained from this global reconstruction through 3D segmentation. The initial $E_{k,i}$ can be roughly estimated as the Euler matrix obtained from the global reconstruction. These initial values can also be obtained by further focused 3D refinement with corresponding local mask applied. The initial location $r_{h,i}$ for V_h can be obtained from its location in the global 3D reconstruction ($r_{3D,h,i}$) and corresponding Euler matrix $E_{h,i}$

$$r_{h,i} = P_{XY} \cdot E_{h,i} \cdot r_{3D,h,i}, h = 1, 2, \qquad (6)$$

where P_{XY} is an operation to project vector to XY plane.

More specifically, we can first subtract V_2 and generate images for V_1. Then we can get an updated volume and Euler matrix for V_1 with which we can generate images for V_2. These procedures can be iterated between V_1 and V_2 for several rounds until convergence (Fig. 1B).

Because the true value of V_k ($V_{k,true}$) is unknown and can only be estimated with V_k at the resolution of the 3D reconstruction, the projection subtracting residual should be:

$$\Delta P_{k,i} \approx F^{-1}\left[CTF_i \cdot A^{E_{k,i}} \cdot F(V_{k,true})\big|_{R > R_k}\right], k = 1, 2, \qquad (7)$$

where R is spatial frequency and R_k is the 3D reconstruction resolution. If the initial estimated volume function of V_k can be of enough high resolution, the intensity of $\Delta P_{k,i}$ can be neglected.

RESULTS

Segmentation algorithm improves the resolution of simulated 20S particle dataset

From the 48 simulated micrographs of 20S particles (Fig. 2, Table 1 for simulating parameters), we extracted the 20S particle images and performed 2D classification and 3D reconstruction of the whole particle images. These showed overall shape of the 20S particle comprising two fuzzy parts corresponding to the SNARE/SNAP (SS) and the D1–D2 domain of NSF (DD), respectively (Supplementary Fig. S1A, Fig. 3A). While the FSC of this overall reconstruction reported a resolution of 5.8 Å, the EM map lacks clear features especially in the SS region. We performed additional 3D reconstruction refinements with local masks around SS or DD, resulting in slightly better-defined SS at 5.7 Å resolution (Fig. 3B) and much better DD at 3.4 Å resolution (Fig. 3C), respectively. The 3D auto-refinements with sub-particles generated with relion_project resulted in similar resolution of 5.45 Å for SS and 3.35 Å for DD (Supplementary Fig. S2, Table 2). Alternatively, we

Fig. 1 Flowchart of particle segmentation and 3D reconstruction. **A** The V_2 part of a particle is re-windowed and centered from the raw particle image according to its location r_2, meanwhile the V_1 part is subtracted from the raw particle image. **B** The flowchart of iterative segmentation and reconstruction. The raw particles are composed of two rigid parts flexible to each other: V_1 and V_2. Firstly, the whole 3D volume of initial model is segmented into V_1 and V_2. Then V_2 is subtracted from raw particle images or micrographs, from which the V_1 particle images are re-windowed and subjected to 3D reconstruction, resulting in a refined V_1. This process is repeated again with V_1 subtracted from raw particle images or micrographs, obtaining V_2 particle images and a refined V_2. The procedure can be repeated until convergence

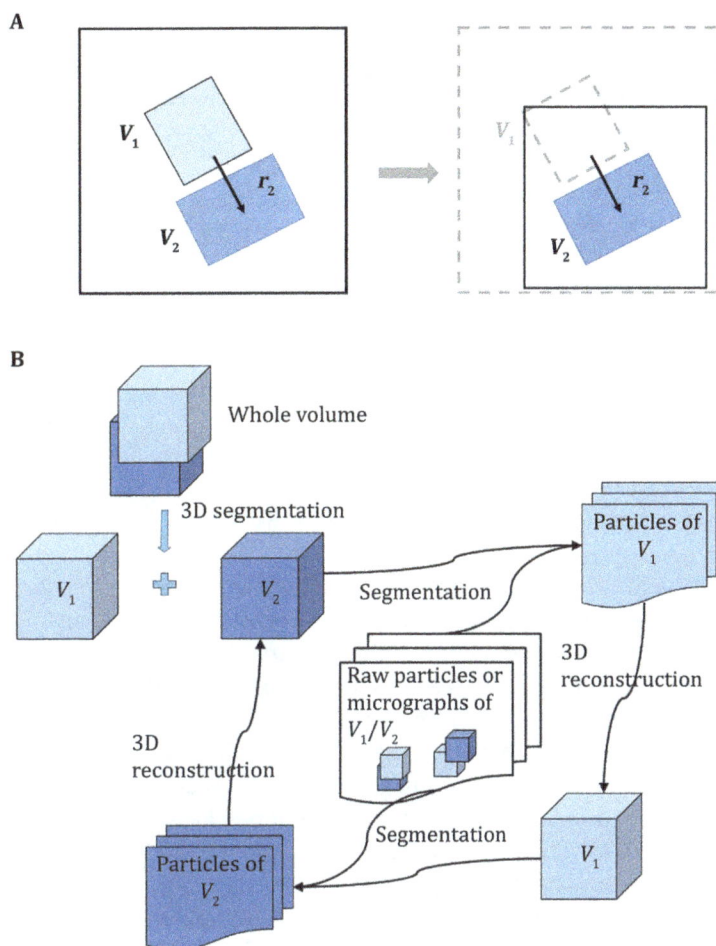

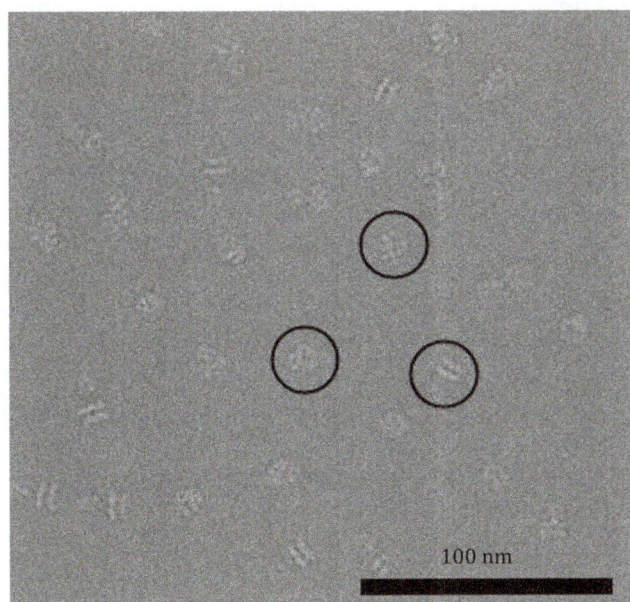

Fig. 2 An area of simulated micrograph. Three simulated 20S particles in various views are marked by circles

applied the segmentation algorithm to the dataset (Supplementary Fig. S1B–D) and obtained a better-defined reconstruction of SS than the previous two SS volumes at 4.59 Å resolution even in the first round of segmentation (Fig. 3D). After second round of segmentation, the map quality was further improved (Fig. 3E, F) although the apparent FSC value didn't change significantly from the first round reconstruction (Fig. 3J). The segmentation algorithm also resulted in a DD (Fig. 3G–I, Supplementary Fig. S1E) better than those in the overall 3D reconstruction.

It is notable that the image box size of the windowed particle has an effect on the reconstruction resolution of DD particles. The 3D reconstruction resolution of the segmented DD with a box size of 160 and 256 pixels was 3.52 Å and 3.41 Å, respectively (Fig. 3G, I, J, Table 2). Because the signal of particles is proportional to the molecular weight and the noise is proportional to the box size (Rosenthal and Henderson 2003), using too large box size will decrease the signal to noise ratio of particles. But on the other hand the too small box size

Table 1 Parameters for micrograph simulation

Cs (mm)	Voltage (kV)	Defocus range (μm)	Astigmatism (Å)	Amplitude contrast	B factor (Å²)	Pixel size (Å/pixel)	σ of translation between SS and DD (pixel)	σ of Euler angle difference between SS and DD (°)
2.7	300	−1 to −3	1000 ± 50	0.1	50 ± 2	1.32	2	10

results in too large reciprocal pixel size, which may limit the CTF correction and interpolation in Fourier space (Penczek *et al.* 2014). The optimal box size used for 3D reconstruction may be variable for particles with different sizes and/or symmetry.

Segmentation algorithm improves the reconstruction quality of influenza RdRP

Our previous work has shown that the influenza RdRP tetramer contains two homo-dimers interacting with each other in a flexible manner (Chang *et al.* 2015). We were able to obtain a 3D reconstruction of the RdRP dimer at resolution of 4.3 Å by applying a mask around one of the dimer density during the refinement (Fig. 4A). In this practice, each particle image lost half of its structural information in the final reconstruction. The segmentation algorithm provides the opportunity to include the other dimer in the final 3D reconstruction thus double the effective dataset. We segmented the RdRP dimers from all the tetramer dataset and performed 2D classification (Supplementary Fig. S3) and 3D refinement. The 3D reconstruction obtained in this way showed a similar apparent resolution as the previous one (Fig. 4B). But closer look at the FSC curves indicated an elevated signal at medium-resolution range from 10 to 5 $Å^{-1}$ in the latter reconstruction (Fig. 4C). The EM density obtained by the segmentation reconstruction algorithm showed better-defined feature and higher local resolution than that obtained by the local masking reconstruction algorithm (Fig. 4D–F). As a control, the 3D auto-refinements with dimer sub-particles generated with relion_project also resulted in similar resolution of 4.45 Å (Supplementary Fig. S4, Table 2).

Segmentation algorithm calculates conformational flexible distribution of 70S ribosome

It is well-known that there is a ratchet motion between the 30S and 50S subunits within a 70S ribosome. Former analysis of 70S ribosomes using supervised classification, maximum likelihood classification, and local masking reconstruction can all separate the different conformers and reconstruct the 30S and 50S portions of

the complex. We tested the segmentation algorithm in separating and reconstructing the two portions of 70S ribosome. As a control, we firstly performed 3D reconstruction of the entire 70S particle images and obtained a structure at 3.4 Å resolution. Using local masking approaches, the 30S and 50S subunits can be further refined to 3.4 Å and 3.2 Å resolutions, respectively (Fig. 5A, B). We applied the segmentation algorithm on the dataset and reconstructed the 30S and 50S subunits separately, resulting in final reconstructions at 3.3 Å and 3.2 Å resolutions, respectively (Fig. 5C, D). The 3D auto-refinements with sub-particles generated with relion_project also resulted in similar resolution of 3.4 Å for 30S and 3.2 Å for 50S (Supplementary Fig. S5, Table 2). In summary, both the local masking refinement and segmentation algorithm improved the resolution than the whole particle refinement procedure (Fig. 5E). For both 30S and 50S subunits, the 3D reconstructions using local masking refinement and segmentation algorithm have very similar resolution (Fig. 5E). The reason that there was no improvement is probably due to the rather small motion between the 30S and 50S subunits for which local masking in an auto-refinement obviously restored the orientation of the subunits effectively.

Because we were using segmentation reconstruction, we could calculate the relative rotating angles between 30S and 50S subunits for each individual particle by comparing their Euler angles after the reconstructions. The distribution of the rotation angles showed two peaks, in agreement to the fact that there are two major populations of conformers in the ratchet switch of the 70S ribosome (Fig. 5F). When we aligned the two classes of 3D reconstructions of 70S ribosome based on the 50S subunit, the 30S subunit has a rotation of about 3.8°(Fig. 5G).

Direct segmentation of particle images from raw micrographs

We noted that the segmentation algorithm can be directly applied to segment particle images from raw micrographs. As we have discussed previously, the segmentation of raw particle images may suffer from the loss of information due to the point spread function

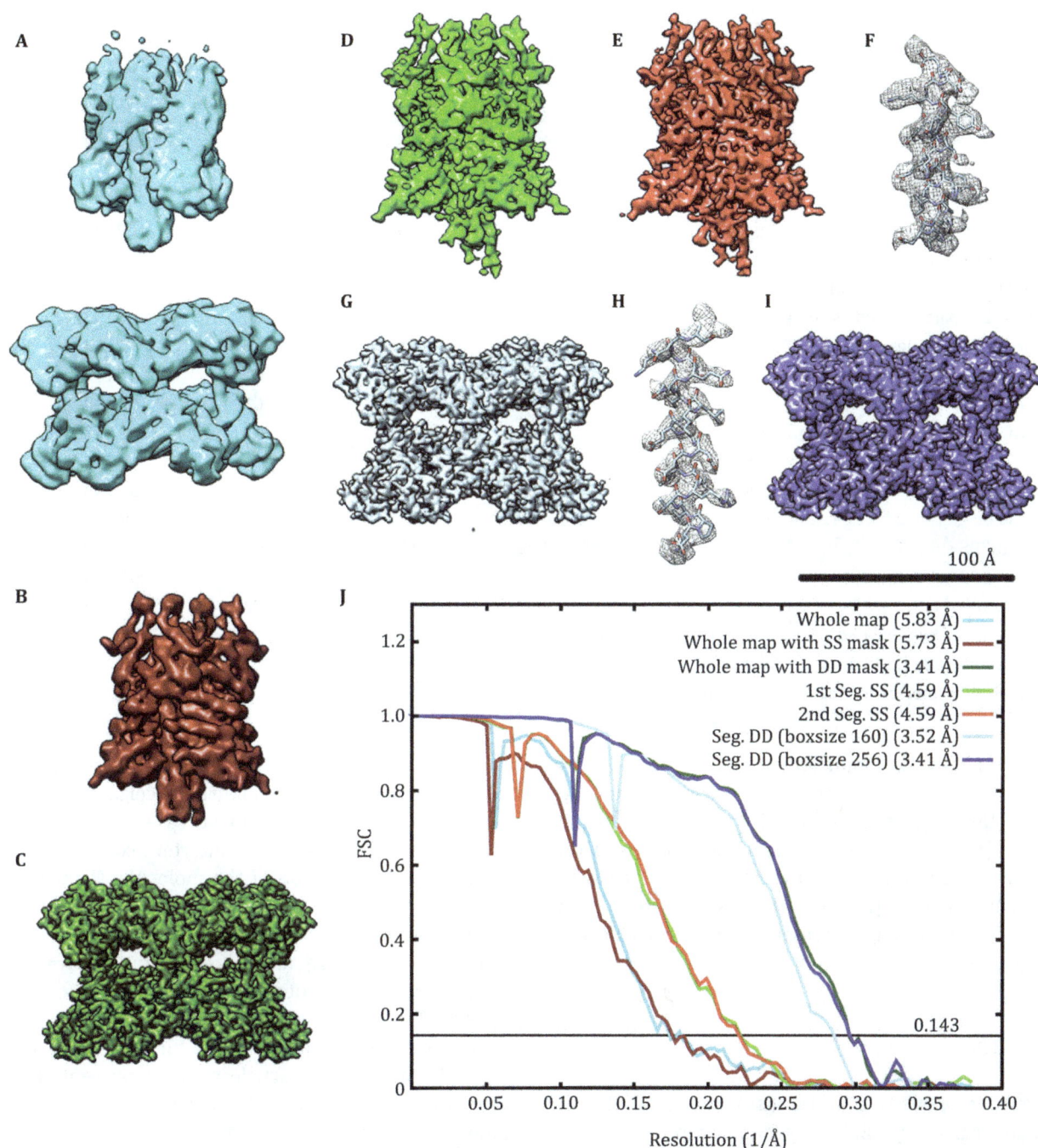

Fig. 3 Comparison of 3D reconstructions from simulated 20S particles. **A** 3D reconstruction of whole particles without local mask. **B** 3D reconstruction of whole particles with a local mask around the SS portion. Only SS is shown. **C** 3D reconstruction of whole particles with a local mask around the DD portion. Only DD is shown. **D** 3D reconstruction of the SS particles after the first round of segmentation. **E** 3D reconstruction of the SS particles after the second round of segmentation. **F** An α-helix from the 3D density of **E** with the corresponding atomic model docked in. This corresponds to the amino acid residues 138–156 of the α-SNAP. **G** 3D reconstruction of the DD particles after the first round of segmentation. **H** An α-helix from the 3D density of **G** with the corresponding atomic model docked in. This corresponds to the amino acid residues 511–531 of the NSF. **I** 3D reconstruction of the DD particles after the first round of segmentation with a box size of 256 pixels. **J** FSC curves of the 3D reconstructions. The FSC curve of segmented SS is the one after the second round of segmentation

caused by the CTF. After aligning each of the raw particle images with the reference calculated from the partial volume, we should be able to subtract reference

projections from the raw micrographs directly. Because there is no cutoff of the CTF fringes around the raw particle images in the whole micrograph, we don't need

Table 2 Summary of 3D reconstruction

	Resolution before post-processing	Resolution after post-processing	Symmetry	# Particles	Box size (pixel)
Whole volume of simulated particles	9.13	5.83	C1	7193	256
Whole volume of simulated particles with SS mask	8.24	5.73	C1	7193	256
Whole volume of simulated particles with DD mask	4.12	3.41	C6	7193	256
SS sub-particles generated with relion_project	7.86	5.45	C1	7193	256
DD sub-particles generated with relion_project	4.12	3.35	C6	7193	256
Segmented SS particles round I	6.40	4.59	C1	7163	160
Segmented DD particles (box size 160)	3.91	3.52	C6	7157	160
Segmented DD particles (box size 256)	4.33	3.41	C6	7157	256
Segmented SS particles round II	6.21	4.59	C1	7157	160
Whole volume of 70S ribosome	3.93	3.45	C1	68,543	280
Whole volume of 70S ribosome with 50S mask	3.81	3.16	C1	68,543	280
Whole volume of 70S ribosome with 30S mask	4.20	3.39	C1	68,543	280
50S ribosome generated with relion_project	3.81	3.16	C1	68,543	280
30S ribosome generated with relion_project	4.25	3.36	C1	68,543	280
Segmented 30S subunit	4.20	3.33	C1	68,543	280
Segmented 50S subunit	3.81	3.19	C1	63,499	280
Whole volume of influenza RdRP tetramer	7.68	7.51	C2	67,066	256
Whole volume of influenza RdRP tetramer with dimer mask	6.38	4.33	C2	67,066	256
Segmented influenza RdRP dimer	4.95	4.32	C2	122,758	180
Influenza RdRP dimer generated with relion_project	6.14	4.45	C2	134,132	256

to worry about the information loss caused by the windowing. In our simulated micrographs, we can easily subtract the projections of DD from each of the 20S particles (Fig. 6A, B). This can also be done in a real electron micrograph that contains relatively crowded 20S particle images (Fig. 6C, D). This provided opportunities for processing of wider range of cryo-electron micrographs.

DISCUSSION

Sample heterogeneity is still a major technical obstacle in single particle cryo-EM 3D reconstruction. The source of heterogeneity includes but is not limited to the following aspects: compositional diversity and conformational flexibility. The conformational variation that molecules undergo can be continuous or discrete. Compositional heterogeneity and conformational heterogeneity with discrete states usually lead to a finite number of classes that current 3D classification algorithms can handle reasonably well. In contrast, continuous conformational change within a molecule would lead to an almost infinite number of classes.

3D refinement and reconstruction with an adaptive local mask around the relatively rigid portion of the molecule has shown to be successful in some cases to solve high-resolution structure of certain part of the whole molecule. But in most cases, the overlapped structures in 2D projections interfere correct alignment of the common portion of the molecule. Using the particle segmentation algorithm, we can separate the relatively mobile portions within a molecule image and thus perform single particle analysis of the separated portions without the interference from each other. The image after segmentation has much cleaner signals for more precise alignment and further analysis. Our example of the 20S particle analysis presented in this work indicates the particular advantage of segmentation algorithm in analyzing complexes with internal symmetry mismatch. The further refinement with local angular searching may result in artifact in some cases. In the example of simulated 20S particle, the asymmetric feature of SS part was lost after local angular searching. However, this feature can be well recovered by the segmentation algorithm.

In our segmentation algorithm, after projecting the 3D partial density, it is critical to subtract the projection from raw particles with correct operation. There have been several attempts (Wang and Sigworth 2009; Bai et al. 2015b; Ilca et al. 2015; Liu and Cheng 2015; Zhang et al. 2015) to subtract the projection of a 3D

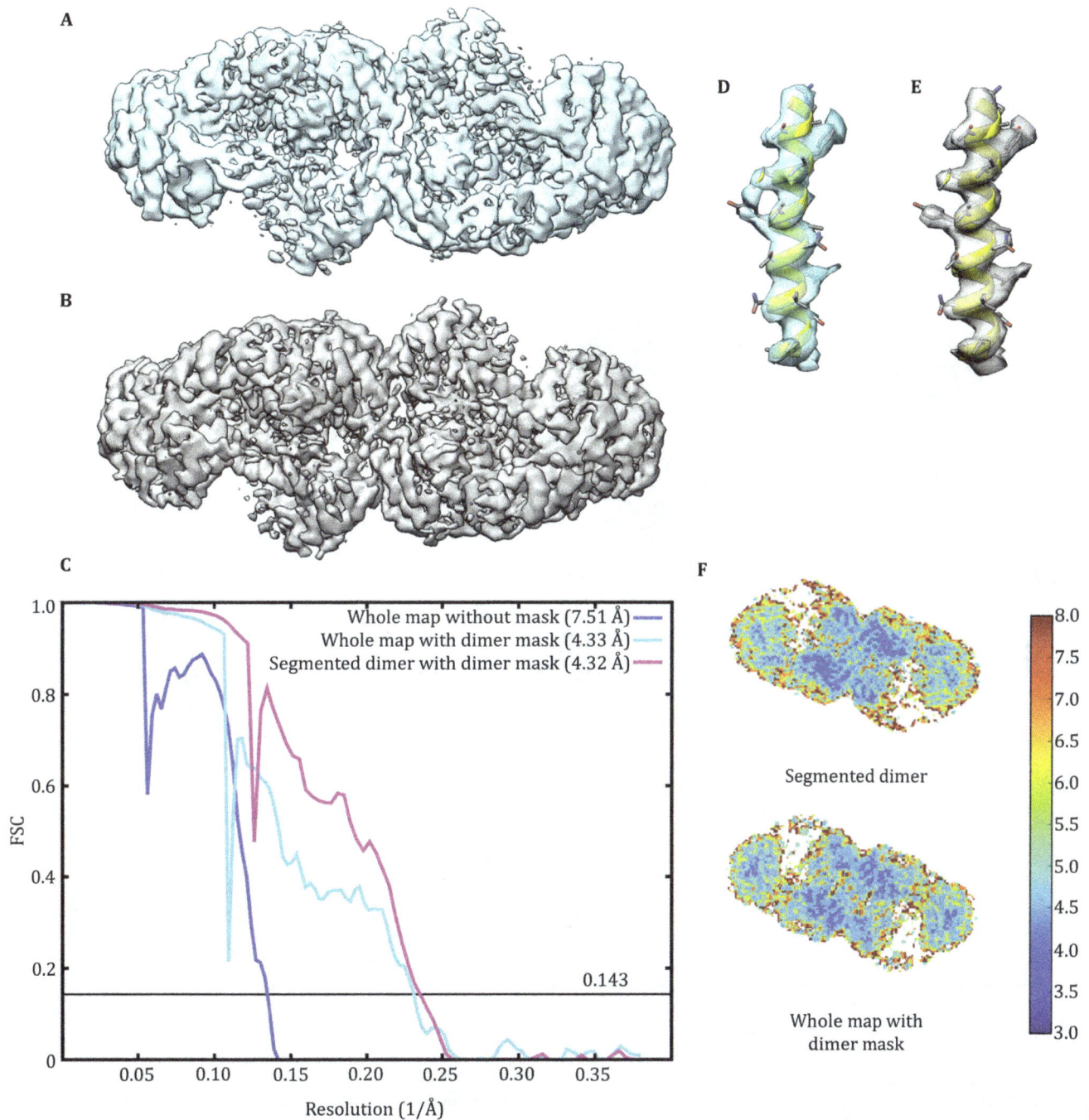

Fig. 4 Comparison of 3D reconstructions of influenza RdRP. **A** 3D reconstruction of influenza RdRP tetramer particles with a local mask around the dimer portion (EMD ID: 6202). **B** 3D reconstruction of the influenza RdRP dimer after the first round of segmentation from the tetramer particle images. **C** FSC curves of 3D reconstructions. **D** and **E** Enlarged views of an α-helix density with the corresponding atomic models from **A** and **B**, respectively. The α-helix corresponds to the amino acid residue 454–476 of polymerase basic protein 1 of RdRP. **F** Central slice of the maps colored by local resolution computed with ResMap

reconstruction or 3D model from raw particles. We found that the absolution gray scale feature of the 3D reconstruction within RELION makes the subtraction easy and intuitive. This operation, which removes most of the low frequency signals of one macromolecule part from the raw particle images, immediately allows the alignment of the other macromolecule part more precisely. This is proved by the fact that reference-free 2D classes of segmented particles show more detailed features than the entire particle but are free of contaminated features from the subtracted references. Furthermore, while we can use the iterative approach

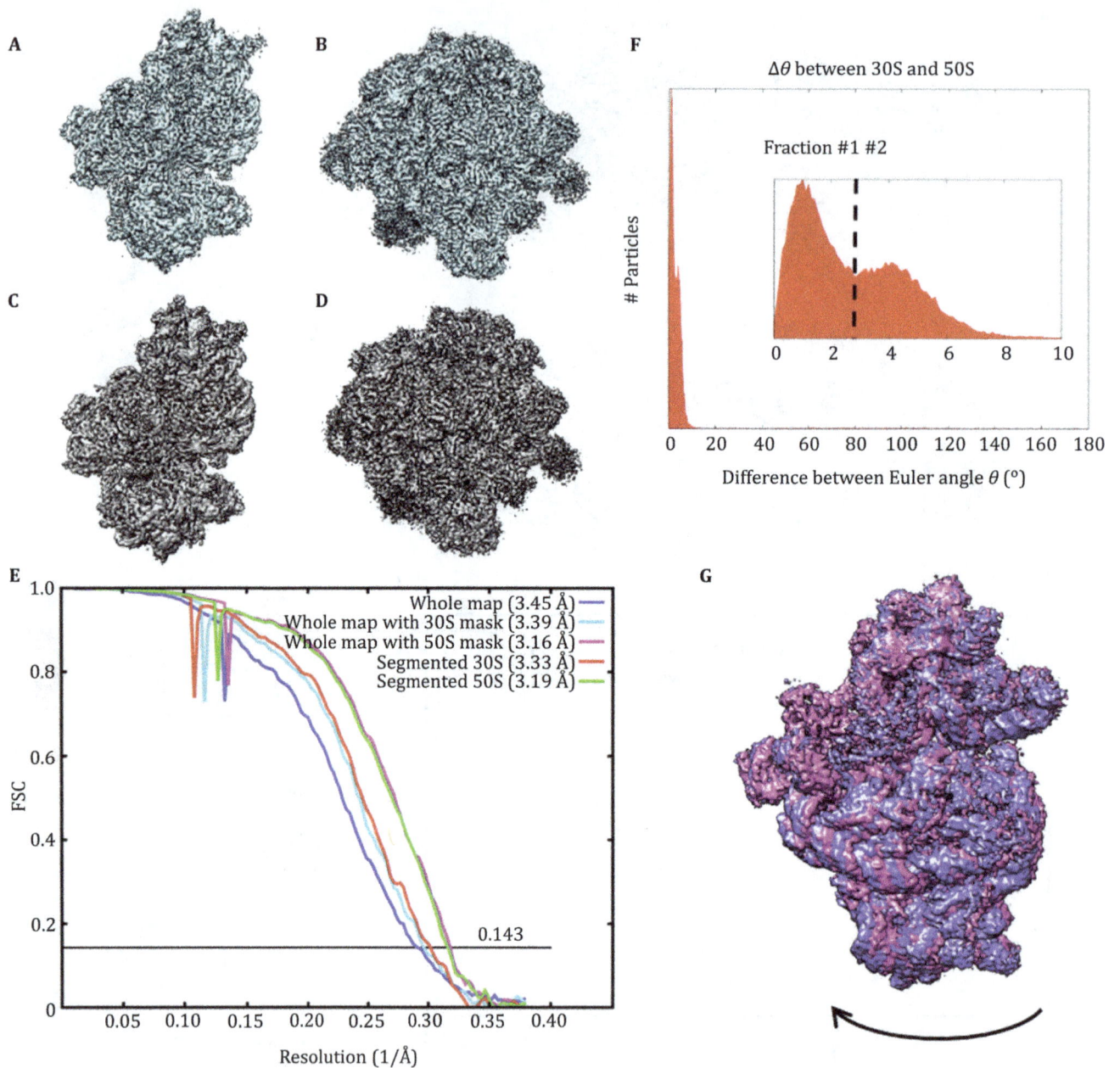

Fig. 5 Comparison of 3D reconstructions of 70S ribosome. **A** and **B** are the 3D reconstruction maps of 70S ribosome particles with a local mask of 30S and 50S, respectively. **C** and **D** are the 3D reconstruction maps of 30S and 50S ribosomes after the particle segmentation, respectively. **E** FSC curves of 3D reconstructions. **F** Distribution of the difference of Euler angle theta between the 30S and 50S subunits. Inset is an enlarged view corresponding to the range of theta from 0° to 10°. **G** Comparison between 30S subunit of the 70S ribosome 3D reconstructed from dataset fraction #1 (*blue*) and fraction #2 (*purple*) using the alignment parameters from the 3D auto-refinement of segmented 50S subunit

(Fig. 1B) to improve the segmentation and alignment of each portion of the molecule, at most two iterations are enough to result the convergence of the solutions in practice (Table 2). This proved that our approximation in Eq. 7 is reasonable for practical purpose.

Besides solving the high-resolution structure of each compositional rigid parts of a complex, the segmentation algorithm provides additional information of the spatial relationship between the rigid parts within each individual particle image. Although in the examples of this work, we mainly focused at the molecules made of two rigid components, the concept can be extended to molecules composed of three or even more rigid bodies that are mobile to each other. Such information of the whole dataset can then be summarized for statistical analysis to reflect the distribution of various

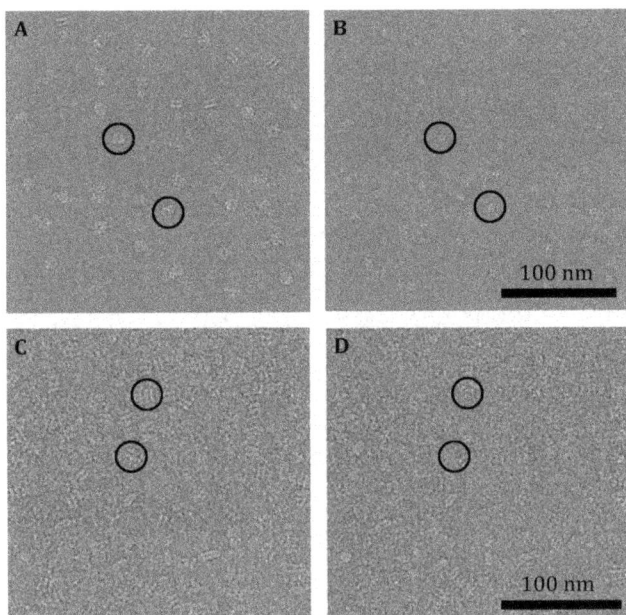

Fig. 6 Particle segmentation from raw micrographs. **A** An area of simulated micrograph of the 20S particles. **B** The same micrograph in **A** from which DD particles were subtracted. **C** An area of a raw micrograph of 20S particles. **D** The same micrograph in **C** from which the 20S particles were subtracted. Some typical particles are marked with black circles

conformational states within the flexible molecule. The conformational distribution is of important biological relevance beyond what the static structure can provide, thus realizing the unique power of single particle analysis.

MATERIALS AND METHODS

Computation implementation

The particle segmentation algorithm described above was implemented as a new program "subtract_micrograph" and its mpi version "subtract_micrograph_mpi" within the RELION 1.4 package. Part of the source code was copied or adapted from RELION 1.3 or 1.4. We also incorporated this program in a GUI version of RELION 1.4 (Fig. 7).

Generation of simulated dataset

Previous works (Zhao *et al.* 2015; Zhou *et al.* 2015) showed that human 20S particle functioning in membrane fusion processes in eukaryotic cells is composed of two parts relatively flexible to each other: the SS complex with pseudo four-fold symmetry and the hexameric NSF complex. We used the 20S particle as a

testing model to generate simulated dataset. For convenience of the simulation, we built a model of the SS complex without symmetry and a hexameric model of DD imposed with a C6 symmetry using the Modeller software package (Eswar *et al.* 2006). The two atomic models were converted to MRC format with e2pdb2mrc.py in EMAN2 package (Tang *et al.* 2007). The two MRC volumes with voxel size of 1.32 Å representing the SS and DD portions of 20S particle were then assembled together to resemble the overall architecture of 20S particle. Heterogeneous conformational states were generated by randomly tilting the two portions independently with a standard deviation of 10° for all three Euler angles and translating the two parts with a standard deviation of 2 pixels in coordinates. Subsequently, we used the full set of simulated 3D MRC volumes to generate simulated electron micrographs using a program genRandomImage.py written with EMAN2 package. A total of 48 simulated electron micrographs each containing 150 particle images at random orientations and locations were generated. In each of these micrographs, CTF-independent Gaussian white noise was superimposed and CTF-dependent water noise was generated by randomizing the Fourier phase of the atomic model of water molecules simulated with NAMD and VMD (Humphrey *et al.* 1996). The noise level and CTF parameters in these simulated micrographs were chosen to mimic the real micrographs obtained by a Gatan K2-Summit electron counting camera on a Titan Krios microscope operated at 300 kV. More details of the parameters for simulation are listed in Table 1.

Processing of simulated dataset

A total of 7200 SS/DD particle images were extracted from simulated micrographs with a box size of 256 pixels. These particle images were first 3D refined with RELION 1.3 against an initial model of 20S particle low-pass filtered at 60 Å resolution. As a control, we refined the 3D reconstruction with local angular search range of 30°, during which a SS or DD mask was applied, resulting in a SS or DD volume, respectively. As another control, we also generated SS or DD sub-particles with relion_project and performed 3D auto-refinement with these sub-particles with a local angular search range of 30°. Alternatively, using our implemented segmentation algorithm, the SS particles were segmented by subtracting the DD density from the whole particle images. The segmented and re-windowed SS particles with a box size of 160 pixels were subjected to 2D classification to select the good SS particle images for further 3D refinement in RELION 1.3. After the 3D refinement of segmented SS particles, DD particles were segmented

Fig. 7 The GUI interface of the segmentation algorithm embedded in RELION package. The segmentation algorithm was embedded in RELION

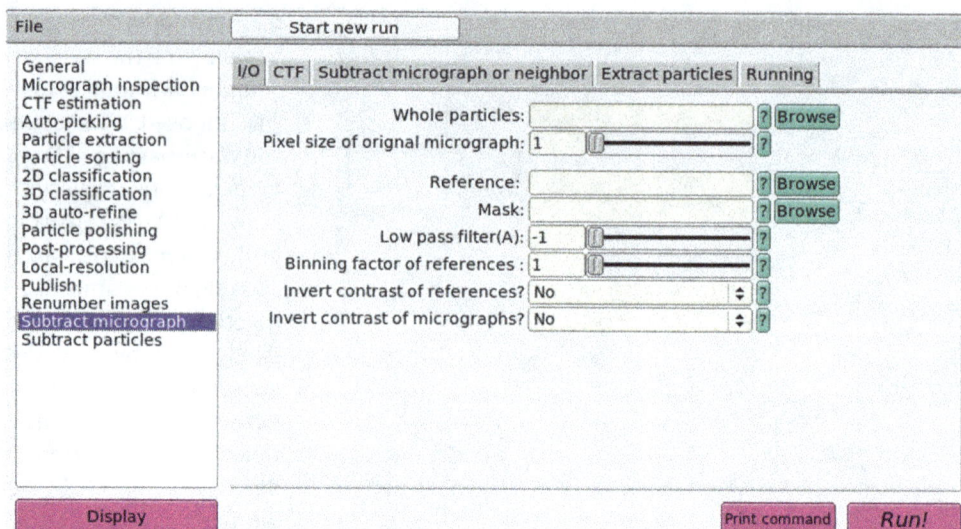

and re-windowed from the whole particle images by subtracting the SS density calculated from the new SS 3D volume. The DD particle images were then subjected to 2D classification and 3D refinement, resulting in an updated DD 3D volume, which was then used for the next cycle of SS segmentation and 3D reconstruction.

Processing of influenza RdRP

The 3D reconstruction of influenza RdRP tetramer and dimer was described previously (Chang *et al.* 2015). The RdRP dataset from the previous work was used in this study. Each raw particle image containing a tetramer has a pixel size of 1.32 Å and a dimension of 256 pixels. Two RdRP dimer particles were segmented and re-windowed from each raw tetramer particle image with a box size of 180 pixels. Therefore, the particle number of RdRP dimer was doubled after segmentation from the tetramers. The segmented RdRP dimer particles were subsequently used for 2D classification and 3D refinement analysis. As a control, we also generated dimer sub-particles with relion_project and performed 3D auto-refinement with all of the dimer sub-particles.

Processing of 70S ribosome

We used a cryo-EM dataset of 70S ribosome comprising 68,543 particle images with box size of 280 pixels and a pixel size of 1.32 Å from Prof. Ning Gao's group. These micrographs were taken from a Titan Krios microscope equipped with a Gatan K2-Summit electron counting camera. We firstly reconstructed a 3D volume of the entire 70S ribosome following the conventional way. This 3D reconstruction was further refined with a local angular search range of 15°, during which a 30S or 50S

mask was applied, resulting in the 3D map of 30S or 50S subunit, respectively. We then segmented the 30S subunit from the dataset with a box size of 280 pixels by subtracting the 50S subunit with the segmentation algorithm. The segmented 30S particles were subjected to 2D classification to select good particles for further 3D auto-refinement. The 50S subunit was subsequently segmented from the 70S ribosome images by subtracting the 30S signal using the segmentation algorithm. The segmented 50S subunit images were then refined to reconstruct a 3D volume. As a control, we also generated 30S or 50S sub-particles with relion_project and performed 3D auto-refinement with these sub-particles. The rotating angles between segmented 30S and 50S subunits were calculated with a program CompareDataStars_data.py written with EMAN2 package.

Other procedures

The micrograph of 20S particle was obtained as described in our previous paper (Zhou *et al.* 2015). 2D classification, 3D reconstruction, and auto-refinement were performed with RELION 1.3. CTF parameters were determined with CTFFIND3 (Mindell and Grigorieff 2003). Reconstruction resolution was estimated with high-frequency noise substituted gold-standard FSC (Scheres and Chen 2012; Chen *et al.* 2013). Local resolution was calculated with ResMap (Kucukelbir *et al.* 2014). Corresponding masks were also applied during the 3D auto-refinement of the segmented particles if not particularly indicated. 3D volume segmentation and atomic model docking were performed with UCSF Chimera (Pettersen *et al.* 2004). The 3D refinements mentioned above are summarized in Table 2.

Acknowledgements The software and scripts used in the work can be accessed via https://github.com/zhouqiang00/Particle-Segmentation. We thank Prof. X. Li, S.-F. Sui for helpful discussions, Dr. D.P. Sun and Dr. J. Wang for kindly providing the RdRP dataset, and Prof. N. Gao and Dr. Y.X. Zhang for kindly providing the ribosome dataset. This work was supported by Grant (2016YFA0501100 to H.W.) from the Ministry of Science and Technology of China and Grant (Z161100000116034 to H.W.) from the Beijing Municipal Science & Technology Commission. Q.Z. was supported by CLS Postdoctoral Fellowship Foundation.

Compliance with ethical standards

Conflict of interest Qiang Zhou, Niyun Zhou, and Hongwei Wang declare that they have no conflict of interest.

Human and animal rights and informed consent This article does not contain any studies with human or animal subjects performed by any of the authors.

References

Amunts A, Brown A, Bai XC, Llacer JL, Hussain T, Emsley P, Long F, Murshudov G, Scheres SH, Ramakrishnan V (2014) Structure of the yeast mitochondrial large ribosomal subunit. Science 343:1485–1489

Anden J, Katsevich E, Singer A (2015) COVARIANCE ESTIMATION USING CONJUGATE GRADIENT FOR 3D CLASSIFICATION IN CRYO-EM. Proceedings. IEEE Int Symp Biomed Imaging 2015:200–204

Bai XC, Fernandez IS, McMullan G, Scheres SH (2013) Ribosome structures to near-atomic resolution from thirty thousand cryo-EM particles. Elife 2:e00461

Bai XC, McMullan G, Scheres SH (2015a) How cryo-EM is revolutionizing structural biology. Trends Biochem Sci 40:49–57

Bai XC, Rajendra E, Yang G, Shi Y, Scheres SH (2015b) Sampling the conformational space of the catalytic subunit of human gamma-secretase. Elife 4:e11182

Bartesaghi A, Merk A, Banerjee S, Matthies D, Wu X, Milne JL, Subramaniam S (2015) 2.2 A resolution cryo-EM structure of beta-galactosidase in complex with a cell-permeant inhibitor. Science 348:1147–1151

Brink J, Ludtke SJ, Kong Y, Wakil SJ, Ma J, Chiu W (2004) Experimental verification of conformational variation of human fatty acid synthase as predicted by normal mode analysis. Structure 12:185–191

Brown A, Amunts A, Bai XC, Sugimoto Y, Edwards PC, Murshudov G, Scheres SH, Ramakrishnan V (2014) Structure of the large ribosomal subunit from human mitochondria. Science 346:718–722

Carazo JM, Sorzano CO, Oton J, Marabini R, Vargas J (2015) Three-dimensional reconstruction methods in single particle analysis from transmission electron microscopy data. Arch Biochem Biophys 581:39–48

Chang S, Sun D, Liang H, Wang J, Li J, Guo L, Wang X, Guan C, Boruah BM, Yuan L, Feng F, Yang M, Wang L, Wang Y, Wojdyla J, Li L, Wang M, Cheng G, Wang HW, Liu Y (2015) Cryo-EM structure of influenza virus RNA polymerase complex at 4.3 A resolution. Mol Cell 57:925–935

Chen S, McMullan G, Faruqi AR, Murshudov GN, Short JM, Scheres SH, Henderson R (2013) High-resolution noise substitution to measure overfitting and validate resolution in 3D structure

determination by single particle electron cryomicroscopy. Ultramicroscopy 135:24–35

Cheng Y (2015) Single-particle cryo-EM at crystallographic resolution. Cell 161:450–457

Cheng Y, Grigorieff N, Penczek PA, Walz T (2015) A primer to single-particle cryo-electron microscopy. Cell 161:438–449

Dashti A, Schwander P, Langlois R, Fung R, Li W, Hosseinizadeh A, Liao HY, Pallesen J, Sharma G, Stupina VA, Simon AE, Dinman JD, Frank J, Ourmazd A (2014) Trajectories of the ribosome as a Brownian nanomachine. Proc Natl Acad Sci 111:17492–17497

Eswar N, Webb B, Marti-Renom MA, Madhusudhan MS, Eramian D, Shen MY, Pieper U, Sali A, (2006). Comparative protein structure modeling using Modeller. Current protocols in bioinformatics/editoral board, Andreas D. Baxevanis... [et al.] Chapter 5, Unit 5.6

Frank J, Ourmazd A (2016) Continuous changes in structure mapped by manifold embedding of single-particle data in cryo-EM. Methods 100:61–67

Humphrey W, Dalke A, Schulten K (1996) VMD: visual molecular dynamics. J Mol Graph 14(33–38):27–38

Ilca S, Kotecha A, Sun X, Poranen M, Stuart D, Huiskonen J (2015) Localized reconstruction of subunits from electron cryomicroscopy images of macromolecular complexes. Nat Commun. doi:10.1038/ncomms9843

Jin Q, Sorzano CO, de la Rosa-Trevin JM, Bilbao-Castro JR, Nunez-Ramirez R, Llorca O, Tama F, Jonic S (2014) Iterative elastic 3D-to-2D alignment method using normal modes for studying structural dynamics of large macromolecular complexes. Structure 22:496–506

Katsevich E, Katsevich A, Singer A (2015) Covariance matrix estimation for the cryo-EM heterogeneity problem. SIAM J Imaging Sci 8:126–185

Kucukelbir A, Sigworth FJ, Tagare HD (2014) Quantifying the local resolution of cryo-EM density maps. Nat Methods 11:63–65

Kuhlbrandt W (2014) Cryo-EM enters a new era. Elife 3:e03678

Leschziner AE, Nogales E (2006) The orthogonal tilt reconstruction method: an approach to generating single-class volumes with no missing cone for ab initio reconstruction of asymmetric particles. J Struct Biol 153:284–299

Liao M, Cao E, Julius D, Cheng Y (2013) Structure of the TRPV1 ion channel determined by electron cryo-microscopy. Nature 504:107–112

Liao HY, Hashem Y, Frank J (2015) Efficient estimation of three-dimensional covariance and its application in the analysis of heterogeneous samples in cryo-electron microscopy. Structure 23:1129–1137

Liu H, Cheng L (2015) Cryo-EM shows the polymerase structures and a nonspooled genome within a dsRNA virus. Science 349:1347–1350

Ma J (2005) Usefulness and limitations of normal mode analysis in modeling dynamics of biomolecular complexes. Structure 13:373–380

Ma J, Karplus M (1997) Ligand-induced conformational changes in ras p21: a normal mode and energy minimization analysis. J Mol Biol 274:114–131

Mindell JA, Grigorieff N (2003) Accurate determination of local defocus and specimen tilt in electron microscopy. J Struct Biol 142:334–347

Nogales E, Scheres SH (2015) Cryo-EM: a unique tool for the visualization of macromolecular complexity. Mol Cell 58:677–689

Penczek PA, Frank J, Spahn CM (2006) A method of focused classification, based on the bootstrap 3D variance analysis, and its application to EF-G-dependent translocation. J Struct Biol 154:184–194

Penczek PA, Kimmel M, Spahn CM (2011) Identifying conformational states of macromolecules by eigen-analysis of resampled cryo-EM images. Structure 19:1582–1590

Penczek PA, Fang J, Li X, Cheng Y, Loerke J, Spahn CM (2014) CTER-rapid estimation of CTF parameters with error assessment. Ultramicroscopy 140:9–19

Pettersen EF, Goddard TD, Huang CC, Couch GS, Greenblatt DM, Meng EC, Ferrin TE (2004) UCSF Chimera–a visualization system for exploratory research and analysis. J Comput Chem 25:1605–1612

Radermacher M, Wagenknecht T, Verschoor A, Frank J (1987) Three-dimensional reconstruction from a single-exposure, random conical tilt series applied to the 50S ribosomal subunit of *Escherichia coli*. J Microsc 146:113–136

Rosenthal PB, Henderson R (2003) Optimal determination of particle orientation, absolute hand, and contrast loss in single-particle electron cryomicroscopy. J Mol Biol 333:721–745

Scheres SH (2012a) A Bayesian view on cryo-EM structure determination. J Mol Biol 415:406–418

Scheres SH (2012b) RELION: implementation of a Bayesian approach to cryo-EM structure determination. J Struct Biol 180:519–530

Scheres SH, Chen S (2012) Prevention of overfitting in cryo-EM structure determination. Nat Methods 9:853–854

Scheres SH, Gao H, Valle M, Herman GT, Eggermont PP, Frank J, Carazo JM (2007) Disentangling conformational states of macromolecules in 3D-EM through likelihood optimization. Nat Methods 4:27–29

Shan H, Wang Z, Zhang F, Xiong Y, Yin CC, Sun F (2016) A local-optimization refinement algorithm in single particle analysis for macromolecular complex with multiple rigid modules. Protein Cell 7:46–62

Tagare HD, Kucukelbir A, Sigworth FJ, Wang H, Rao M (2015) Directly reconstructing principal components of heterogeneous particles from cryo-EM images. J Struct Biol 191:245–262

Tang G, Peng L, Baldwin PR, Mann DS, Jiang W, Rees I, Ludtke SJ (2007) EMAN2: an extensible image processing suite for electron microscopy. J Struct Biol 157:38–46

van Heel M, Frank J (1981) Use of multivariate statistics in analysing the images of biological macromolecules. Ultramicroscopy 6:187–194

Wang HW, Nogales E (2005) An iterative Fourier-Bessel algorithm for reconstruction of helical structures with severe Bessel overlap. J Struct Biol 149:65–78

Wang L, Sigworth FJ (2009) Structure of the BK potassium channel in a lipid membrane from electron cryomicroscopy. Nature 461:292–295

Yan C, Hang J, Wan R, Huang M, Wong CC, Shi Y (2015) Structure of a yeast spliceosome at 3.6-angstrom resolution. Science (New York, N.Y.) 349:1182–1191

Zhang W, Kimmel M, Spahn CM, Penczek PA (2008) Heterogeneity of large macromolecular complexes revealed by 3D cryo-EM variance analysis. Structure 16:1770–1776

Zhang X, Ding K, Yu X, Chang W, Sun J, Zhou ZH (2015) In situ structures of the segmented genome and RNA polymerase complex inside a dsRNA virus. Nature 527:531–534

Zhao M, Wu S, Zhou Q, Vivona S, Cipriano D, Cheng Y, Brunger A (2015) Mechanistic insights into the recycling machine of the SNARE complex. Nature 518:61–67

Zhou Q, Huang X, Sun S, Li XM, Wang HW, Sui SF (2015) Cryo-EM structure of SNAP-SNARE assembly in 20S particle. Cell Res 25:551–560

19

Hessian single-molecule localization microscopy using sCMOS camera

Fudong Xue[1,2], Wenting He[1], Fan Xu[3], Mingshu Zhang[1], Liangyi Chen[4], Pingyong Xu[1,2]✉

[1] Key Laboratory of RNA Biology, Institute of Biophysics, Chinese Academy of Sciences, Beijing 100101, China
[2] College of Life Sciences, University of Chinese Academy of Sciences, Beijing 100049, China
[3] Weldon School of Biomedical Engineering, Purdue University, West Lafayette, IN 47907, USA
[4] State Key Laboratory of Membrane Biology, Beijing Key Laboratory of Cardiometabolic Molecular Medicine, Institute of Molecular Medicine, Peking University, Beijing 100871, China

Abstract   Single-molecule localization microscopy (SMLM) has the highest spatial resolution among the existing super-resolution imaging techniques, but its temporal resolution needs further improvement. An sCMOS camera can effectively increase the imaging rate due to its large field of view and fast imaging speed. Using an sCMOS camera for SMLM imaging can significantly improve the imaging time resolution, but the unique single-pixel-dependent readout noise of sCMOS cameras severely limits their application in SMLM imaging. This paper develops a Hessian-based SMLM (Hessian-SMLM) method that can correct the variance, gain, and offset of a single pixel of a camera and effectively eliminate the pixel-dependent readout noise of sCMOS cameras, especially when the signal-to-noise ratio is low. Using Hessian-SMLM to image mEos3.2-labeled actin was able to significantly reduce the artifacts due to camera noise.

Keywords   Hessian, SMLM, Super-resolution, sCMOS

INTRODUCTION

Super-resolution (SR) microscopy enables biological researchers to see nanoscale images of intracellular structures. Single-molecule localization microscopy (SMLM), such as PALM/STORM (Betzig *et al.* 2006; Hell 2007; Rust *et al.* 2006), uses photocontrollable fluorescent proteins to label target proteins and image a small number of single molecules that are randomly activated at different times. And then the central position of a single fluorescent molecule is precisely located by a Gaussian fitting algorithm. Next, a final image with diffraction-unlimited resolution is reconstructed if a sufficient number of single molecules are obtained. This SMLM technology can achieve a spatial resolution of approximately 20–30 nm. However, this method requires thousands of image frames to reconstruct SR structures, and it is difficult to meet the researchers' requirements for temporal resolution.

To achieve higher temporal resolution, a faster signal acquisition with a sufficiently high signal-to-noise ratio is important. Increasing the exposure frequency (*i.e.*, decreasing the exposure time) provides faster signal acquisition, but for back-illuminated electron multiplying charge-coupled devices (EMCCDs), the read speed is currently limited to 70 frames per second, limiting the temporal resolution of SR techniques that use multiple sequential frames to reconstruct a diffraction-unlimited image, especially SMLM microscopy. Compared with EMCCD cameras, the newly developed sCMOS camera with the ability to detect single molecules has the advantages of a large field of view, fast readout speed,

✉ Correspondence: pyxu@ibp.ac.cn (P. Xu)

etc., which can significantly increase the data acquisition rate and improve the temporal resolution. However, unlike an EMCCD, an sCMOS camera has a unique single-pixel-dependent readout noise, and the random noise appearing in each single frame at different time points will be incorrectly fitted to the fluorescence signal of a single molecule. Since an SMLM image is a superposition of a single molecule localized in a large number of single-frame (tens of thousands to several tens of thousands) images, the superposition of randomly occurring noise in a single-frame image generates a pseudostructure in the final SR image.

The unique noise of the sCMOS camera limits its application in fast SMLM imaging. Therefore, there is an urgent need for noise calibration and for the development of a corresponding single-molecule localization algorithm. Huang *et al.* developed an algorithm to calibrate the independent readout noise by addressing three important characteristics (Huang *et al.* 2013): (1) Remove single-pixel-dependent offsets; (2) correct single-pixel-dependent gain values; and (3) remove the systematic fluctuation noise of each pixel. This method first calculates the brightness offset and the fluctuation level of each pixel by acquiring the dark image of the camera and then calculates the gain per pixel by illuminating the camera with different intensities. Then, these three parameters are used to calibrate the actual fluorescence image. This method is a good way to distinguish the source of readout noise and to calibrate and eliminate the single-pixel-dependent systematic readout noise of the camera. However, Huang *et al.* assumed Gaussian noise model of readout noise for all pixels on the sensor, which is based on the data obtained when the camera is dark, may not reflect the underlying fluctuation pattern for each pixel on the camera at different time points during the actual sampling, because during the actual sampling the fluctuation level of each pixel at a certain time point varies greatly. In SMLM experiments, these fluctuating noises can easily be treated as signals by single-molecule detection algorithms, resulting in artifacts in the reconstruction results.

The Hessian matrix is a square matrix composed of second-order partial derivatives of multivariate functions and is often used in the field of boundary detection and denoising (Lefkimmiatis *et al.* 2013; Sun *et al.* 2015). In 2018, Huang *et al.* developed a Hessian-SIM method (Huang *et al.* 2018). The Hessian penalty matrix was used to remove the scattering noise after SIM reconstruction, and the SR images with minimal artifacts were obtained. In this paper, we develop the new technique Hessian-SMLM, an SMLM imaging technology for rapid imaging using an sCMOS camera and combining the Hessian and Huang's sCMOS calibrating algorithms to effectively remove sCMOS pixel-dependent readout noise. The new method can significantly reduce the reconstruction artifacts due to noise when used for actin PALM imaging.

OVERVIEW OF HESSIAN-SMLM FOR SR IMAGING

First, we collected the image when the sCMOS camera was in the dark or illuminated by a series of white lights with different intensities and obtained the calibration parameters of the sCMOS camera: offset and gain. Then, a fluorescently labeled biological sample was imaged by the sCMOS camera. The corresponding pixel offset obtained previously was subtracted from each pixel in the acquired single-molecule data, and the fluorescence of each pixel was then divided by the gain. A Hessian denoising algorithm was then used to remove the fluctuation noise caused by variance. Next, the noise-corrected single-molecule image was used for single-molecule extraction. Last, the precise single-molecule position was obtained by Gaussian fitting, and the SR image was reconstructed. The specific process of Hessian-SMLM is shown in Fig. 1.

Acquisition of offset and gain of sCMOS camera

Offset is a constant luminance term in the readout process of sCMOS that is added in advance to avoid the

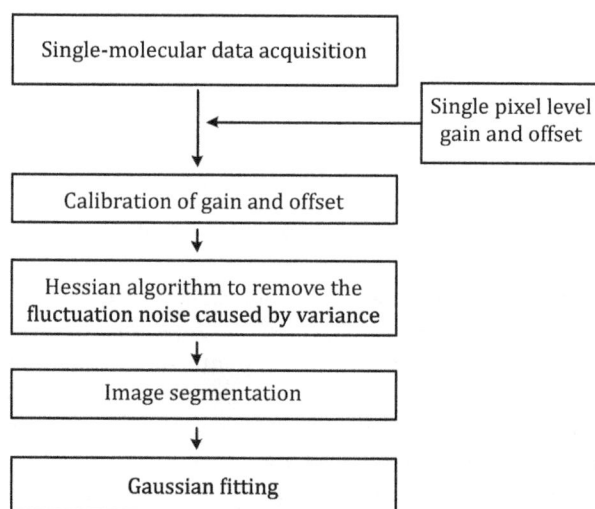

Fig. 1 Diagram of Hessian-SMLM SR imaging. First, single-pixel level characterization of sCMOS is obtained, and the single-molecule data of each pixel are calibrated by the gain and offset values of the sCMOS camera. Then, the calibrated data are treated with the Hessian algorithm to remove the noise caused by variance. In the localization step, the images are segmented and fitted by Gaussian fitting to obtain the final SR image

negative luminance caused by the readout noise in the sCMOS camera. By taking a series of dark images with an sCMOS camera and calculating the average value of each pixel, the offset of each pixel can be measured. First, we turned off the autocalibration mode of the Prime 95B sCMOS camera (Photometrics, Tucson, AZ USA) and recorded 60,000 frames of dark images (Fig. 2A). The offset of each pixel in the sCMOS camera can be derived from the following statistical analysis:

$$Offset_{ij} = \frac{1}{N} \sum_{s=1}^{N} pix_{ij}^s, \tag{1}$$

where N is the total number of dark frames and pix_{ij}^s is the photon number at frame s for pixel ij.

Because the gain of each pixel of the sCMOS camera is different, the amplification of the fluorescence signal is different for each pixel during actual imaging, so it is necessary to measure the gain of each pixel for calibration. To calculate the gain for each pixel, we first took a series of dark images and calculated the variance as follows:

$$Var_{ij} = \frac{1}{N} \sum_{s=1}^{N} \left(pix_{ij}^s - Offset_{ij} \right)^2. \tag{2}$$

Then, we uniformly illuminated the camera with white light of different light intensities ranging from ~ 20 to 200 photons, which were evenly divided into 15 groups from low to high according to the light intensity. In total, 20,000 frames of images were captured for each group (Fig. 2B). The variance and mean value of brightness under different light intensities were then obtained, and linear fitting by the least-squares method was performed, as in Eq. 3 (Fig. 2C).

$$\hat{g}_{ij} = \arg\ \min \sum_{k=1}^{N} \left(\left(var_{ij}^k - var_{ij} \right) - g_{ij} \left(Mean_{ij}^k - Offset_{ij} \right) \right)^2. \tag{3}$$

Hessian algorithm removes variance noise in single-molecule images

For Prime 95B sCMOS imaging, the pixel size is 110 nm/pixel for a 100× magnification microscope. Due to the optical diffraction limit, the full width at half maximum (FWHM) of a diffraction spot is greater than 250 nm, so when a single molecule on the sample is imaged through an optical system, the molecule occupies at least two pixels on the camera, which means that spatial correlation exists for two pixels. At the same time, if the speed of single-molecule blinking is lower than the camera's readout speed, there should also be correlations across time series for the same pixel. However, there should be no correlation of random Gaussian noise, especially sCMOS pixel-dependent noise, between different pixels and/or across time series for the same pixel. Therefore, we refer to a reference (Huang *et al.* 2018) and introduce the Hessian penalty to remove the temporally and spatially random noise. The optimization function is as follows:

$$H(r,t) = \arg\ \min \left[m\|H(r,t) - D(r,t)\|_2^2 + R_{\text{Hessain}}(H(r,t)) \right], \tag{4}$$

where D is the raw data collected by the camera; H is the optimized data; m is the relative weight between the first term and the Hessian penalty; and R_{Hessian} is the Hessian penalty (Lefkimmiatis *et al.* 2012), which is expressed by Eq. 5,

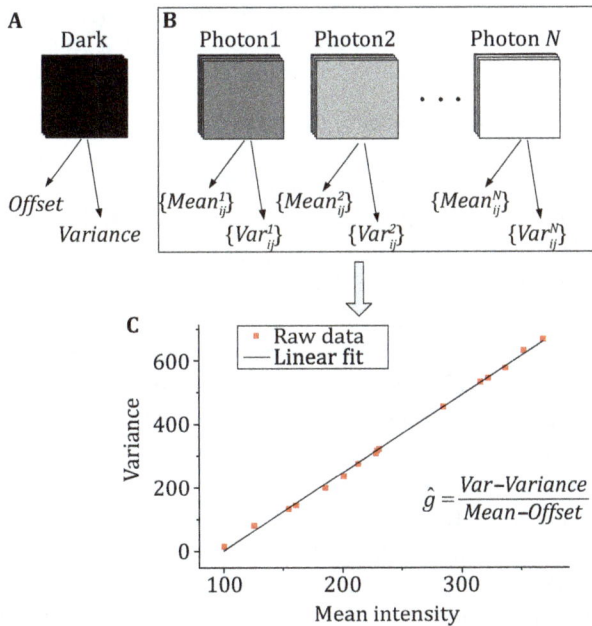

Fig. 2 sCMOS characterization at the single-pixel level. **A** To measure the offset and variance of sCMOS, 60,000 dark frames were used. **B** A series of image sequences were recorded at different average intensity levels to measure the mean values and variances. **C** The gain for each pixel can then be calculated with the help of the previously obtained variance and mean value. The *red dashed line* is the variance at different mean intensities of one pixel of the sCMOS camera, and the *solid line* shows the linear fit result

$$R_{\text{Hession}}(H(r,t))$$

$$= \int_\Omega \left\| \begin{pmatrix} H_{xx}(r,t) & H_{xy}(r,t) & \sqrt{\sigma}H_{xt}(r,t) \\ H_{yx}(r,t) & H_{yy}(r,t) & \sqrt{\sigma}H_{yt}(r,t) \\ \sqrt{\sigma}H_{tx}(r,t) & \sqrt{\sigma}H_{ty}(r,t) & \sigma H_{tt}(r,t) \end{pmatrix} \right\|_1 drdt,$$

(5)

where r is the position of each pixel; Ω represents the integral area that contains all pixels within the image H; and H_{xy} is the second-order partial derivative of H versus xy; σ is a parameter that was introduced to enforce the continuity of structures along the time axis.

We thus finally obtain the optimal noise-removed single-molecule data by calculating the minimum value of Eq. 4 from the single-molecule data, and use these data for single-molecule extraction and reconstruction as previous reference (Olivo-Marin 2002; Smith *et al.* 2010).

Cell culture

U2OS cells were cultured using complete MCMM (McCoy's 5A Medium Modified, Gibco) supplemented with 10% inactivated fetal bovine serum (FBS, Gibco) in a 37 °C incubator with 5% CO_2. The construct of Life-Act-mEos3.2 for labeling actin structures and the Hessian-SMLM SR imaging were described in Zhang *et al.* (2012). The constructs were transfected using Lipofectamine™ 2000 Transfection Reagent (Invitrogen, USA). After a 48-h transfection, the cells were then fixed by 4% (*w/v*) paraformaldehyde (NOVON Scientific, Pleasanton, CA USA) and 0.2% glutaraldehyde (Electron Microscopy Sciences, Hatfield, PA USA) and imaged by a custom built PALM microscope.

Simulation of Hessian-SMLM imaging

To verify the Hessian-SMLM's ability to remove variance noise, we simulated single-molecule data containing fluctuating noise and performed PALM imaging simulations. We simulated and generated the original data shown in Fig. 3B and used the Thunder STORM imaging plug-in to generate single-molecule images (Ovesný *et al.* 2014). Simulated wave noise was added (Fig. 3A) so that we obtained simulated single-molecule data containing fluctuating noise. Then, the PALM single-molecule localization algorithm was used to reconstruct the image with (Fig. 3D) and without (Fig. 3C) Hessian denoising methods. As shown in Fig. 3C and D, the resulting reconstruction without the Hessian denoising method contains more artifacts, while Hessian denoising efficiently removed the artifacts caused by the fluctuating noise.

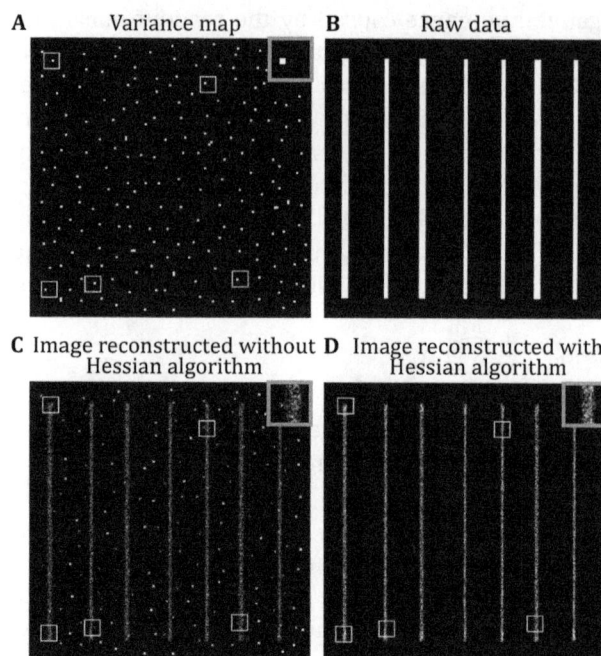

Fig. 3 Simulation of single emitters on parallel lines based on the shown variance map with localization using single-molecule localization microscopy with and without the Hessian algorithm. **A** Simulated variance map. **B** Raw data used to simulate the single-molecule image. **C**, **D** Simulated single-molecule image reconstructed without (**C**) and with (**D**) the Hessian algorithm. The *green box* denotes the enlarged section shown in the inset

Hessian-SMLM imaging of actin

To verify the effectiveness of Hessian-SMLM in SR imaging of biological samples, we transfected LifeAct-mEos3.2 in U2OS cells and acquired single-molecule data using the Prime 95B sCMOS camera. We treated the single-molecule data to reconstruct an SR image using different algorithms separately as follows: (1) Single-molecule localization algorithm of PALM; (2) the sCMOS-calibrated single-molecule localization algorithm by Huang *et al.*; and (3) the Hessian-SMLM algorithm in the current study.

We found that when using the Prime 95B sCMOS camera to acquire single-molecule data, there is substantial noise in the variance map of the fluorescence signal. Most of the noise is completely colocalized with the variance map of the dark camera (Fig. 4C), suggesting that the noise comes mainly from the variance when the camera is dark. Moreover, when the single-molecule data collected by the sCMOS camera are directly used in the PALM single-molecule localization algorithm, the reconstruction results contain more artifacts (Fig. 4D, indicated by boxes). Obviously, these artifacts come from the variance of the sCMOS camera because they are exactly the same noise points in the

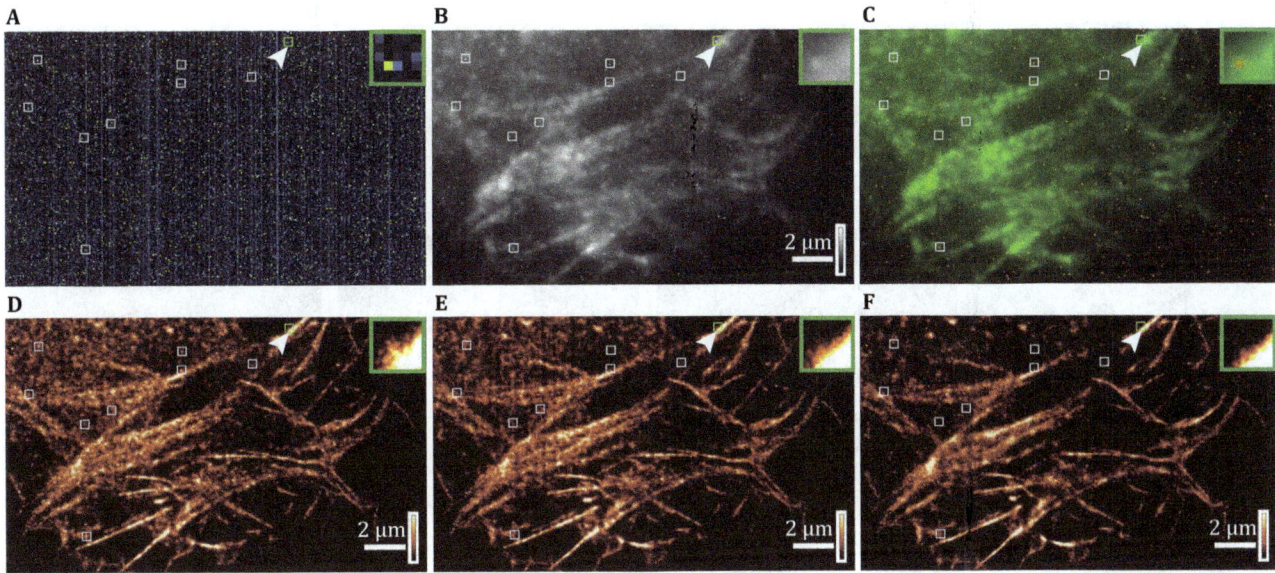

Fig. 4 Hessian-SMLM imaging of actin with three SMLM algorithms. **A** Readout variance map of the sCMOS camera. **B** Standard deviation (STD) projection image of single-molecule data. **C** Merged image of the variance map (**A**) and the STD projection image (**B**). *Green* pixel-level readout variance of the dark camera; *Red* STD image. **D–F** Reconstructed super-resolution images of actin analyzed using the conventional algorithm (**D**), sCMOS camera-specific algorithm (**E**), and Hessian-SMLM algorithm (**F**). The *green box* denotes the enlarged section shown in the inset. Color scales, bottom to top: 0–65 ADU^2, where ADU means analog-to-digital units (**A**), minimum-to-maximum signal (**B**), same upper bound chosen for best visualization for **D–F**. Scale bars 2 μm

variance map when the camera is dark (Fig. 4A). Using Huang *et al.*'s sCMOS correction method to correct the variance, gain, and offset greatly improved the reconstruction results and significantly reduced the artifacts due to single-pixel fluctuations. However, for some areas, especially those where the variance is smaller, the calibration effect of this method is not ideal (Fig. 4E). One possible reason is that the variance we obtained with the dark camera can only represent the statistical fluctuation level of each pixel during the sampling process. In the actual sampling, the averaged variance of each pixel from the dark sCMOS camera does not accurately characterize the true fluctuation level of each frame at different time points. However, as shown in Fig. 4F, Hessian-SMLM can efficiently remove the variance noise in each frame produced by the fluctuation at different time points, and the artifacts generated in the reconstruction result are significantly reduced.

Next, we further compared the effect of the above three methods on a LifeAct-mEos3.2 reconstruction with a low exposure time and laser intensity. The low exposure time and laser intensity significantly reduced the signal-to-noise ratio of the single-molecule data. The structure directly reconstructed with the PALM localization algorithm had substantial noise and many artifactual structures (Fig. 5B). Using the sCMOS calibration method removed some of the noise (Fig. 5C), the effect was not as good as the effect when the signal-to-noise

ratio was high (Fig. 4E). However, with the Hessian-SMLM method, the reconstruction artifacts were basically removed, and the structural continuity of actin was much better than the continuity with the other two methods (Fig. 5D compared to Fig. 5B and C, and enlarged regions shown in Fig. 5E). We chose four pixels with different variance and measured the pixel intensity before and after Hessian algorithm (Fig. 5F). The fluctuating value of the four pixels decreased significantly after Hessian algorithm (Fig. 5F, blue and red solid lines), while the mean pixel values are almost the same before and after Hessian algorithm (Fig. 5F, blue and red dash lines). Therefore, Hessian-SMLM is able to efficiently remove the variance noise of each frame produced by the fluctuation at different time points, and reduce the artifacts.

DISCUSSION

Although sCMOS cameras are inferior to EMCCDs for single-molecule imaging, sCMOS cameras have the advantages of fast speed and large field of view and have potential applications in live-cell SMLM imaging. However, the unique single-pixel-dependent readout noise of sCMOS cameras severely limits their application in live SMLM, especially when the signal-to-noise ratio of a single-molecule signal is low. For SMLM, there are at

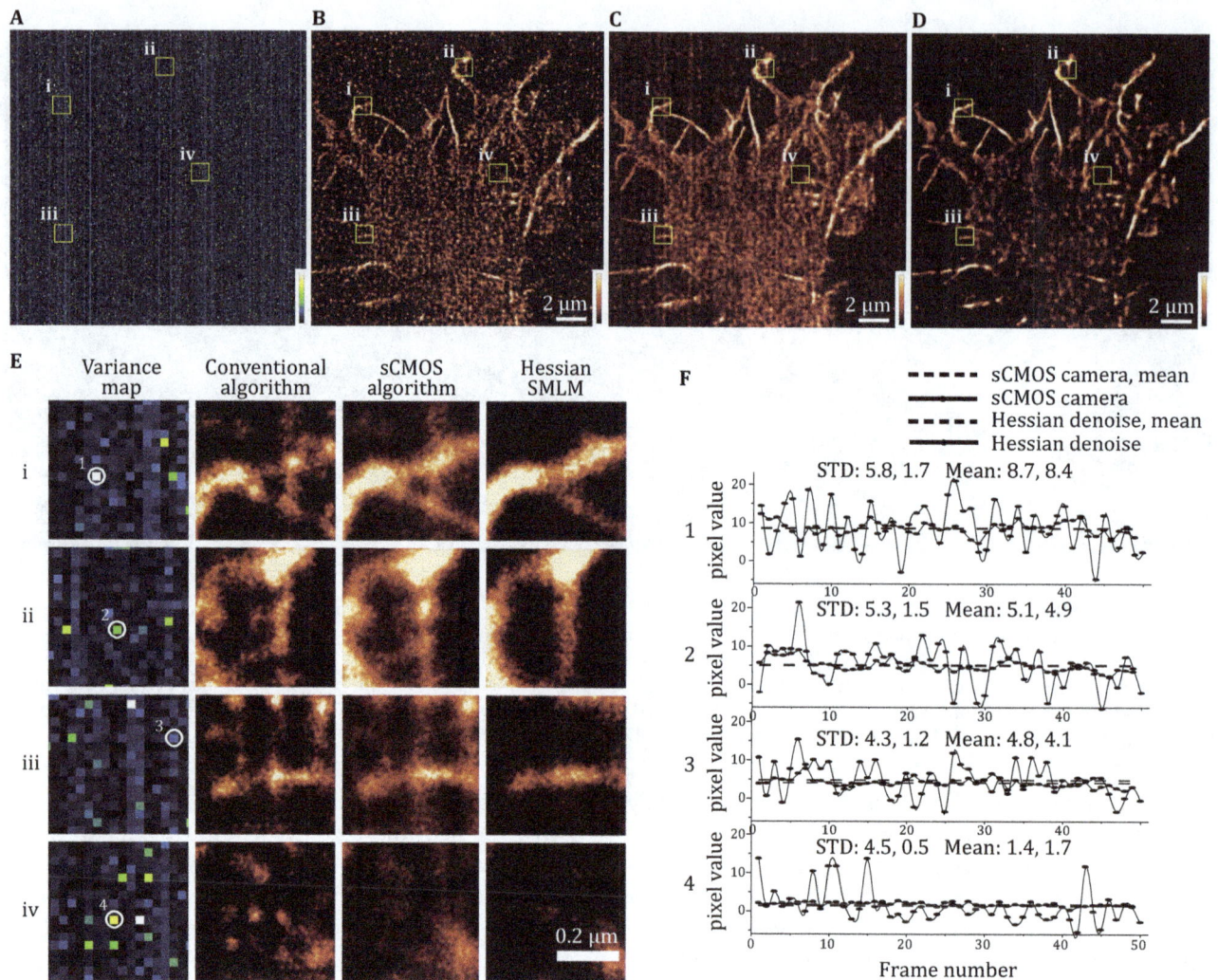

Fig. 5 Hessian-SMLM significantly removes the noise of sCMOS and artifacts in SMLM images even when the signal-to-noise ratio is low. **A** Readout variance map of the sCMOS camera. **B–D** Reconstructed super-resolution images of actins analyzed using the conventional algorithm (**B**), sCMOS algorithm (**C**), and Hessian-SMLM algorithm (**D**). **E** Zoomed in regions (i–iv) from (**A–D**) show actin structures by the PALM localization algorithm, the sCMOS calibration method, and Hessian-SMLM algorithm. **F** The pixel intensity traces of the selected pixels were from cropped regions (i–iv) over 50 frames. Color scales, bottom to top: 0–50 ADU^2 (**A**); same upper bound chosen for best visualization for **B–D**. Scale bars 2 μm (**B–D**), 0.2 μm (**E**)

least two situations that produce low signal-to-noise ratios for single-molecule signals. The first situation is the imaging of fixed cells by SMLM when the signal-to-noise ratio becomes worse as the sampling time increases. The other one is live SMLM. For live SMLM, on the one hand, the exposure time must be decreased to increase the sampling frequency and thus improve the temporal resolution. On the other hand, it is necessary to reduce the irradiation light intensity to reduce phototoxicity. However, reducing the exposure time and reducing the intensity of the illumination light significantly reduce the signal-to-noise ratio of the fluorescence signal. Under these conditions, the camera's readout noise will be particularly notable. In particular,

the camera's variance noise will be Gaussian fitted as a fake single-molecule signal, which will produce false structures or artifacts in the final reconstructed SR image. In this study, our new Hessian-SMLM SR imaging method can well remove the single-pixel-dependent readout noise and effectively reduce the artifacts caused by the readout noise of the sCMOS camera. This method is especially suitable for SMLM imaging when the signal-to-noise ratio is low and has greater advantages over other SMLM imaging methods under conditions of low exposure time and laser intensity. Therefore, Hessian-SMLM is potentially more suitable than the other methods for live-cell SMLM imaging.

Acknowledgements This work was supported by the National Key R&D Program of China (2016YFA0501500 and 2017YFA0505300), the National Natural Science Foundation of China (31421002, 21778069 and 31670870), Project of the Chinese Academy of Sciences (XDB08030203), and CAS-Peking University Joint Team Project. We thank Dr. Fang Huang from Purdue University for very helpful comments on the manuscript.

Compliance with Ethical Standards

Conflict of interest Fudong Xue, Wenting He, Fan Xu, Mingshu Zhang, Liangyi Chen, and Pingyong Xu declare that they have no conflict of interest.

Human and animal rights and informed consent This article does not contain any studies with human or animal subjects performed by any of the authors.

References

Betzig E, Patterson GH, Sougrat R, Lindwasser OW, Olenych S, Bonifacino JS, Davidson MW, Lippincott-Schwartz J, Hess HF (2006) Imaging intracellular fluorescent proteins at nanometer resolution. Science 313(5793):1642–1645

Hell SW (2007) Far-field optical nanoscopy. Science 316(5828):1153–1158

Huang F, Hartwich TMP, Rivera-Molina FE, Lin Y, Duim WC, Long JJ, Uchil PD, Myers JR, Baird MA, Mothes W, Davidson MW, Toomre D, Bewersdorf J (2013) Video-rate nanoscopy using sCMOS camera—specific single-molecule localization algorithms. Nat Methods 10(7):653–658

Huang X, Fan J, Li L, Liu H, Wu R, Wu Y, Wei L, Mao H, Lal A, Xi P, Tang L, Zhang Y, Liu Y, Tan S, Chen L (2018) Fast, long-term, super-resolution imaging with Hessian structured illumination microscopy. Nat Biotechnol 36(5):451–459

Lefkimmiatis S, Bourquard A, Unser M (2012) Hessian-based regularization for 3D microscopy image restoration. In *ISBI*, pp 1731–1734

Lefkimmiatis S, Ward JP, Unser M (2013) Hessian Schatten-norm regularization for linear inverse problems. IEEE Trans Image Process 22(5):1873–1888

Olivo-Marin J (2002) Extraction of spots in biological images using multiscale products. Pattern Recogn 35(9):1989–1996

Ovesný M, Křížek P, Borkovec J, Švindrych Z, Hagen GM (2014) ThunderSTORM: a comprehensive ImageJ plug-in for PALM and STORM data analysis and super-resolution imaging. Bioinformatics 30(16):2389–2390

Rust MJ, Bates M, Zhuang X (2006) Sub-diffraction-limit imaging by stochastic optical reconstruction microscopy (STORM). Nat Methods 3(10):793

Smith CS, Joseph N, Rieger B, Lidke KA (2010) Fast, single-molecule localization that achieves theoretically minimum uncertainty. Nat Methods 7(5):373–375

Sun T, Sun N, Wang J, Tan S (2015) Iterative CBCT reconstruction using Hessian penalty. Phys Med Biol 60(5):1965

Zhang M, Chang H, Zhang Y, Yu J, Wu L, Ji W, Chen J, Liu B, Lu J, Liu Y (2012) Rational design of true monomeric and bright photoactivatable fluorescent proteins. Nat Methods 9(7):727

20

PI4KIIα regulates insulin secretion and glucose homeostasis via a PKD-dependent pathway

Lunfeng Zhang[1,2], **Jiangmei Li**[1,3], **Panpan Zhang**[3], **Zhen Gao**[1,2],
Yingying Zhao[1,3], **Xinhua Qiao**[1,2], **Chang Chen**[1,2,4✉]

[1] National Laboratory of Biomacromolecules, CAS Center for Excellence in Biomacromolecules, Institute of Biophysics, Chinese Academy of Sciences, Beijing 100101, China
[2] University of Chinese Academy of Sciences, Beijing 100049, China
[3] Shanghai Institute for Advanced Immunochemical Studies, ShanghaiTech University, Shanghai 201210, China
[4] Beijing Institute for Brain Disorders, Beijing 100069, China

Abstract   Insulin release by pancreatic β cells plays a key role in regulating blood glucose levels in humans, and to understand the mechanism for insulin secretion may reveal therapeutic strategies for diabetes. We found that PI4KIIα transgenic (TG) mice have abnormal glucose tolerance and higher serum glucose levels than wild-type mice. Glucose-stimulated insulin secretion was significantly reduced in both PI4KIIα TG mice and PI4KIIα-overexpressing pancreatic β cell lines. A proximity-based biotin labeling technique, BioID, was used to identify proteins that interact with PI4KIIα, and the results revealed that PI4KIIα interacts with PKD and negatively regulates its activity. The effect of PI4KIIα on insulin secretion was completely rescued by altering PKD activity. PI4KIIα overexpression also worsened glucose tolerance in streptozotocin/high-fat diet-induced diabetic mice by impairing insulin secretion. Our study has shed new light on PI4KIIα function and mechanism in diabetes and identified PI4KIIα as an important regulator of insulin secretion.

Keywords   Phosphatidylinositol 4-kinase IIα (PI4KIIα), Insulin secretion, Protein kinase D (PKD), Carriers of the trans-Golgi network to the cell surface (CARTS)

INTRODUCTION

To maintain the balance of the glucose homeostasis, β cells of pancreas adapt their insulin secretory capability in response to various physiological and pathological demands (Zhang *et al.* 2012). The deterioration of insulin secretion can lead to the hyperglycemic environment that promotes loss of β cell mass and β cell dysfunction. Although insulin resistance has been received as the key character of type II diabetes (T2DM) for a long time, the development of obvious hyperglycemia requires a decrease in β cell function (Pimenta *et al.* 1995; Vauhkonen *et al.* 1998). β Cells are distinct endocrine cells that can respond positively by secreting insulin in response to changes of glucose concentration in the extracellular and to activators of phospholipase C, such as acetylcholine or cholecystokinin, and adenylate cyclase, such as glucagon, glucagon-like peptide-1, or gastric inhibitory polypeptide (Radosavljevic *et al.* 2004). The crucial regulators that can mediate glucose-stimulated insulin release are Ca^{2+}, adenosine

---

Lunfeng Zhang, Jiangmei Li, and Panpan Zhang have contributed equally to this work.

✉ Correspondence: changchen@moon.ibp.ac.cn (C. Chen)

triphosphate (ATP), and diacylglycerol (DAG) (Radosavljevic *et al.* 2004; Rorsman and Renstrom 2003). In addition, there are many direct regulators of each step of insulin release, such as the packaging of insulin in small secretory granules, the trafficking of these granules to the plasma membrane, the exocytotic fusion of the granules with the plasma membrane, and the eventual retrieval of the secreted membranes by endocytosis (Easom 2000; Rorsman and Renstrom 2003). However, the regulation of insulin secretion is not precisely understood (Rorsman and Renstrom 2003).

Phosphatidylinositol kinases and phosphatidylinositol phosphates (PIPs) have recently been strongly associated with insulin secretion by pancreatic β cells. Researchers have indicated that phosphatidylinositol-4-phosphate (PI4P) and phosphatidylinositol-4,5-biphosphate [PI(4,5)P$_2$] increase the insulin secretory response triggered by 10 μmol/L Ca^{2+}, and insulin secretion was diminished by inhibiting the expression of type III PI4-kinase β (PI4KIIIβ) or type I phosphatidylinositol-4-phosphate 5-kinase γ (PI4P5Kγ) (Olsen *et al.* 2003; Waselle *et al.* 2005). Huang *et al.* showed that PI4P5Kα-knockout mice have increased first-phase insulin release and resist the high-fat diet (HFD)-induced development of type 2-like diabetes and obesity. In addition, they concluded that PI4P5Kα regulates insulin release from pancreatic β cells by helping maintain plasma membrane PI(4,5)P$_2$ levels and the integrity of the actin cytoskeleton under both basal and stimulated conditions (Huang *et al.* 2011). Phosphatidylinositol-4-kinase IIα (PI4KIIα), the most abundant PI4K in mammalian cells (Balla and Balla 2006), localizes to the trans-Golgi network (TGN), endosomes and secreted vesicles and has been implicated in the regulation of protein sorting (Balla 2013; Guo *et al.* 2003; Minogue *et al.* 2006; Wang *et al.* 2003). Recently, Ketel *et al.* reported that depleting PI4KIIα causes defects in endosomal exocytosis and that PI(4)P produced by PI4KIIα on Rab11 endosomes is required for the recruitment of the exocyst to enable endosomal exocytosis (Ketel *et al.* 2016). Studies have also indicated that PI4KIIα is involved in recycling and retrograde transport (Jovic *et al.* 2014). Ryder *et al.* showed that PI4KIIα interacts with and regulates the WASH complex and influences vesicle transport (Ryder *et al.* 2013). In addition, PI4KIIα dysfunction contributes to several secretory diseases, such as breast cancer (Chu *et al.* 2010; Lang 2003; Li *et al.* 2010, 2014), spastic paraplegia (Simons *et al.* 2009), Gaucher's disease (Jovic *et al.* 2012), and Alzheimer's disease (Kang *et al.* 2013; Wu *et al.* 2004). However, nothing is known about PI4KIIα in diabetes, the most common disease associated with secretion.

In this study, we demonstrate that PI4KIIα transgenic (TG) mice have impaired glucose tolerance due to abolished insulin secretion under both physiological and pathological conditions. Mechanistic studies indicated that PI4KIIα influences insulin and CARTS complex secretion by regulating PKD activity. The above results suggest that PI4KIIα plays an important role in diabetes and insulin secretion.

RESULTS

Generation and characterization of PI4KIIα TG mice

To study the function of PI4KIIα in T2DM and insulin secretion, we first investigated PI4KIIα expression levels in mouse models of diabetes. As shown in Supplementary materials (Figs. S1A, S1B), PI4KIIα expression in pancreatic islets was markedly increased in KK mice and *db/db* mice compared to wild-type (WT) C57BL/6 mice. These results demonstrate that PI4KIIα is upregulated with diabetes. To determine whether upregulated PI4KIIα expression plays a role in T2DM, a transgenic (TG) PI4KIIα-overexpressing BALB/c mouse model was generated (Fig. 1A). Four independent TG lines (lines 9, 11, 12, and 17) that expressed WT PI4KIIα protein were obtained. Lines 12 and 17 were chosen for further analysis (Fig. 1A). We then detected PI4KIIα expression levels by Western blot and found that it was upregulated in all detected tissues, including brain, pancreas, lung, stomach, fat, liver, intestine, spleen, heart, and muscle (Fig. S2A). Islet and acinar cells were isolated, and PI4KIIα expression was detected by Western blot. As shown in Fig. 1B, both islets and other neighboring cells in the pancreas expressed PI4KIIα, and immunohistochemistry studies indicated that PI4KIIα is highly colocalized with insulin-staining positive cells (Fig. 1C). Therefore, the data confirm that PI4KIIα is expressed in pancreatic β cells. We also analyzed the activity of overexpressed PI4KIIα; the PI4P content in the TG mouse pancreas was twofold higher than that in the WT mouse pancreas (Fig. 1D). PI4KIIα TG mice exhibited normal general health, viability, fertility, and body composition (data not shown). Although the TG mouse body weight was reduced (Fig. S2B), PI4KIIα overexpression had little effect on food intake (Fig. S2C) or plasma cholesterol, triglyceride, LDL-c, and HDL-c content compared to WT littermates (Fig. S2D).

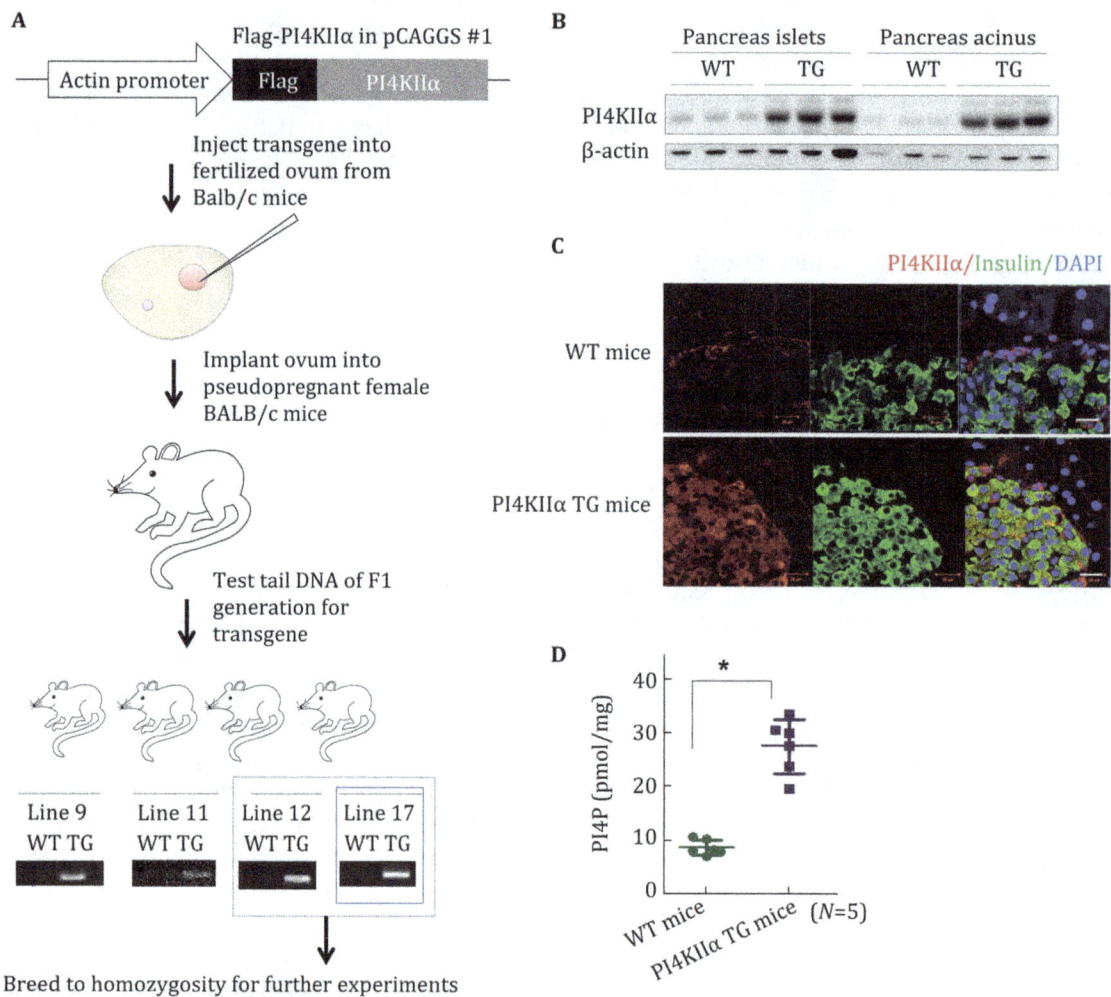

Fig. 1 Generation of PI4KIIα transgenic (TG) mice. **A** Workflow to generate PI4KIIα TG BALB/c mice. **B** Islets and acinus were isolated from the mouse pancreas, and PI4KIIα expression was measured by Western blot. **C** The pancreases of WT and PI4KIIα TG littermates were sectioned at 10 μm using a cryostat. Protein expression in pancreas sections was determined using antibodies against insulin and PI4KIIα. Images were obtained using a laser confocal fluorescence microscope. Scale bar, 40 μm. **D** Islets were isolated from the pancreas of 25-week-old male PI4KIIα TG mice (line 17) and age-matched WT littermates ($N = 5$ for each line), and PI4P content was measured using a PI(4)P Mass Strip Kit. The data are presented as the mean ± SD. All the experiments except **A** were performed three times in triplicate. *$p < 0.05$

PI4KIIα overexpression abolishes glucose tolerance and insulin secretion

The above results indicated that PI4KIIα overexpression has no effect on the blood lipid profile. We then further tested its effect on blood glucose. Under normal chow-feeding conditions, PI4KIIα TG mice had slightly higher blood glucose levels (Fig. 2A) and impaired glucose tolerance (Fig. 2B) compared to their respective WT littermates after 16 h of fasting. However, insulin tolerance has no significant difference between PI4KIIα TG and WT mice, and the blood glucose level after insulin injection was not different among these four lines of mice (Fig. 2C), indicating no effect on insulin resistance. The observed results prompted us to evaluate the influence of PI4KIIα on insulin secretion. As shown in Fig. 2D, PI4KIIα overexpression significantly reduced insulin secretion during hyperglycemic stimulation; both the first and second phases were impaired in both lines (12 and 17) of PI4KIIα TG mice.

To further confirm that PI4KIIα regulates insulin secretion, we overexpressed GFP-PI4KIIα and the kinase-dead mutant GFP-PI4KIIαK152A (Minogue *et al.* 2006) in MIN6 cells (murine insulinoma-derived pancreatic β cell line), an insulin-secreting cell line (Ishihara *et al.* 1993). As shown in Fig. 3A, both WT and kinase-dead PI4KIIα reduced insulin secretion in response to high glucose (33 mmol/L) stimulation, indicating that PI4KIIα kinase activity is not necessary for its regulation of insulin secretion. Consistent with

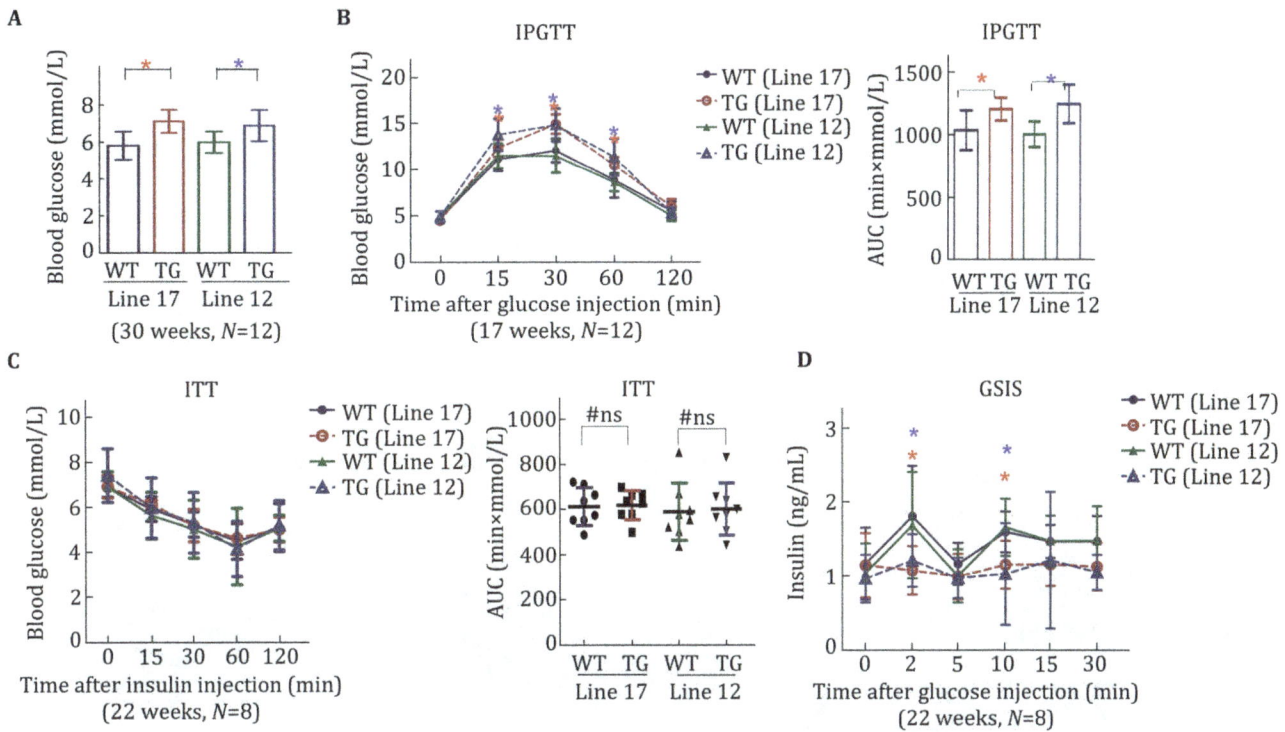

Fig. 2 PI4KIIα overexpression impairs glucose tolerance and insulin secretion. **A** Fasting blood glucose was measured in 30-week-old male PI4KIIα TG mice (lines 12 and 17) and their respective WT littermates ($N = 12$ for each line). **B** Intraperitoneal glucose tolerance test (IPGTT) was performed in overnight-fasted 17-week-old male PI4KIIα TG mice (lines 12 and 17) and age-matched WT littermates ($N = 12$ for each line). **C** ITT was performed in 6-h fasted 22-week-old male PI4KIIα TG mice (lines 12 and 17) and age-matched WT littermates ($N = 8$ for each line). **D** GSIS was performed in overnight-fasted 22-week-old male PI4KIIα TG mice (lines 12 and 17) and age-matched WT littermates ($N = 8$ for each line). The data are presented as the mean $\pm$ SD. All the experiments were performed three times in triplicate. *$p < 0.05$

the above results, siRNA-mediated suppression of PI4KIIα expression in MIN6 cells significantly increased insulin secretion in response to stimulation with 33 mmol/L glucose (Fig. 3B), and this upregulation could not be rescued by adding PI4P, the product of PI4KIIα, to the cell culture (Fig. S3). These findings are consistent with the above result that overexpression of the kinase-dead PI4KIIα suppressed insulin secretion (Fig. 3A). Together, the results indicated that PI4KIIα can regulate insulin secretion independent of kinase activity. To study the regulatory mechanism, we first investigated whether pancreas islet mass was affected in PI4KIIα TG mice. As shown in Figs. S4A and B, neither islet size nor β cell mass were different between PI4KIIα TG mice and WT mice fed a normal chow diet. With the 4′,6-diamidino-2-phenylindole (DAPI), insulin, and BrdU triple staining of pancreatic sections, we discovered that there was no difference in BrdU incorporation into β cells between PI4KIIα TG mice and WT mice (Fig. S4C). Terminal deoxynu-cleotidyl transferase-mediated dUTP-biotin nick end labeling (TUNEL) assay results also indicated that

PI4KIIα overexpression did not induce β cell apoptosis (Fig. S4D). Based on these results, we concluded that PI4KIIα overexpression reduced insulin secretion but did not affect β cell mass. We then questioned whether PI4KIIα directly regulates insulin secretion. To answer this question, we investigated the presence of insulin in PI4KIIα-positive granules by visualizing EGFP-tagged PI4KIIα and endogenous insulin. Insulin was juxtaposed with the PI4KIIα signal (Fig. 3C). A qualitative assessment was performed using more complex sections by viewing Z-stacks of images sequentially in a movie, which made it easier to follow particular structures in three dimensions (3D). Some of these Z-stacks were then converted to surface-rendered 3D objects by Imaris software (Imaris 8 with colocalization; Bitplane, Belfast, UK) to investigate spatial relationships and visualize colocalization. The results indicated that insulin was surrounded by PI4KIIα-containing organelles (Fig. 3C, Supplemental movie). All the above results indicated that PI4KIIα may regulate insulin secretion via protein–protein interactions.

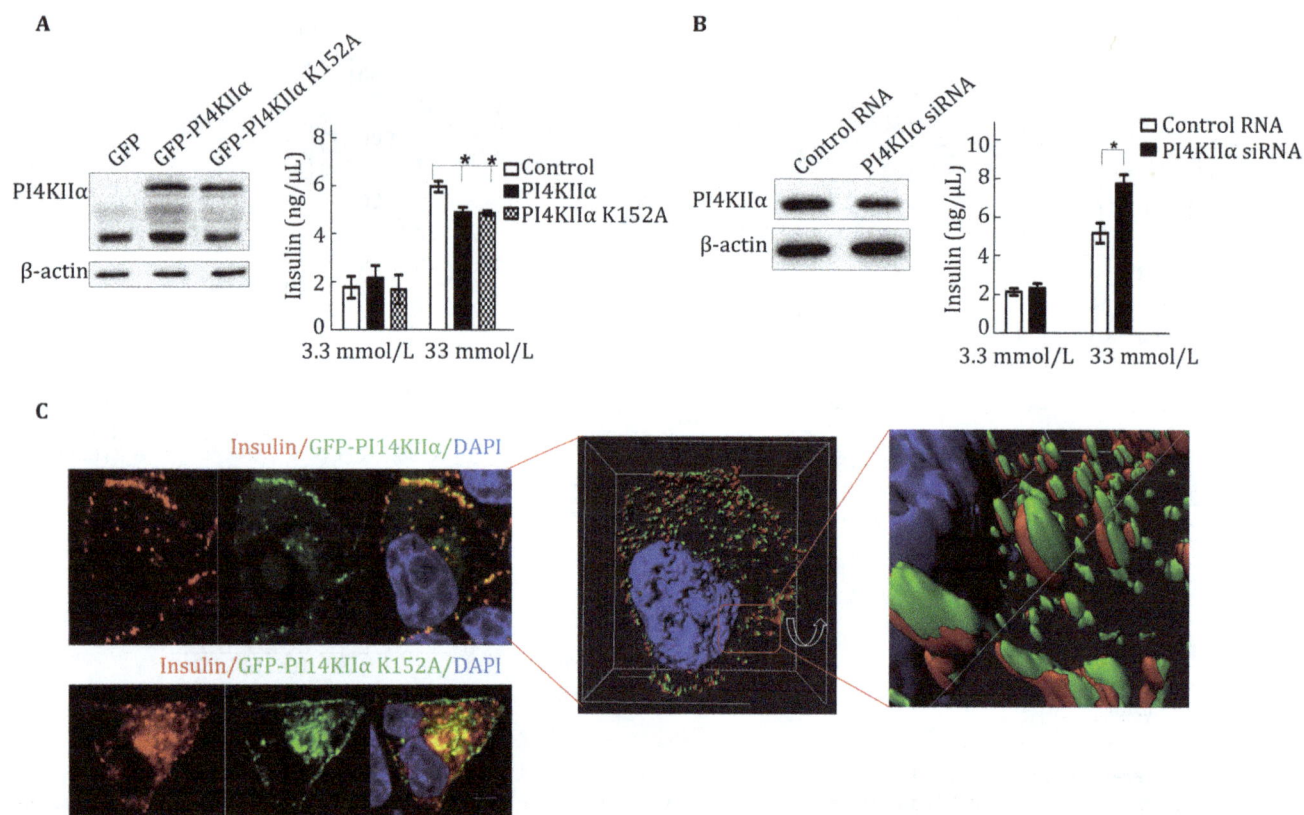

Fig. 3 PI4KIIα regulates insulin secretion in MIN6 cells. **A** MIN6 cells overexpressing GFP, GFP-PI4KIIα, or GFP-PI4KIIα K152A. **B** MIN6 cells with siRNA-mediated knockdown of PI4KIIα. Insulin secretion in response to 3.3 or 33 mmol/L glucose was measured using an insulin ELISA kit. **C** Immunostaining of the nucleus (*blue*), insulin (*red*), and GFP-PI4KIIα or GFP-PI4KIIα K152A in MIN6 cells. The 3D cell model was built using Imaris software (Imaris 8 with colocalization; Bitplane, Belfast, UK). Scale bar, 2 μm. The data are presented as the mean ± SD of three independent experiments, and all the experiments were performed three times in triplicate. *$p < 0.05$

PI4KIIα regulates CARTS complex secretion

We next checked a possible molecular mechanism by which PI4KIIα makes insulin exocytosis decline using BioID, which was an unbiased proteomic method and was developed for the characterization of protein–protein interaction networks recently. It was a kind of proximity-based biotin labeling (Roux *et al.* 2012). We ectopically expressed PI4KIIα fused to a mutant *Escherichia coli* biotin ligase (BirA R118G, or BirA*) in MCF-7 cells. BirA* efficiently activates biotin to label PI4KIIα proximate targets (Kwon *et al.* 2002). We used PI4KIIIβ as a control in this BioID experiment. All the hits are presented in Fig. 4A. We then analyzed these proximate proteins by Gene Ontology (GO) biological process analysis, and the result showed that proteins involved in translation, intracellular transport, protein folding, and metabolic process were enriched (Fig. 4B). Based on the observation that PI4KIIα can regulate insulin secretion, we then carefully analyzed the targets involved in intracellular transport, which are listed in Table 1. Interestingly, three individual components of

the CARTS complex (Rab8a, p115, and clathrin heavy chain 1) (Wakana *et al.* 2012) are included in the PI4KIIα interaction target list but not that of PI4KIIIβ (Fig. 4A, B; Table 1). CARTS forms at the TGN and it is a class of transport carriers. Protein kinase D (PKD) is required for the trafficking of these carriers that contain Rab8a, p115, and a number of secretory and plasma membrane-specific cargos, such as pancreatic adeno-carcinoma upregulated factor (PAUF) (Wakana *et al.* 2015). To verify the relationship between PI4KIIα and the CARTS complex, we ascertained the localization of PI4KIIα, an important component of CARTS (Rab8) and the most classical cargo of CARTS, PAUF, by immunofluorescence. As shown in Fig. 4C, PI4KIIα partially colocalized with both Rab8 and PAUF. Thus, we compared PAUF secretion in PI4KIIα-knockout cells and WT cells. Monoclonal MCF-7 knockout cell lines (Fig. S5) were generated by the CRISPR-CAS9 method. As shown in Fig. 4D and E, PAUF secretion was highly upregulated in both monoclonal PI4KIIα-knockout cell lines. These data are consistent with the finding that PI4KIIα sup-pression increases insulin secretion (Fig. 3B).

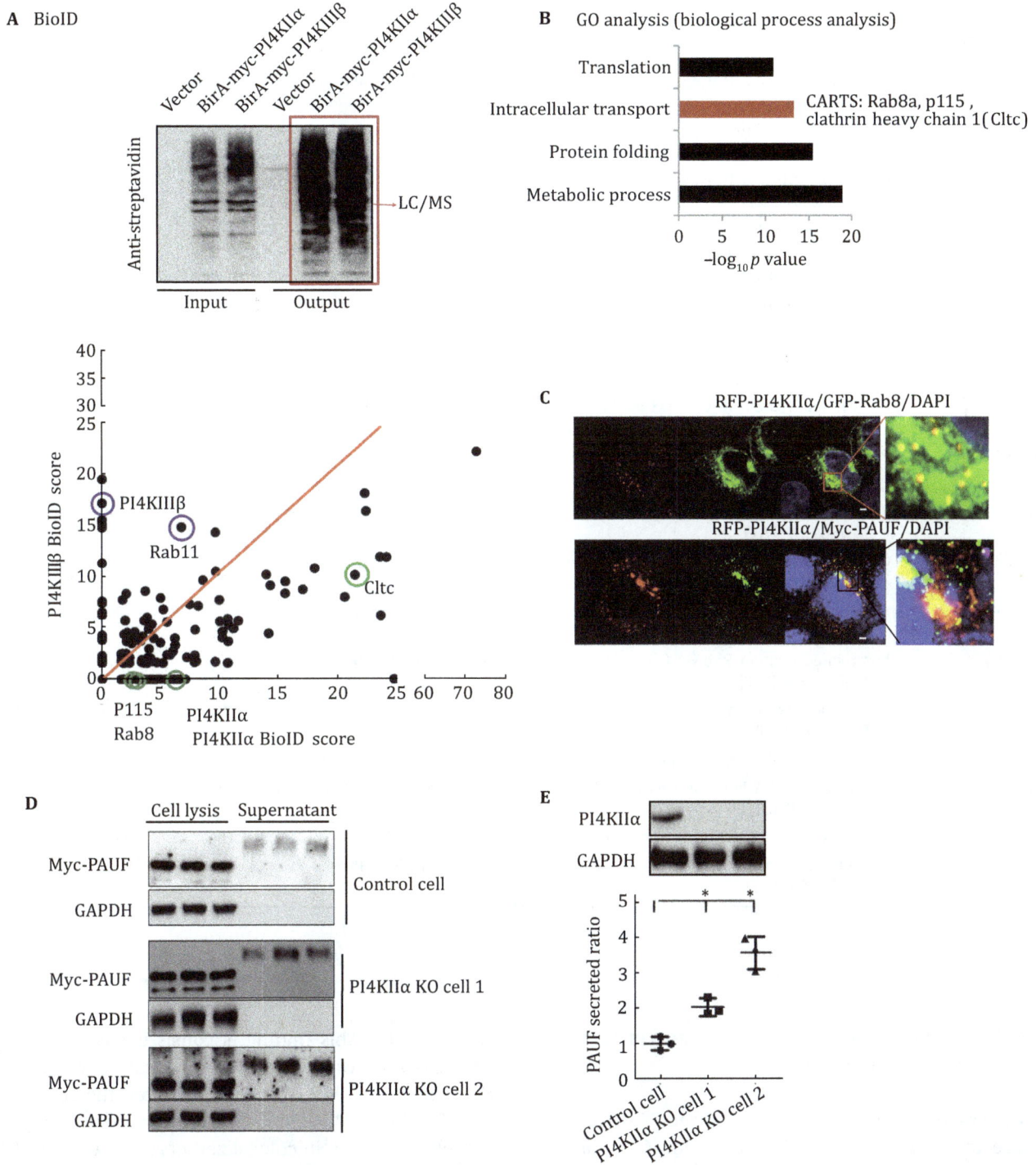

Fig. 4 PI4KIIα interactome reveals its regulation of the CARTS complex. **A** Proteins biotinylated by BirA*-PI4KIIα or BirA*-PI4KIIIβ in MCF-7 cells were identified by mass spectrometry (LTQ-Orbitrap XL). **B** Gene ontology (GO) biological process analysis was performed to characterize these PI4KIIα proximate proteins. **C** RFP-PI4KIIα was cotransfected with GFP-Rab8 or Myc-PAUF into MCF-7 cells, and the nucleus (*blue*) or Myc-PAUF (*green*) were immunostained using DAPI or a Myc mouse monoclonal antibody, respectively. Scale bar, 2 μm. **D**, **E** Myc-PAUF was overexpressed in WT and PI4KIIα-knockout MCF-7 cells. After 30 h, cells were incubated in serum-free DMEM for another 8 h. Cell lysates and culture media were collected, PAUF and GAPDH were detected by Western blot (**D**), and the bands were analyzed using ImageJ (**E**). PI4KIIα expression levels were also detected by Western blot, with GAPDH as a control. The data are presented as the mean ± SD of three independent experiments, except for those from the LC–MS/MS experiments

Table 1 Protein targets related to intracellular transport identified by LC–MS/MS in the PI4KIIα and PI4KIIIβ BioID experiments

Accession number	Description	Score 2α	Score 3β
P34058	Heat shock protein HSP 90-beta OS = *Rattus norvegicus* GN = Hsp90ab1 PE = 1 SV = 4—[HS90B_RAT]	51.98	35.20
P15999	ATP synthase subunit alpha, mitochondrial OS = *Rattus norvegicus* GN = Atp5a1 PE = 1 SV = 2—[ATPA_RAT]	41.16	16.54
P82995	Heat shock protein HSP 90-alpha OS = *Rattus norvegicus* GN = Hsp90aa1 PE = 1 SV = 3—[HS90A_RAT]	30.15	15.51
F1M779	Clathrin heavy chain 1 OS = *Rattus norvegicus* GN = Cltc PE = 2 SV = 1—[F1M779_RAT]	21.36	10.17
P62494	Ras-related protein Rab-11A OS = *Rattus norvegicus* GN = Rab11a PE = 1 SV = 3—[RB11A_RAT]	6.75	14.85
P46462	Transitional endoplasmic reticulum ATPase OS = *Rattus norvegicus* GN = Vcp PE = 1 SV = 3—[TERA_RAT]	6.57	0.00
Q99M64	Phosphatidylinositol 4-kinase type 2-alpha OS = *Rattus norvegicus* GN = Pi4k2a PE = 1 SV = 1—[P4K2A_RAT]	6.09	0.00
P09527	Ras-related protein Rab-7a OS = *Rattus norvegicus* GN = Rab7a PE = 1 SV = 2—[RAB7A_RAT]	5.19	1.62
B0BNK1	Protein Rab5c OS = *Rattus norvegicus* GN = Rab5c PE = 2 SV = 1—[B0BNK1_RAT]	4.31	1.77
Q63716	Peroxiredoxin-1 OS = *Rattus norvegicus* GN = Prdx1 PE = 1 SV = 1—[PRDX1_RAT]	3.98	0.00
Q4KM74	Vesicle-trafficking protein SEC22b OS = *Rattus norvegicus* GN = Sec22b PE = 1 SV = 3—[SC22B_RAT]	3.00	4.00
D4ABY2	Coatomer subunit gamma OS = *Rattus norvegicus* GN = Copg2 PE = 2 SV = 2—[D4ABY2_RAT]	2.49	0.00
G3V8T9	Apoptosis regulator BAX OS = *Rattus norvegicus* GN = Bax PE = 4 SV = 1—[G3V8T9_RAT]	1.88	0.00
P35280	Ras-related protein Rab-8A OS = *Rattus norvegicus* GN = Rab8a PE = 2 SV = 2—[RAB8A_RAT]	1.85	0.00
P41542	General vesicular transport factor p115 OS = *Rattus norvegicus* GN = Uso1 PE = 1 SV = 1—[USO1_RAT]	1.81	0.00
B2RYP4	Protein Snx2 OS = *Rattus norvegicus* GN = Snx2 PE = 2 SV = 1—[B2RYP4_RAT]	1.66	0.00
Q63413	Spliceosome RNA helicase Ddx39b OS = *Rattus norvegicus* GN = Ddx39b PE = 1 SV = 3—[DX39B_RAT]	0.00	3.95
B5DEP2	Protein Rab25 OS = *Rattus norvegicus* GN = Rab25 PE = 2 SV = 1—[B5DEP2_RAT]	0.00	1.70
O08561	Phosphatidylinositol 4-kinase beta OS = Rattus norvegicus GN = Pi4 kb PE = 1 SV = 1—[PI4KB_RAT]	0.00	17.18

PI4KIIα-regulated insulin and CARTS secretion is dependent on PKD activity

PKD is essential for the biogenesis of the "TGN-to-cell-surface transport carriers" and is the most important common regulator of CARTS complex and insulin secretion (Sumara *et al.* 2009; Wakana *et al.* 2012). Therefore, we speculated that PI4KIIα-mediated regulation of insulin and CARTS secretion is dependent on the interaction between PI4KIIα and PKD. To address this hypothesis, we first determined whether PKD could be identified in a PI4KIIα BioID assay. As shown in Fig. 5A, PKD was labeled by both BirA*-PI4KIIα and

BirA*-PI4KIIIβ; this finding is consistent with previous data showing that PI4KIIIβ is a substrate of PKD (Hausser *et al.* 2005). We then analyzed the colocalization of PI4KIIα, PKD, and insulin. As shown in Fig. 5B, both PKD and insulin colocalized perfectly with PI4KIIα in MIN6 cells. In addition, the interaction between PKD and PI4KIIα was confirmed by GST pull-down assay; human GFP-tagged PKD (GFP-PKD) was captured by both WT PI4KIIα and K152A-mutant PI4KIIα but not by the GST tag alone (Fig. 5C). This is consistent with what we observed previously, *i.e.*, that both WT and kinase-dead (K152A) PI4KIIα reduce insulin secretion. To clarify whether this interaction contributes to the

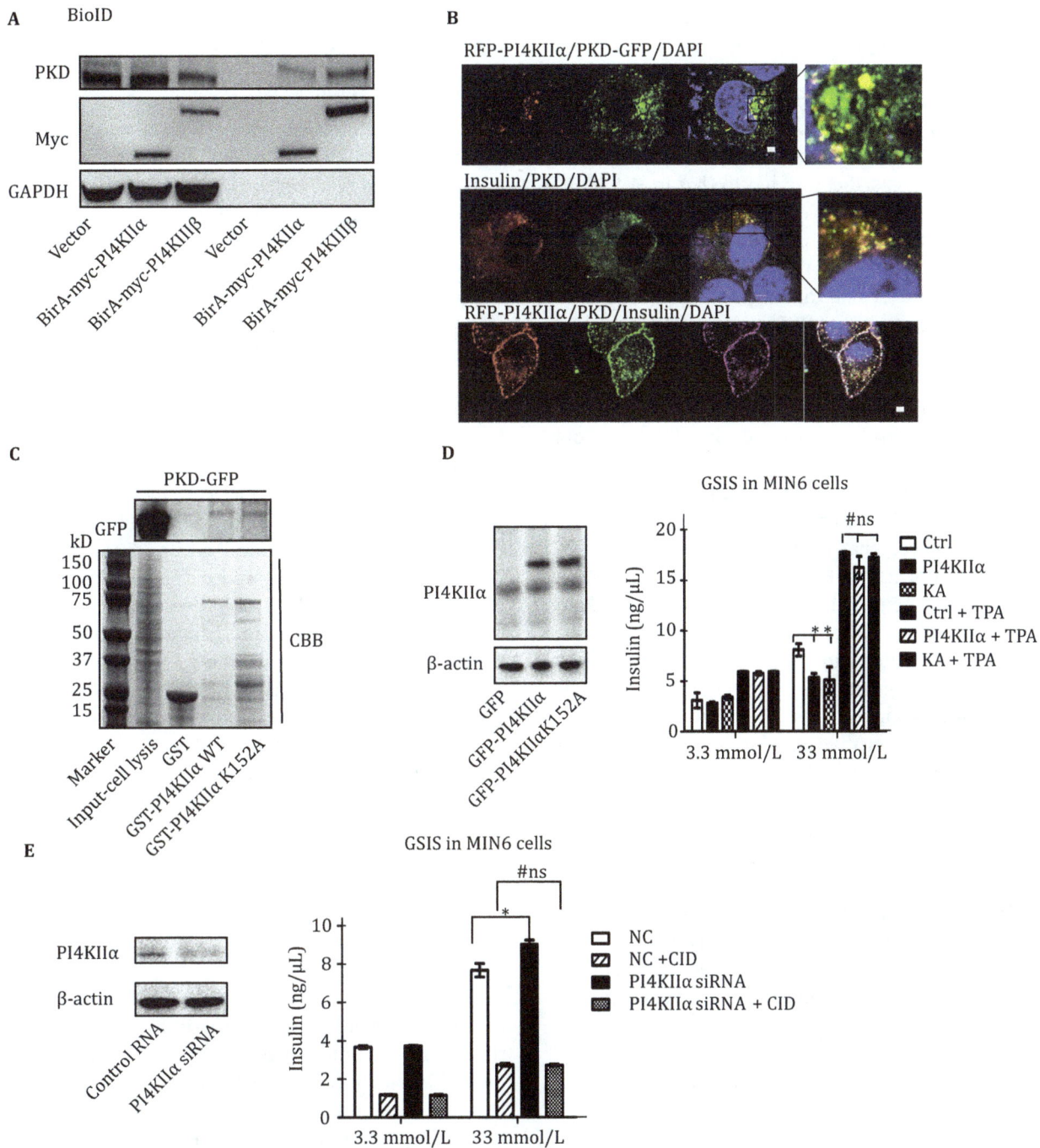

Fig. 5 PI4KIIα regulates insulin secretion via a PKD-dependent pathway. **A** Proteins biotinylated by BirA*-PI4KIIα or BirA*-PI4KIIIβ in MCF-7 cells were purified using streptavidin agarose. The immunoprecipitates were immunoblotted with antibodies against PKD, Myc, and GAPDH. **B** MIN6 cells were transfected with or without RFP-PI4KIIα, and the nucleus (*blue*), PKD (*green*), and insulin (*red* or *magenta* as indicated) were immunostained using DAPI or the respective antibodies. Scale bar, 2 μm. **C** MCF-7 cells were transfected with GFP-PKD. After 24 h, the cells were lysed in RIPA buffer and subjected to pull-down with exogenous GST, GST-PI4KIIα WT, or GST-PI4KIIαK152A expressed in *E. coli*. **D** MIN6 cells were transfected with GFP, GFP-PI4KIIα, or GFP-PI4KIIαK152A. After 30 h, insulin secretion in response to 3.3 or 33 mmol/L glucose stimulation with or without 0.2 μmol/L TPA was measured using an insulin ELISA kit. **E** MIN6 cells were transfected with control siRNA or mouse PI4KIIα siRNA for 60 h; then, insulin secretion in response to 3.3 or 33 mmol/L glucose stimulation with or without 10 μmol/L CID755673 was measured using an insulin ELISA kit. The data are presented as the mean ± SD of three independent experiments, and all the experiments were performed three times in triplicate. *p < 0.05

regulation of PKD activity, we evaluated that PKD activity in pancreatic islets isolated from WT mice or PI4KIIα TG mice. As shown in Fig. S6A, there was an obvious decrease in autophosphorylated PKD and phosphorylated PKD substrates in PI4KIIα TG mice. However, in PI4KIIα-knockout MCF-7 cells, PKD activity was markedly increased compared to that in WT MCF-7 cells (Fig. S6B). Thus, we can conclude that PI4KIIα regulates PKD activity via a protein–protein interaction. In addition, we performed rescue experiments to validate whether the regulation of insulin secretion by PI4KIIα is dependent on PKD activity. As shown in Fig. 5D, overexpression of either WT and kinase-dead (K152A) PI4KIIα obviously inhibited glucose-induced insulin secretion in MIN6 cells, while TPA (PKD agonist) treatment significantly increased insulin secretion. However, there was no difference in insulin secretion between WT MIN6 cells and PI4KIIα-overexpressing MIN6 cells upon TPA treatment. Consistent with this result, we observed that a PKD inhibitor (CID755673) blocked the PI4KIIα siRNA-mediated increasing of insulin secretion (Fig. 5E). The PKD inhibitor markedly reduced insulin secretion in response to stimulation with either low or high glucose, and PI4KIIα knockdown increased insulin secretion in only control MIN6 cells, not CID755673-treated cells. Together, these results indicated that the negative regulation of insulin secretion by PI4KIIα is dependent on PKD activity. To further confirm this regulatory pathway, we ascertained the effect of CID755673 on PAUF secretion induced by PI4KIIα knockout. As shown in Fig. S6C, inhibiting PKD activity obviously abolished the increased secretion of PAUF (traditional cargo for CARTS complex) induced by suppressing PI4KIIα. Based on these results, we concluded that PI4KIIα negatively regulates insulin and CARTS complex secretion and that this effect is dependent on PKD activity.

PI4KIIα overexpression worsens glucose tolerance and insulin secretion in streptozotocin/high-fat diet-induced diabetic mice

To determine whether PI4KIIα upregulation increases susceptibility to diabetes, we investigated the effect of a HFD and streptozotocin (STZ) treatment on PI4KIIα TG mice and WT littermates. BALB/c mice are insensitive to a HFD (Schreyer et al. 1998); thus, we first constructed PI4KIIα TG C57BL/6 mice by breeding PI4KIIα TG BALB/c mice with C57BL/6 WT mice, which are sensitive to a HFD. After 8 homozygous generations, PI4KIIα TG C57BL/6 mice were successful obtained (Fig. 6A). We first tested whether PI4KIIα overexpression

increases serum glucose levels in C57BL/6 mice; fasting blood glucose level (Fig. S7A) and glucose tolerance (Fig. S7B) were impaired in PI4KIIα TG C57BL/6 mice, there was no significant difference in insulin tolerance (Fig. S7C), while insulin secretion was significantly weakened (Fig. S7D). These results were consistent with those in model mice on the BALB/c genetic background. Then, three-week-old male PI4KIIα TG and WT mice on the C57BL/6 genetic background were fed a HFD. After 3 weeks on a HFD, a single dose of STZ (80 mg/kg in 0.1 mol/L citrate buffer, pH 4.5) was administered by intraperitoneal injection. At 2 and 3 weeks after the injection, fasting blood glucose and glucose-stimulated insulin secretion (GSIS) were measured, and the intraperitoneal glucose tolerance test (IPGTT) and insulin tolerance test (ITT) were administered (Fig. 6B). The treatment highly raised serum glucose levels over time, and the STZ/HFD-induced hyperglycemic effect was extremely pronounced in the PI4KIIα TG mice (Fig. 6C). In addition, male PI4KIIα TG mice displayed relatively worse glucose tolerance after STZ/HFD treatment, with more rapid progression of diabetes compared to WT mice; the phenotype of PI4KIIα TG mice after 2 weeks was similar to that of WT mice after 3 weeks (Fig. 6D). Consistent with previous findings, the ITT results were not different between these two strains of mice (Fig. 6E), but PI4KIIα TG mice had lower glucose-induced insulin secretion compared to WT littermates (Fig. 6F). These results indicated that PI4KIIα TG mice are more sensitive to STZ/HFD treatment, and overexpressing PI4KIIα increased the susceptibility to diabetes.

DISCUSSION

Insulin secretion from pancreatic β cells is critical for the proper maintenance of blood glucose levels, and perturbations in this process lead to diabetes (Del Prato et al. 2002; Gupta et al. 2012). We provide new evidence that PI4Kα is a key regulator of β cell function in pancreas. The work of us uncovered a negative regulatory role for PI4Kα as shown in Fig. 6G of the hypothetic model: PI4KIIα can negatively regulate PKD activity via protein–protein interaction, while PKD activity is essential for insulin exocytosis.

PKD is a serine/threonine kinase that is activated by DAG signaling pathways to control fission and transport of Golgi vesicles, mediate survival responses to oxidative stress, regulate antigen-activated signaling in T and B cells, inhibit JNK-dependent proliferation, modulate adhesion, and elicit nuclear export of histone deacetylases (Ellwanger and Hausser 2013; Fu and Rubin

A

PI4KIIα TG BALB/c mice   Wild type C57BL/6 mice

Breed

PI4KIIα TG F1 mice   Wild type C57BL/6 mice

Breed for 8 generations homozygosity

PI4KIIα TG C57BL/6 mice

B

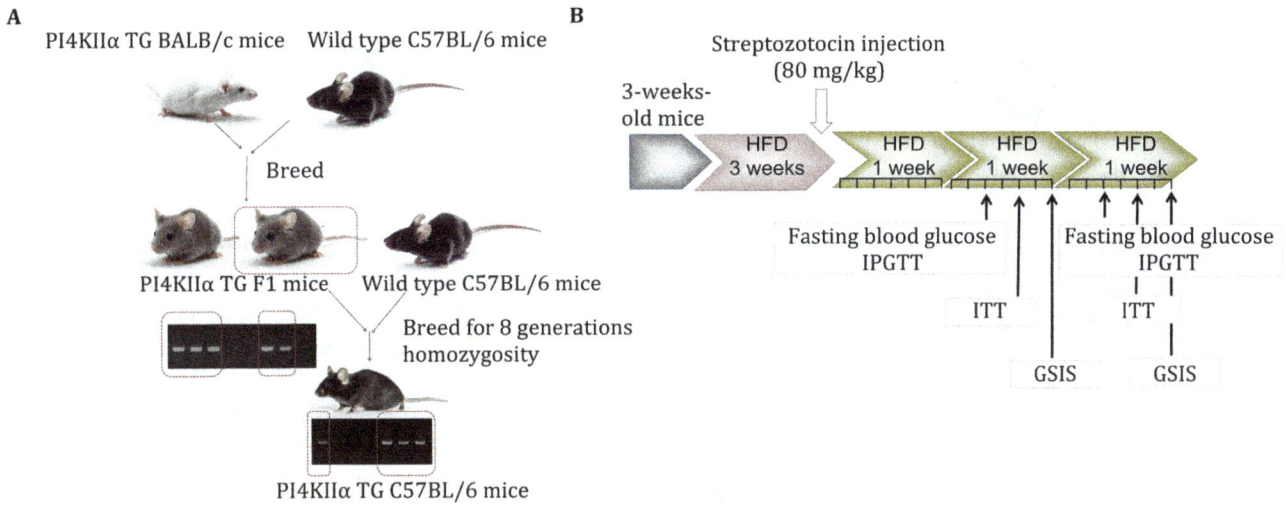

Streptozotocin injection (80 mg/kg)

3-weeks-old mice

HFD 3 weeks | HFD 1 week | HFD 1 week | HFD 1 week

Fasting blood glucose IPGTT

Fasting blood glucose IPGTT

ITT

ITT

GSIS

GSIS

C

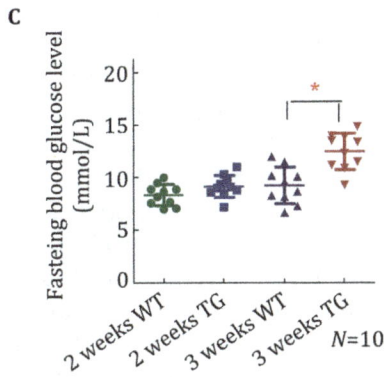

D   IPGTT   IPGTT

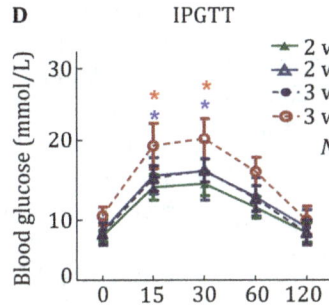

- ▲ 2 weeks WT
- ▲ 2 weeks TG
- ● 3 weeks WT
- ○ 3 weeks TG

N=10

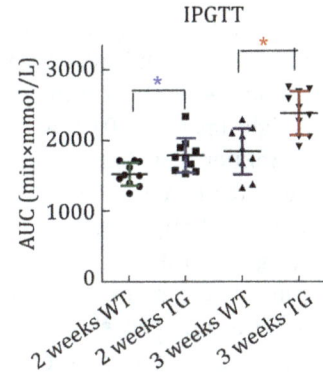

E   ITT   ITT   **F**   GSIS

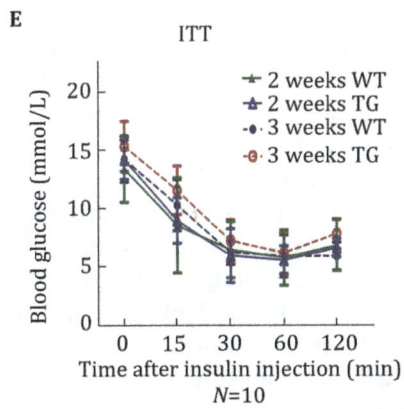

Time after insulin injection (min)
N=10

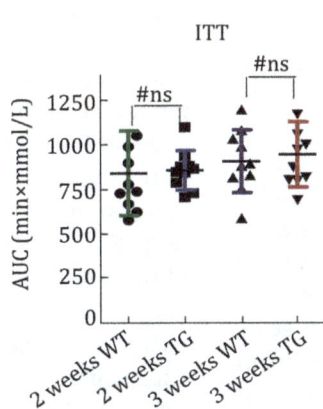

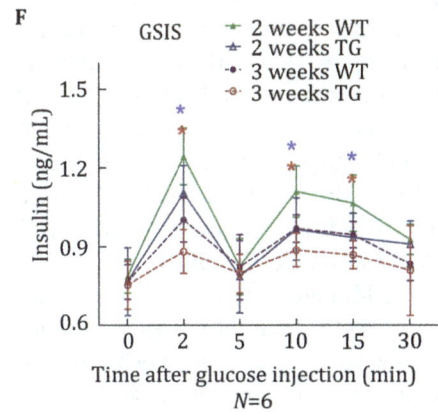

Time after glucose injection (min)
N=6

- 2 weeks WT
- 2 weeks TG
- 3 weeks WT
- 3 weeks TG

G

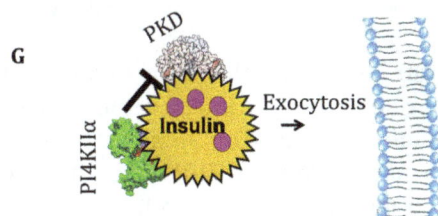

PKD

PI4KIIα   Insulin   Exocytosis

◀**Fig. 6** PI4KIIα overexpression enhances the sensitivity to STZ/HFD-induced diabetes in mice. **A** Workflow to generate PI4KIIα TG C57BL/6 mice from PI4KIIα TG BALB/c mice. **B** Workflow to generate diabetic mice by STZ/HFD treatment. **C** Fasting blood glucose was measured in STZ/HFD-induced mice. **D** IPGTT was performed in overnight-fasted PI4KIIα TG C57BL/6 mice and age-matched WT C57BL/6 littermates ($N = 10$ for each line). **E** ITT was performed in 6 h-fasted PI4KIIα TG C57BL/6 mice and age-matched WT C57BL/6 littermates ($N = 10$ for each line). **F** GSIS was performed in overnight-fasted PI4KIIα TG C57BL/6 mice and age-matched WT littermates ($N = 6$ for each line). **G** Hypothetic model: PI4KIIα can negatively regulate PKD activity via protein-protein interaction, while PKD activity is essential for insulin exocytosis. The data are presented as the mean ± SD. All the experiments were performed three times in triplicate. *$p < 0.05$

2011). Recently, researchers identified PKD as a pivotal regulator of stimulated insulin exocytosis (Sumara *et al.* 2009). In addition to its function in the TGN, PKD is thought to play an important role in priming insulin vesicles for transport and immediate fusion (Li *et al.* 2004; Sumara *et al.* 2009). Several studies have indicated that G protein-coupled receptor (GPR) 40 (Ferdaoussi *et al.* 2012; Iglesias *et al.* 2012) and MAPK p38δ (Sumara *et al.* 2009) influence insulin secretion by regulating PKD activity. Here, we revealed that PI4KIIα is a novel regulator of PKD activity by direct interaction, not by the DAG pathway (Figs. 5, S6). As shown in Fig. 5B, PKD and PI4KIIα colocalized at insulin-positive granules rather than at the TGN. Lu *et al.* showed that PKD localized at vesicular structures and promoted the recruitment of VAMP2 vesicles to the targeted membrane(Lu *et al.* 2007). Meanwhile, PI4KIIα was reported to have a similar function as PKD in regulating the association of VAMP3 with its cognate Q-SNARE Vti1a (Jovic *et al.* 2014). Therefore, we hypothesized that the PI4KIIα/PKD complex may have a role in the insulin and CARTS sorting process, which merits further investigation.

PI4KIIα is involved in various essential cellular functions, including membrane trafficking (Salazar *et al.* 2005; Wang *et al.* 2007, 2003), signal transduction (Li *et al.* 2010; Minogue *et al.* 2006; Pan *et al.* 2008), and the exo-endocytic cycle of synaptic vesicles (Guo *et al.* 2003). However, the precise mechanism of PI4KIIα in the cell is not yet completely deciphered because it engages in low-affinity interactions with dynamic cellular signaling pathways (Gokhale *et al.* 2016). Gokhale *et al.* identified novel interactors of PI4KIIα using a chemical cross-linker, DSP, combined with immunoprecipitation and immunoaffinity purification (Gokhale *et al.* 2016). Here, we used another transient and dynamic interaction method, BioID proximity-based biotin labeling, to identify proteins that interact with PI4KIIα. As shown in Table 1 and Fig. 4B, PI4KIIα participates in transient, low-affinity and dynamic

interactions that are difficult to identify by direct pull-down or coimmunoprecipitation assays. This could explain that why the interaction between PI4KIIα and PKD identified by pull-down is quite weak; both proteins are highly dynamic in membrane trafficking and signal transduction (Balla and Balla 2006; Ellwanger and Hausser 2013). In addition, Rab8, Rab5, and Rab7 were also detected as PI4KIIα proximity targets (Table 1), which accord with previous results that PI4KIIα has an important role in late endosome (Salazar *et al.* 2005), early endosome, and sorting endosome (Henmi *et al.* 2016; Ketel *et al.* 2016) functions. Our results indicated that BioID could be an ideal tool for detecting dynamic PI4KIIα interactions and could provide valuable assistance in determining its functional role in physiologically and pathologically processes.

Recent studies indicated that PI4KIIα is essential for endosomal trafficking of transferrin and certain receptors (Henmi *et al.* 2016; Jovic *et al.* 2014; Ketel *et al.* 2016; Minogue *et al.* 2006). Therefore, we ascertained the effect of PI4KIIα knockout on transferrin recycling. As described by Jovic *et al.* (2014), suppressing PI4KIIα induced a significant delay in transferrin delivery to the recycling compartment (data not shown). Studies indicated that PI4KIIα is required for the production of endosomal PtdIns(4)P on early endosomes and for the sorting of transferrin and EGFR into the recycling and degradation pathways; both knocking down PI4KIIα and inhibiting its kinase activity influence the surface delivery of endosomal cargos (Henmi *et al.* 2016; Jovic *et al.* 2014; Ketel *et al.* 2016). However, in our study, we found that PI4KIIα is a negative regulator of insulin and PAUF secretion and that this regulation is completely independent of kinase activity: both WT and kinase-dead PI4KIIα reduced insulin secretion (Fig. 3A), and PI4P, the product of PI4KIIα, could not rescue the increase in insulin secretion upon PI4KIIα knockdown (Fig. S3). Together, the above results indicated that PI4KIIα has a different effect on different cargos, chiefly because of different regulatory mechanisms. The complexities of cargo classification and the intricate positive and negative feedback mechanisms among different cargos make it impossible to state an exact rule about the positive or negative regulation of various cargos by PI4KIIα; however, we will address this issue in the future.

To the best of our knowledge, this is the first study to reveal the pivotal role of PI4KIIα in regulating diabetes via insulin secretion and PKD. Our findings indicated that PI4KIIα is a new player in T2DM and that high PI4KIIα expression increases the susceptibility to HFD-induced hyperglycemia. Because PI4KIIα regulation of PKD and insulin secretion is independent of kinase activity, it is hard to evaluate its therapeutic effect in animal models of diabetes using inhibitors. However, the cellular assays indicated that suppressing PI4KIIα expression markedly increased insulin secretion (Figs. 3B, 5E). Therefore, it is worth developing tools to suppress PI4KIIα expression or disrupt the interaction between PI4KIIα and PKD and exploring the therapeutic effect against type 1 and type 2 diabetes; this will be the main direction of our future work.

MATERIALS AND METHODS

Reagents, plasmids, and antibodies

PI(4)P Mass ELISA Kit (K-4000E) was purchased from Echelon Biosciences. The original full-length human PI4KIIα plasmid was a kind gift from Shane Minogue (Minogue *et al.* 2001, University College London). pSpCas9(BB)-2A-GFP (PX458) (Addgene plasmid #48138) and lentiCRISPR V2 (Addgene plasmid #52963) were gifts from Feng Zhang (Ran *et al.* 2013). Antibodies to c-Myc, GAPDH, and β-actin were purchased from Santa Cruz Biotechnology (TX, USA). Antibodies to PKD, p-PKD (916), p-PKD (744/748), and PKD substrates were from Cell Signaling Technology (Herts, UK). Rabbit polyclonal PI4KIIα antibody was a kind gift from Pietro De Camilli (Guo *et al.* 2003, Yale University, HHMI). Insulin (Mouse) Ultrasensitive EIA was from Alpco (NH, USA). Other reagents were purchased from Sigma (Dorset, UK) unless otherwise stated.

Generation of PI4KIIα transgenic mice

All animals were housed in the specific facilities which were pathogen-free and maintained on a 12-h light/dark cycle, and fed standard rodent chow at the Laboratory Animal Resources in the Institute of Biophysics, Chinese Academy of Science. Human PI4KIIα tagged with a flag epitope was subcloned into pCAGGS. The DNA was eluted in filtered microinjection buffer and injected into zygotes from BALB/c mice (purchased from Weitonglihua, Beijing, China). For genotyping, mouse tail DNA was isolated (by alkaline lysis) and analyzed by PCR (Forward primer: tctttcccgagcgcatctaccag; Reverse primer: agcagcaaggacagcacagcttc).

To study the function of PI4KIIα in STZ/HFD-induced diabetes, we generated PI4KIIα TG mice on the C57BL/6 genetic background. The first generation of heterozygous PI4KIIα TG mice was obtained by crossing WT C57BL/6 mice with PI4KIIα TG mice on the BALB/c genetic background; the resulting mice were the first (F1) generation. The identified PI4KIIα TG F1 mice were

backcrossed with C57BL/6 WT mice for eight generations. Finally, we obtained heterozygous PI4KIIα TG mice on a pure C57BL/6 genetic background. At each generation, the genotype was confirmed by PCR.

STZ/HFD-induced diabetic mouse model

Three-week-old male PI4KIIα TG mice ($N = 10$) on a C57BL/6 genetic background were fed a HFD (26.2% protein, 26.3% carbohydrate, 34.9% fat), and the control group ($N = 10$) comprised their WT littermates. After 3 weeks on a HFD, a single dose of STZ (80 mg/kg in 0.1 mol/L citrate buffer, pH 4.5) was administered by intraperitoneal injection to induce partial insulin deficiency. Three weeks after the STZ injection, the majority of animals fed a HFD and treated with STZ exhibited hyperglycemia. To monitor disease progression on STZ/HFD treatment, we tested fasting blood glucose, IPGTT, ITT, and GSIS at 2 and 3 weeks after STZ injection.

BioID, on-bead protein digestion, and mass spectrometry

BioID was performed according to the previously described procedures (Roux *et al.* 2012). In brief, transfected cells were incubated with 50 μmol/L biotin for 6 h before harvest. Cells lysed as described above were incubated at 4 °C for 3 h with 500 μl of streptavidin conjugated to beads (New England Biolabs, Ipswich, MA). Beads were washed once with 1.5 ml of wash buffer 1 (2% SDS in H_2O), once with wash buffer 2 (0.1% deoxycholate, 1% Triton X-100, 500 mmol/L NaCl, 1 mmol/L EDTA, and 50 mmol/L 171 HEPES, pH 7.5), once with wash buffer 3 (250 mmol/L LiCl, 0.5% NP-40, 0.5% deoxycholate, 1 mmol/L EDTA, and 10 mmol/L Tris, pH 8.1), and then twice with wash buffer 4 (50 mmol/L Tris, pH 7.4, and 50 mmol/L NaCl). To evaluate sample integrity, 10% of the total was retained for immunoblots. The remaining beads were centrifuged at 2000 *g* and resuspended in 50 μl of 50 mmol/L ammonium bicarbonate for mass spectrometry (LTQ-Orbitrap XL) as previously described (Roux *et al.* 2012).

Islet isolation and Western blot analysis

Mouse islets were isolated by collagenase digestion of the pancreas according to previously described procedures (Martinez *et al.* 2006). In brief, overnight-fasted mice were anesthetized with an intraperitoneal injection of pentobarbital sodium (80 mg/kg body weight). The pancreas of mouse was inflated by the injection of 3 ml of a collagenase P solution (Sigma Chemical, St. Louis, MO; 0.5 mg/mL in Hank's buffered salt solution).

Pancreases were removed and incubated at 37 °C for approximately 20 min to make the digestion complete, which was stopped by the addition of 10 ml of Hank's buffered salt solution containing 5% fetal bovine serum. The pancreases were washed three times with 10 ml of RPMI-1640 medium. Isolated islets were selected from the medium with the aid of a pipette under a stereoscopic microscope. The isolated islets were subjected to Western blot analysis or further incubated in RPMI-1640 with or without TPA (PKD agonist) at 37 °C for 30 min. Islets, cells or tissue were lysed and analyzed by Western blot using specific antibodies.

Statistics

Statistical analysis was performed using the two-tailed paired Student's t test. Differences were considered statistically significant at $p < 0.05$ or $p < 0.01$, as indicated in the legends. All data are presented as the mean $\pm$ SD.

Acknowledgements We thank Pietro De Camilli for providing the PI4KIIα antibody and Shane Minogue for providing the full-length human PI4KIIα cDNA. This work was supported by the National Key Research and Development Program of China (2017YFA0504000, 2016YFC0903100), the National Natural Science Foundation of China (31570857, 31101021, and 81472839); the "863" National High-Technology Development Program of China (0A200202D03); the Novo Nordisk—Chinese Academy of Sciences Research Fund (NNCAS-2012-2); the Beijing Natural Science Foundation (7132156); Science and Technology Commission of Shanghai Municipality (15431903100); Personalized Medicines—Molecular Signature-based Drug Discovery and Development, the Strategic Priority Research Program of the Chinese Academy of Sciences (XDA12020316).

Compliance with Ethical Standards

Conflict of interest Jiangmei Li, Lunfeng Zhang, Panpan Zhang, Zhen Gao, Yingying Zhao, Xinhua Qiao, and Chang Chen declare that they have no conflict of interest.

Human and animal rights and informed consent All institutional and national guidelines for the care and use of laboratory animals were followed.

References

Balla T (2013) Phosphoinositides: tiny lipids with giant impact on cell regulation. Physiol Rev 93:1019–1137

Balla A, Balla T (2006) Phosphatidylinositol 4-kinases: old enzymes with emerging functions. Trends Cell Biol 16:351–361

Chu KM, Minogue S, Hsuan JJ, Waugh MG (2010) Differential effects of the phosphatidylinositol 4-kinases, PI4KIIalpha and PI4KIIbeta, on Akt activation and apoptosis. Cell Death Dis 1:e106

Del Prato S, Marchetti P, Bonadonna RC (2002) Phasic insulin release and metabolic regulation in type 2 diabetes. Diabetes 51(Suppl 1):S109–S116

Easom RA (2000) Beta-granule transport and exocytosis. Semin Cell Dev Biol 11:253–266

Ellwanger K, Hausser A (2013) Physiological functions of protein kinase D in vivo. IUBMB Life 65:98–107

Ferdaoussi M, Bergeron V, Zarrouki B, Kolic J, Cantley J, Fielitz J, Olson EN, Prentki M, Biden T, MacDonald PE et al (2012) G protein-coupled receptor (GPR)40-dependent potentiation of insulin secretion in mouse islets is mediated by protein kinase D1. Diabetologia 55:2682–2692

Fu Y, Rubin CS (2011) Protein kinase D: coupling extracellular stimuli to the regulation of cell physiology. EMBO Rep 12:785–796

Gokhale A, Ryder PV, Zlatic SA, Faundez V (2016) Identification of the interactome of a palmitoylated membrane protein, phosphatidylinositol 4-kinase type II alpha. Methods Mol Biol 1376:35–42

Guo J, Wenk MR, Pellegrini L, Onofri F, Benfenati F, De Camilli P (2003) Phosphatidylinositol 4-kinase type IIalpha is responsible for the phosphatidylinositol 4-kinase activity associated with synaptic vesicles. Proc Natl Acad Sci USA 100:3995–4000

Gupta D, Krueger CB, Lastra G (2012) Over-nutrition, obesity and insulin resistance in the development of beta-cell dysfunction. Curr Diabetes Rev 8:76–83

Hausser A, Storz P, Martens S, Link G, Toker A, Pfizenmaier K (2005) Protein kinase D regulates vesicular transport by phosphorylating and activating phosphatidylinositol-4 kinase IIIbeta at the Golgi complex. Nat Cell Biol 7:880–886

Henmi Y, Morikawa Y, Oe N, Ikeda N, Fujita A, Takei K, Minogue S, Tanabe K (2016) PtdIns4KIIalpha generates endosomal PtdIns(4)P and is required for receptor sorting at early endosomes. Mol Biol Cell 27:990–1001

Huang P, Yeku O, Zong H, Tsang P, Su W, Yu X, Teng S, Osisami M, Kanaho Y, Pessin JE, Frohman MA (2011) Phosphatidylinositol-4-phosphate-5-kinase alpha deficiency alters dynamics of glucose-stimulated insulin release to improve glucohomeostasis and decrease obesity in mice. Diabetes 60:454–463

Iglesias J, Barg S, Vallois D, Lahiri S, Roger C, Yessoufou A, Pradevand S, McDonald A, Bonal C, Reimann F, Gribble F, Debril MB, Metzger D, Chambon P, Herrera P, Rutter GA, Prentki M, Thorens B, Wahli W (2012) PPARbeta/delta affects pancreatic beta cell mass and insulin secretion in mice. J Clin Investig 122:4105–4117

Ishihara H, Asano T, Tsukuda K, Katagiri H, Inukai K, Anai M, Kikuchi M, Yazaki Y, Miyazaki JI, Oka Y (1993) Pancreatic beta cell line MIN6 exhibits characteristics of glucose metabolism and glucose-stimulated insulin secretion similar to those of normal islets. Diabetologia 36:1139–1145

Jovic M, Kean MJ, Szentpetery Z, Polevoy G, Gingras AC, Brill JA, Balla T (2012) Two phosphatidylinositol 4-kinases control lysosomal delivery of the Gaucher disease enzyme, beta-glucocerebrosidase. Mol Biol Cell 23:1533–1545

Jovic M, Kean MJ, Dubankova A, Boura E, Gingras AC, Brill JA, Balla T (2014) Endosomal sorting of VAMP3 is regulated by PI4K2A. J Cell Sci 127:3745–3756

Kang MS, Baek SH, Chun YS, Moore AZ, Landman N, Berman D, Yang HO, Morishima-Kawashima M, Osawa S, Funamoto S, Ihara Y, Di Paolo G, Park JH, Chung S, Kim TW (2013) Modulation of lipid kinase PI4KIIalpha activity and lipid raft association of presenilin 1 underlies gamma-secretase inhibition by ginsenoside(20S)Rg3. J Biol Chem 288(29):20868–20882

Ketel K, Krauss M, Nicot AS, Puchkov D, Wieffer M, Muller R, Subramanian D, Schultz C, Laporte J, Haucke V (2016) A phosphoinositide conversion mechanism for exit from endosomes. Nature 529:408–412

Kwon K, Streaker ED, Beckett D (2002) Binding specificity and the ligand dissociation process in the *E. coli* biotin holoenzyme synthetase. Protein Sci 11:558–570

Lang J (2003) PIPs and pools in insulin secretion. Trends Endocrinol Metab 14:297–299

Li J, O'Connor KL, Hellmich MR, Greeley GH Jr, Townsend CM Jr, Evers BM (2004) The role of protein kinase D in neurotensin secretion mediated by protein kinase C-alpha/-delta and rho/rho kinase. J Biol Chem 279:28466–28474

Li J, Lu Y, Zhang J, Kang H, Qin Z, Chen C (2010) PI4KIIalpha is a novel regulator of tumor growth by its action on angiogenesis and HIF-1alpha regulation. Oncogene 29:2550–2559

Li J, Zhang L, Gao Z, Kang H, Rong G, Zhang X, Chen C (2014) Dual inhibition of EGFR at protein and activity level via combinatorial blocking of PI4KIIalpha as anti-tumor strategy. Protein Cell 5:457–468

Lu G, Chen J, Espinoza LA, Garfield S, Toshiyuki S, Akiko H, Huppler A, Wang QJ (2007) Protein kinase D3 is localized in vesicular structures and interacts with vesicle-associated membrane protein 2. Cell Signal 19:867–879

Martinez SC, Cras-Meneur C, Bernal-Mizrachi E, Permutt MA (2006) Glucose regulates Foxo1 through insulin receptor signaling in the pancreatic islet beta-cell. Diabetes 55:1581–1591

Minogue S, Anderson JS, Waugh MG, dos Santos M, Corless S, Cramer R, Hsuan JJ (2001) Cloning of a human type II phosphatidylinositol 4-kinase reveals a novel lipid kinase family. J Biol Chem 276:16635–16640

Minogue S, Waugh MG, De Matteis MA, Stephens DJ, Berditchevski F, Hsuan JJ (2006) Phosphatidylinositol 4-kinase is required for endosomal trafficking and degradation of the EGF receptor. J Cell Sci 119:571–581

Olsen HL, Hoy M, Zhang W, Bertorello AM, Bokvist K, Capito K, Efanov AM, Meister B, Thams P, Yang SN, Rorsman P, Berggren PO, Gromada J (2003) Phosphatidylinositol 4-kinase serves as a metabolic sensor and regulates priming of secretory granules in pancreatic beta cells. Proc Natl Acad Sci USA 100:5187–5192

Pan W, Choi SC, Wang H, Qin Y, Volpicelli-Daley L, Swan L, Lucast L, Khoo C, Zhang X, Li L, Abrams CS, Sokol SY, Wu D (2008) Wnt3a-mediated formation of phosphatidylinositol 4,5-bisphosphate regulates LRP6 phosphorylation. Science 321:1350–1353

Pimenta W, Korytkowski M, Mitrakou A, Jenssen T, Yki-Jarvinen H, Evron W, Dailey G, Gerich J (1995) Pancreatic beta-cell dysfunction as the primary genetic lesion in NIDDM. Evidence from studies in normal glucose-tolerant individuals with a first-degree NIDDM relative. JAMA 273:1855–1861

Radosavljevic T, Todorovic V, Sikic B (2004) Insulin secretion: mechanisms of regulation. Med Pregl 57:249–253

Ran FA, Hsu PD, Wright J, Agarwala V, Scott DA, Zhang F (2013) Genome engineering using the CRISPR-Cas9 system. Nat Protoc 8:2281–2308

Rorsman P, Renstrom E (2003) Insulin granule dynamics in pancreatic beta cells. Diabetologia 46:1029–1045

Roux KJ, Kim DI, Raida M, Burke B (2012) A promiscuous biotin ligase fusion protein identifies proximal and interacting proteins in mammalian cells. J Cell Biol 196:801–810

Ryder PV, Vistein R, Gokhale A, Seaman MN, Puthenveedu MA, Faundez V (2013) The WASH complex, an endosomal Arp2/3 activator, interacts with the Hermansky–Pudlak syndrome complex BLOC-1 and its cargo phosphatidylinositol-4-kinase type IIalpha. Mol Biol Cell 24:2269–2284

Salazar G, Craige B, Wainer BH, Guo J, De Camilli P, Faundez V (2005) Phosphatidylinositol-4-kinase type II alpha is a component of adaptor protein-3-derived vesicles. Mol Biol Cell 16:3692–3704

Schreyer SA, Wilson DL, LeBoeuf RC (1998) C57BL/6 mice fed high fat diets as models for diabetes-accelerated atherosclerosis. Atherosclerosis 136:17–24

Simons JP, Al-Shawi R, Minogue S, Waugh MG, Wiedemann C, Evangelou S, Loesch A, Sihra TS, King R, Warner TT, Hsuan JJ (2009) Loss of phosphatidylinositol 4-kinase 2alpha activity causes late onset degeneration of spinal cord axons. Proc Natl Acad Sci USA 106:11535–11539

Sumara G, Formentini I, Collins S, Sumara I, Windak R, Bodenmiller B, Ramracheya R, Caille D, Jiang H, Platt KA, Meda P, Aebersold R, Rorsman P, Ricci R (2009) Regulation of PKD by the MAPK p38delta in insulin secretion and glucose homeostasis. Cell 136:235–248

Vauhkonen I, Niskanen L, Vanninen E, Kainulainen S, Uusitupa M, Laakso M (1998) Defects in insulin secretion and insulin action in non-insulin-dependent diabetes mellitus are inherited. Metabolic studies on offspring of diabetic probands. J Clin Investig 101:86–96

Wakana Y, van Galen J, Meissner F, Scarpa M, Polishchuk RS, Mann M, Malhotra V (2012) A new class of carriers that transport selective cargo from the trans Golgi network to the cell surface. EMBO J 31:3976–3990

Wakana Y, Kotake R, Oyama N, Murate M, Kobayashi T, Arasaki K, Inoue H, Tagaya M (2015) CARTS biogenesis requires VAP-lipid transfer protein complexes functioning at the endoplasmic reticulum-Golgi interface. Mol Biol Cell 26:4686–4699

Wang YJ, Wang J, Sun HQ, Martinez M, Sun YX, Macia E, Kirchhausen T, Albanesi JP, Roth MG, Yin HL (2003) Phosphatidylinositol 4 phosphate regulates targeting of clathrin adaptor AP-1 complexes to the Golgi. Cell 114:299–310

Wang J, Sun HQ, Macia E, Kirchhausen T, Watson H, Bonifacino JS, Yin HL (2007) PI4P promotes the recruitment of the GGA adaptor proteins to the trans-Golgi network and regulates their recognition of the ubiquitin sorting signal. Mol Biol Cell 18:2646–2655

Waselle L, Gerona RR, Vitale N, Martin TF, Bader MF, Regazzi R (2005) Role of phosphoinositide signaling in the control of insulin exocytosis. Mol Endocrinol 19:3097–3106

Wu B, Kitagawa K, Zhang NY, Liu B, Inagaki C (2004) Pathophysiological concentrations of amyloid beta proteins directly inhibit rat brain and recombinant human type II phosphatidylinositol 4-kinase activity. J Neurochem 91:1164–1170

Zhang J, Zhang N, Liu M, Li X, Zhou L, Huang W, Xu Z, Liu J, Musi N, DeFronzo RA, Cunningham JM, Zhou Z, Lu XY, Liu F (2012) Disruption of growth factor receptor-binding protein 10 in the pancreas enhances beta-cell proliferation and protects mice from streptozotocin-induced beta-cell apoptosis. Diabetes 61:3189–3198

PERMISSIONS

LIST OF CONTRIBUTORS

Jian-Sheng Kang
The First Affiliated Hospital of Zhengzhou University, Zhengzhou 450052, China

Xiaohua Wan and Fa Zhang
Key Laboratory of Intelligent Information Processing, Institute of Computing Technology, Chinese Academy of Sciences, Beijing 100190, China

Yu Chen, Zihao Wang and Jingrong Zhang
Key Laboratory of Intelligent Information Processing, Institute of Computing Technology, Chinese Academy of Sciences, Beijing 100190, China
University of Chinese Academy of Sciences, Beijing 100049, China

Lun Li
Key Laboratory of Intelligent Information Processing, Institute of Computing Technology, Chinese Academy of Sciences, Beijing 100190, China
School of Mathematical Sciences, University of Chinese Academy of Sciences, Beijing 100049, China

Fei Sun
University of Chinese Academy of Sciences, Beijing 100049, China
National Key Laboratory of Biomacromolecules, CAS Center for Excellence in Biomacromolecules, Institute of Biophysics, Chinese Academy of Sciences, Beijing 100101, China
Center for Biological Imaging, Institute of Biophysics, Chinese Academy of Sciences, Beijing 100101, China

Xiaomin Li
Technology Center for Protein Sciences, Ministry of Education Key Laboratory of Protein Sciences, School of Life Sciences, Tsinghua University, Beijing 100084, China

Jianlin Lei
Technology Center for Protein Sciences, Ministry of Education Key Laboratory of Protein Sciences, School of Life Sciences, Tsinghua University, Beijing 100084, China
Beijing Advanced Innovation Center for Structural Biology, School of Life Sciences, Tsinghua University, Beijing 100084, China

Hong-Wei Wang
Beijing Advanced Innovation Center for Structural Biology, School of Life Sciences, Tsinghua University, Beijing 100084, China

Jesús G. Galaz-Montoya and Steven J. Ludtke
National Center for Macromolecular Imaging, Verna and Marrs McLean Department of Biochemistry and Molecular Biology, Baylor College of Medicine, Houston, TX 77030, USA

Zheng Liu
Department of Biophysics, Health Science Centre, Peking University, Beijing 100191, China
Department of Biochemistry and Molecular Biophysics, Columbia University, New York 10032, USA

Tom S. Y. Guu and Yizhi Jane Tao
Department of BioSciences, Rice University, Houston, TX 77005, USA

Jingqiang Zhang and Yinyin Li
School of Life sciences, Sun Yat-sen University, Guangzhou 510275, China

Lingpeng Cheng
School of Life Science, Tsinghua University, Beijing 100084, China

Jianhao Cao
School of Life sciences, Sun Yat-sen University, Guangzhou 510275, China
State Key Laboratory of Organ Failure Research, Institute of Antibody Engineering, School of Biotechnology, Southern Medical University, Guangzhou 510515, China

Yiming Jiang, Haiying Huang, Xuan Yu, Yong Chen and Xiangkai Li
Ministry of Education Key Laboratory of Cell Activities and Stress Adaptations, School of Life Science, Lanzhou University, Lanzhou 730000, China

Mengru Wu
State Key Laboratory of Microbial Resources, Institute of Microbiology, Chinese Academy of Sciences, Beijing 100101, China

Pu Liu
Department of Development Biology Sciences, School of Life Science, Lanzhou University, Lanzhou 730000, China

Dongsheng Liu
iHuman Institute, ShanghaiTech University, Shanghai 201203, China
Department of Biochemistry, Albert Einstein College of Medicine, Bronx, NY 10461, USA

David Cowburn
Department of Biochemistry, Albert Einstein College of Medicine, Bronx, NY 10461, USA

Fanlei Ran and Shihua Wang
Key Laboratory of Pathogenic Fungi and Mycotoxins of Fujian Province, Key Laboratory of Biopesticide and Chemical Biology of Education Ministry, School of Life Sciences, Fujian Agriculture and Forestry University, Fuzhou 350002, China

Lili An, Yingjun Fan and Haiying Hang
Key Laboratory for Protein and Peptide Pharmaceuticals, Institute of Biophysics, Chinese Academy of Sciences, Beijing 100101, China

Yanqing Liu and Xuejun C. Zhang
National Laboratory of Biomacromolecules, National CAS Center for Excellence in Biomacromolecules, Institute of Biophysics, Chinese Academy of Sciences, Beijing 100101, China
College of Life Science, University of Chinese Academy of Sciences, Beijing 100049, China

Yue Liu, Lingli He and Yongfang Zhao
National Laboratory of Biomacromolecules, National CAS Center for Excellence in Biomacromolecules, Institute of Biophysics, Chinese Academy of Sciences, Beijing 100101, China

Jiyun Shi and Fan Wang
Interdisciplinary Laboratory, Institute of Biophysics, Chinese Academy of Sciences, Beijing 100101, China
Medical Isotopes Research Center, Peking University, Beijing 100191, China

Shuang Liu
School of Health Sciences, Purdue University, West Lafayette, IN 47907, USA

Yao Peng
National Laboratory of Biomacromolecules, Institute of Biophysics, Chinese Academy of Sciences, Beijing 100101, China
Institute of Molecular and Clinical Medicine, Kunming Medical University, Kunming 650500, China
iHuman Institute, ShanghaiTech University, Shanghai 201210, China
University of Chinese Academy of Sciences, Beijing 100049, China

Guisheng Zhong, Yiran Wu, Wenqing Shui, Suwen Zhao, Ling Shen and Raymond C. Stevens
iHuman Institute, ShanghaiTech University, Shanghai 201210, China
School of Life Science and Technology, ShanghaiTech University, Shanghai 201210, China

Zhi-Jie Liu
National Laboratory of Biomacromolecules, Institute of Biophysics, Chinese Academy of Sciences, Beijing 100101, China
Institute of Molecular and Clinical Medicine, Kunming Medical University, Kunming 650500, China
iHuman Institute, ShanghaiTech University, Shanghai 201210, China
School of Life Science and Technology, ShanghaiTech University, Shanghai 201210, China

Simeng Zhao, Yueming Xu, Xiaoyan Liu and Jianjun Cheng
iHuman Institute, ShanghaiTech University, Shanghai 201210, China

Haijie Cao
College of Pharmacy, Nankai University, Tianjin 300071, China
High-throughput Molecular Drug Discovery Center, Tianjin Joint Academy of Biotechnology and Medicine, Tianjin 300457, China

Jun Ma and Ronald J. Quinn
Eskitis Institute for Drug Discovery, Griffith University, Brisbane, QLD 4111, Australia

Shengliu Wang
National Key Laboratory of Biomacromolecules, CAS Center for Excellence in Biomacromolecules, Institute of Biophysics, Chinese Academy of Sciences, Beijing 100101, China

Shuoguo Li, Gang Ji and Xiaojun Huang
Center for Biological Imaging, Institute of Biophysics, Chinese Academy of Sciences, Beijing 100101, China

Fei Sun
National Key Laboratory of Biomacromolecules, CAS Center for Excellence in Biomacromolecules, Institute of Biophysics, Chinese Academy of Sciences, Beijing 100101, China
Center for Biological Imaging, Institute of Biophysics, Chinese Academy of Sciences, Beijing 100101, China
University of Chinese Academy of Sciences, Beijing, China

Hua Peng
Key Laboratory of Infection and Immunity, Institute of Biophysics, Chinese Academy of Sciences, Beijing 100101, China

Zhichen Sun
Key Laboratory of Infection and Immunity, Institute of Biophysics, Chinese Academy of Sciences, Beijing 100101, China
University of Chinese Academy of Sciences, Beijing 100049, China

Yang-Xin Fu
Key Laboratory of Infection and Immunity, Institute of Biophysics, Chinese Academy of Sciences, Beijing 100101, China
Department of Pathology, University of Texas Southwestern Medical Center, Dallas, TX 75390, USA

Tao-Rong Xie, Chun-Feng Liu and Jian-Sheng Kang
CAS Key Laboratory of Nutrition and Metabolism, Institute for Nutritional Sciences, Shanghai Institutes for Biological Sciences, Graduate School of the Chinese Academy of Sciences, Chinese Academy of Sciences, Shanghai 200031, China

Jan J. T. M. Swartjes
University of Groningen and University Medical Center Groningen, Department of Biomedical Engineering, Antonius Deusinglaan 1, 9713 AV Groningen, the Netherlands

Deepak H. Veeregowda
Ducom Instruments Europe B.V, Center for Innovation, 9713 GX Groningen, the Netherlands

Xuejun C. Zhang, Min Liu and Lei Han
National Laboratory of Biomacromolecules, CAS Center for Excellence in Biomacromolecules, Institute of Biophysics, Chinese Academy of Sciences, Beijing 100101, China
College of Life Science, University of Chinese Academy of Sciences, Beijing 100049, China

Cheng Zhong, Liping Wang and Yi Lu
Shenzhen Key Lab of Neuropsychiatric Modulation and Collaborative Innovation Center for Brain Science, CAS Center for Excellence in Brain Science and Intelligence Technology, The Brain Cognition and Brain Disease Institute, Shenzhen Institutes of Advanced Technology, Chinese Academy of Sciences, Shenzhen 518055, China

Lulu Wang and Kang Huang
Shenzhen Key Lab of Neuropsychiatric Modulation and Collaborative Innovation Center for Brain Science, CAS Center for Excellence in Brain Science and Intelligence Technology, The Brain Cognition and Brain Disease Institute, Shenzhen Institutes of Advanced Technology, Chinese Academy of Sciences, Shenzhen 518055, China
Shenzhen College of Advanced Technology, University of Chinese Academy of Sciences, Shenzhen 518055, China

Qiang Zhou
State Key Laboratory of Biomembrane and Membrane Biotechnology, Center for Structural Biology, School of Life Sciences, Tsinghua University, Beijing 100084, China

Niyun Zhou and Hong-Wei Wang
Ministry of Education Key Laboratory of Protein Science, Tsinghua-Peking Joint Center for Life Sciences, Center for Structural Biology, School of Life Sciences, Tsinghua University, Beijing 100084, China

Fudong Xue and Pingyong Xu
Key Laboratory of RNA Biology, Institute of Biophysics, Chinese Academy of Sciences, Beijing 100101, China
College of Life Sciences, University of Chinese Academy of Sciences, Beijing 100049, China

Wenting He and Mingshu Zhang
Key Laboratory of RNA Biology, Institute of Biophysics, Chinese Academy of Sciences, Beijing 100101, China

Fan Xu
Weldon School of Biomedical Engineering, Purdue University, West Lafayette, IN 47907, USA

Liangyi Chen
State Key Laboratory of Membrane Biology, Beijing Key Laboratory of Cardiometabolic Molecular Medicine, Institute of Molecular Medicine, Peking University, Beijing 100871, China

Lunfeng Zhang, Zhen Gao and Xinhua Qiao
National Laboratory of Biomacromolecules, CAS Center for Excellence in Biomacromolecules, Institute of Biophysics, Chinese Academy of Sciences, Beijing 100101, China
University of Chinese Academy of Sciences, Beijing 100049, China

Jiangmei Li and Yingying Zhao
National Laboratory of Biomacromolecules, CAS Center for Excellence in Biomacromolecules, Institute of Biophysics, Chinese Academy of Sciences, Beijing 100101, China
Shanghai Institute for Advanced Immunochemical Studies, ShanghaiTech University, Shanghai 201210, China

Panpan Zhang
Shanghai Institute for Advanced Immunochemical Studies, ShanghaiTech University, Shanghai 201210, China

Chang Chen
National Laboratory of Biomacromolecules, CAS Center for Excellence in Biomacromolecules, Institute of Biophysics, Chinese Academy of Sciences, Beijing 100101, China
University of Chinese Academy of Sciences, Beijing 100049, China
Beijing Institute for Brain Disorders, Beijing 100069, China

Index

www.ingramcontent.com/pod-product-compliance
Lightning Source LLC
Chambersburg PA
CBHW080533200326

41458CB00012B/4419